Phase Transformations in Metals and Alloys

THIRD EDITION

Phase Transformations in Metals and Alloys

THIRD EDITION

DAVID A. PORTER,
KENNETH E. EASTERLING, and
MOHAMED Y. SHERIF

CRC Press is an imprint of the
Taylor & Francis Group, an **informa** business

CRC Press
Taylor & Francis Group
6000 Broken Sound Parkway NW, Suite 300
Boca Raton, FL 33487-2742

Printed in the United States of America on acid-free paper
10 9 8 7 6 5 4 3 2 1

International Standard Book Number-13: 978-1-4200-6210-6 (0)

Library of Congress Cataloging-in-Publication Data

Porter, David A.
Phase transformations in metals and alloys. -- 3rd ed. / David A. Porter, Kenneth E. Easterling, and Mohamed Y. Sherif.
p. cm.
Includes bibliographical references and index.
ISBN 978-1-4200-6210-6 (alk. paper)
1. Physical metallurgy. 2. Phase rule and equilibrium. I. Easterling, K. E. II. Sherif, Mohamed Y. III. Title.

TN690.P597 2008
669'.94--dc22 2008037684

Visit the Taylor & Francis Web site at
http://www.taylorandfrancis.com

and the CRC Press Web site at
http://www.crcpress.com

This book is dedicated

to the memory of

Kenneth E. Easterling

Contents

Preface to the Third Edition

The fundamental theories of phase transformations in metals and alloys remain largely unchanged, so the third edition is essentially an expanded version of the second edition with additional material covering some of the more important application developments of the last 17 years. A section addressing the computation of phase diagrams has been added to Chapter 1 and recent developments in metallic glasses have been treated in Chapter 4. Chapter 5 contains most new material: the Scheil method of calculating a CCT diagram from a TTT diagram is now given; the treatment of the nucleation and growth of polygonal ferrite and bainite has been expanded to include new theories, while new case studies cover copper precipitation hardening of very low-carbon bainitic steel and very fine carbide-free bainite were added. In Chapter 6, a more detailed treatment of stress-assisted and strain-induced martensite is included to provide a theoretical background to transformation-induced plasticity (TRIP) steels.

David A. Porter and Mohamed Y. Sherif
August 2008

Preface to the Second Edition

In the ten years since the book was first published, there have been many new developments in the metallurgical field. Rapidly solidified metals and glasses have come of age; new Al–Li alloys are now used in modern aircraft; microalloyed (structural) and high-purity (pipeline) steels have become more sophisticated; radically new oxide-dispersed steels have appeared; a number of new memory metals have been developed; the list could go on. In spite of this, the underlying principles governing all of these developments have obviously not changed over the years. This is really the strength of the present book. From the beginning, our objective was to present these principles in an organized, comprehensive manner, so that undergraduates could appreciate them and carry out their own research. The present book is thus deliberately little changed from the original. We have, however, hopefully corrected any errors, expanded lists of further reading, and perhaps, most importantly, included a complete set of solutions to exercises. We hope that the revised edition continues to be enjoyed and appreciated in the many schools of metallurgy, materials science, and engineering materials who use our book throughout the world.

In completing this revised edition we are grateful to all the students and professors who have written to us over the last decade. We would particularly like to thank Dr. Wen-Bin Li (University of Luleå, Sweden) for using a fine-tooth comb in bringing out both obvious and less obvious errors in the original text. There remain, inevitably, a few "points of contention" concerning our description of certain phenomena, as raised by some of our correspondents, but there is nothing unhealthy about that. Finally, we would like to thank Dr. John Ion (University of Lappeenranta, Finland) for helping us compile the "Solutions to Exercises" chapter.

David A. Porter and Kenneth E. Easterling
September 1991

Preface to the First Edition

This book has been written as an undergraduate course in phase transformations for final year students specializing in metallurgy, materials science, or engineering materials. It should also be useful for research students interested in revising their knowledge of the subject. The book is based on lectures originally given by the authors at the University of Luleå for engineering students specializing in engineering materials. Surprisingly, we found no modern treatments of this important subject in a form suitable for a course book, the most recent probably being P.G. Shewmon's *Transformations in Metals* (McGraw-Hill, 1969). There have, however, been some notable developments in the subject over the last decade, particularly in studies of interfaces between phases and interface migration, as well as the kinetics of precipitate growth and the stability of precipitates. There have also been a number of important new practical developments based on phase transformations, including the introduction of TRIP steels (transformation induced by plastic deformation), directionally aligned eutectic composites, and sophisticated new structural steels with superior weldability and forming properties, to mention just a few. In addition, continuous casting and high-speed, high-energy fusion welding have emerged strongly in recent years as important production applications of solidification. It was the objective of this course to present a treatment of phase transformations in which these and other new developments could be explained in terms of the basic principles of thermodynamics and atomic mechanisms.

The book is effectively divided in two parts. Chapters 1 through 3 contain the background material necessary for understanding phase transformations: thermodynamics, kinetics, diffusion theory, and the structure and properties of interfaces. Chapters 4 through 6 deal with specific transformations: solidification, diffusional transformations in solids, and diffusionless transformations. At the end of the chapters on solidification, diffusion-controlled transformations, and martensite, we give a few selected case studies of engineering alloys to illustrate some of the principles discussed earlier. In this way, we hope that the text will provide a useful link between theory and the practical reality. It should be stated that we found it necessary to give this course in conjunction with a number of practical laboratory exercises and worked examples. Sets of problems are also included at the end of each chapter.

In developing this course and writing the text we have had continuous support and encouragement from our colleagues and students in the Department of Engineering Materials. Particular thanks are due to Agneta Engfors for her patience and skill in typing the manuscript as well as assisting with the editing.

David A. Porter and Kenneth E. Easterling
February 1980

Authors

David A. Porter received his BA and PhD in materials science from the University of Cambridge, United Kingdom. He has since been involved in materials' research and development in the University of Luleå in Sweden, the research center of the aluminum producer Årdal and Sunndal Verk in Norway, and the steel producer Fundia Special Bar also in Sweden. He is currently a product development manager responsible for the development of hot-rolled steel in Rautaruukki Oyj in Finland.

The late Professor Kenneth E. Easterling received his DrTechSc in physical metallurgy from the Institute of Technology, Helsinki. After lecturing at Chalmers University of Technology, Gothenburg, he became the professor and head of the Department of Engineering Materials at the University of Luleå, Sweden. He later became a professor of engineering (materials science) at the University of Exeter, United Kingdom.

Mohamed Y. Sherif received his BSc in mechanical engineering with honors from al-Azhar University in Cairo, Egypt. He then received his MPhil in materials modeling and PhD from the Department of Materials Science and Metallurgy, University of Cambridge, United Kingdom. His work on this new edition started while he was a research associate in the Phase Transformations and Complex Properties Group in the same department in Cambridge. He is now a research engineer at SKF Group, Engineering and Research Centre, Nieuwegein, the Netherlands.

1

Thermodynamics and Phase Diagrams

This chapter deals with some of the basic thermodynamic concepts that are required for a more fundamental appreciation of phase diagrams and phase transformations. It is assumed that the reader is already acquainted with elementary thermodynamics and only a summary of the most important results as regards phase transformations will be given here. Fuller treatment can be found in the books listed at the end of this chapter.

The main use of thermodynamics in physical metallurgy is to allow the predication of whether an alloy is in equilibrium. In considering phase transformations we are always concerned with changes toward equilibrium, and thermodynamics is therefore a very powerful tool. It should be noted, however, that rate at which equilibrium is reached cannot be determined by thermodynamics alone, as will become apparent in later chapters.

1.1 Equilibrium

It is useful to begin this chapter on thermodynamics by defining a few of the terms that will be frequently used. In the study of phase transformations, we will be dealing with the changes that can occur within a given system, e.g., an alloy that can exist as a mixture of one or more phases. A phase can be defined as a portion of the system whose properties and composition are homogeneous and which is physically distinct from other parts of the system. The components of a given system are the different elements or chemical compounds which make up the system, and the composition of a phase or the system can be described by giving the relative amounts of each component.

The study of phase transformations, as the name suggests, is concerned with how one or more phases in an alloy (the system) change into a new phase or mixture of phases. The reason why a transformation occurs at all is because the initial state of the alloy is unstable relative to the final state. But how is phase stability measured? The answer to this question is provided by thermodynamics. For transformations that occur at constant temperature and pressure, the relative stability of a system is determined by its Gibbs free energy (G).

The Gibbs free energy of a system is defined by the equation

$$G = H - TS \tag{1.1}$$

where

H is the enthalpy
T is the absolute temperature
S is the entropy of the system

Enthalpy is a measure of the heat content of the system and is given by

$$H = E + PV \tag{1.2}$$

where

E is the internal energy of the system
P is the pressure
V is the volume

The internal energy arises from the total kinetic and potential energies of the atoms within the system. Kinetic energy can arise from atomic vibration in solids or liquids and from translational and rotational energies for the atoms and molecules within a liquid or gas; whereas potential energy arises from the interactions, or bonds, between the atoms within the system. If a transformation or reaction occurs the heat that is absorbed or evolved will depend on the change in the internal energy of the system. However, it will also depend on changes in the volume of the system and the term PV takes this into account, so that at constant pressure the heat absorbed or evolved is given by the change in H. When dealing with condensed phases (solids and liquids) the PV term is usually very small in comparison to E, that is, $H \approx E$. This approximation will be made frequently in the treatments given in this book. The other function that appears in the expression for G is entropy (S) which is a measure of the randomness of the system.

A system is said to be in equilibrium when it is in the most stable state, i.e., shows no desire to change ad infinitum. An important consequence of the laws of classical thermodynamics is that at constant temperature and pressure a closed system (i.e., one of fixed mass and composition) will be in stable equilibrium if it has the lowest possible value of the Gibbs free energy, or in mathematical terms

$$\mathrm{d}G = 0 \tag{1.3}$$

It can be seen from the definition of G (Equation 1.1) that the state with the highest stability will be that with the best compromise between low enthalpy and high entropy. Thus at low temperatures, solid phases are most stable since they have the strongest atomic binding and therefore the lowest internal energy (enthalpy). At high temperatures, however, the $-TS$ term

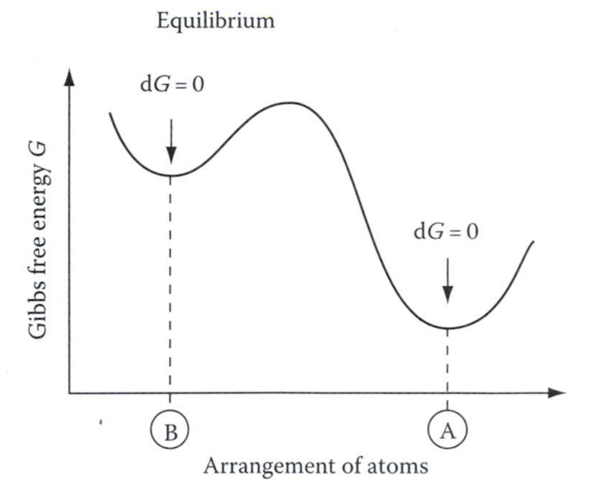

FIGURE 1.1
Schematic variation of Gibbs free energy with the arrangement of atoms. Configuration "A" has the lowest free energy and is therefore the arrangement when the system is at stable equilibrium. Configuration "B" is a metastable equilibrium.

dominates and phases with more freedom of atom movement, liquids and gases, become most stable. If pressure changes are considered it can be seen from Equation 1.2 that phases with small volumes are favored by high pressures.

The definition of equilibrium given by Equation 1.3 can be illustrated graphically as follows. If it were possible to evaluate the free energy of a given system for all conceivable configurations the stable equilibrium configuration would be found to have the lowest free energy. This is illustrated in Figure 1.1 where it is imagined that the various atomic configurations can be represented by points along the abscissa. Configuration A would be the stable equilibrium state. At this point small changes in the arrangement of atoms to a first approximation produce no change in G, i.e., Equation 1.3 applies. However, there will always be other configurations, e.g., B, which lie at a local minimum in free energy and therefore also satisfy Equation 1.3, but which do not have the lowest possible value of G. Such configurations are called metastable equilibrium states to distinguish them from the stable equilibrium state. The intermediate states for which $dG \neq 0$ are unstable and are only ever realized momentarily in practice. If, as result of thermal fluctuations, the atoms become arranged in an intermediate state they will rapidly rearrange into one of the free energy minima. If by a change of temperature or pressure, for example, a system is moved from a stable to a metastable state it will, given time, transform to the new stable equilibrium state.

Graphite and diamond at room temperature and pressure are examples of stable and metastable equilibrium states. Given time, therefore, all diamond under these conditions will transform to graphite.

Any transformation that results in a decrease in Gibbs free energy is possible. Therefore, a necessary criterion for any phase transformation is

$$\Delta G = G_2 - G_1 < 0 \tag{1.4}$$

where G_1 and G_2 are the free energies of the initial and final states, respectively. The transformation need not go directly to the stable equilibrium state but can pass through a whole series of intermediate metastable states.

The answer to the question "How fast does a phase transformation occur?" is not provided by classical thermodynamics. Sometimes metastable states can be very short-lived; at other times they can exist almost indefinitely as in the case of diamond at room temperature and pressure. The reason for these differences is the presence of the free energy hump between the metastable and stable states in Figure 1.1. The study of transformation rates in physical chemistry belongs to the realm of kinetics. In general, higher humps or energy barriers lead to slower transformation rates. Kinetics obviously plays a central role in the study of phase transformations and many examples of kinetic processes will be found throughout this book.

The different thermodynamic functions that have been mentioned in this section can be divided into two types called intensive and extensive properties. The intensive properties are those which are independent of the size of the system such as T and P, whereas the extensive properties are directly proportional to the quantity of material in the system, e.g., V, E, H, S, and G. The usual way of measuring the size of the system is by the number of moles of material it contains. The extensive properties are then molar quantities, i.e., expressed in units per mole. The number of moles of a given component in the system is given by the mass of the component in grams divided by its atomic or molecular weight.

The number of atoms or molecules within 1 mol of material is given by Avogadro's number (N_a) and is 6.023×10^{23}.

1.2 Single-Component Systems

Let us begin by dealing with the phase changes that can be induced in a single component by changes in temperature at a fixed pressure, say 1 atm. A single-component system could be one containing a pure element or one type of molecule that does not dissociate over the range of temperature of interest. In order to predict the phases that are stable or mixtures that are in equilibrium at different temperatures it is necessary to be able to calculate the variation of G with T.

1.2.1 Gibbs Free Energy as a Function of Temperature

The specific heat of most substances is easily measured and readily available. In general, it varies with temperature as shown in Figure 1.2a. The specific heat is the quantity of heat (in J) required to raise the temperature of the substance by 1 K. At constant pressure this is denoted by C_p and is given by

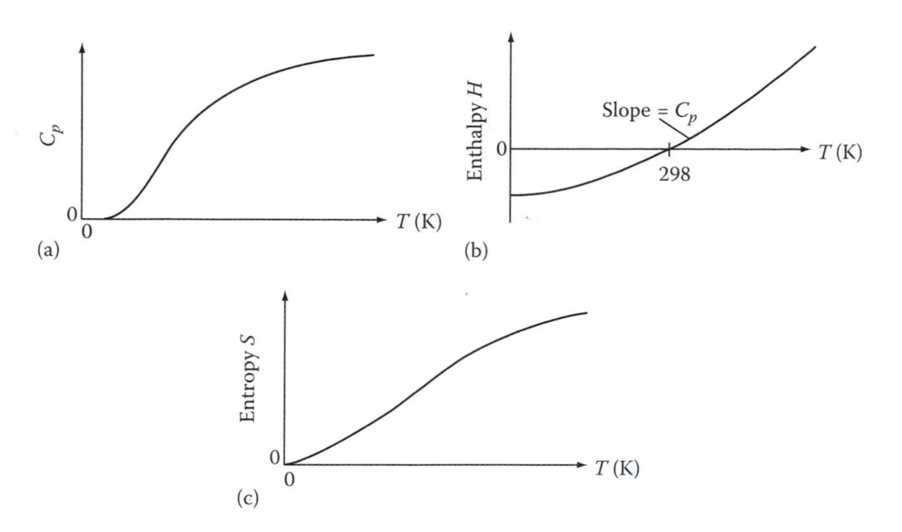

FIGURE 1.2
(a) Variation of C_p with temperature, C_p tends to a limit of ~3*R*. (b)Variation of enthalpy (*H*) with absolute temperature for a pure metal. (c) Variation of entropy (*S*) with absolute temperature.

$$C_p = \left(\frac{\partial H}{\partial T}\right)_p \tag{1.5}$$

Therefore, the variation of *H* with *T* can be obtained from a knowledge of the variation of C_p with *T*. In considering phase transformations or chemical reactions it is only changes in thermodynamic functions that are of interest. Consequently, *H* can be measured relative to any reference level which is usually done by defining $H=0$ for a pure element in its most stable state at 298 K (25°C). The variation of *H* with *T* can then be calculated by integrating Equation 1.5, i.e.,

$$H = \int_{298}^{T} C_p \, dT \tag{1.6}$$

The variation is shown schematically in Figure 1.2b. The slope of the $H-T$ curve is C_p.

The variation of entropy with temperature can also be derived from the specific heat C_p. From classical thermodynamics

$$\frac{C_p}{T} = \left(\frac{\partial S}{\partial T}\right)_p \tag{1.7}$$

Taking entropy at 0 K as zero, Equation 1.7 can be integrated to give

$$S = \int_{0}^{T} \frac{C_p}{T} \, dT \tag{1.8}$$

as shown in Figure 1.2c.

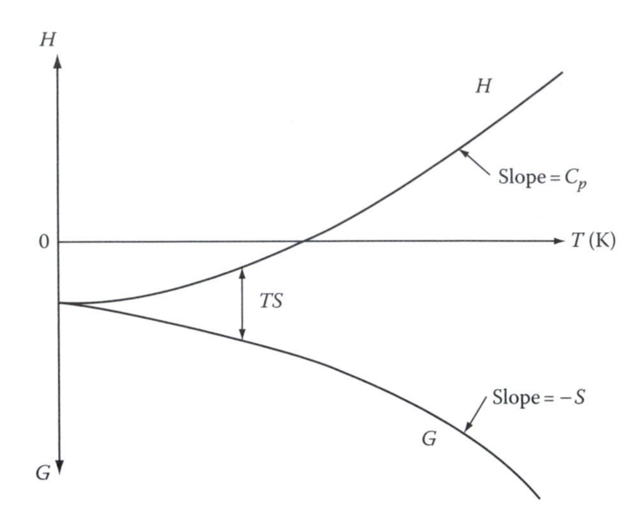

FIGURE 1.3
Variation of Gibbs free energy with temperature.

Finally the variation of G with temperature shown in Figure 1.3 is obtained by combining Figure 1.2b and c using Equation 1.1. When temperature and pressure vary the change in Gibbs free energy can be obtained from the following result of classical thermodynamics: for a system of fixed mass and composition:

$$dG = -S\,dT + V\,dP \tag{1.9}$$

At constant pressure $dP = 0$ and

$$\left(\frac{\partial G}{\partial T}\right)_p = -S \tag{1.10}$$

This means that G decreases with increasing T at a rate given by $-S$. The relative positions of the free energy curves of solid and liquid phases are illustrated in Figure 1.4. At all temperatures the liquid has a higher enthalpy (internal energy) than the solid. Therefore, at low temperatures $G^L > G^S$. However, the liquid phase has a higher entropy than the solid phase and the Gibbs free energy of the liquid therefore decreases more rapidly with increasing temperature than that of the solid. For temperatures up to T_m the solid phase has the lowest free energy and is therefore the stable equilibrium phase, whereas above T_m the liquid phase is the equilibrium state of the system. At T_m both phases have the same value of G and both solid and liquid can exist in equilibrium. T_m is therefore the equilibrium melting temperature at the pressure concerned.

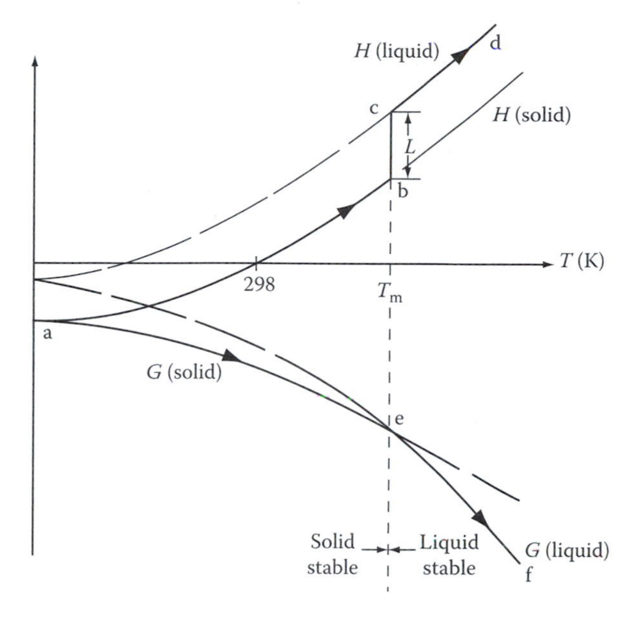

FIGURE 1.4
Variation of enthalpy (H) and free energy (G) with temperature for the solid and liquid phases of a pure metal. L is the latent heat of melting. T_m is the equilibrium melting temperature.

If a pure component is heated from absolute zero the heat supplied will raise the enthalpy at a rate determined by C_p (solid) along the line ab in Figure 1.4. Meanwhile the free energy will decrease along ae. At T_m the heat supplied to the system will not raise its temperature but will be used in supplying the latent heat of melting (L) that is required to convert solid into liquid (bc in Figure 1.4). Note that at T_m the specific heat appears to be infinite since the addition of heat does not appear as an increase in temperature. When all solid has transformed into liquid the enthalpy of the system will follow the line cd while the Gibbs free energy decreases along ef. At still higher temperatures than shown in Figure 1.4 the free energy of the gas phase (at atmospheric pressure) becomes lower than that of the liquid and the liquid transforms to a gas. If the solid phase can exist in different crystal structures (allotropes or polymorphs) free energy curves can be constructed for each of these phases and the temperature at which they intersect will give the equilibrium temperature for the polymorphic transformation. For example, at atmospheric pressure iron can exist as either body-centered cubic (bcc) ferrite below 910°C or face-centered cubic (fcc) austenite above 910°C, and at 910°C both phases can exist in equilibrium.

1.2.2 Pressure Effects

The equilibrium temperatures discussed so far only apply at a specific pressure (1 atm, say). At other pressures the equilibrium temperatures will differ. For example, Figure 1.5 shows the effect of pressure on the equilibrium temperatures for pure iron. Increasing pressure has the effect of depressing the α/γ equilibrium temperature and raising the equilibrium melting

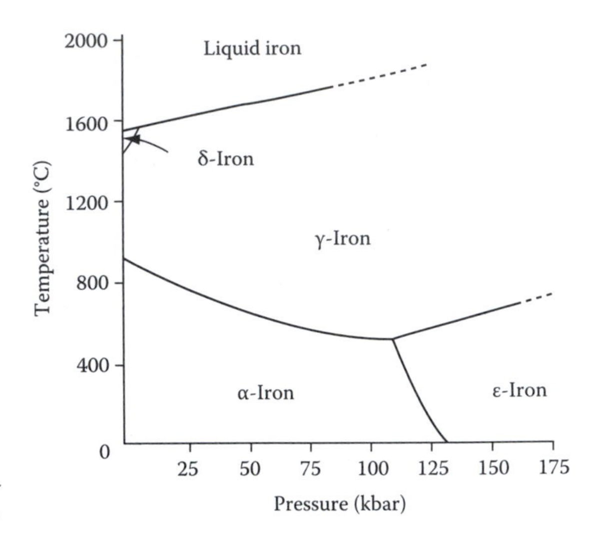

FIGURE 1.5
Effect of pressure on the equilibrium phase diagram for pure iron.

temperature. At very high pressures hexagonal close-packed (hcp) ε-Fe becomes stable. The reason for these changes derives from Equation 1.9. At constant temperature the free energy of a phase increases with pressure such that

$$\left(\frac{\partial G}{\partial P}\right)_T = V \tag{1.11}$$

If the two phases in equilibrium have different molar volumes their respective free energies will not increase by the same amount at a given temperature and equilibrium will, therefore, be disturbed by changes in pressure. The only way to maintain equilibrium at different pressure is by varying the temperature.

If the two phases in equilibrium are α and β, application of Equation 1.9 to 1 mol of both gives

$$\begin{aligned} \mathrm{d}G^{\alpha} &= V_{\mathrm{m}}^{\alpha}\,\mathrm{d}P - S^{\alpha}\,\mathrm{d}T \\ \mathrm{d}G^{\beta} &= V_{\mathrm{m}}^{\beta}\,\mathrm{d}P - S^{\beta}\,\mathrm{d}T \end{aligned} \tag{1.12}$$

If α and β are in equilibrium $G^{\alpha} = G^{\beta}$ therefore $\mathrm{d}G^{\alpha} = \mathrm{d}G^{\beta}$ and

$$\left(\frac{\mathrm{d}P}{\mathrm{d}T}\right)_{\mathrm{eq}} = \frac{S^{\beta} - S^{\alpha}}{V_{\mathrm{m}}^{\beta} - V_{\mathrm{m}}^{\alpha}} = \frac{\Delta S}{\Delta V} \tag{1.13}$$

This equation gives the change in temperature of dT required to maintain equilibrium between α and β if pressure is increased by dP. The equation can be simplified as follows. From Equation 1.1

$$G^\alpha = H^\alpha - TS^\alpha$$
$$G^\beta = H^\beta - TS^\beta$$

Therefore, putting $\Delta G = G^\beta - G^\alpha$, etc., gives

$$\Delta G = \Delta H - T\Delta S$$

But since at equilibrium $G^\beta - G^\alpha$, $\Delta G = 0$, and

$$\Delta H - T\Delta S = 0$$

Consequently, Equation 1.13 becomes

$$\left(\frac{\mathrm{d}P}{\mathrm{d}T_{\mathrm{eq}}}\right) = \frac{\Delta H}{T_{\mathrm{eq}}\Delta V} \tag{1.14}$$

which is known as the Clausius–Clapeyron equation. Since close-packed γ-Fe has a smaller molar volume than α-Fe, $\Delta V = V_{\mathrm{m}}^{\gamma} - V_{\mathrm{m}}^{\alpha} < 0$ whereas $\Delta H = H^\gamma - H^\alpha > 0$ (for the same reason that a liquid has a higher enthalpy than a solid), so that $(\mathrm{d}P/\mathrm{d}T)$ is negative, i.e., an increase in pressure lowers the equilibrium transition temperature. On the other hand the δ/L equilibrium temperature is raised with increasing pressure due to the larger molar volume of the liquid phase. It can be seen that the effect of increasing pressure is to increase the area of the phase diagram over which the phase with the smallest molar volume is stable (γ-Fe in Figure 1.5). It should also be noted that ε-Fe has the highest density of the three allotropes, consistent with the slopes of the phase boundaries in the Fe phase diagram.

1.2.3 Driving Force for Solidification

In dealing with phase transformations, we are often concerned with the difference in free energy between two phases at temperatures away from the equilibrium temperature. For example, if a liquid metal is undercooled by ΔT below T_{m} before it solidifies, solidification will be accompanied by a decrease in free energy ΔG (J mol^{-1}) as shown in Figure 1.6. This free energy decrease provides the driving force for solidification. The magnitude of this change can be obtained as follows.

The free energies of the liquid and solid at a temperature T are given by

$$G^{\mathrm{L}} = H^{\mathrm{L}} - TS^{\mathrm{L}}$$
$$G^{\mathrm{S}} = H^{\mathrm{S}} - TS^{\mathrm{S}}$$

Therefore, at a temperature T

$$\Delta G = \Delta H - T\Delta S \tag{1.15}$$

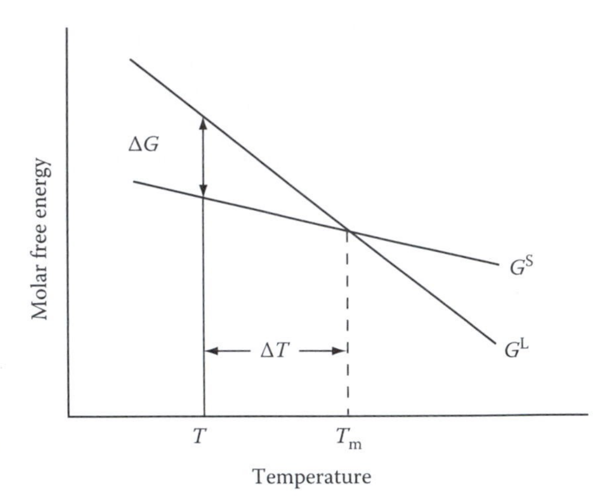

FIGURE 1.6
Difference in free energy between liquid and solid close to the melting point. The curvature of the G^S and G^L lines has been ignored.

where

$$\Delta H = H^L - H^S \quad \text{and} \quad \Delta S = S^L - S^S$$

At the equilibrium melting temperature T_m the free energies of solid and liquid are equal, i.e., $\Delta G = 0$. Consequently

$$\Delta G = \Delta H - T_m \Delta S = 0$$

and therefore at T_m

$$\Delta S = \frac{\Delta H}{T_m} = \frac{L}{T_m} \tag{1.16}$$

This is known as the entropy of fusion. It is observed experimentally that the entropy of fusion is a constant $\simeq R$ (8.3 J mol^{-1} K^{-1}) for most metals (Richard's rule). This is not unreasonable as metals with high bond strengths can be expected to have high values for both L and T_m.

For small undercoolings (ΔT) the difference in the specific heats of the liquid and solid ($C_p^L - C_p^S$) can be ignored. ΔH and ΔS are therefore approximately independent of temperature. Combining Equations 1.15 and 1.16 thus gives

$$\Delta G \simeq L - T\frac{L}{T_m}$$

i.e., for small ΔT

$$\Delta G \simeq \frac{L\Delta T}{T_{\mathrm{m}}} \tag{1.17}$$

This is a very useful result which will frequently recur in subsequent chapters.

1.3 Binary Solutions

In single-component systems all phases have the same composition, and equilibrium simply involves pressure and temperature as variables. In alloys, however, composition is also variable and to understand phase changes in alloys requires an appreciation of how the Gibbs free energy of a given phase depends on composition as well as temperature and pressure. Since the phase transformations described in this book mainly occur at a fixed pressure of 1 atm most attention will be given to changes in composition and temperature. In order to introduce some of the basic concepts of the thermodynamics of alloys a simple physical model for binary solid solutions will be described.

1.3.1 Gibbs Free Energy of Binary Solutions

The Gibbs free energy of a binary solution of A and B atoms can be calculated from the free energies of pure A and B in the following way.

It is assumed that A and B have the same crystal structures in their pure states and can be mixed in any proportions to make a solid solution with the same crystal structure. Imagine that 1 mol of homogeneous solid solution is made by mixing together X_{A} mol of A and X_{B} mol of B. Since there is a total of 1 mol of solution

$$X_{\mathrm{A}} + X_{\mathrm{B}} = 1 \tag{1.18}$$

where X_{A} and X_{B} are the mole fractions of A and B, respectively, in the alloy. In order to calculate the free energy of the alloy, the mixing can be made in two steps (Figure 1.7). They are

1. Bring together X_{A} mol of pure A and X_{B} mol of pure B
2. Allow the A and B atoms to mix together to make a homogeneous solid solution

After step 1 the free energy of the system is given by

$$G_1 = X_{\mathrm{A}}G_{\mathrm{A}} + X_{\mathrm{B}}G_{\mathrm{B}} \ (\mathrm{J\ mol^{-1}}) \tag{1.19}$$

where G_{A} and G_{B} are the molar free energies of pure A and B at the temperature and pressure of the above experiment. G_1 can be most

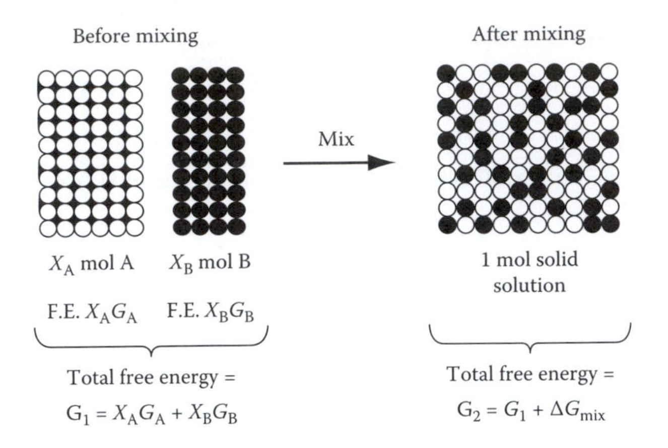

FIGURE 1.7
Free energy of mixing.

conveniently represented on a molar free energy diagram (Figure 1.8) in which molar free energy is plotted as a function of X_B or X_A. For all alloy compositions G_1 lies on the straight line between G_A and G_B.

The free energy of the system will not remain constant during the mixing of the A and B atoms and after step 2 the free energy of the solid solution G_2 can be expressed as

$$G_2 = G_1 + \Delta G_{mix} \tag{1.20}$$

where ΔG_{mix} is the change in Gibbs free energy caused by the mixing.

Since

$$G_1 = H_1 - TS_1$$

and

$$G_2 = H_2 - TS_2$$

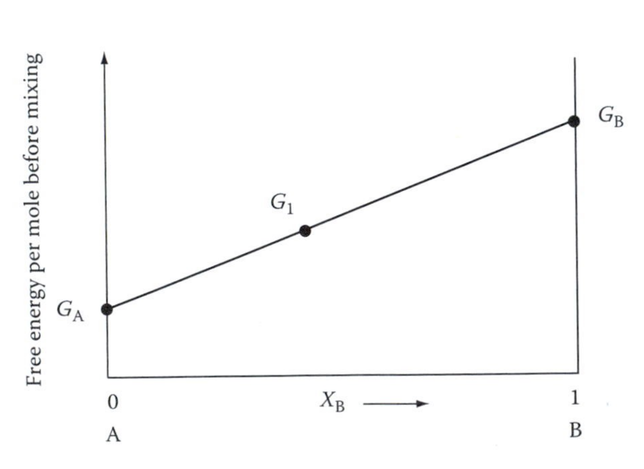

FIGURE 1.8
Variation of G_1 (the free energy before mixing) with alloy composition (X_A or X_B).

putting

$$\Delta H_{mix} = H_2 - H_1$$

and

$$\Delta S_{mix} = S_2 - S_1$$

gives

$$\Delta G_{mix} = \Delta H_{mix} - T\Delta S_{mix} \tag{1.21}$$

where ΔH_{mix} is the heat absorbed or evolved during step 2, i.e., it is the heat of solution, and ignoring volume changes during the process, it represents only the difference in internal energy (E) before and after mixing. ΔS_{mix} is the difference in entropy between the mixed and unmixed states.

1.3.2 Ideal Solutions

The simplest type of mixing to treat first is when $\Delta H_{mix} = 0$, in which case the resultant solution is said to be ideal and the free energy change on mixing is only due to the change in entropy

$$\Delta G_{mix} = -T\Delta S_{mix} \tag{1.22}$$

In statistical thermodynamics, entropy is quantitatively related to randomness by the Boltzmann equation, i.e.,

$$S = k \ln \omega \tag{1.23}$$

where
 k is Boltzmann's constant
 ω is a measure of randomness

There are two contributions to the entropy of a solid solution—a thermal contribution S_{th} and configurational contribution S_{config}.

In the case of thermal entropy, ω is the number of ways in which the thermal energy of the solid can be divided among the atoms, that is, the total number of ways in which vibrations can be set up in the solid. In solutions, additional randomness exists due to the different ways in which the atoms can be arranged. This gives extra entropy S_{config} for which ω is the number of distinguishable ways of arranging the atoms in the solution.

If there is no volume change or heat change or heat change during mixing then the only contribution to ΔS_{mix} is the change in configurational entropy. Before mixing, the A and B atoms are held separately in the system and there is only one distinguishable way in which the atoms can be arranged. Consequently $S_1 = k \ln 1 = 0$ and therefore $\Delta S_{mix} = S_2$.

Assuming that A and B mix to form a substitutional solid solution and that all configurations of A and B atoms are equally probable, the number of distinguishable ways of arranging the atoms on the atom site is

$$\omega_{\text{config}} = \frac{(N_A + N_B)!}{N_A!N_B!} \tag{1.24}$$

where
N_A is the number of A atoms
N_B is the number of B atoms

Since we are dealing with 1 mol of solution, i.e., N_a atoms (Avogadro's number),

$$N_A = X_A N_a$$

and

$$N_B = X_B N_a$$

By substituting into Equations 1.23 and 1.24, using Stirling's approximation ($\ln N! \simeq N \ln N - N$) and the relationship $N_a k = R$ (the universal gas constant) gives

$$\Delta S_{\text{mix}} = -R(X_A \ln X_A + X_B \ln X_B) \tag{1.25}$$

Note that, since X_A and X_B are less than unity, ΔS_{mix} is positive, i.e., there is an increase in entropy on mixing, as expected. The free energy of mixing, ΔG_{mix}, is obtained from Equation 1.22 as

$$\Delta G_{\text{mix}} = RT(X_A \ln X_A + X_B \ln X_B) \tag{1.26}$$

Figure 1.9 shows ΔG_{mix} as a function of composition and temperature.

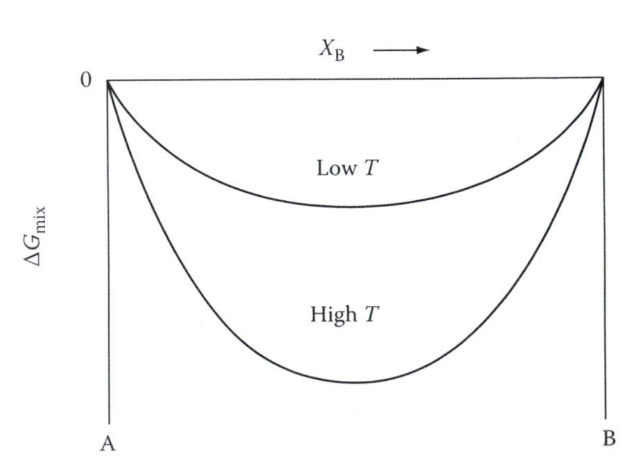

FIGURE 1.9
Free energy of mixing for an ideal solution.

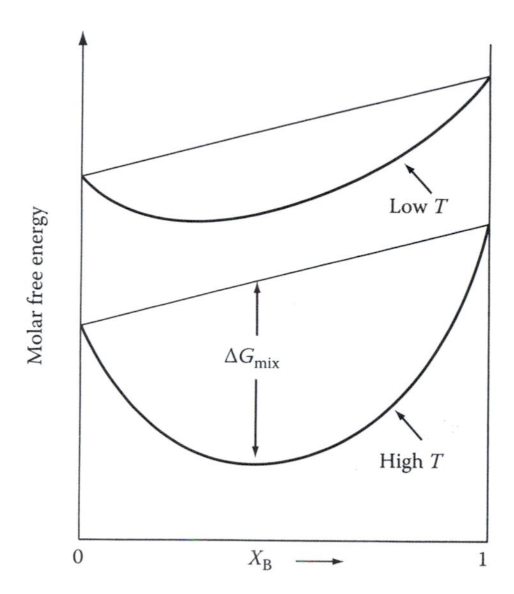

FIGURE 1.10
The molar free energy (free energy per mole of solution) for an ideal solid solution. A combination of Figures 1.8 and 1.9.

The actual free energy of the solution G will also depend on G_A and G_B. From Equations 1.19, 1.20, and 1.26

$$G = G_2 = X_A G_A + X_B G_B + RT(X_A \ln X_A + X_B \ln X_B) \tag{1.27}$$

This is shown schematically in Figure 1.10. Note that, as the temperature increases, G_A and G_B decrease and the free energy curves assume a greater curvature. The decrease in G_A and G_B is due to the thermal entropy of both components and is given by Equation 1.10.

It should be noted that all of the free energy–composition diagrams in this book are essentially schematic; if properly plotted the free energy curves must end asymptotically at the vertical axes of the pure components, i.e., tangential to the vertical axes of the diagrams. This can be shown by differentiating Equation 1.26 or Equation 1.27.

1.3.3 Chemical Potential

In alloys it is of interest to know how the free energy of a given phase will change when atoms are added or removed. If a small quantity of A, dn_A mol, is added to a large amount of a phase at constant temperature and pressure, the size of the system will increase by dn_A and therefore the total free energy of the system will also increase by a small amount dG'. If dn_A is small enough dG' will be proportional to the amount of A added. Thus we can write

$$dG' = \mu_A dn_A \ (T, P, n_B \text{ constant}) \tag{1.28}$$

The proportionality constant μ_A is called the partial molar free energy of A or alternatively the chemical potential of A in the phase. μ_A depends on the composition of the phase, and therefore, dn_A must be so small that the composition is not significantly altered. If Equation 1.28 is rewritten it can be seen that a definition of the chemical potential of A is

$$\mu_A = \left(\frac{\partial G'}{\partial n_A}\right)_{T,P,n_B} \tag{1.29}$$

The symbol G' has been used for the Gibbs free energy to emphasize the fact that it refers to the whole system. The usual symbol G will be used to denote the molar free energy and is therefore independent of the size of the system.

Equations similar to Equations 1.28 and 1.29 can be written for the other components in the solution. For a binary solution at constant temperature and pressure the separate contributions can be summed:

$$dG' = \mu_A dn_A + \mu_B dn_B \tag{1.30}$$

This equation can be extended by adding further terms for solutions containing more than two components. If T and P changes are also allowed Equation 1.9 must be added giving the general equation

$$dG' = -SdT + VdP + \mu_A dn_A + \mu_B dn_B + \mu_C dn_C + \cdots$$

If 1 mol of the original phase contained X_A mol A and X_B mol B, the size of the system can be increased without altering its composition if A and B are added in the correct proportions, i.e., such that dn_A:$dn_B = X_A$:X_B. For example, if the phase contains twice as many A as B atoms ($X_A = 2/3$, $X_B = 1/3$) the composition can be maintained constant by adding two A atoms for every one B atom (dn_A:$dn_B = 2$). In this way, the size of the system can be increased by 1 mol without changing μ_A and μ_B. To do this X_A mol A and X_B mol B must be added and the free energy of the system will increase by the molar free energy G. Therefore from Equation 1.30

$$G = \mu_A X_A + \mu_B X_B \ (\text{J mol}^{-1}) \tag{1.31}$$

When G is known as a function of X_A and X_B, as in Figure 1.10 for example, μ_A and μ_B can be obtained by extrapolating the tangent to the G curve to the sides of the molar free energy diagram as shown in Figure 1.11. This can be obtained from Equations 1.30 and 1.31, remembering that $X_A + X_B = 1$, i.e., $dX_A = -dX_B$, and this is left as an exercise for the reader. It is clear from Figure 1.11 that μ_A and μ_B vary systematically with the composition of the phase.

Comparison of Equations 1.27 and 1.31 gives μ_A and μ_B for an ideal solution as

$$\begin{aligned} \mu_A &= G_A + RT \ln X_A \\ \mu_B &= G_B + RT \ln X_B \end{aligned} \tag{1.32}$$

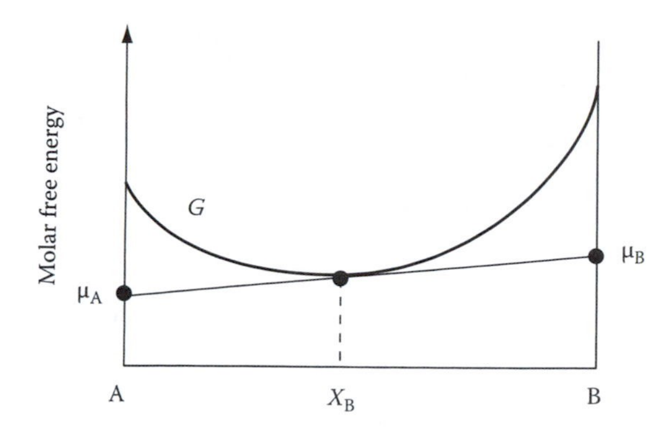

FIGURE 1.11
The relationship between the free energy curve for a solution and the chemical potentials of the components.

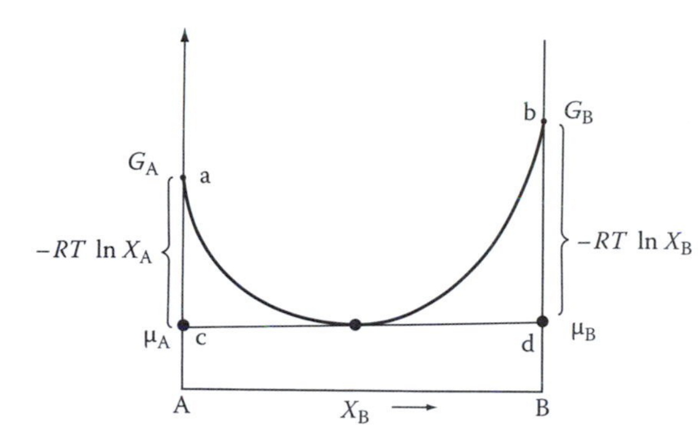

FIGURE 1.12
The relationship between the free energy curve and chemical potentials for an ideal solution.

which is a much simpler way of presenting Equation 1.27. These relationships are shown in Figure 1.12. The distances ac and bd are simply $-RT \ln X_A$ and $-RT \ln X_B$.

1.3.4 Regular Solutions

Returning to the model of a solid solution, so far it has been assumed that $\Delta H_{mix} = 0$; however, this type of behavior is exceptional in practice and usually mixing is endothermic (heat absorbed) or exothermic (heat evolved). The simple model used for an ideal solution can, however, be extended to include the ΔH_{mix} term by using the so-called quasichemical approach.

In the quasichemical model it is assumed that the heat of mixing, ΔH_{mix}, is only due to the bond energies between adjacent atoms. For this assumption to be valid it is necessary that the volumes of pure A and B are equal and do not change during mixing so that the interatomic distance and bond energies are independent of composition.

The structure of an ordinary solid solution is shown schematically in Figure 1.13. Three types of interatomic bonds are present:

1. A–A bonds each with an energy ε_{AA}
2. B–B bonds each with an energy ε_{BB}
3. A–B bonds each with an energy ε_{AB}

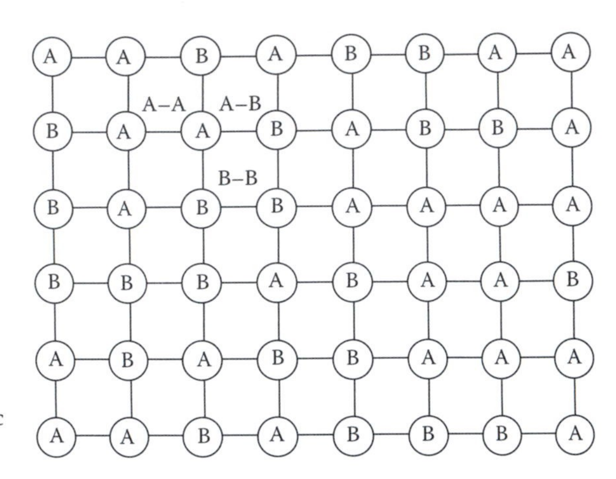

FIGURE 1.13
Different types of interatomic bond in solid solution.

By considering zero energy to be the state where the atoms are separated to infinity ε_{AA}, ε_{BB}, and ε_{AB} are negative quantities, and become increasingly more negative as the bonds become stronger. The internal energy of the solution E will depend on the number of bonds of each type P_{AA}, P_{BB}, and P_{AB} such that

$$E = P_{AA}\varepsilon_{AA} + P_{BB}\varepsilon_{BB} + P_{AB}\varepsilon_{AB}$$

Before mixing pure A and B contain only A–A and B–B bonds, respectively, and by considering the relationships between P_{AA}, P_{BB}, and P_{AB} in the solution it can be shown[1] that the change in internal energy on mixing is given by

$$\Delta H_{mix} = P_{AB}\varepsilon \tag{1.33}$$

where

$$\varepsilon = \varepsilon_{AB} - \frac{1}{2}(\varepsilon_{AA} + \varepsilon_{BB}) \tag{1.34}$$

that is, ε is the difference between the A–B bond energy and the average of the A–A and B–B bond energies.

If $\varepsilon = 0$, $\Delta H_{mix} = 0$ and the solution is ideal, as considered in Section 1.3.2. In this case, the atoms are completely randomly arranged and the entropy of mixing is given by Equation 1.25. In such a solution it can also be shown[1] that

$$P_{AB} = N_a z X_A X_B \text{ (bonds mol}^{-1}\text{)} \tag{1.35}$$

where
N_a is Avogadro's number
z is the number of bonds per atom

If $\varepsilon < 0$ the atoms in the solution will prefer to be surrounded by atoms of the opposite type and this will increase P_{AB}, whereas, if $\varepsilon > 0$, P_{AB} will tend to be less than in a random solution. However, provided ε is not too different from zero, Equation 1.35 is still a good approximation in which case

$$\Delta H_{mix} = \Omega X_A X_B \tag{1.36}$$

where

$$\Omega = N_a z \varepsilon \tag{1.37}$$

Real solutions that closely obey Equation 1.36 are known as regular solutions. The variation of ΔH_{mix} composition is parabolic and is shown in Figure 1.14 for $\Omega > 0$. Note that the tangents at $X_A = 0$ and 1 are related to Ω as shown.

The free energy change on mixing a regular solution is given by Equations 1.21, 1.25, and 1.36 as

$$\Delta G_{mix} = \underbrace{\Omega X_A X_B}_{\Delta H_{mix}} + \underbrace{RT(X_{mix} \ln X_A + X_B \ln X_B)}_{-T\Delta S_{mix}} \tag{1.38}$$

This is shown in Figure 1.15 for different values of Ω and temperature. For exothermic solutions $\Delta H_{mix} < 0$ and mixing results in a free energy decrease at all temperatures (Figure 1.15a and b). When $\Delta H_{mix} > 0$, however, the situation is more complicated. At high temperatures $T\Delta S_{mix}$ is greater than ΔH_{mix} for all compositions and the free energy curve has a positive curvature at all points (Figure 1.15c). At low temperatures, on the other hand, $T\Delta S_{mix}$ is smaller and ΔG_{mix} develops a negative curvature in the middle (Figure 1.15d).

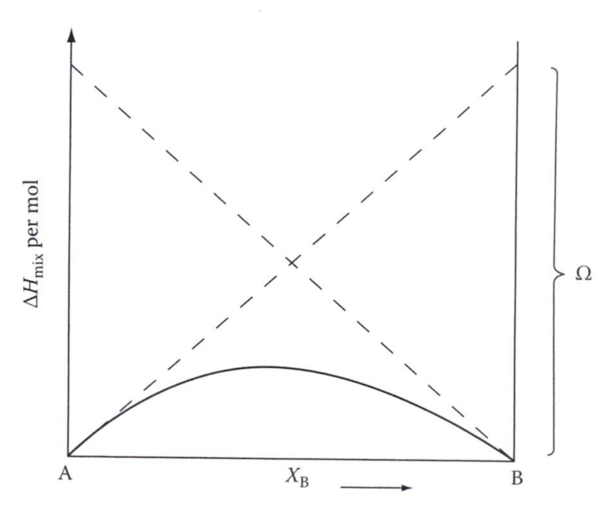

FIGURE 1.14
The variation of ΔH_{mix} with composition for a regular solution.

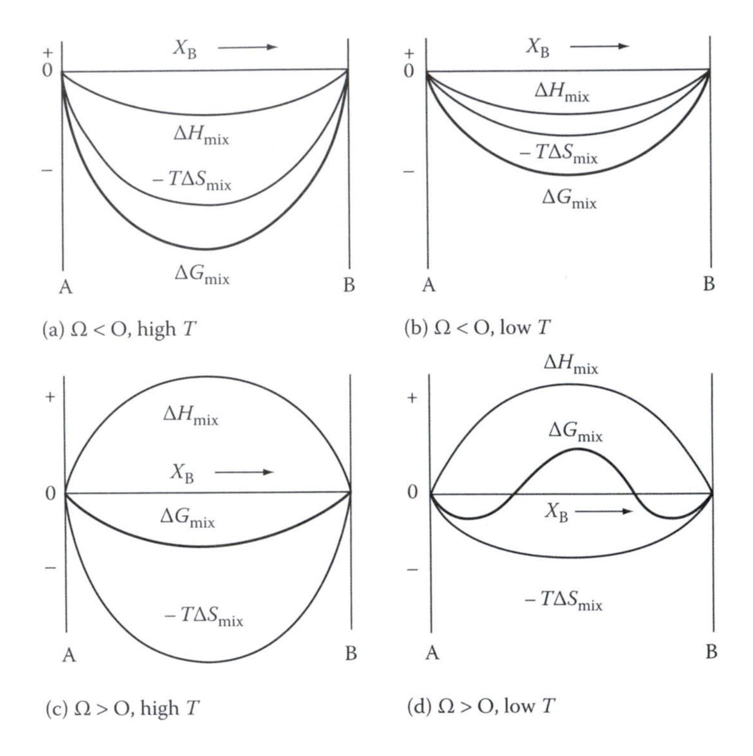

FIGURE 1.15
The effect of ΔH_{mix} and T on ΔG_{mix}.

Differentiating Equation 1.25 shows that, as X_A or $X_B \rightarrow 0$, the $-T\Delta S_{mix}$ curve becomes vertical whereas the slope of the ΔH_{mix} curve tends to a finite value Ω (Figure 1.14). This means that, except at absolute zero, ΔG_{mix} always decreases on addition of a small amount of solute.

The actual free energy of the alloy depends on the values chosen for G_A and G_B and is given by Equations 1.19, 1.20, and 1.38 as

$$G = X_A G_A + X_B G_B + \Omega X_A X_B + RT(X_A \ln X_A + X_B \ln X_B) \tag{1.39}$$

This is shown in Figure 1.16 along with the chemical potentials of A and B in the solution. Using the relationship $X_A X_B = X_A^2 X_B + X_B^2 X_A$ and comparing Equations 1.31 and 1.39 shows that for a regular solution

$$\mu_A = G_A + \Omega(1 - X_A)^2 + RT \ln X_A \tag{1.40}$$

and

$$\mu_B = G_B + \Omega(1 - X_B)^2 + RT \ln X_B$$

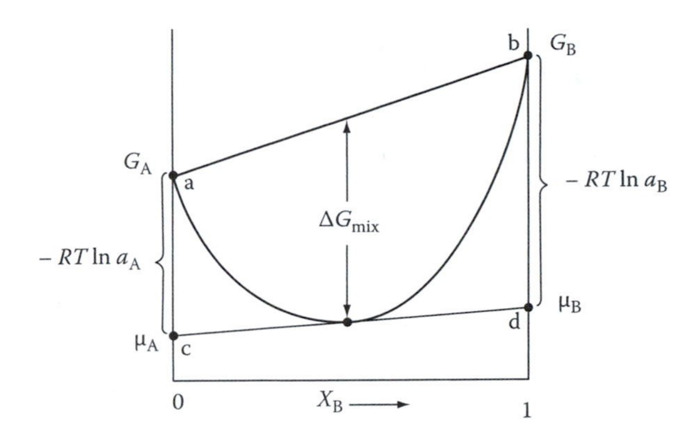

FIGURE 1.16
The relationship between molar free energy and activity.

1.3.5 Activity

Equation 1.32 for the chemical potential of an ideal alloy was simple and it is convenient to retain a similar expression for any solution. This can be done by defining the activity of a component, a, such that the distances ac and bd in Figure 1.16 are $-RT \ln a_A$ and $-RT \ln a_B$. In this case,

$$\mu_A = G_A + RT \ln a_A \tag{1.41}$$

and

$$\mu_B = G_B + RT \ln a_B$$

In general a_A and a_B will be different from X_A and X_B and the relationship between them will vary with the composition of the solution. For a regular solution, comparison of Equations 1.40 and 1.41 gives

$$\ln\left(\frac{a_A}{X_A}\right) = \frac{\Omega}{RT}(1 - X_A)^2 \tag{1.42}$$

and

$$\ln\left(\frac{a_B}{X_B}\right) = \frac{\Omega}{RT}(1 - X_B)^2$$

Assuming pure A and pure B have the same crystal structure, the relationship between a and X for any solution can be represented graphically as illustrated in Figure 1.17. Line 1 represents an ideal solution for which $a_A = X_A$ and $a_B = X_B$. If $\Delta H_{mix} < 0$ the activity of the components in solution will be less in an ideal solution (line 2) and vice versa when $\Delta H_{mix} > 0$ (line 3).

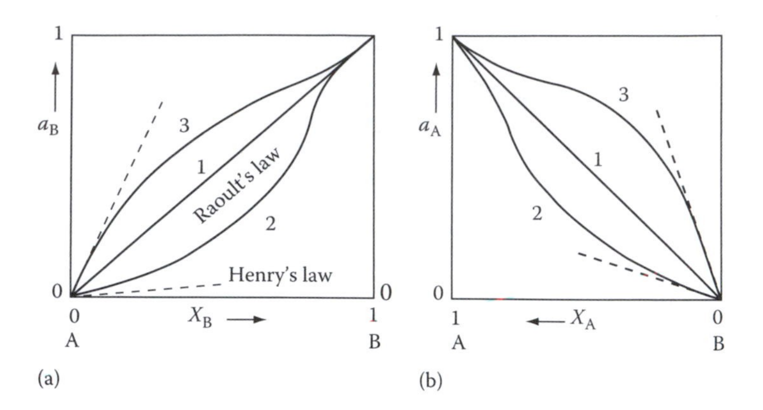

FIGURE 1.17
The variation of activity with composition (a) a_B (b) a_A. Line 1, ideal solution (Raoult's law); line 2, $\Delta H_{mix} < 0$; line 3, $\Delta H_{mix} > 0$.

The ratio (a_A/X_A) is usually referred to as γ_A, the activity coefficient of A, that is

$$\gamma_A = a_A/X_A \tag{1.43}$$

For a dilute solution of B in A, Equation 1.42 can be simplified by letting $X_B \rightarrow 0$ in which case,

$$\gamma_B = \frac{a_B}{X_B} \simeq \text{constant (Henry's law)} \tag{1.44}$$

and

$$\gamma_A = \frac{a_A}{X_A} \simeq 1 \text{ (Raoult's law)} \tag{1.45}$$

Equation 1.44 is known as Henry's law and Equation 1.45 as Raoult's law; they apply to all solutions when sufficiently dilute.

Since activity is simply related to chemical potential via Equation 1.41 the activity of a component is just another means of describing the state of the component in a solution. No extra information is supplied and its use is simply a matter of convenience as it often leads to simpler mathematics.

Activity and chemical potential are simply a measure of the tendency of the atom to leave a solution. If the activity or chemical potential is low the atoms are reluctant to leave the solution which means, for example, that the vapor pressure of the component in equilibrium with the solution will be relatively low. It will also be apparent later that the activity or chemical potential of a component is important when several condensed phases are in equilibrium.

1.3.6 Real Solutions

While the previous model provides a useful description of the effects of configurational entropy and interatomic bonding on the free energy of binary solutions its practical use is rather limited. For many systems the model is an oversimplification of reality and does not predict the correct dependence of ΔG_{mix} on composition and temperature.

As already indicated, in alloys where the enthalpy of mixing is not zero (ε and $\Omega \neq 0$) the assumption that a random arrangement of atoms is the equilibrium, or most stable arrangement is not true, and the calculated value for ΔG_{mix} will not give the minimum free energy. The actual arrangement of atoms will be a compromise that gives that lowest internal energy consistent with sufficient entropy, or randomness, to achieve the minimum free energy. In systems with $\varepsilon < 0$ the internal energy of the system is reduced by increasing the number of A–B bonds, i.e., by ordering the atoms as shown in Figure 1.18a. If $\varepsilon > 0$ the internal energy can be reduced by increasing the number of A–A and B–B bonds, i.e., by the clustering of the atoms into A-rich and B-rich groups (Figure 1.18b). However, the degree of ordering or clustering will decrease as temperature increases due to the increasing importance of entropy.

In systems where there is a size difference between the atoms the quasi-chemical model will underestimate the change in internal energy on mixing since no account is taken of the elastic strain fields which introduce a strain energy term into ΔH_{mix}. When the size difference is large this effect can dominate over the chemical term.

When the size difference between the atoms is very large then interstitial solid solutions are energetically most favorable (Figure 1.18c). New mathematical models are needed to describe these solutions.

In systems where there is a strong chemical bonding between the atoms there is a tendency for the formation of intermetallic phases. These are distinct from solutions based on the pure components since they have

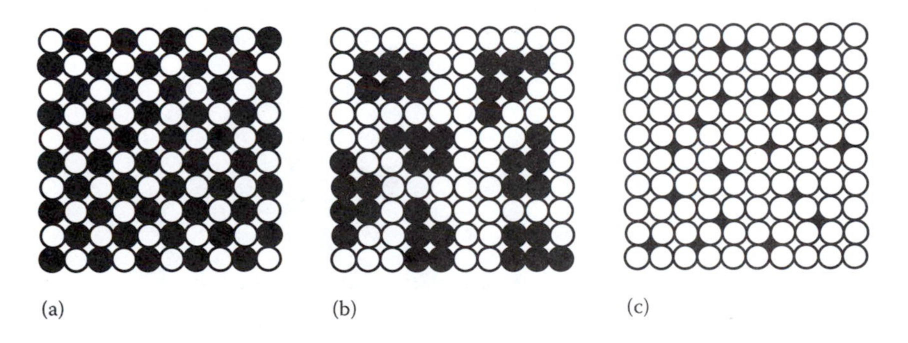

FIGURE 1.18
Schematic representation of solid solutions: (a) ordered substitutional, (b) clustering, and (c) random interstitial.

a different crystal structure and may also be highly ordered. Intermediate phases and ordered phases are discussed further in the next two sections.

1.3.7 Ordered Phases

If the atoms in a substitutional solid solution are completely randomly arranged each atom position is equivalent and the probability that any given site in the lattice will contain an A atom will be equal to the fraction of A atoms in the solution X_A, similarly X_B for the B atoms. In such solutions P_{AB}, the number of A–B bonds, is given by Equation 1.35. If $\Omega < 0$ and the number of A–B bonds is greater than this, the solution is said to contain short-range order (SRO). The degree of ordering can be quantified by defining a SRO parameter s such that

$$s = \frac{P_{AB} - P_{AB}(\text{random})}{P_{AB}(\max) - P_{AB}(\text{random})}$$

where $P_{AB}(\text{max})$ and $P_{AB}(\text{random})$ refer to the maximum number of bonds possible and the number of bonds for a random solution, respectively. Figure 1.19 illustrates the difference between random and short-range ordered solutions.

In solutions with compositions that are close to a simple ratio of A:B atoms another type of order can be found as shown schematically in Figure 1.18a. This is known as long-range order (LRO). Now the atom sites are no longer equivalent but can be labeled as A-sites and B-sites. Such a solution can be considered to be a different (ordered) phase separate from the random or nearly random solution.

Consider Cu–Au alloys as a specific example. Cu and Au are both fcc and totally miscible. At high temperatures Cu or Au atoms can occupy any site

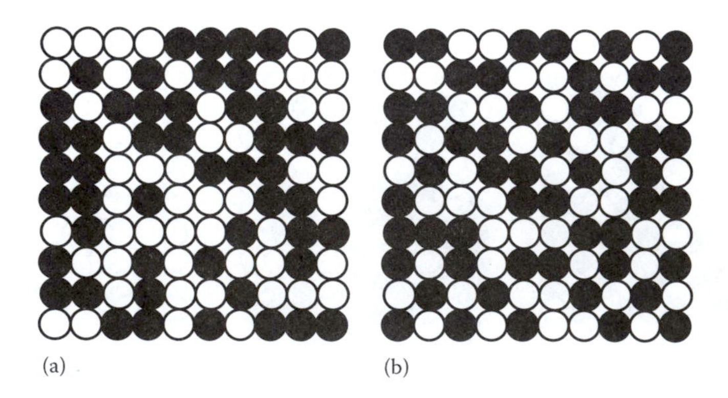

FIGURE 1.19
(a) Random A–B solution with total of 100 atoms and $X_A = X_B = 0.5$. $P_{AB} \sim 100$, $S = 0$. (b) Same alloy with SRO $P_{AB} = 132$. $P_{AB(\max)} \sim 200$, $S = (132 - 100)/(200 - 100) = 0.32$.

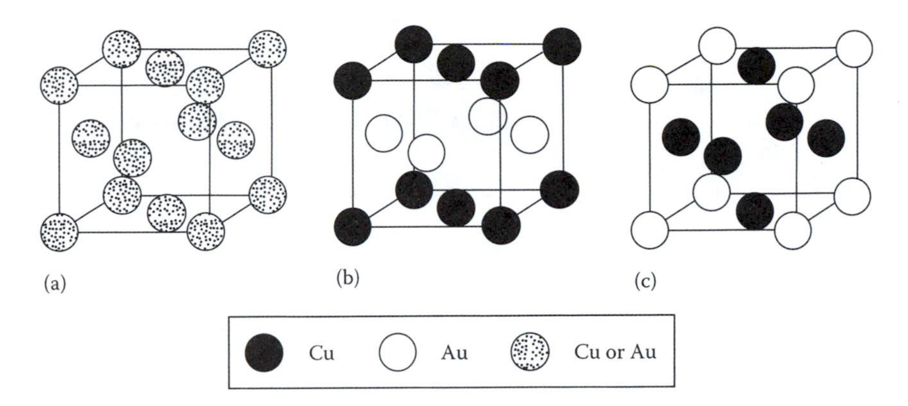

FIGURE 1.20
Ordered substitutional structures in the Cu–Au system: (a) high-temperature disordered structure, (b) CuAu superlattice, and (c) Cu_3Au superlattice.

and the lattice can be considered as fcc with a ''random'' atom at each lattice point as shown in Figure 1.20a. At low temperatures, however, solutions with $X_{Cu} = X_{Au} = 0.5$, i.e., a 50/50 Cu/Au mixture, form an ordered structure in which the Cu and Au atoms are arranged in alternate layers (Figure 1.20b). Each atom position is no longer equivalent and the lattice is described as a CuAu superlattice. In alloys with the composition Cu_3Au another superlattice is found (Figure 1.20c).

The entropy of mixing of structures with LRO is extremely small and with increasing temperature the degree order decreases until above some critical temperature there is no LRO at all. This temperature is a maximum when the composition is the ideal required for the superlattice. However, LRO can still be obtained when the composition deviates from the ideal if some of the atom sites are left vacant of if some atoms sit on wrong sites. In such cases it can be easier to disrupt the order with increasing temperature and critical temperature is lower (Figure 1.21).

The most common ordered lattices in other systems are summarized in Figure 1.22 along with their Structurbericht notation and examples of alloys

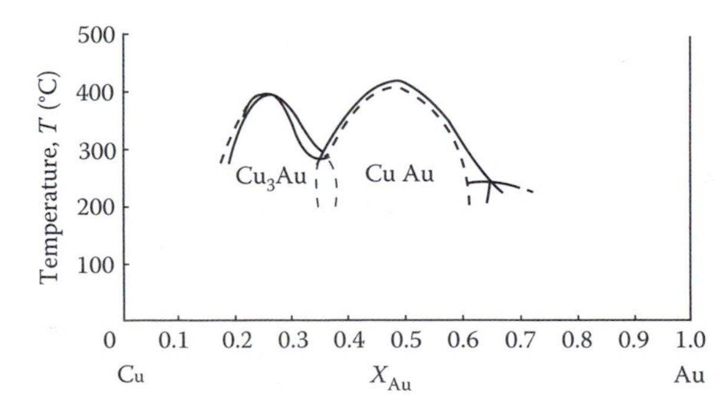

FIGURE 1.21
Part of the Cu–Au phase diagram showing the regions where the Cu_3Au and CuAu superlattices are stable.

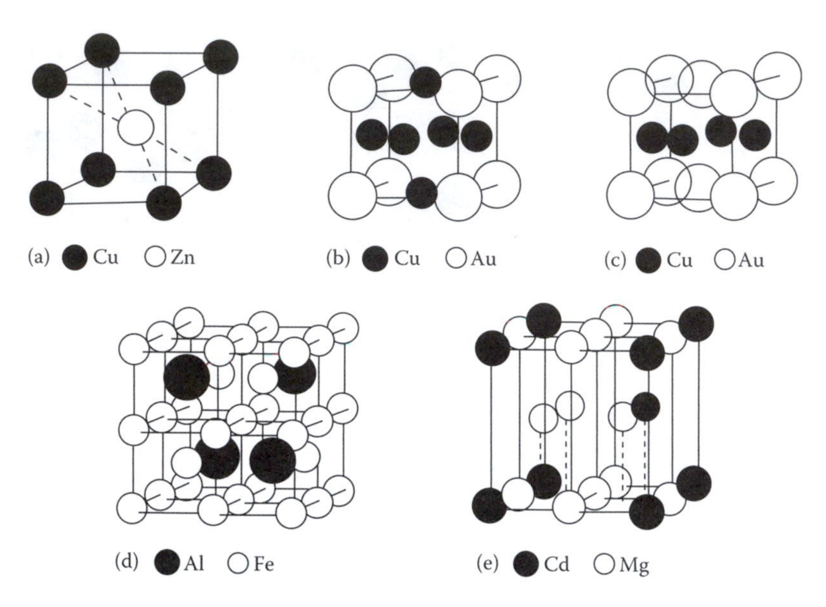

FIGURE 1.22
The five common ordered lattices, examples of which are (a) $L2_0$: CuZn, FeCo, NiAl, FeAl, AgMg; (b) $L1_2$: Cu_3Au, Au_3Cu, Ni_3Mn, Ni_3Fe, Ni_3Al, Pt_3Fe; (c) $L1_0$: CuAu, CoPt, FePt; (d) $D0_3$: Fe_3Al, Fe_3Si, Fe3Be, Cu_3Al; (e) $D0_{19}$: Mg_3Cd, Cd_3Mg, Ti_3Al, Ni_3Sn. (After Smallman, R.E., *Modern Physical Metallurgy*, 3rd edn., Butterworths, London, 1970.)

in which they are found. Finally, note that the critical temperature for loss of LRO increases with increasing Ω, or ΔH_{mix}, and in many systems the ordered phase is stable up to the melting point.

1.3.8 Intermediate Phases

Often the configuration of atoms that has the minimum free energy after mixing does not have the same crystal structure as either of the pure components. In such cases the new structure is known as an intermediate phase.

Intermediate phases are often based on an ideal atom ratio that results in a minimum Gibbs free energy. For compositions that deviate from the ideal, the free energy is higher giving a characteristic "U" shape to the G curve, as in Figure 1.23. The range of compositions over which the free energy curve has a meaningful existence depends on the structure of the phase and the type of interatomic bonding—metallic, covalent, or ionic. When small composition deviations cause a rapid rise in G the phase is referred to as an intermetallic compound and is usually stoichiometric, i.e., has a formula A_mB_n where m and n are integers (Figure 1.23a). In other structures fluctuations in composition can be tolerated by some atoms occupying "wrong" positions or by atom sites being left vacant, and in these cases the curvature of the G curve is much less (Figure 1.23b).

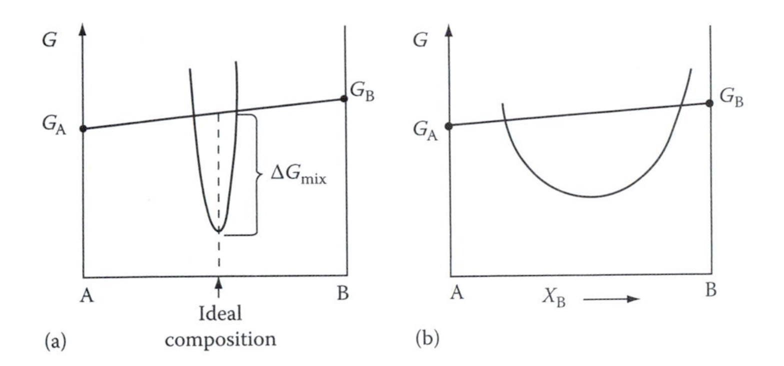

FIGURE 1.23
Free energy curves for intermediate phases: (a) for an intermetallic compound with a very narrow stability range, (b) for an intermediate phase with a wide stability range.

Some intermediate phases can undergo order–disorder transformations in which an almost random arrangement of the atoms is stable at high temperatures and an ordered structure is stable below some critical temperature. Such a transformation occurs in the β-phase in the Cu–Zn system for example (see Section 5.10).

The structure of intermediate phases is determined by three main factors: relative atomic size, valency, and electronegativity. When the component atoms differ in size by a factor of about 1.1–1.6 it is possible for the atoms to fill space most efficiently if the atoms order themselves into one of the so-called Laves phases based on $MgCu_2$, $MgZn_2$, and $MgNi_2$ (Figure 1.24). Another example where atomic size determines the structure is in the formation of the interstitial compounds MX, M_2X, MX_2, and M_6X where M can be Zr, Ti, V, Cr, etc. and X can be H, B, C, and N. In this case, the M atoms form a

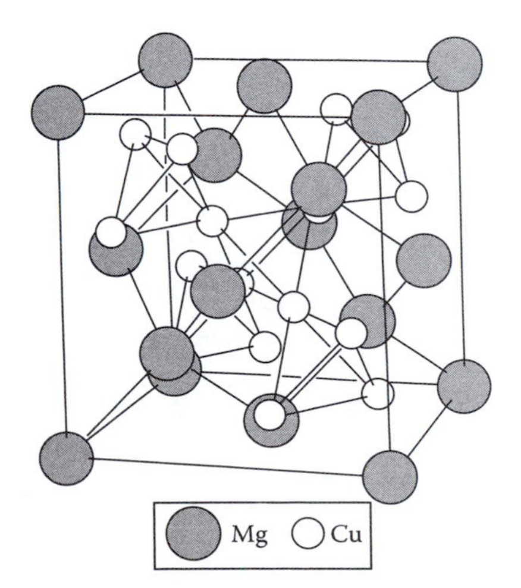

FIGURE 1.24
Structure of $MgCu_2$ (A Laves phase). (From Wernick, J.H., in *Physical Metallurgy*, 2nd edn., Chapter 5, Cahn, R.W. (Ed.), North Holland, Amsterdam, 1974. With permission.)

cubic or hcp arrangement and the X atoms are small enough to fit into the interstices between them.

The relative valency of the atoms becomes important in the so-called electron phases, e.g., α and β phases. The free energy of these phases depends on the number of valency electrons per unit cell, and this varies with composition due to the valency difference.

The electronegativity of an atom is a measure of how strongly it attracts electrons and in systems where the two components have very different electronegativities ionic bonds can be formed producing normal valency compounds, e.g., Mg^{2+} and Sn^{4-} are ionically bonded in Mg_2Sn.[2]

1.4 Equilibrium in Heterogeneous Systems

It is usually the case that A and B do not have the same crystal structure in their pure states at a given temperature. In such cases two free energy curves must be drawn, one for each structure. The stable forms of pure A and B at a given temperature (and pressure) can be denoted as α and β, respectively. For the sake of illustration let α be fcc and β bcc. The molar free energies of fcc A and bcc B are shown in Figure 1.25a as points a and b. The first step in drawing the free energy curve of the fcc α-phase is, therefore, to convert the stable bcc arrangement of B atoms into an unstable fcc arrangement. This requires an increase in free energy, bc. The free energy curve for the α-phase can now be constructed as before by mixing fcc A and fcc B as shown in Figure 1.25. $-\Delta G_{mix}$ for α of composition X is given by the distance de as usual.

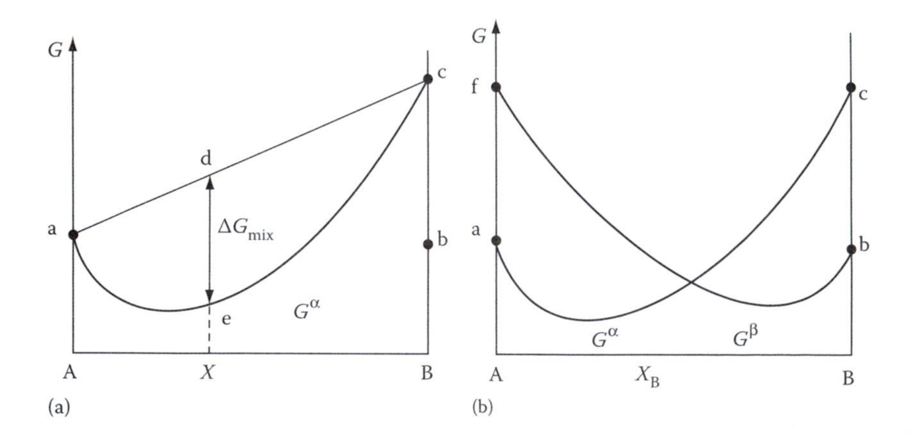

FIGURE 1.25
(a) Molar free energy curve for the α-phase. (b) Molar free energy curves for α- and β-phases.

A similar procedure produces the molar free energy curve for the β-phase (Figure 1.25b). The distance af is now the difference in free energy between bcc A and fcc A.

It is clear from Figure 1.25b that A-rich alloys will have the lowest free energy as a homogeneous α-phase and B-rich alloys as β-phase. For alloys with compositions near the cross-over in G curves the situation is not so straightforward. In this case, it can be shown that the total free energy can be minimized by the atoms separating into two phases.

It is first necessary to consider a general property of molar free energy diagrams when phase mixtures are present. Suppose an alloy consists of two phases α and β each of which has a molar free energy given by G^{α} and G^{β} (Figure 1.26). If the overall composition of the phase mixture is X_B^0 the lever rule gives the relative number of moles of α and β that must be present, and the molar free energy of the phase mixture G is given by the point on the straight line between α and β as shown in the figure. This result can be proven most readily using the geometry of Figure 1.26. The lengths ad and cf, respectively, represents the molar free energies of the α- and β-phases present in the alloy. Point g is obtained by the intersection of be and dc so that bcg and acd, as well as deg and dfc, form similar triangles. Therefore, bg/ad = bc/ac and ge/cf = ab/ac. According to the lever rule, 1 mol of alloy will contain bc/ac mol of α and ab/ac mol of β. It follows that bg and ge represent the separate contributions from the α- and β-phases to the total free energy of 1 mol of alloy. Therefore, the length be represents the molar free energy of the phase mixture.

Consider now alloy X^0 in Figure 1.27a. If the atoms are arranged as a homogeneous phase, the free energy will be lowest as α, i.e., G_0^{α} per mole. However, from the above it is clear that the system can lower its free energy if the atoms separate into two phases with compositions α_1 and β_1 for example. The free energy of the system will then be reduced to G_1. Further reductions in free energy can be achieved if the A and B atoms interchange between the α- and β-phases until the compositions α_e and β_e are reached (Figure 1.27b). The free energy of the system G_e is now a minimum and there is no desire for further change. Consequently the system is in equilibrium and α_e and β_e are the equilibrium compositions of the α- and β-phases.

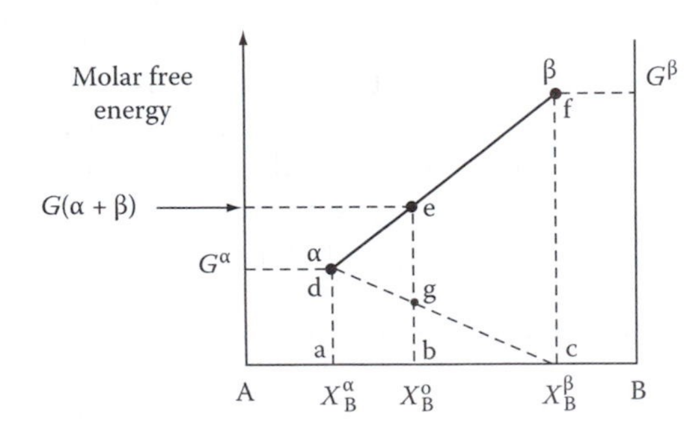

FIGURE 1.26
The molar free energy of a two-phase mixture (α + β).

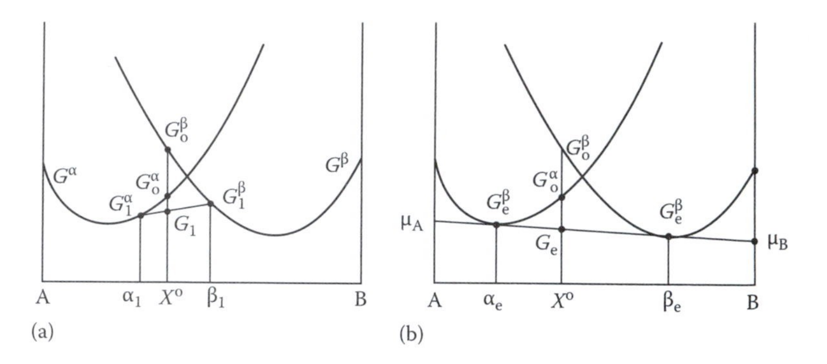

FIGURE 1.27
(a) Alloy X^0 has a free energy G_1 as a mixture of $\alpha_1 + \beta_1$. (b) At equilibrium, alloy X^0 has a minimum free energy G_e when it is a mixture of $\alpha_e + \beta_e$.

This result is quite general and applies to any alloy with an overall composition between α_e and β_e: only the relative amounts of the two phases change, as given by the lever rule. When the alloy composition lies outside this range, however, the minimum free energy lies on the G^α or G^β curves and the equilibrium state of the alloy is a homogeneous single phase.

From Figure 1.27 it can be seen that equilibrium between two phases requires that the tangents to each G curve at the equilibrium compositions lie on a common line. In other words, each component must have the same chemical potential in the phased, i.e., for heterogeneous equilibrium:

$$\mu_A^\alpha = \mu_A^\beta, \quad \mu_\beta^\alpha = \mu_B^\beta \tag{1.46}$$

The condition for equilibrium in a heterogeneous system containing two phases can also be expressed using the activity concept defined for homogeneous systems in Figure 1.16. In heterogeneous systems containing more than one phase the pure components can, at least theoretically, exist in different crystal structures. The most stable state, with the lowest free energy, is usually defined as the state in which the pure component has unit activity. In the present example this would correspond to defining the activity of A in pure α-A as unity, i.e., when $X_A = 1$, $a_A^\alpha = 1$. Similarly when $X_B = 1$, $a_B^\beta = 1$. This definition of activity is shown graphically in Figure 1.28a; Figure 1.28b and c shows how the activities of B and A vary with the composition of the α- and β-phases. Between A and α_e, and β_e and B, where single phases are stable, the activities (or chemical potentials) vary and for simplicity ideal solutions have been assumed in which case there is a straight line relationship between a and X. Between α_e and β_e the phase compositions in equilibrium do not change and the activities are equal and given by points q and r. In other words, when two phases exist in equilibrium, the activities of the components in the system must be equal in the two phases, i.e.,

$$a_A^\alpha = a_A^\beta, \quad a_B^\alpha = a_B^\beta \tag{1.47}$$

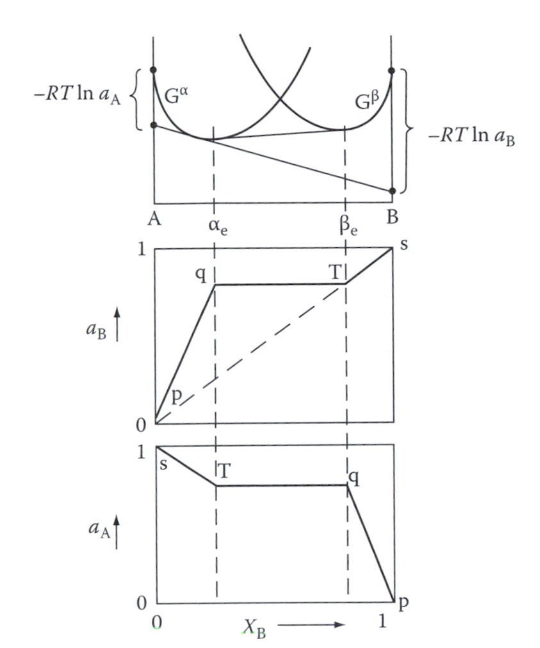

FIGURE 1.28
The variation of a_A and a_B with composition for a binary system containing two ideal solutions, α and β.

1.5 Binary Phase Diagrams

In Section 1.4, it has been shown how the equilibrium state of an alloy can be obtained from the free energy curves at a given temperature. The next step is to see how equilibrium is affected by temperature.

1.5.1 Simple Phase Diagram

The simplest case to start with is when A and B are completely miscible in both the solid and liquid states and both are ideal solutions. The free energy of pure A and pure B will vary with temperature as shown schematically in Figure 1.4. The equilibrium melting temperatures of the pure components occur when $G^S = G^L$, i.e., at $T_m(A)$ and $T_m(B)$. The free energy of both phases decreases as temperature increases. These variations are important for A–B alloys also since they determine the relative positions of G_A^S, G_A^L, G_B^S, and G_B^L on the molar free energy diagrams of the alloy at different temperatures (Figure 1.29).

At a high temperature $T_1 > T_m(A) > T_m(B)$ the liquid will be the stable phase for pure A and pure B, and for the simple case we are considering that the liquid also has a lower free energy than the solid at all the intermediate compositions as shown in Figure 1.29a.

Decreasing the temperature will have two effects: first G_A^L and G_B^L will increase more rapidly than G_A^S and G_B^S, second the curvature of the G

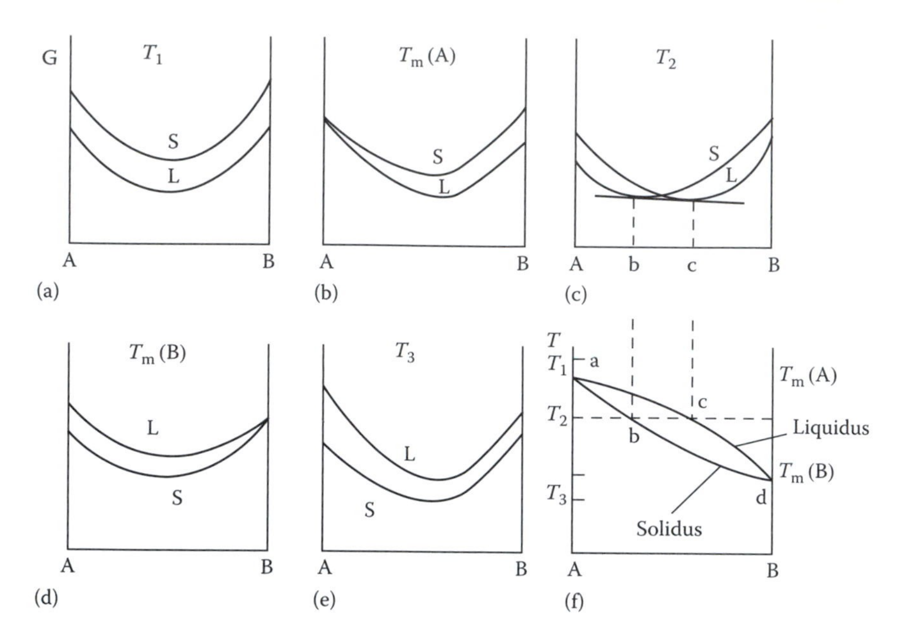

FIGURE 1.29
Derivation of a simple phase diagram from the free energy curves for the liquid (L) and solid (S).

curves will be reduced due to the smaller contribution of $-T\Delta S_{mix}$ to the free energy.

At T_m(A), Figure 1.29b, $G_A^S = G_A^L$, and this corresponds to point a on the A–B phase diagram, Figure 1.29f. At a lower temperature T_2 the free energy curves cross, Figure 1.29c, and common tangent construction indicates that alloys between A and b are solid at equilibrium, between c and B they are liquid, and between b and c equilibrium consists of a two-phase mixture (S + L) with compositions b and c. These points are plotted on the equilibrium phase diagram at T_2.

Between T_2 and T_m(B) G^L continues to rise faster than G^S so that points b and c in Figure 1.29c will both move to the right tracing out the solidus and liquidus lines in the phase diagram. Eventually the T_m(B) b and c will meet at a single point, d in Figure 1.29f. Below T_m(B) the free energy of the solid phase is everywhere below that of the liquid and all alloys are stable as a single-phase solid.

1.5.2 Systems with a Miscibility Gap

Figure 1.30 shows the free energy curves for a system in which the liquid phase is approximately ideal, but for the solid phase $\Delta H_{mix} > 0$, i.e., the A and B atoms dislike each other. Therefore, at low temperatures (T_3) the free energy curve for the solid assumes a negative curvature in the middle, Figure 1.30c, and the solid solution is most stable as a mixture of two phases α' and α''

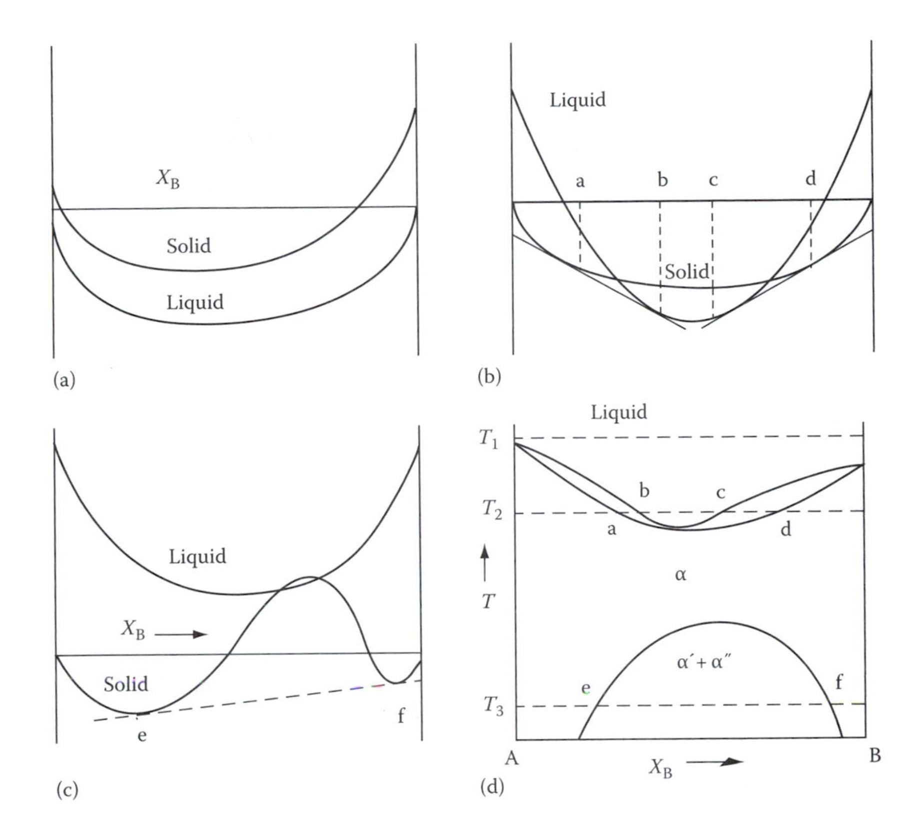

FIGURE 1.30
Derivation of a phase diagram (d) where $\Delta H^{S}_{mix} > \Delta H^{L}_{mix} = 0$. Free energy versus composition curves for (a) T_1, (b) T_2, and (c) T_3.

with composition e and f. At higher temperatures, when $-T\Delta S_{mix}$ becomes larger, e and f approach each other and eventually disappear as shown in the phase diagram (Figure 1.30d). The $\alpha' + \alpha''$ region is known as the miscibility gap.

The effect of a positive ΔH_{mix} in the solid is already apparent at higher temperatures where it gives rise to a minimum melting point mixture. The reason why all alloys should melt at temperatures below the melting points of both components can be qualitatively understood since the atoms in the alloy repel each other making the disruption of the solid into a liquid phase possible at lower temperatures than in either pure A or pure B.

1.5.3 Ordered Alloys

The opposite type of effect arises when $\Delta H_{mix} < 0$. In these systems melting will be more difficult in the alloys and a maximum melting point mixture may appear. This type of alloy also has a tendency to order at low temperatures as shown in Figure 1.31a. If the attraction between unlike atoms is very strong the ordered phase may extend as far as the liquid (Figure 1.31b).

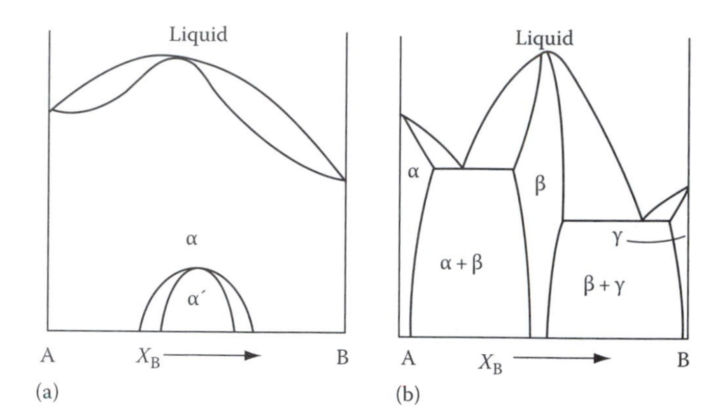

FIGURE 1.31
(a) Phase diagram when $\Delta H_{mix}^{S} < 0$; (b) as (a) but even more negative ΔH_{mix}^{S}. (After Swalin, R.A., *Thermodynamics of Solids*, Wiley, New York, 1972.)

1.5.4 Simple Eutectic Systems

If ΔH_{mix}^{S} is much larger than zero the miscibility gap in Figure 1.30d can extend into the liquid phase. In this case, a simple eutectic phase diagram results as shown in Figure 1.32. A similar phase diagram can result when A and B have different crystal structures as illustrated in Figure 1.33.

1.5.5 Phase Diagrams Containing Intermediate Phases

When stable intermediate phases can form, extra free energy curves appear in the phase diagram. An example is shown in Figure 1.34, which also illustrates how a peritectic transformation is related to the free energy curves.

An interesting result of the common tangent construction is that the stable composition range of the phase in the phase diagram need not include the composition with the minimum free energy, but is determined by the relative free energies of adjacent phases (Figure 1.35). This can explain why the composition of the equilibrium phase appears to deviate from that which would be predicated from the crystal structure. For example, the θ-phase in the Cu–Al system is usually denoted as $CuAl_2$ although the composition $X_{Cu} = 1/3$, $X_{Al} = 2/3$ is not covered by the θ field on the phase diagram.

1.5.6 Gibbs Phase Rule

The condition for equilibrium in a binary system containing two phases is given by Equation 1.46 or Equation 1.47. A more general requirement for systems containing several components and phases is that the chemical potential of each component must be identical in every phase, i.e.,

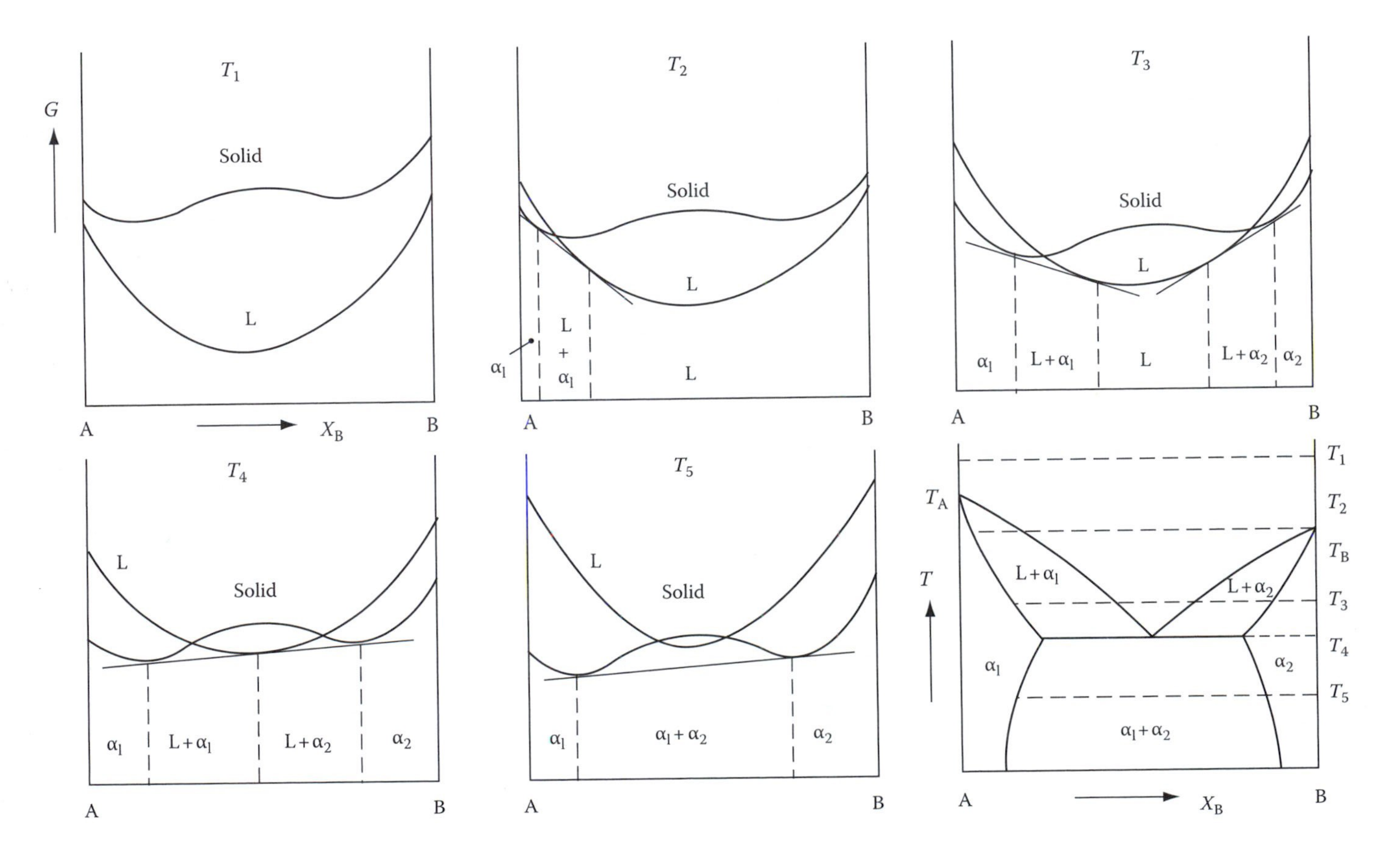

FIGURE 1.32
Derivation of a eutectic phase diagram where both solid phases have the same crystal structure. (After Cottrell, A.H., *Theoretical Structural Metallurgy*, Edward Arnold, London, 1995.)

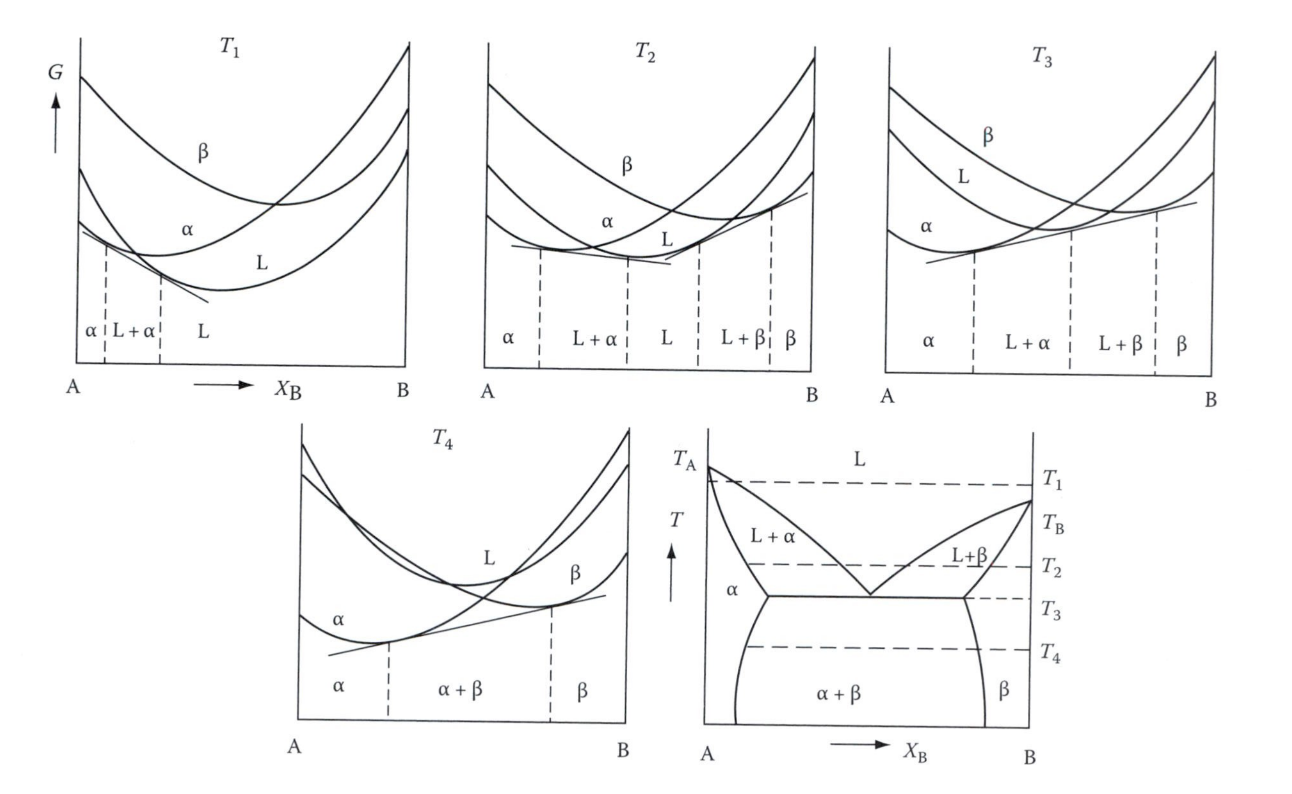

FIGURE 1.33
Derivation of a eutectic phase diagram where each solid phase has a different crystal structure. (After Prince, A., *Alloy Phase Equilibria*, Elsevier, Amsterdam, 1966.)

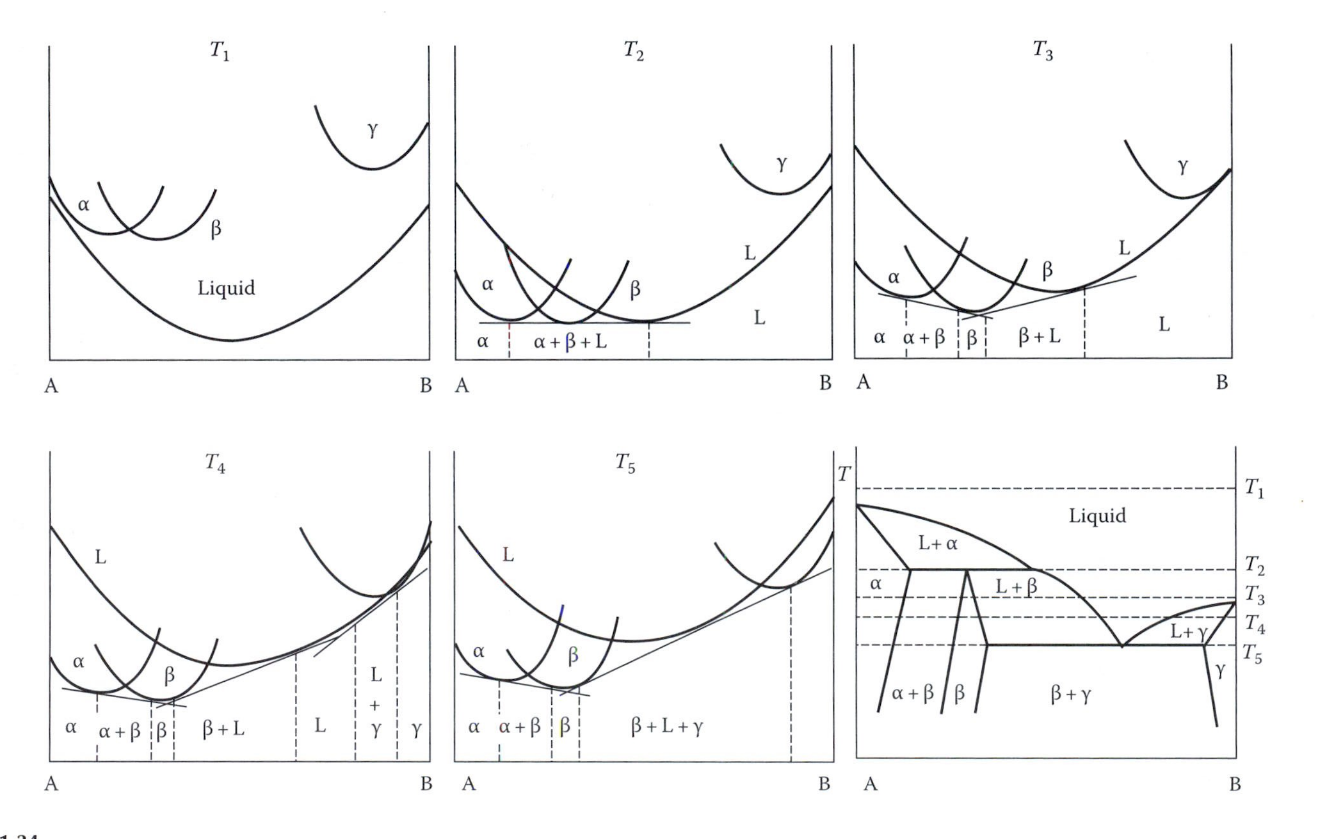

FIGURE 1.34
Derivation of a complex phase diagram. (After Cottrell, A.H., *Theoretical Structural Metallurgy*, Edward Arnold, London, 1955.)

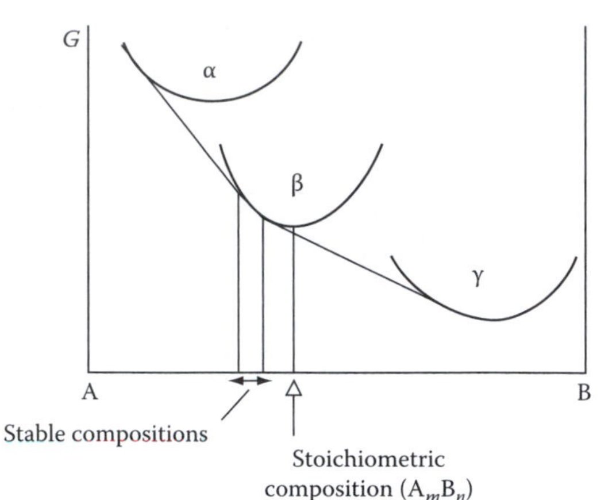

FIGURE 1.35
Free energy diagram to illustrate that the range of compositions over which a phase is stable depends on the free energies of the other phases in equilibrium.

$$
\begin{aligned}
\mu_A^\alpha &= \mu_A^\beta = \mu_A^\gamma = \cdots \\
\mu_B^\alpha &= \mu_B^\beta = \mu_B^\gamma = \cdots \\
\mu_C^\alpha &= \mu_C^\beta = \mu_C^\gamma = \cdots
\end{aligned}
\tag{1.48}
$$

The proof of this relationship is left as an exercise for the reader (see Exercise 1.10). A consequence of this general condition is the Gibbs phase rule. This states that if a system containing C components and P phases is in equilibrium the number of degrees of freedom F is given by

$$P + F = C + 2 \tag{1.49}$$

A degree of freedom is an intensive variable such as T, P, X_A, $X_B, \ldots$ that can be varied independently while still maintaining equilibrium. If pressure is maintained constant one degree of freedom is lost and the phase rule becomes

$$P + F = C + 1 \tag{1.50}$$

At present we are considering binary alloys so that $C = 2$ therefore

$$P + F = 3$$

This means that a binary system containing one phase has two degrees of freedom, i.e., T and X_B can be varied independently. In a two-phase region of a phase diagram $P = 2$ and therefore $F = 1$, which means that if the temperature is chosen independently the compositions of the phases are fixed. When three phases are in equilibrium, such as at a eutectic or peritectic temperature, there are no degrees of freedom and the compositions of the phases and the temperature of the system are all fixed.

1.5.7 Effect of Temperature on Solid Solubility

The equations for free energy and chemical potential can be used to derive the effect of temperature on the limits of solid solubility in a terminal solid solution. Consider for simplicity the phase diagram shown in Figure 1.36a where B is soluble in A, but A is virtually insoluble in B. The corresponding free energy curves for temperature T_1 are shown schematically in Figure 1.36b. Since A is almost insoluble in B the G^{β} curve rises rapidly as shown. Therefore, the maximum concentration of B soluble in A(X_B^e) is given by the condition

$$\mu_B^{\alpha} = \mu_B^{\beta} \simeq G_B^{\beta}$$

For a regular solid solution Equation 1.40 gives

$$\mu_B^{a} = G_B^{\alpha} + \Omega(1 - X_B)^2 + RT \ln X_B$$

But from Figure 1.36b, $G_B^{\alpha} - \mu_B^{\alpha} = \Delta G_B$, the difference in free energy between pure B in the stable β-form and the unstable α-form. Therefore for $X_B = X_B^c$

$$-RT \ln X_B^e - \Omega(1 - X_B^e)^2 = \Delta G_B \tag{1.51}$$

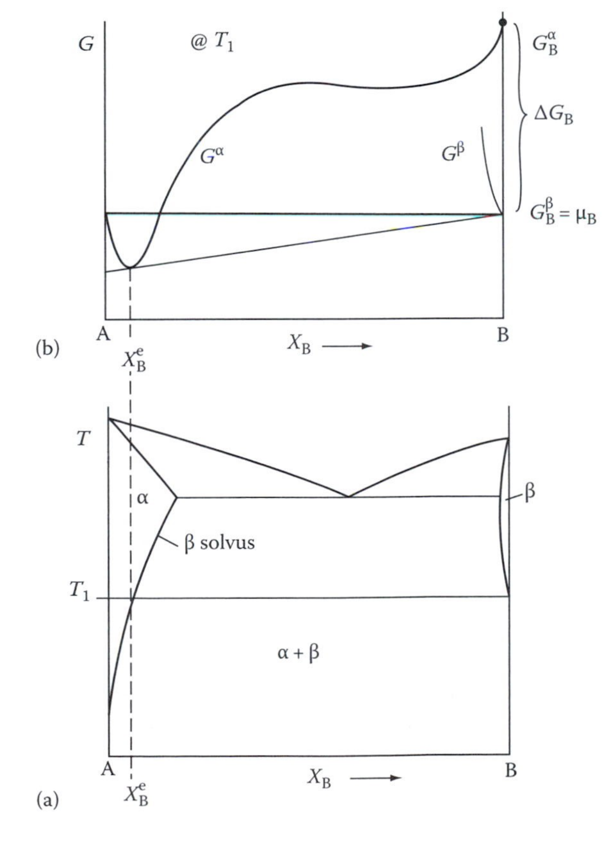

FIGURE 1.36
Solubility of B in A.

If the solubility is low $X_B^e \ll 1$ and this gives

$$X_B^e = \exp\left\{-\frac{\Delta G_B + \Omega}{RT}\right\} \quad (1.52)$$

Putting

$$\Delta G_B = \Delta H_B - T\Delta S_B$$

gives

$$X_B^e = A\exp\frac{-Q}{RT} \quad (1.53)$$

where A is a constant equal to $\exp(\Delta S_B/R)$ and

$$Q = \Delta H_B + \Omega \quad (1.54)$$

ΔH_B is the difference in enthalpy between the β-form of B and the α-form in J mol^{-1}. Ω is the change in energy when 1 mol of B with the α-structure dissolves in A to make a dilute solution. Therefore Q is just the enthalpy change, or heat absorbed, when 1 mol of B with the β-structure dissolves in A to make a dilute solution.

ΔS_B is the difference in entropy between β-B and α-B and is approximately independent of temperature. Therefore, the solubility of B in α increases exponentially with temperature at a rate determined by Q. It is interesting to note that, except at absolute zero, X_B^e can never be equal to zero, that is, no two components are ever completely insoluble in each other.

1.5.8 Equilibrium Vacancy Concentration

So far it has been assumed that in a metal lattice every atom site is occupied. However, let us now consider the possibility that some sites remain without atoms, that is, there are vacancies in the lattice. The removal of atoms from their sites not only increases the internal energy of the metal, due to the broken bonds around the vacancy, but also increases the randomness or configurational entropy of the system. The free energy of the alloy will depend on the concentration of vacancies, and the equilibrium concentration X_v^e will be that which gives the minimum free energy.

If, for simplicity, we consider vacancies in a pure metal the problem of calculating X_v^e is almost identical to the calculation of ΔG_{mix} for A and B atoms when ΔH_{mix} is positive. Because the equilibrium concentration of vacancies is small the problem is simplified because vacancy–vacancy interactions can be ignored and the increase in enthalpy of the solid (ΔH) is directly proportional to the number of vacancies added, i.e.,

$$\Delta H \simeq \Delta H_v X_v$$

where

X_v is the mole fraction of vacancies

ΔH_v is the increase in enthalpy per mole of vacancies added

Each vacancy causes an increase of $\Delta H_v/N_a$ where N_a is Avogadro's number.

There are two contributions to the entropy change ΔS on adding vacancies. There is a small change in the thermal entropy of ΔS_v per mole of vacancies added due to changes in the vibrational frequencies of the atoms around a vacancy. The largest contribution, however, is due to the increase in configurational entropy given by Equation 1.25. The total entropy change is thus

$$\Delta S = X_v \Delta S_v - R(X_v \ln X_v + (1 - X_v)\ln(1 - X_v))$$

The molar free energy of the crystal containing X_v mol of vacancies is therefore given by

$$G = G_A + \Delta G = G_A + \Delta H_v X_v - T\Delta S_v X_v + RT(X_v \ln X_v + (1 - X_v)\ln(1 - X_v)) \tag{1.55}$$

This is shown schematically in Figure 1.37. Given time the number of vacancies will adjust so as to reduce G to a minimum. The equilibrium concentration of vacancies X_v^e is therefore given by the condition

$$\left.\frac{dG}{dX_v}\right|_{X_v = X_v^e} = 0$$

Differentiating Equation 1.55 and making the approximation $X_v \ll 1$ gives

$$\Delta H_v - T\Delta S_v + RT \ln X_v^e = 0$$

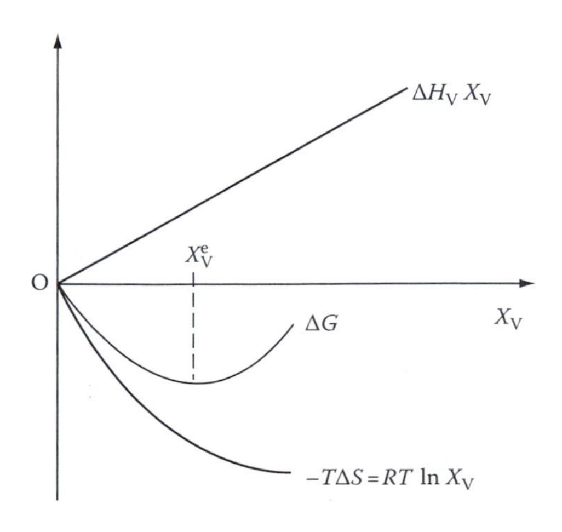

FIGURE 1.37
Equilibrium vacancy concentration.

Therefore, the expression for X_v^e is

$$X_v^e = \exp\frac{\Delta S_v}{R} \cdot \exp\frac{-\Delta H_v}{RT} \quad (1.56)$$

or, putting $\Delta G_v = \Delta H_v - T\Delta S_v$ gives

$$X_v^e = \exp\frac{-\Delta G_v}{RT} \quad (1.57)$$

The first term on the right-hand side of Equation 1.56 is a constant ~3, independent of T, whereas the second term increases rapidly with increasing T. In practice, ΔH_v is of the order of 1 eV per atom and X_v^e reaches a value of about 10^{-4} to 10^{-3} at the melting point of the solid.

1.6 Influence of Interfaces on Equilibrium

The free energy curves that have been drawn so far have been based on the molar free energies of infinitely large amounts of material of a perfect single crystal. Surfaces, grain boundaries, and interphase interfaces have been ignored. In real situations these and other crystal defects such as dislocations do exist and raise the free energies of the phases. Therefore the minimum free energy of an alloy, i.e., the equilibrium state, is not reached until virtually all interfaces and dislocations have been annealed out. In practice such a state is unattainable within reasonable periods of time.

Interphase interfaces can become extremely important in the early stages of phase transformations when one phase, β, say, can be present as very fine particles in the other phase, α, as shown in Figure 1.38a. If the α-phase is acted on by a pressure of 1 atm the β-phase is subjected to an extra pressure ΔP due to the curvature of the α/β interface, just as a soap bubble exerts an extra pressure ΔP on its contents. If γ is the α/β interfacial energy and the particles are spherical with a radius r, ΔP is given approximately by

$$\Delta P = \frac{2\gamma}{r}$$

By definition, the Gibbs free energy contains a PV term and an increase of pressure P therefore causes an increase in free energy G. From Equation 1.9 at constant temperature

$$\Delta G = \Delta P \cdot V$$

Therefore, the β curve on the molar free energy–composition diagram in Figure 1.38b will be raised by an amount

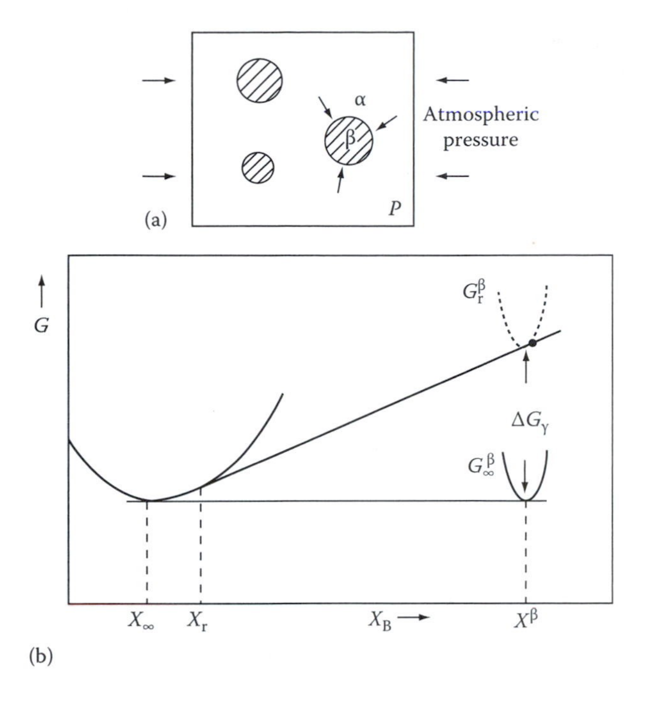

FIGURE 1.38
The effect of interfacial energy on the solubility of small particles.

$$\Delta G_\gamma = \frac{2\gamma V_m}{r} \tag{1.58}$$

where V_m is the molar volume of the β-phase. This free energy increase due to interfacial energy is known as a capillarity effect or the Gibbs–Thomson effect.

The concept of a pressure difference is very useful for spherical liquid particles, but it is less convenient in solids. This is because, as will be discussed in Chapter 3, finely dispersed solid phases are often nonspherical. For illustration, therefore, consider an alternative derivation of Equation 1.58 which can be more easily modified to deal with nonspherical cases.[3]

Consider a system containing two β particles, one with a spherical interface of radius r and the other with a planar interface ($r = \infty$) embedded in an α matrix as shown in Figure 1.39. If the molar free energy difference between the two particles is ΔG_γ, the transfer of a small quantity (dn mol) of β from the large to the small particle will increase the free energy of the system by a small amount (dG) given by

$$\mathrm{d}G = \Delta G_\gamma \mathrm{d}n$$

If the surface area of the large particle remains unchanged the increase in free energy will be due to the increase in the interfacial area of the spherical particle (dA). Therefore, assuming γ is constant

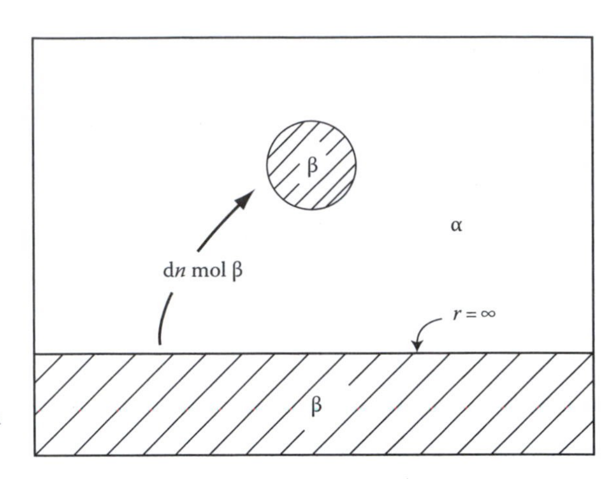

FIGURE 1.39
Transfer of dn mol of β from a large to a small particle.

$$dG = \gamma dA$$

Equating these two expressions gives

$$\Delta G_\gamma = \gamma \frac{dA}{dn} \tag{1.59}$$

Since $n = 4\pi r^3/3V_m$ and $A = 4\pi r^2$ it can easily be shown that

$$\frac{dA}{dn} = \frac{dA/dr}{dn/dr} = \frac{2V_m}{r}$$

from which Equation 1.58 can be obtained.

An important practical consequence of the Gibbs–Thomson effect is that the solubility of β in α is sensitive to the size of the β particles. From the common tangent construction in Figure 1.38b, it can be seen that the concentration of solute B in α in equilibrium with β across curved interface (X_r) is greater than X_∞, the equilibrium concentration for a planar interface. Assuming for simplicity that the α-phase is a regular solution and that the β-phase is almost pure B, i.e., $X_B^\beta \sim 1$, Equation 1.52 gives

$$X_\infty = \exp\left\{-\frac{\Delta G_B + \Omega}{RT}\right\}$$

Similarly X_r can be obtained by using $(\Delta G_B - 2\gamma V_m/r)$ in place of ΔG_B

$$X_r = \exp\left\{-\frac{\Delta G_B + \Omega - 2\gamma V_m/r}{RT}\right\}$$

Therefore,

$$X_r = X_\infty \exp\frac{2\gamma V_m}{RTr} \tag{1.60}$$

and for small values of the exponent

$$X_r \simeq X_\infty\left(1 + \frac{2\gamma V_m}{RTr}\right) \tag{1.61}$$

Taking the following typical values: $\gamma = 200$ mJ m^{-2}, $V_m = 10^{-5}$ m^3, $R = 8.31$ J mol^{-1} K^{-1}, $T = 500$ K gives

$$\frac{X_r}{X_\infty} \simeq 1 + \frac{1}{r\ (\text{nm})}$$

e.g., for $r = 10$ nm $X_r/X_\infty \sim 1.1$. It can be seen therefore that quite large solubility differences can arise for particles in the range $r = 1-100$ nm. However, for particles visible in the light microscope ($r > 1$ μm) capillarity effects are very small.

1.7 Ternary Equilibrium

Since most commercial alloys are based on at least three components, an understanding of ternary phase diagrams is of great practical importance. The ideas that have been developed for binary systems can be extended to systems with three or more components.[4]

The composition of a ternary alloy can be indicated on an equilateral triangle (the Gibbs triangle) whose corners represent 100% A, B, or C as shown in Figure 1.40. The triangle is usually divided by equidistant lines parallel to the sides marking 10% intervals in atomic or weight percent. All points on lines parallel to BC contain the same percentage of A, the lines parallel to AC represent constant B concentration, and lines parallel to AB constant C concentrations. Alloys on PQ for example contain 60% A, on RS 30% B, and TU 10% C. Clearly the total percentage must sum to 100%, or expressed as mole fractions

$$X_A + X_B + X_C = 1 \tag{1.62}$$

The Gibbs free energy of any phase can now be represented by a vertical distance from the point in the Gibbs triangle. If this is done for all possible compositions, the points trace out the free energy surfaces for all the possible phases, as shown in Figure 1.41a. The chemical potentials of A, B, and C in

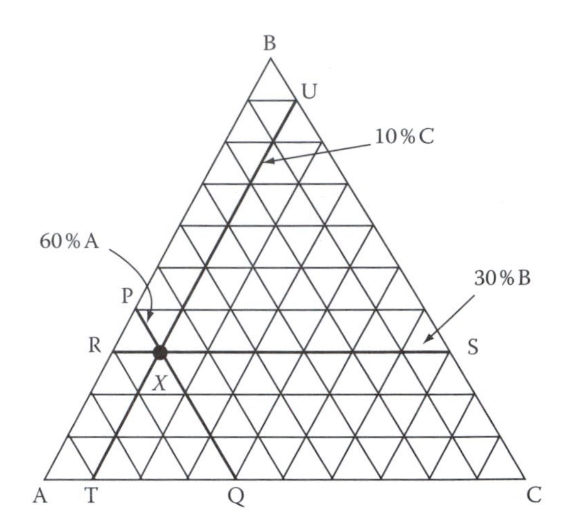

FIGURE 1.40
The Gibbs triangle.

any phase are then given by the points where the tangential plane to the free energy surfaces intersects the A, B, and C axes. Figure 1.41a is drawn for a system in which the three binary systems AB, BC, and CA are simple eutectics. Free energy surfaces exist for three solid phases α, β, and γ and the liquid phase, L. At this temperature the liquid phase is most stable for all alloy compositions. At lower temperatures the G^L surface moves upward and eventually intersects the G^α surface as shown in Figure 1.41b. Alloys with compositions in the vicinity of the intersection of the two curves consist of a + L at equilibrium. In order for the chemical potentials to be equal in both phases the compositions of the two phases in equilibrium must be given by points connected by a common tangential plane, for example, *s* and *l* in Figure 1.41b. These points can be marked on an isothermal section of the equilibrium phase diagram as shown in Figure 1.41c. The lines joining the compositions in equilibrium are known as tie-lines. By rolling the tangential plane over the two free energy surfaces a whole series of tie-lines will be generated, such as pr and qt, and the region covered by these tie-lines pqtr is a two-phase region on the phase diagram. An alloy with composition *x* in Figure 1.41c will therefore minimize its free energy by separating into solid α with composition *s* and liquid with composition *l*. The relative amounts of α and L are simply given by the lever rule. Alloys with compositions within Apq will be a homogeneous α-phase at this temperature, whereas alloys within BCrt will be liquid.

On further cooling the free energy surface for the liquid will rise through the other free energy surfaces producing the sequence of isothermal sections shown in Figure 1.42. In Figure 1.42f, for example, the liquid is stable near the center of the diagram whereas at the corners the α, β, and γ solid phases are stable. In between are several two-phase regions containing bundles of tie-lines. In addition there are three-phase regions known as tie-triangles. The L + α + β triangle for example arises because the common tangential plane simultaneously touches the G^α, G^β, and G^L surfaces. Therefore, any alloy

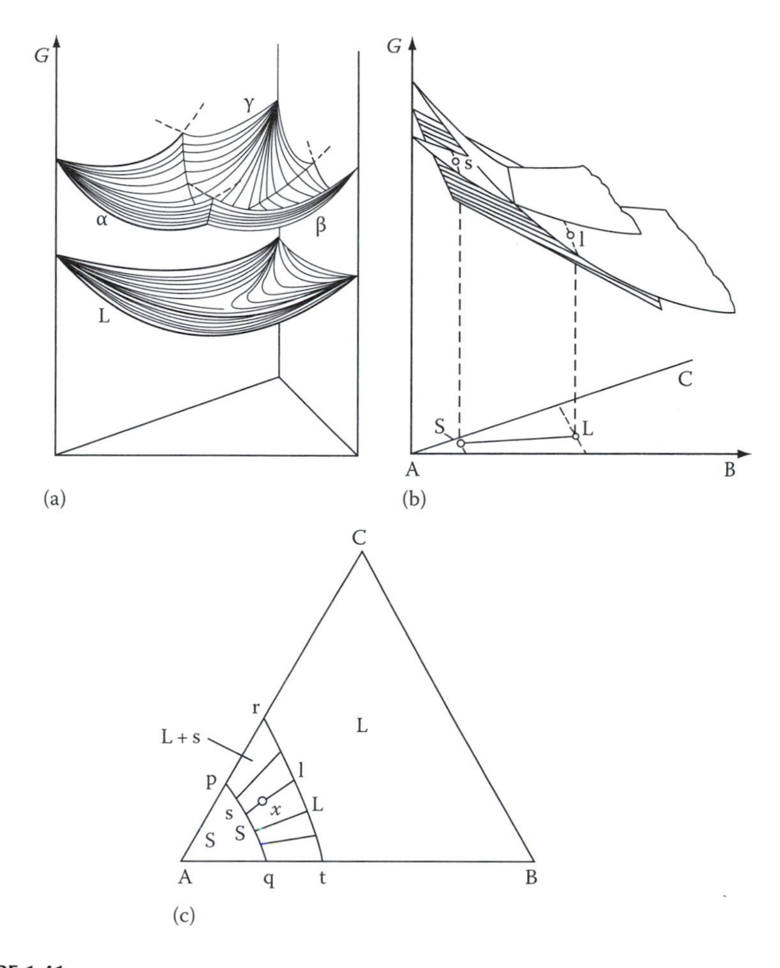

FIGURE 1.41
(a) Free energies of a liquid and three solid phases of a ternary system. (b) A tangential plane construction to the free energy surfaces defines equilibrium between *s* and *l* in the ternary system. (c) Isothermal section through a ternary phase diagram obtained in this way with a two-phase region (L + S) and various tie-lines. The amounts of *l* and *s* at point *x* are determined by the lever rule. (After Haasen, P., *Physical Metallurgy*, Cambridge University Press, Cambridge, 1978.)

with a composition within the $L + \alpha + \beta$ triangle at this temperature will be in equilibrium as a three-phase mixture with compositions given by the corners of the triangle. If the temperature is lowered still further the L region shrinks to a point at which four phases are in equilibrium $L + \alpha + \beta + \gamma$. This is known as the ternary eutectic point and the temperature at which it occurs is the ternary eutectic temperature (Figure 1.42g). Below this temperature the liquid is no longer stable and an isothermal section contains three two-phase regions and one three-phase tie triangle $\alpha + \beta + \gamma$ as shown in Figure 1.42h. If isothermal sections are constructed for all temperatures they can be combined into a three-dimensional ternary phase diagram as shown in Figure 1.44.

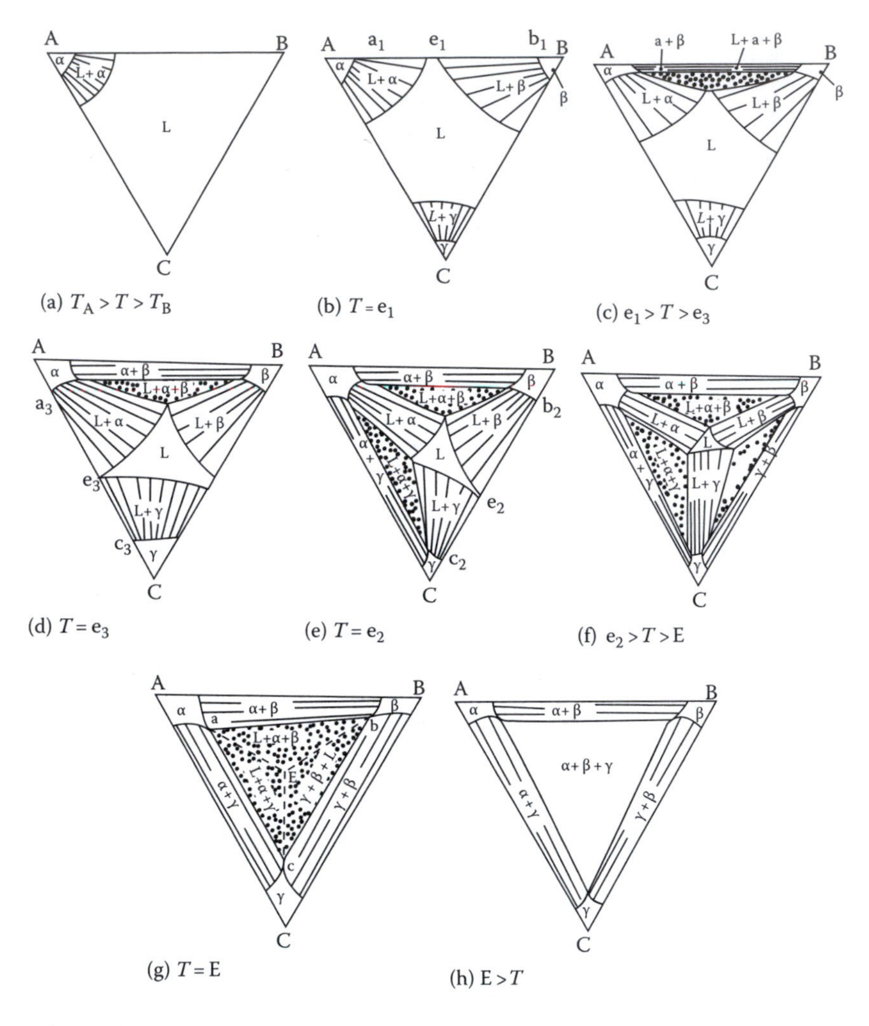

FIGURE 1.42
Isothermal sections through Figure 1.44. (After Prince, A., *Alloy Phase Equilibria*, Elsevier, Amsterdam, 1966.)

In order to follow the course of solidification of a ternary alloy, assuming equilibrium is maintained at all temperatures, it is useful to plot the liquidus surface contours as shown in Figure 1.43. During equilibrium freezing of alloy X the liquid composition moves approximately along the line Xe (drawn through A and X) as primary α-phase is solidified; then along the eutectic valley eE as both α and β solidify simultaneously. Finally at E, the ternary eutectic point, the liquid transforms simultaneously into $\alpha + \beta + \gamma$. This sequence of events is also illustrated in the perspective drawing in Figure 1.44.

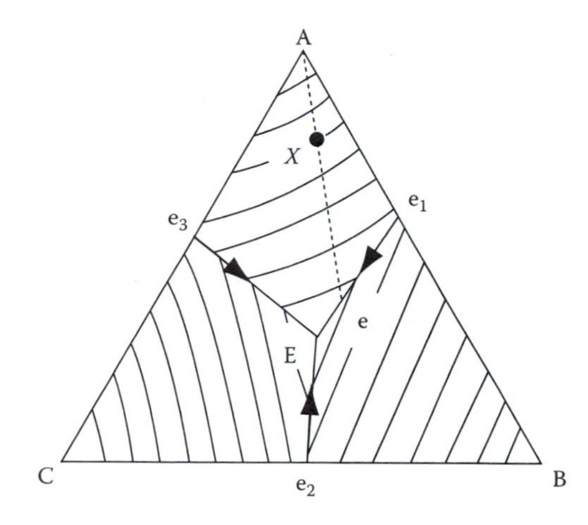

FIGURE 1.43
A projection of the liquidus surfaces of Figure 1.44 onto the Gibbs triangle.

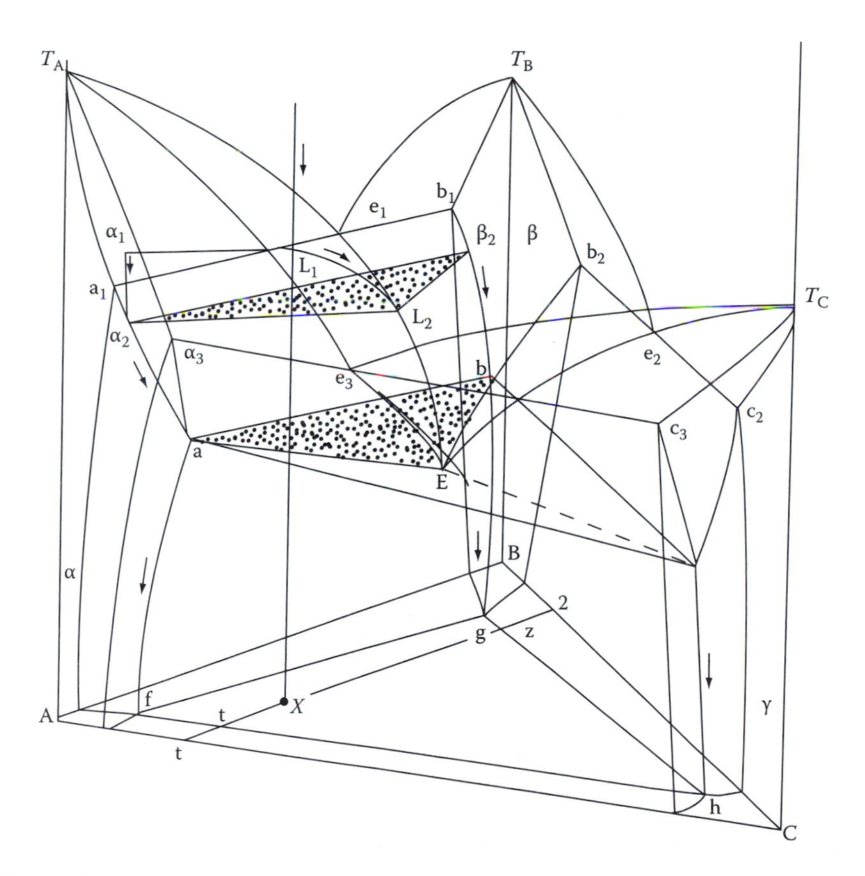

FIGURE 1.44
Equilibrium solidification of alloy *X*. (After Prince, A., *Alloy Phase Equilibria*, Elsevier, Amsterdam, 1966.)

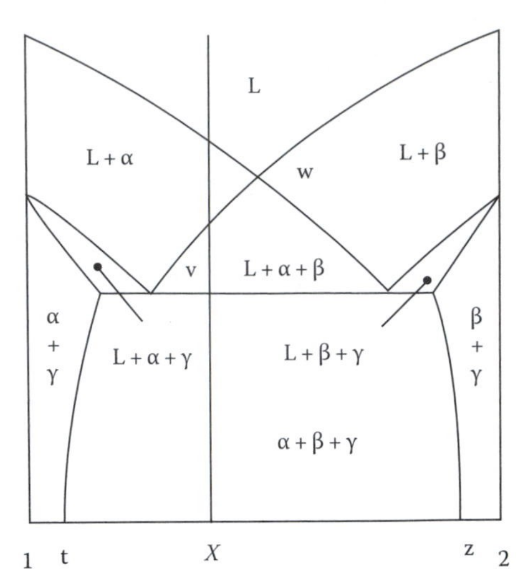

FIGURE 1.45
A vertical section between points 1, 2, and X in Figure 1.44. (After Prince, A., *Alloy Phase Equilibria*, Elsevier, Amsterdam, 1966.)

The phases that form during solidification can also be represented on a vertical section through the ternary phase diagram. Figure 1.45 shows such a section taken through X parallel to AB in Figure 1.44. It can be seen that on cooling from the liquid phase the alloy first passes into the $L + \alpha$ region, then into $L + \alpha + \beta$, and finally all liquid disappears and the $\alpha + \beta + \gamma$ region is entered, in agreement with the above.

An important limitation of vertical sections is that in general the section will not coincide with the tie-lines in the two-phase regions and so the diagram only shows the phases that exist in equilibrium at different temperatures and not their compositions. Therefore, they cannot be used like binary phase diagrams, despite the superficial resemblance.

1.8 Additional Thermodynamic Relationships for Binary Solutions

It is often of interest to be able to calculate the change in chemical potential ($d\mu$) that results from a change in alloy composition (dX). Considering Figure 1.46 and comparing triangles it can be seen that

$$-\frac{d\mu_A}{X_B} = \frac{d\mu_B}{X_A} = \frac{d(\mu_B - \mu_A)}{1} \tag{1.63}$$

and that the slope of the free energy–composition curve is given by

$$\frac{dG}{dX_B} = \frac{\mu_B - \mu_A}{1} \tag{1.64}$$

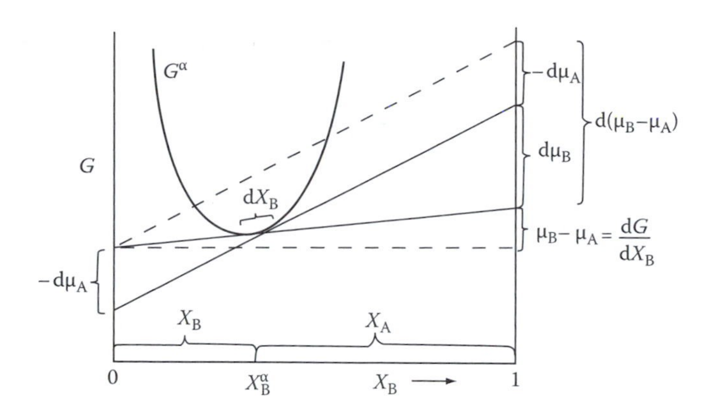

FIGURE 1.46
Evaluation of the change in chemical potential due to a change in composition. (After Hillert, M., in *Lectures on the Theory of Phase Transformations*, Aaronson, H.I. (Ed.), The American Society for Metals and The Metallurgical Society of AIME, New York, 1969.)

Substituting this expression into Equation 1.63 and multiplying throughout by $X_A X_B$ leads to the following equalities:

$$-X_A d\mu_A = X_B d\mu_B = X_A X_B \frac{d^2G}{dX^2} dX_B \tag{1.65}$$

which are the required equations relating $d\mu_A$, $d\mu_B$, and dX_B. The first equality in this equation is known as the Gibbs–Duhem relationship for a binary solution. Note that the subscript "B" has been dropped from d^2G/dX^2 as $d^2G/dX_B^2 = d^2G/dX_A^2$. For a regular solution differentiation of Equation 1.39 gives

$$\frac{d^2G}{dX^2} = \frac{RT}{X_A X_B} - 2\Omega \tag{1.66}$$

For an ideal solution $\Omega = 0$ and

$$\frac{d^2G}{dX^2} = \frac{RT}{X_A X_B} \tag{1.67}$$

Equation 1.65 can be written in a slightly different form by making use of activity coefficients. Combining Equations 1.41 and 1.43 gives

$$\mu_B = G_B + RT \ln \gamma_B X_B \tag{1.68}$$

Therefore,

$$\frac{d\mu_B}{dX_B} = \frac{RT}{X_B}\left\{1 + \frac{X_B}{\gamma_B}\frac{d\gamma_B}{dX_B}\right\} = \frac{RT}{X_B}\left\{1 + \frac{d\ln\gamma_B}{d\ln X_B}\right\} \tag{1.69}$$

A similar relationship can be derived for $d\mu_A/dX_B$. Equation 1.65 therefore becomes

$$-X_A d\mu_A = X_B d\mu_B = RT\left\{1+\frac{d\ln\gamma_A}{d\ln X_A}\right\}dX_B = RT\left\{1+\frac{d\ln\gamma_B}{d\ln X_B}\right\}dX_B \quad (1.70)$$

Comparing Equations 1.65 and 1.70 gives

$$X_A X_B \frac{d^2G}{dX^2} = RT\left\{1+\frac{d\ln\gamma_A}{d\ln X_A}\right\} = RT\left\{1+\frac{d\ln\gamma_B}{d\ln X_B}\right\} \quad (1.71)$$

1.9 Computation of Phase Diagrams

The determination of Gibbs energy of phases, with various properties, pure stoichiometric substances for instance, as a function of temperature, composition, and pressure is necessary for establishing a thermodynamic databank. As described earlier, if Gibbs energy is known, other thermodynamic properties like enthalpy and entropy could easily be derived. Thermodynamic information gathered by conducting experiments are usually compared against existing physical models the purpose of which is to quantify and give predictions of the measured thermodynamic properties. However, it is important to realize that such models could yield unreliable predictions if used outside the range over which they were verified.

1.9.1 Pure Stoichiometric Substances

Thermodynamic data of pure stoichiometric substances are stored as the enthalpy of formation and entropy measured at room temperature (298.15 K) and 1 bar atmospheric pressure (called henceforth the standard conditions), in addition to heat capacity[5]

$$H = H^{\text{ref}} + \int_{T_{\text{ref}}}^{T} C_p dT \quad (1.72)$$

$$S = S^{\text{ref}} + \int_{T_{\text{ref}}}^{T} \frac{C_p}{T} dT \quad (1.73)$$

where the new values of H and S can be determined as a function of temperature compared with a reference state. Note that the second term in the enthalpy and entropy equations is essentially the same as shown in Equations 1.6 and 1.8, respectively.

Gibbs energy could then be estimated from Gibbs–Helmholtz relationship $G = H - TS$. Therefore, obviously it is necessary to describe the heat capacity as a function of temperature.

The Debye function expresses the heat capacity as a function of temperature as follows:

$$C = D\left(\frac{\Theta}{T}\right) \tag{1.74}$$

where Θ is the Debye temperature which can be written as

$$\Theta = \frac{h\omega_D}{2\pi k} \tag{1.75}$$

where

h and k are the Planck and Boltzmann constants, respectively

ω_D is defined as the Debye frequency

The Debye temperature is a material-dependent constant. As atoms vibrate, such vibration increases with temperature yet not all atoms vibrate with the same frequency. When deriving the internal energy due to lattice vibrations, a spectrum is considered, the maximum frequency of which is called the Debye frequency. However, such an approach may not be suitable for determining all the experimental thermodynamic values of solid substances.[6]

According to Hack,[5] a system of thermochemical data was introduced based on the so-called standard element reference state (SER). That is to say, the enthalpy of a state of elements stable under the standard conditions is set to zero, by convention, at these conditions. The entropy, however, takes the absolute value under these conditions.

After Mayer and Kelley,[7] a polynomial is then used to represent the relation between specific heat capacity at constant pressure, C_p, and temperature as follows:

$$C_p = C_1 + C_2T + C_3T^2 + \frac{C_4}{T^2} \tag{1.76}$$

The above relation can be used to describe most experimental data adequately by adjusting its empirical fitting parameters. However, as shown in Figure 1.47, the use of more than one function is sometimes necessary.

Consequently, under standard conditions, the Gibbs energy can be written as follows:

$$G = C_1 + C_2T + C_3T\ln(T) + C_4T^2 + C_5T^3 + \frac{C_6}{T} \tag{1.77}$$

These fitting parameters, C_i, are then stored in the Gibbs energy databank.

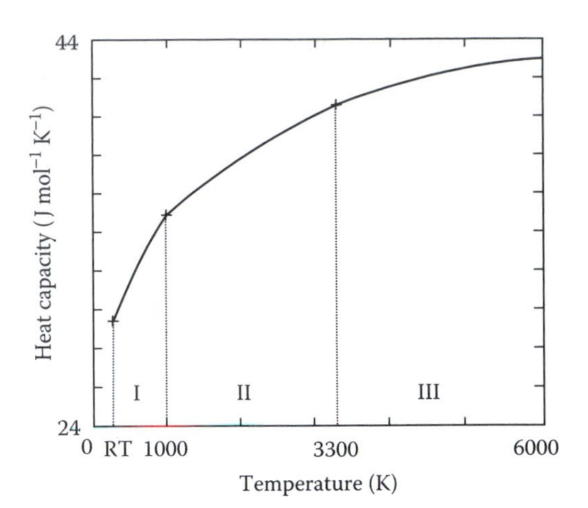

FIGURE 1.47
Heat capacity of oxygen gas versus temperature.[5] Note the use of several polynomials over different temperature ranges to best fit the measured data. (After Hack, K., in *The SGTE Casebook: Thermodynamics at Work*, Hack, K. (Ed.), The Institute of Materials, London, 1996.)

Contributions to Gibbs energy of pure stoichiometric phases such as metallic elements or stoichiometric oxides, SiO_2 for example, include those arising from the lattice, magnetic as well as pressure contributions

$$G_{\mathrm{m}} = G_{\mathrm{lattice}} + G_{\mathrm{magnetic}} + G_{\mathrm{pressure}} \tag{1.78}$$

where G_{lattice} is a function of ΔH and S under the standard conditions, as well as the polynomial representing C_p. Other contributions to the Gibbs energy are discussed in the following section.

When applying the above scheme to substances exhibiting second-order magnetic phase transitions, where the shape of the heat capacity curve shows a rather sudden change, many polynomials are used around Curie temperature.[8] However, this approach has numerical limitations given the large number of fitting parameters needed, and also their large values. Inden suggested that exceptional changes in C_p should be treated separately.[9,10]

For the case of a magnetic phase transition, the magnetic contribution to Gibbs energy can be written as

$$G_{\mathrm{magnetic}} = RTf\left(\frac{T}{T_{\mathrm{c}}}\right)\ln(\beta + 1) \tag{1.79}$$

where

- f is a structure-dependent function of temperature
- T_{c} is a critical temperature which could be either Curie or Néel temperature
- β is the magnetic moment per atom in the lattice

Note that f will differ in the temperature ranges below and above the critical temperature.

For a complete description of the Gibbs free energy, it is also necessary to consider the contribution caused by the effect of pressure on the molar volume. The SGTE consortium[5] followed the approach suggested by Murnaghan[11]

$$G_{\text{pressure}} = V^0 \exp\left[\int_{298}^{T} \alpha(T)\mathrm{d}T\right] \frac{[1 + nK(T)P]^{(1-(1/n))} - 1}{(n-1)K(T)} \tag{1.80}$$

where
- V^0 is the molar volume at room temperature
- $\alpha(T)$ is the thermal expansion of that volume
- n is the pressure derivative of the bulk modulus which is the inverse of the compressibility

$K(T)$ is a polynomial function of temperature which stands for the compressibility at a pressure of 1 bar

$$K(T) = K_0 + K_1 T + K_2 T^2 \tag{1.81}$$

The thermal expansion function, $\alpha(T)$, may be written as

$$\alpha(T) = A_0 + A_1 T + A_2 T^2 + \frac{A_3}{T^2} \tag{1.82}$$

An example of a calculated pressure–temperature phase diagram is shown in Figure 1.48.

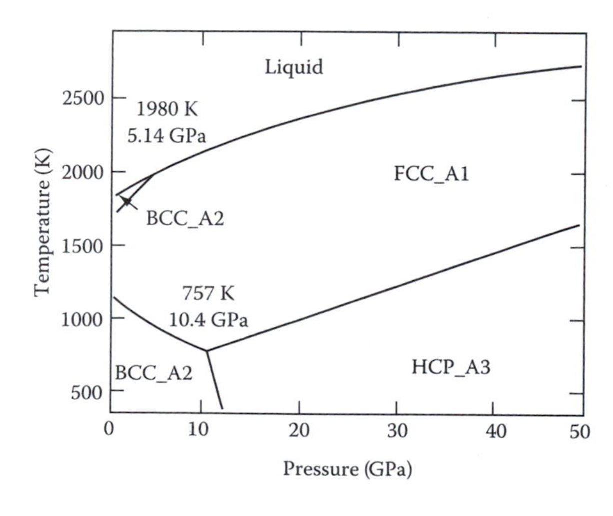

FIGURE 1.48
Pressure–temperature phase diagram of iron, calculated. (After Hack, K., in *The SGTE Casebook: Thermodynamics at Work*, Hack, K. (Ed.), The Institute of Materials, London, 1996.)

1.9.2 Solution Phases

1.9.2.1 Substitutional Solutions

In Section 1.3.2, we discussed ideal solutions where the free energy of such solutions was presented in Equation 1.27. For a multicomponent ideal solution, Equation 1.27 may be written as follows:

$$G^{\mathrm{id}} = RT\sum X_i \ln(X_i) + \sum X_i G_i \tag{1.83}$$

where X_i is the mole fraction of constituent i. However, there is a need for mathematical expressions describing the deviation from the ideal solution, i.e., the excess Gibbs energy. These expressions should allow the estimation of thermodynamic properties of higher order systems from those measured on lower order systems.

Redlich and Kister[12] suggested determining the properties of a ternary system, say A–B–C, from the three binaries (A–B, A–C, and B–C). Therefore, for a binary solution with constituents A and B, say, the excess Gibbs energy may be described as

$$G^{\mathrm{ex}}_{\mathrm{AB}} = X_{\mathrm{A}}X_{\mathrm{B}}\sum L^{(i)}_{\mathrm{AB}}(X_{\mathrm{A}} - X_{\mathrm{B}})^i \tag{1.84}$$

So by fitting experimental data of a binary A–B to this relation, the parameters can be determined. When $i = 0$, the excess energy is $X_{\mathrm{A}}X_{\mathrm{B}}L^{(0)}_{\mathrm{AB}}$, a term which is that of the regular solution model, see Equation 1.36. Here the parameter $L^{(0)}_{\mathrm{AB}}$ is independent from chemical composition. Other $L^{(i)}_{\mathrm{AB}}$ parameters, when $i > 0$, describe the composition dependence of parameter L_{AB}. Parameter L_{AB} is an energy parameter which is called the interaction energy. So the excess Gibbs energy for a ternary solution A–B–C may be written as a first approximation:

$$\begin{aligned} G^{\mathrm{ex}}_{\mathrm{ABC}} = {} & X_{\mathrm{A}}X_{\mathrm{B}}\sum L^{(i)}_{\mathrm{AB}}(X_{\mathrm{A}} - X_{\mathrm{B}})^i + X_{\mathrm{A}}X_{\mathrm{C}}\sum L^{(i)}_{\mathrm{AC}}(X_{\mathrm{A}} - X_{\mathrm{C}})^i \\ & + X_{\mathrm{B}}X_{\mathrm{C}}\sum L^{(i)}_{\mathrm{BC}}(X_{\mathrm{B}} - X_{\mathrm{C}})^i \end{aligned} \tag{1.85}$$

Logically, when experimental data are available for ternary solutions the measured properties should be compared against those predicted by Redlich–Kister approach. Any possible deviation should first be dealt with by adding a term $X_{\mathrm{A}}X_{\mathrm{B}}X_{\mathrm{C}}L^{(0)}_{\mathrm{ABC}}$ to the excess Gibbs energy. If the deviation is not adequately corrected for, the term could be formulated as

$$\begin{aligned} X_{\mathrm{A}}X_{\mathrm{B}}X_{\mathrm{C}}\Bigg[& L^{(0)}_{\mathrm{ABC}} + \frac{1}{3}(1 + 2X_{\mathrm{A}} - X_{\mathrm{B}} - X_{\mathrm{C}})L^{(1)}_{\mathrm{ABC}} + \frac{1}{3}(1 + 2X_{\mathrm{B}} - X_{\mathrm{C}} - X_{\mathrm{A}})L^{(1)}_{\mathrm{BCA}} \\ & + \frac{1}{3}(1 + 2X_{\mathrm{C}} - X_{\mathrm{A}} - X_{\mathrm{B}})L^{(1)}_{\mathrm{CAB}}\Bigg] \end{aligned} \tag{1.86}$$

For quaternary systems, often there is no information regarding all four ternary systems and sometimes the data of the quaternary system itself are not available. Nevertheless, a quaternary term $X_A X_B X_C X_D L^{(0)}_{ABCD}$ may be added to the excess Gibbs energy in this case.

For other phases with different properties such as interstitial solutions, the reader is referred to, for example, Hack,[5] for further information.

1.10 Kinetics of Phase Transformations

The thermodynamic functions that have been described in this chapter apply to systems that are in stable or metastable equilibrium. Thermodynamics can therefore be used to calculate the driving force for a transformation (Equation 1.4), but it cannot say how fast a transformation will proceed. The study of how fast processes occur belongs to the science of kinetics.

Let us redraw Figure 1.1 for the free energy of a single atom as it takes part in a phase transformation from an initially metastable state into a state of lower free energy (Figure 1.49). If G_1 and G_2 are the free energies of the initial and final states, the driving force for the transformation will be $\Delta G = G_2 - G_1$. However, before the free energy of the atom can decrease from G_1 to G_2 the atom must pass through a so-called transition or activated state with a free energy ΔG^a above G_1. The energies shown in Figure 1.49 are average energies associated with large numbers of atoms. As a result of the random thermal motion of the atoms the energy of any particular atom will vary with time and occasionally it may be sufficient for the atom to reach the activated state. This process is known as thermal activation.

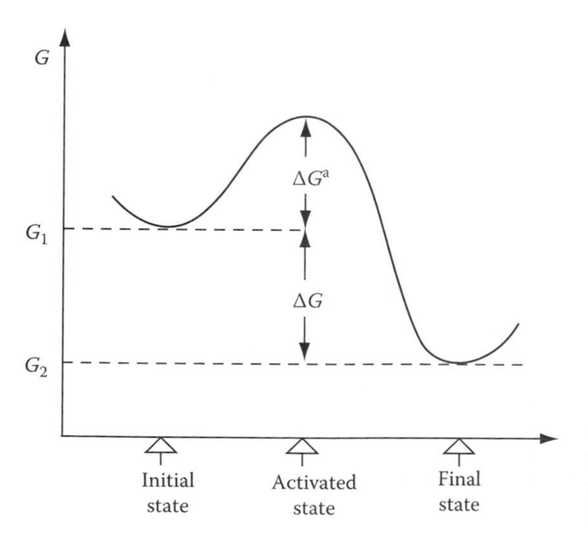

FIGURE 1.49
Transformation from initial to final state through an activated state of higher free energy.

According to kinetic theory, the probability of an atom reaching the activated state is given by $\exp(-\Delta G^a/kT)$ where k is Boltzmann's constant (R/N_a) and ΔG^a is known as the activation free energy barrier. The rate at which a transformation occurs will depend on the frequency with which atoms reach the activated state. Therefore, we can write

$$\text{Rate} \propto \exp\left(-\frac{\Delta G^a}{kT}\right)$$

Putting $\Delta G^a = \Delta H^a - T\Delta S^a$ and changing from atomic to molar quantities enables this equation to be written as

$$\text{Rate} \propto \exp\left(\frac{\Delta H^a}{RT}\right) \tag{1.87}$$

This equation was first derived empirically from the observed temperature dependence of the rate of chemical reactions and is known as the Arrhenius rate equation. It is also found to apply to a wide range of processes and transformations in metals and alloys, the simplest of these is the process of diffusion which is discussed in Chapter 2.

Exercises

1.1 The specific heat of solid copper above 300 K is given by

$$C_p = 22.64 + 6.28 \times 10^{-3}T \text{ J mol}^{-1}\text{K}^{-1}$$

By how much does the entropy of copper increase on heating from 300 to 1358 K?

1.2 With the aid of Equation 1.11 and Figure 1.5, draw schematic free energy–pressure curves for pure Fe at 1600°C, 800°C, 500°C, and 300°C.

1.3 Estimate the change in the equilibrium melting point of copper caused by a change of pressure of 10 kbar. The molar volume of copper is 8.0×10^{-6} m^3 for the liquid, and 7.6×10^{-6} m^3 for the solid phase. The latent heat of fusion of copper is 13.05 kJ mol^{-1}. The melting point is 1085°C.

1.4 For a single-component system, why do the allotropes stable at high temperatures have higher enthalpies than allotropes stable at low temperatures, e.g., $H(\gamma\text{-Fe}) > H(\alpha\text{-Fe})$?

1.5 Determine, by drawing, the number of distinguishable ways of arranging two black balls and two white balls in a square array. Check your answer with Equation 1.24.

1.6 By using Equations 1.30 and 1.31, show that the chemical potentials of A and B can be obtained by extrapolating the tangent to the *G–X* curve to $X_A = 0$ and $X_B = 0$.

1.7 Derive Equation 1.40 from Equations 1.31 and 1.39.

1.8 15 g gold and 25 g of silver are mixed to form a single-phase ideal solid solution.

   a. How many moles of solution are there?
   b. What are the mole fractions of gold and silver?
   c. What is the molar entropy of mixing?
   d. What is the total entropy of mixing?
   e. What is the molar free energy change at 500°C?
   f. What are the chemical potentials of Au and Ag at 500°C taking the free Au atom is added? Express your answer in eV atom^{-1}.
   g. By how much will the free energy of the solution change at 500°C if one Au atom is added? Express your answer in eV atom^{-1}.

1.9 In the Fe–C system Fe_3C is only a metastable phase, while graphite is the most stable carbon-rich phase. By drawing schematic free energy–composition diagrams show how the Fe-graphite phase diagram compares to the Fe–Fe_3C phase diagram from 0 to 2 wt% Fe. Check your answer with the published phase diagram in the *Metals Handbook* for example.

1.10 Consider a multicomponent system A, B, C,... containing several phases α, β, γ,... at equilibrium. If a small quantity of A (dn_A mol) is taken from the α-phase and added to the β-phase at constant *T* and *P* what are the changes in the free energies of the α- and β-phases, dG^α and dG^β? Since the overall mass and composition of the system is unchanged by the above process the total free energy change $dG = dG^\alpha + dG^\beta = 0$. Show, therefore, that $\mu_A^\alpha = \mu_A^\beta$. Repeating for other pairs of phases and other components gives the general equilibrium conditions, Equation 1.48.

1.11 For aluminum $\Delta H_v = 0.8$ eV atom^{-1} and $\Delta S_v/R = 2$. Calculate the equilibrium vacancy concentration at 660°C (T_m) and 25°C.

1.12 The solid solubility of silicon in aluminum is 1.25 atomic % at 550°C and 0.46 atomic % at 450°C. What solubility would you expect at 200°C? Check your answer by reference to the published phase diagram.

1.13 The metal A and metal B form an ideal liquid solution but are almost immiscible in the solid state. The entropy of fusion of both A and B is 8.4 J mol^{-1} K^{-1} and the melting temperatures are 1500 and 1300 K, respectively. Assuming that the specific heats of the solid and liquid are identical calculate the eutectic composition and temperature in the A–B phase diagram.

1.14 Write down an equation that shows by how much the molar free energy of solid Cu is increased when it is present as a small sphere of radius *r* in liquid Cu. By how much must liquid Cu be cooled below T_m before a solid particle of Cu can grow if the particle diameter is (1) 2 μm, (2) 2 nm (20 Å)? (Cu: $T_m = 1085°C = 1358$ K. Atomic weight 63.5.

Density 8900 kg m^{-3}. Solid/liquid interfacial energy $\gamma = 0.144$ J m^{-2}. Latent heat of melting $L = 13{,}300$ J mol^{-1}.)

1.15 Suppose a ternary alloy containing 40 atomic % A, 20 atomic % B, 40 atomic % C solidifies through a ternary eutectic reaction to a mixture of α, β, and γ with the following compositions: 80 atomic % A, 5 atomic % B, 15 atomic % C; 70 atomic % B, 10 atomic % A, 20 atomic % C; and 20 atomic % B, 10 atomic % A, 70 atomic % C. What will be the mole fractions of α, β, and γ in the microstructure?

1.16 Show that a general expression for the chemical potential of a component in solution is given by

$$\mu_A = G_A^0 + S_A(T_0 - T) + RT \ln \gamma_A X_A + (P - P_0)V_m$$

where

G_A^0 is the free energy of pure A at temperature T_0 and pressure P_0
S_A is the entropy of A
R is the gas constant
γ_A is the activity coefficient for A
X_A is the mole fraction in solution
V_m is the molar volume which is assumed to be constant

Under what conditions is the above equation valid?

References

1. D.R. Gaskell, *Introduction to Metallurgical Thermodynamics*, McGraw-Hill, New York, 1973, p. 342.
2. R.W. Cahn (Ed.), *Physical Metallurgy*, 2nd edn., Chapters 4 and 5, North-Holland, Amsterdam, 1974.
3. M. Ferrante and R.D. Doherty, *Acta Metall.*, 27: 1603, 1979.
4. M. Hillert, *Phase Transformations*, Chapter 5, American Society for Metals, Metals Park, OH, 1970.
5. K. Hack (Ed.), *The SGTE Casebook: Thermodynamics at Work*, Part 1, Institute of Materials, London, 1996.
6. C. Gerthsen and H.O. Kneser, *Physik*, Springer Verlag, Berlin, Heidelberg, New York, 1969.
7. K.K. Kelley, *US Bureau of Mines Bulletin 476*, 1949.
8. I. Barin and O. Knache, Springer-Verlag, Berlin, Heidelberg, and Verlag Stahleisen, Düsseldorf, 1973.
9. G. Inden, *Proc. of Calphad V*, Düsseldorf, III.4, 1–13, 1976.
10. G. Inden, *Proc. of Calphad V*, Düsseldorf, IV.1, 1–33, 1976.
11. F.D. Murnaghan, *Proc. Natl. Acad. Sci. (USA)*, 30: 244, 1944.
12. O. Redlich and A.T. Kister, *Ind. Eng. Chem.*, 40: 345–348, 1948.

Further Reading

A.H. Cottrell, Alloys (Chapter 14), The phase diagram (Chapter 15), in *An Introduction to Metallurgy*, Edward Arnold, London, 1967.

D.R. Gaskell, *Introduction to Metallurgical Thermodynamics*, McGraw-Hill, New York, 1973.

P. Gordon, *Principles of Phase Diagrams in Materials Systems*, McGraw-Hill, New York, 1968.

M. Hillert, Calculation of phase equilibria, in *Phase Transformations*, Chapter 5, American Society for Metals, Metals Park, OH, 1970.

M. Hillert, The uses of the Gibbs free energy–composition diagrams, in H.I. Aaronson (Ed.), *Lectures on the Theory of Phase Transformations*, Chapter 1, The Metallurgical Society of AIME, New York, 1975.

C.H.P. Lupis, *Chemical Thermodynamics of Materials*, North Holland, Amsterdam, 1983.

A.D. Pelton, Phase diagrams, in R.W. Cahn and P. Haasen (Eds.), *Physical Metallurgy*, Chapter 7, North-Holland, Amsterdam, 1983.

A. Prince, *Alloy Phase Equilibria*, Elsevier, London, 1966.

G.V. Raynor, Phase diagrams and their determination, in R.W. Cahn (Ed.), *Physical Metallurgy*, Chapter 7, North-Holland, Amsterdam, 1970.

F.N. Rhines, *Phase Diagrams in Metallurgy*, McGraw-Hill, New York, 1956.

P.G. Shewmon, Metallurgical thermodynamics, in R.W. Cahn and P. Haasen (Eds.), *Physical Metallurgy*, Chapter 6, North-Holland, Amsterdam, 1983.

R.A. Swalin, *Thermodynamics of Solids*, 2nd edn., Wiley, New York, 1972.

D.R.F. West, *Ternary Equilibrium Diagrams*, 2nd edn., Chapman & Hall, London, U.K., 1982.

2

Diffusion

Chapter 1 was mainly concerned with stable or equilibrium arrangements of atoms in an alloy. The study of phase transformations concerns those mechanisms by which a system attempts to reach this state and how long it takes. One of the most fundamental processes that control the rate at which many transformations occur is the diffusion of atoms.

The reason why diffusion occurs is always so as to produce a decrease in Gibbs free energy. As a simple illustration of this consider Figure 2.1. Two blocks of the same A–B solid solution, but with different compositions, are welded together and held at a temperature high enough for long-range diffusion to occur. If the molar free energy diagram of the alloy is as shown in Figure 2.1b, the molar free energy of each part of the alloy will be given by G_1 and G_2, and initially the total free energy of the welded block will be G_3. However, if diffusion occurs as indicated in Figure 2.1a so as to eliminate the concentration differences, the free energy will decrease toward G_4, the free energy of a homogeneous alloy. Thus, in this case, a decrease in free energy is produced by A and B atoms diffusing away from the regions of high concentration to that of low concentration, i.e., down the concentration gradients. However, this need not always be the case as was indicated in Section 1.4. In alloy systems that contain a miscibility gap the free energy curves can have a negative curvature at low temperatures. If the free energy curve and composition for the A–B alloy shown in Figure 2.1a were as drawn in Figure 2.1d the A and B atoms would diffuse toward the regions of high concentration, i.e., up the concentration gradients, as shown in Figure 2.1c. However, this is still the most natural process as it reduces the free energy from G_3 toward G_4 again.

As can be seen in Figure 2.1e and f the A and B atoms are diffusing from regions where the chemical potential is high to regions where it is low, i.e., down the chemical potential gradient in both cases. In practice the first case mentioned above is far more common than the second case, and it is usually assumed that diffusion occurs down concentration gradients.

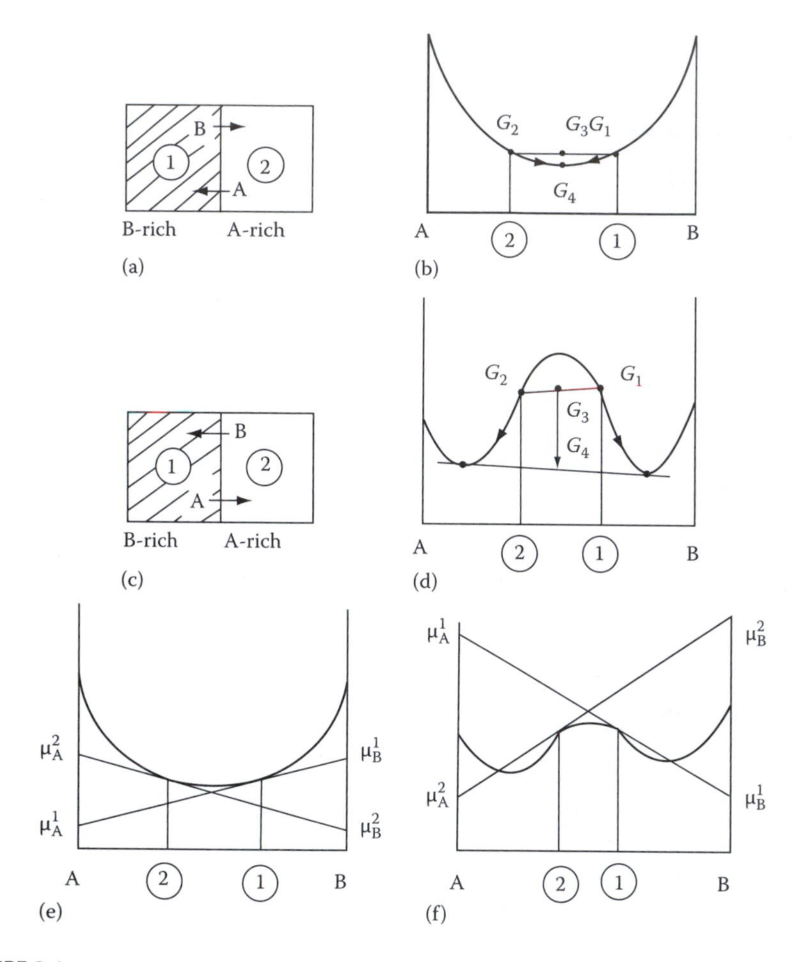

FIGURE 2.1
Free energy and chemical potential changes during diffusion. (a) and (b) "down-hill" diffusion. (c) and (d) "up-hill" diffusion. (e) $\mu_A^2 > \mu_A^1$ therefore A atoms move from (2) to (1), $\mu_B^1 > \mu_B^2$ therefore B atoms move from (1) to (2). (f) $\mu_A^1 > \mu_A^2$ therefore A atoms move from (1) to (2), $\mu_B^2 > \mu_B^1$ therefore B atoms move from (2) to (1).

However, it can be seen that this is only true under special circumstances and for this reason it is strictly speaking better to express the driving force for diffusion in terms of a chemical potential gradient. Diffusion ceases when the chemical potentials of all atoms are everywhere the same and the system is in equilibrium. However, since case 1 above is mainly encountered in practice and because concentration differences are much easier to measure than chemical potential differences, it is nevertheless more convenient to relate diffusion to concentration gradients. The remainder of this chapter will thus be mainly concerned with this approach to diffusion.

2.1 Atomic Mechanisms of Diffusion

There are two common mechanisms by which atoms can diffuse through a solid and the operative mechanism depends on the type of site occupied in the lattice. Substitutional atoms usually diffuse by a vacancy mechanism whereas the smaller interstitial atoms migrate by forcing their way between the larger atoms, i.e., interstitially.

Normally is substitutional atom in a crystal oscillates about a given site and is surrounded by neighboring atoms on similar sites. The mean vibrational energy possessed by each atom is given by $3kT$, and therefore increases in proportion to the absolute temperature. Since the mean frequency of vibration is approximately constant the vibrational energy is increased by increasing the amplitude of the oscillations. Normally the movement of a substitutional atom is limited by its neighbors and the atom cannot move to another site. However, if an adjacent site is vacant it can happen that a particularly violent oscillation results in the atom jumping over on to the vacancy. This is illustrated in Figure 2.2. Note that in order for the jump to occur the shaded atoms in Figure 2.2b must move apart to create enough space for the migrating atom to pass between. Therefore the probability that any atom will be able to jump into a vacant site depends on the probability that it can acquire sufficient vibrational energy. The rate at which any given atom is able to migrate through the solid will clearly be determined by the frequency with which it encounters a vacancy and this in turn depends on the concentration of vacancies in the solid. It will be shown that both the probability of jumping and the concentration of vacancies are extremely sensitive to temperature.

When a solute atom is appreciably smaller in diameter than the solvent, it occupies one of the interstitial sites between the solvent atoms. In face-centered cubic (fcc) materials the interstitial sites are midway along the cube edges or, equivalently, in the middle of the unit cell, Figure 2.3a.

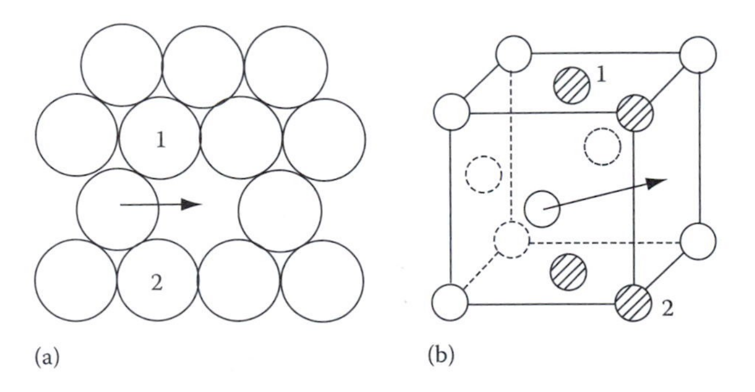

FIGURE 2.2
Movement of an atom into an adjacent vacancy in an fcc lattice. (a) A close-packed plane. (b) A unit cell showing the four atoms (shaded) which must move before the jump can occur. (After Shewmon, P.G., *Diffusion in Solids*, McGraw-Hill, New York, 1963.)

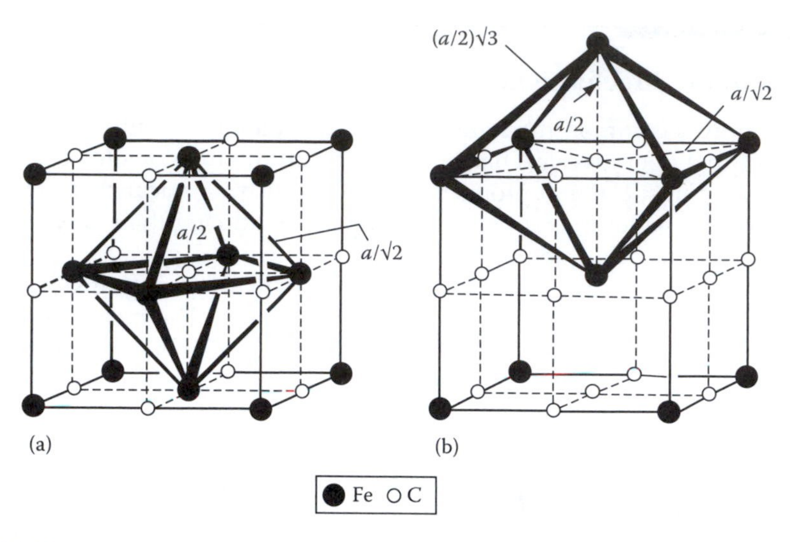

FIGURE 2.3
(a) Octahedral interstices (0) in an fcc crystal. (b) Octahedral interstices in a bcc crystal. (After Haasen, P., *Physical Metallurgy*, Cambridge University Press, Cambridge, 1978.)

These are known as octahedral sites since the six atoms around the site form an octahedron. In the base-centered cubic (bcc) lattice the interstitial atoms also often occupy the octahedral sites which are now located at edge-centering or face-centering positions as shown in Figure 2.3b.

Usually the concentration of interstitial atoms is so low that only a small fraction of the available sites is occupied. This means that each interstitial atom is always surrounded by vacant sites and can jump to another position as often as its thermal energy permits it to overcome the strain energy barrier to migration, Figure 2.4.

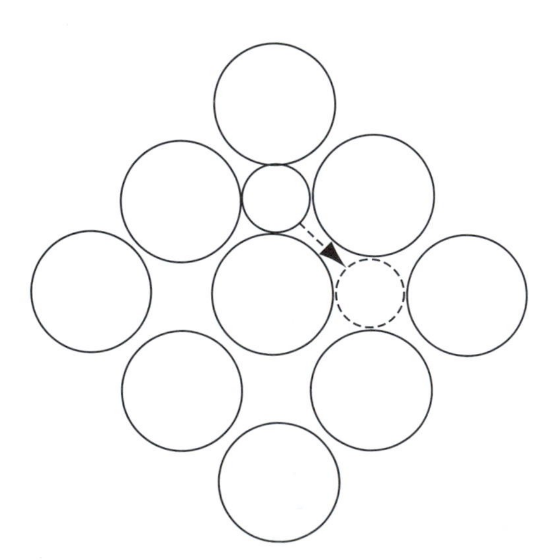

FIGURE 2.4
A {100} plane in an fcc lattice showing the path of an interstitial atom diffusing by the interstitial mechanism.

2.2 Interstitial Diffusion

2.2.1 Interstitial Diffusion as a Random Jump Process

Let us consider first a simple model of a dilute interstitial solid solution where the parent atoms are arranged on a simple cubic lattice and the solute B atoms fit perfectly into the interstices without causing any distortion of the parent lattice. We assume that the solution is so dilute that every interstitial atom is surrounded by six vacant interstitial sites. If the concentration of B varies in one dimension (x) through the solution (Figure 2.5) the B atoms can diffuse throughout the material until their concentration is the same everywhere. The problem to be considered then, concerns how this diffusion is related to the random jump characteristics of the interstitial atoms.

To answer this question consider the exchange of atoms between two adjacent atomic planes such as (1) and (2) in Figure 2.5a. Assume that on average an interstitial atom jumps Γ_B times per second (Γ = Greek capital gamma) and that each jump is in a random direction, i.e., there is an equal probability of the atom jumping to every one of the six adjacent sites. If plane (1) contains n_1 B-atoms m^{-2} the number of atoms that will jump from plane (1) to (2) in 1 s (J) will be given by

$$\vec{J}_B = \frac{1}{6}\Gamma_B n_1 \text{ atoms m}^{-2}\text{ s}^{-1} \tag{2.1}$$

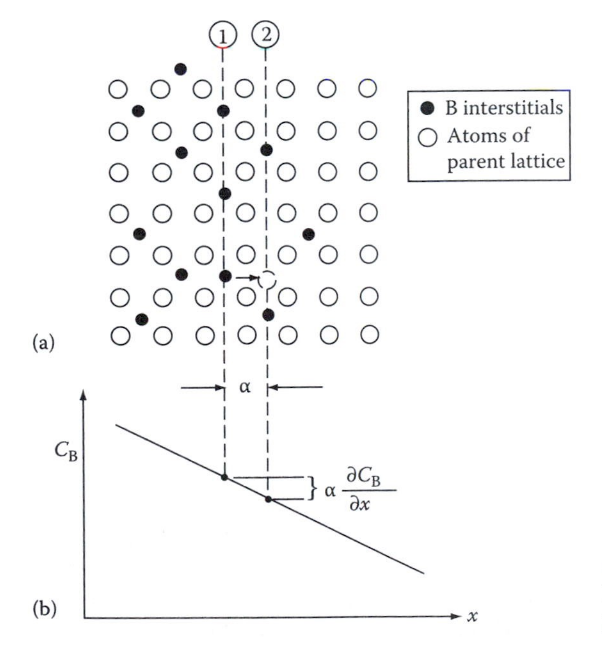

FIGURE 2.5
Interstitial diffusion by random jumps in a concentration gradient.

During the same time the number of atoms that jump from plane (2) to (1), assuming Γ_B is independent of concentration, is given by

$$\overleftarrow{J}_B = \frac{1}{6}\Gamma_B n_2 \text{ atoms m}^{-2}\text{ s}^{-1}$$

Since $n_1 > n_2$ there will be a net flux of atoms from left to right given by

$$J_B = \overrightarrow{J}_B - \overleftarrow{J}_B = \frac{1}{6}\Gamma_B(n_1 - n_2) \tag{2.2}$$

where n_1 and n_2 are related to the concentration of B in the lattice. If the separation of planes (1) and (2) is α the concentration of B at the position of plane (1) $C_B(1) = n_1/\alpha$ atoms m^{-3}. Likewise $C_B(2) = n_2/\alpha$. Therefore $(n_1 - n_2) = \alpha(C_B(1) - C_B(2))$ and from Figure 2.5b it can be seen that $C_B(1) - C_B(2) = -(\partial C_B/\partial x)$. Substituting these equations into Equation 2.2 gives:

$$J_B = -\left(\frac{1}{6}\Gamma_B\alpha^2\right)\frac{\partial C_B}{\partial x} \text{ atoms m}^{-2}\text{ s}^{-1}$$

The partial derivative $\partial C_B/\partial x$ has been used to indicate that the concentration gradient can change with time. Thus in the presence of a concentration gradient the random jumping of individual atoms produces a net flow of atoms down the concentration gradient.

Substituting

$$D_B = \frac{1}{6}\Gamma_B\alpha^2 \tag{2.3}$$

yields

$$J_B = -D_B\frac{\partial C_B}{\partial x} \tag{2.4}$$

This equation is identical to that proposed by Fick in 1855 and is usually known as Fick's first law of diffusion. D_B is known as the intrinsic diffusivity or the diffusion coefficient of B, and has units [m^2 s^{-1}]. The units for J are [quantity m^{-2} s^{-1}] and for $\partial C/\partial x$ [quantity m^{-4}], where the unit of quantity can be in terms of atoms, moles, kg, etc., long it is the same for J and C.

When the jumping of B atoms is truly random with a frequency independent of concentration, D_B is given by Equation 2.3 and is also a constant independent of concentration. Although this equation for D_B was derived for interstitial diffusion in a simple cubic lattice it is equally applicable to any randomly diffusing atom in any cubic lattice provided the correct substitution for the jump distance α is made. In noncubic lattices the probability of jumps in different crystallographic directions is not equal and D varies with direction. Atoms in hexagonal lattices, for example, diffuse at different rates parallel and perpendicular to the basal plane.

The condition that the atomic jumps occur completely randomly and independently of concentration is usually not fulfilled in real alloys. Nevertheless it is found from experiment that Fick's first law is still applicable, though only if the diffusion coefficient D is made to vary with composition. For example the diffusion coefficient for carbon in fcc-Fe at 1000°C is 2.5×10^{-11} m^2 s^{-1} at 0.15 wt% C, but it rises to 7.7×10^{-11} m^2 s^{-1} in solutions containing 1.4 wt% C. The reason for the increase of D_C^γ with concentration is that the C atoms strain the Fe lattice thereby making diffusion easier as the amount of strain increases.

As an example of the use of Equation 2.3 the following data can be used to estimate the jump frequency of a carbon atom in γ-Fe at 1000°C. The lattice parameter of γ-Fe is ~0.37 nm thus the jump distance $\alpha = 0.37/\sqrt{2} = 0.26$ nm (2.6 Å). Assuming $D = 2.5 \times 10^{-11}$ m^2 s^{-1}, leads to the result that $\Gamma = 2 \times 10^9$ jumps s^{-1}. If the vibration frequency of the carbon atoms is ~10^{13}, then only about one attempt in 10^4 results in a jump from one site to another.

It is also interesting to consider the diffusion process from the point of view of a single diffusing atom. If the direction of each new jump is independent of the direction of the previous jump the process is known as a random walk. For a random walk in three dimensions it can be shown[1] that after n steps of length α the "average" atom will be displaced by a net distance $\alpha\sqrt{n}$ from its original position. (This is more precisely that root mean square displacement after n steps.) Therefore after a time t the average atom will have advanced a radial distance r from the origin, where

$$r = \alpha\sqrt{(\Gamma t)} \tag{2.5}$$

Substituting Equation 2.3 for Γ gives

$$r = 2.4\sqrt{(Dt)} \tag{2.6}$$

It will be seen that the distance $\sqrt{(Dt)}$ is a very important quantity in diffusion problems.

For the example of carbon diffusing in γ-Fe above, in 1 s each carbon atom will move a total distance of ~0.5 m but will only reach in net displacement of ~10 μm. It is obvious that very few of the atom jumps provide a useful contribution to the diffusion distance.

2.2.2 Effect of Temperature—Thermal Activation

Let us now take a closer look at the actual jump process for an interstitial atom as in Figure 2.6a. Due to the thermal energy of the solid all the atoms will be vibrating about their rest positions and occasionally a particularly violent oscillation of an interstitial atom, or some chance coincidence of the

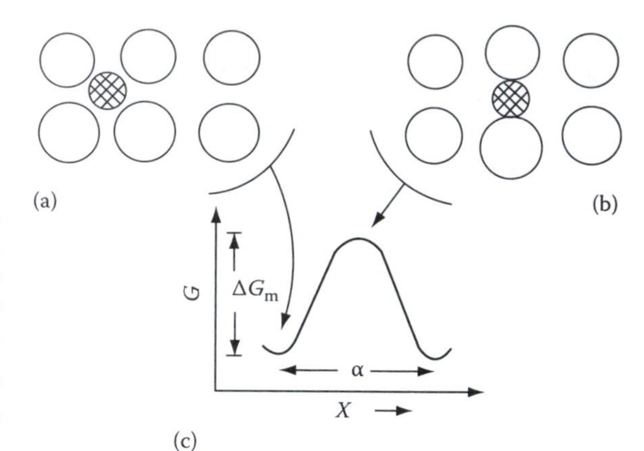

FIGURE 2.6
Interstitial atom, (a) in equilibrium position, (b) at the position of maximum lattice distortion. (c) Variation of the free energy of the lattice as a function of the position of interstitial. (After Shewmon, P.G., *Physical Metallurgy*, 2nd edn., North-Holland, Amsterdam, 1974.)

movements of the matrix and interstitial atoms, will result in a jump. Since the diffusion coefficient is closely related to the frequency of such jumps, Γ, it is of interest to know the factors controlling Γ and the effect of raising the temperature of the system.

The rest positions of the interstitial atoms are positions of minimum potential energy. In order to move an interstitial atom of the parent lattice must be forced apart into higher energy positions as shown in Figure 2.6b. The work that must be done to accomplish this process causes an increase in the free energy of the system by ΔG_m (m refers to migration) as shown in Figure 2.6c. ΔG_m is known as the activation energy for the migration of the interstitial atom. In any system in thermal equilibrium the atoms are constantly colliding with one another and changing their vibrational energy. On average, the fraction of atoms with energy of ΔG or more than the mean energy is given by $\exp(-\Delta G/RT)$. Thus if the interstitial atom in Figure 2.6a is vibrating with a mean frequency v in the x direction it makes v attempts per second to jump into the next site and the fraction of these attempts that are successful is given by $\exp(-\Delta G_m/RT)$. Now the atom is randomly vibrating in three-dimensional space, and if it is surrounded by z sites to which it can jump the jump frequency is given by

$$\Gamma_B = zv \exp \frac{-\Delta G_m}{RT} \tag{2.7}$$

ΔG_m can be considered to be the sum of a large activation enthalpy ΔH_m a small activation entropy term $-T\Delta S_m$.

Combining this expression with Equation 2.3 gives the diffusion coefficient as

$$D_B = \left[\frac{1}{6}\alpha^2 zv \exp \frac{\Delta S_m}{R}\right] \exp \frac{-\Delta H_m}{RT} \tag{2.8}$$

This can be simplified to an Arrhenius-type equation, that is

$$D_{\mathrm{B}} = D_{\mathrm{B0}} \exp \frac{-Q_{\mathrm{ID}}}{RT} \tag{2.9}$$

where

$$D_{\mathrm{B0}} = \frac{1}{6} \alpha^2 z v \exp \frac{\Delta S_{\mathrm{m}}}{R} \tag{2.10}$$

and

$$Q_{\mathrm{ID}} = \Delta H_{\mathrm{m}} \tag{2.11}$$

The terms that are virtually independent of temperature have been grouped into a single material constant D_0. Therefore D or Γ increases exponentially with temperature at a rate determined by the activation enthalpy Q_{ID} (ID refers to interstitial diffusion). Equation 2.9 is found to agree with experimental measurements of diffusion coefficients in substitutional as well as interstitial diffusion. In the case of interstitial diffusion it has been shown that the activation enthalpy Q is only dependent on the activation energy barrier to the movement of interstitial atoms from one site to another.

Some experimental data for the diffusion of various interstitials in bcc-Fe are given in Table 2.1. Note that the activation enthalpy for interstitial diffusion increases as the size of the interstitial atom increases. (The atomic diameters decrease in the order C, N, H.) This is to be expected since smaller atoms cause less distortion of the lattice during migration.

A convenient graphical representation of D as a function of temperature can be obtained writing Equation 2.9 in the form

$$\log D = \log D_0 - \frac{Q}{2.3R}\left(\frac{1}{T}\right) \tag{2.12}$$

Thus if log D is plotted against $(1/T)$ a straight line is obtained with a slope equal to $-(Q/2.3R)$ and an intercept on the log D axis at log D_0, see Figure 2.7.

TABLE 2.1

Experimental Diffusion Data for Interstitials in Ferritic (bcc) Iron

Solute	D (mm^2 s^{-1})	Q (kJ mol^{-1})	Reference
C	2.0	84.1	[2]
N	0.3	76.1	[3]
H	0.1	13.4	[4]

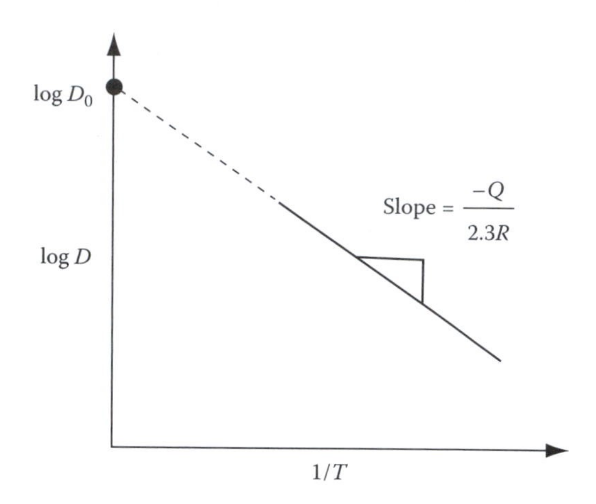

FIGURE 2.7
The slope of log D versus $1/T$ gives the activation energy for diffusion Q.

2.2.3 Steady-State Diffusion

The simplest type of diffusion to deal with is when a steady state exists, that is when the concentration at every point does not change with time. For example consider a thin-walled pressure vessel containing hydrogen. The concentration of hydrogen at the inner surface of the vessel will be maintained at a level C_H depending on the pressure in the vessel, while the concentration at the outer surface is reduced to zero by the escape of hydrogen to the surroundings. A steady state will eventually be reached when the concentration everywhere reaches a constant value. Provided D_H is independent of concentration there will be a single concentration gradient in the wall given by

$$\frac{\partial C}{\partial x} = \frac{0 - C_H}{l}$$

where l is the wall thickness. On this basic the flux through the wall is given by

$$J_H = \frac{D_H C_H}{l} \tag{2.13}$$

2.2.4 Nonsteady-State Diffusion

In most practical situations steady-state conditions are not established, i.e., concentration varies with both distance and time, and Fick's first law can no longer be used. For simplicity let us consider the situation shown in Figure 2.8a where a concentration profile exists along one dimension (x) only. The flux at any point along the x-axis will depend on the local value of D_B and $\partial C_B/dx$ as shown in Figure 2.8b. In order to calculate how the

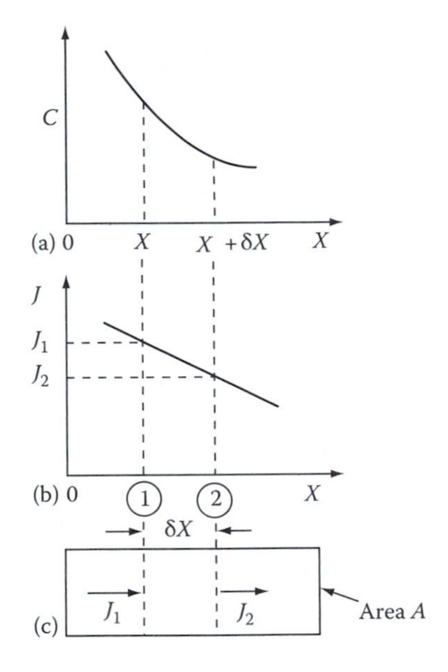

FIGURE 2.8
The derivation of Fick's second law.

concentration of B at any point varies with time consider a narrow slice of material with an area A and a thickness δx as shown in Figure 2.8c.

The number of interstitial B atoms that diffuse into the slice across plane (1) in a small time interval δt will be $J_1 A\delta t$. The number of atoms that leave the thin slice during this time, however, is only $J_2 A\delta$. Since $J_2 < J_1$ the concentration of B within the slice will have increased by

$$\delta C_B = \frac{(J_1 - J_2)A\delta t}{A\delta x} \tag{2.14}$$

But since δx is small,

$$J_2 = J_1 + \frac{\partial J}{\partial x}\delta x \tag{2.15}$$

and in the limit as $\delta t \to 0$ these equations give

$$\frac{\partial C_B}{\partial t} = -\frac{\partial J_B}{\partial x} \tag{2.16}$$

Substituting Fick's first law gives

$$\frac{\partial C_B}{\partial t} = \frac{\partial}{\partial x}\left(D_B \frac{\partial C_B}{\partial x}\right) \tag{2.17}$$

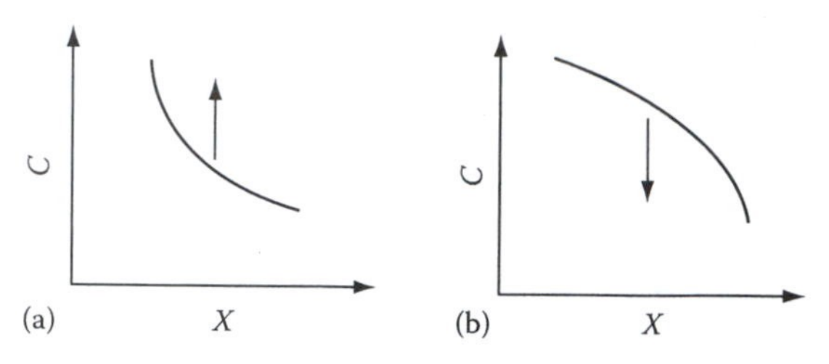

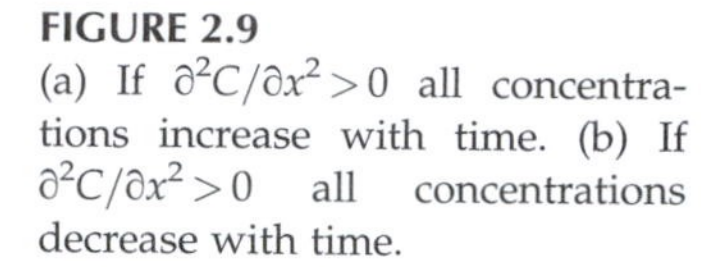

FIGURE 2.9
(a) If $\partial^2 C/\partial x^2 > 0$ all concentrations increase with time. (b) If $\partial^2 C/\partial x^2 > 0$ all concentrations decrease with time.

which is referred to as Fick's second law. If variations of D_B with concentration can be ignored this equation can be simplified to

$$\frac{\partial C_B}{\partial t} = D_B \frac{\partial^2 C_B}{\partial x^2} \tag{2.18}$$

These equations relate the rate of change of composition with time to the concentration profile $C_B(x)$. Equation 2.18 has a simple graphical interpretation as $\partial^2 C_B/\partial x^2$ is the curvature of the C_B versus x curve. If the concentration profile appears as shown in Figure 2.9a it has a positive curvature everywhere and the concentration at all points on such a curve will increase with time ($\partial C_B/\partial t$ positive). When the curvature is negative as in Figure 2.9b C_B decreases with time ($\partial C_B/\partial t$ negative).

2.2.5 Solutions to the Diffusion Equation

Two solutions will be considered which are of practical importance. One concerns the situation which is encountered in homogenization heat treatments, and the other is encountered, for example, in the carburization of steel.

2.2.5.1 Homogenization

It is often of interest to be able to calculate the time taken for an inhomogeneous alloy to reach complete homogeneity, as for example in the elimination of segregation in castings.

The simplest composition variation that can be solved mathematically is if C_B varies sinusoidally with distance in one dimension as shown in Figure 2.10. In this case B atoms diffuse down the concentration gradients, and regions with negative curvature, such as between $x=0$ and $x=l$, decrease in concentration, while regions between $x=l$ and $2l$ increase in concentration. The curvature is zero at $x=0$, l, $2l$, so the concentrations at these points remain unchanged with time. Consequently the concentration profile after a certain time reduces to that indicated by the dashed line in Figure 2.10.

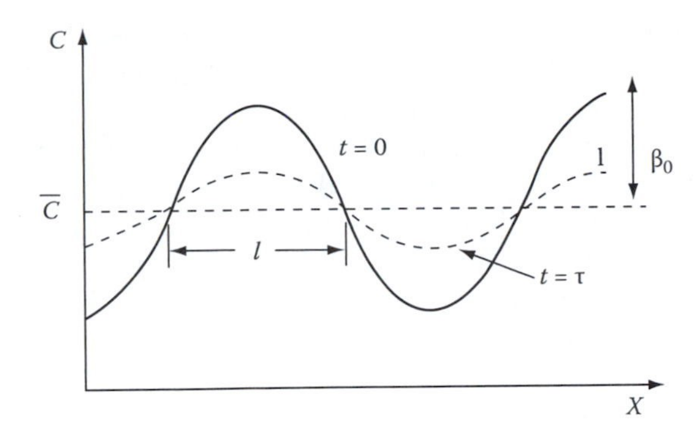

FIGURE 2.10
The effect of diffusion on a sinusoidal variation of composition.

At time $t = 0$ the concentration profile is given by

$$C = \overline{C} + \beta_0 \sin\frac{\pi x}{l} \tag{2.19}$$

where
$\overline{C}$ is the mean composition
β_0 is the amplitude of the initial concentration profile

Assuming D_B is independent of concentration the solution of Equation 2.18 that satisfies this initial condition is

$$C = \overline{C} + \beta_0 \sin\left(\frac{\pi x}{l}\right) \exp\frac{-t}{\tau} \tag{2.20}$$

where τ is a constant called the relaxation time and is given by

$$\tau = \frac{l^2}{\pi^2 D_B} \tag{2.21}$$

Thus the amplitude of the concentration profile after a time $t(\beta)$ is given by C at $x = l/2$, i.e.,

$$\beta = \beta_0 \exp\frac{-t}{\tau} \tag{2.22}$$

In other words, the amplitude of the concentration profile decreases exponentially with time and after a sufficiently long time approaches zero so that $C = \overline{C}$ everywhere. The rate at which this occurs is determined by the relaxation time τ. After a time $t = \tau$, $\beta = \beta_0/e$, that is, the amplitude has decreased to 1/2.72 of its value at $t = 0$. The solute distribution at this stage would therefore appear as shown by the dashed line in Figure 2.10. After at time

$t = 2\tau$ the amplitude is reduced by a total of $1/e^2$, i.e., by about one order of magnitude. From Equation 2.21 it can be seen that the rate of homogenization increases rapidly as the wavelength of the fluctuations decreases.

The initial concentration profile will not usually be sinusoidal, but in general any concentration profile can be considered as the sum of an infinite series of sine waves of varying wavelength and amplitude, and each wave decays at a rate determined by its own τ. Thus the short wavelength terms die away very rapidly and the homogenization will ultimately be determined by τ for the longest wavelength component.

2.2.5.2 Carburization of Steel

The aim of carburization is to increase the carbon concentration in the surface layers of a steel product in order to achieve a harder wear-resistant surface. This is usually done by holding the steel in a gas mixture containing CH_4 and/or CO at a temperature where it is austenitic. By controlling the relative proportions of the two gases the concentration of carbon at the surface of the steel in equilibrium with the gas mixture can be maintained at a suitable constant value. At the same time carbon continually diffuses from the surface into the steel.

The concentration profiles that are obtained after different times are shown in Figure 2.11. An analytical expression for these profiles can be obtained by solving Fick's second law using the boundary conditions: C_B (at $x = 0$) $= C_S$ and $C_B(\infty) = C_0$, the original carbon concentration of the steel. The specimen is considered to be infinitely long. In reality the diffusion coefficient of carbon in austenite increases with increasing concentration, but an approximate solution can be obtained by taking an average value and this gives the simple solution.

$$C = C_S - (C_S - C_0)\,\mathrm{erf}\left(\frac{x}{2\sqrt{(Dt)}}\right) \tag{2.23}$$

where "erf" stands for error function which is an indefinite integral defined by the equation.

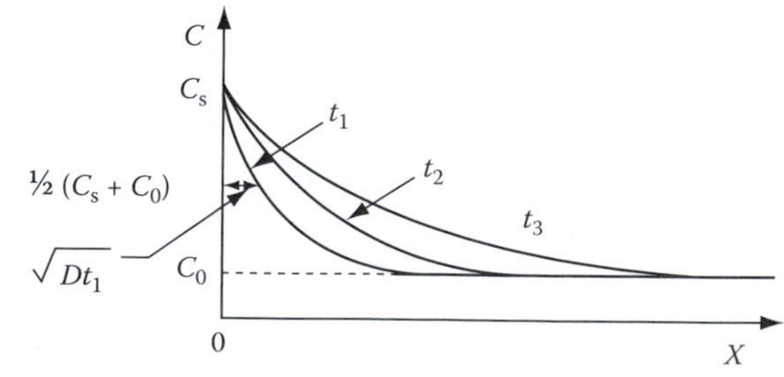

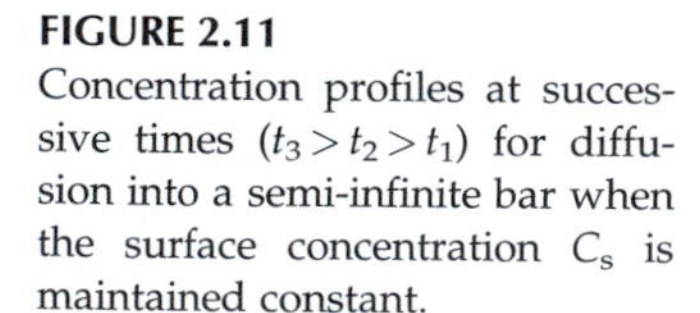
FIGURE 2.11
Concentration profiles at successive times ($t_3 > t_2 > t_1$) for diffusion into a semi-infinite bar when the surface concentration C_s is maintained constant.

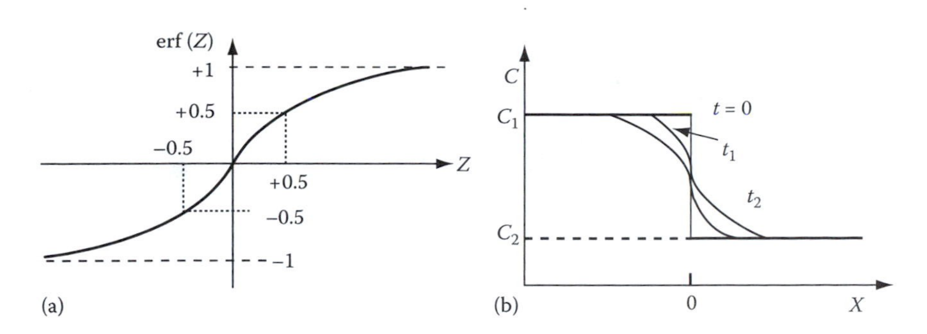

FIGURE 2.12
(a) Schematic diagram illustrating the main features of the error function. (b) Concentration profiles at successive times ($t_2 > t_1 > 0$) when two semi-infinite bars of different composition are annealed after welding.

$$\text{erf}(z) = \frac{2}{\sqrt{\pi}} \int_0^z \exp(-y^2)\text{d}y$$

The function is shown graphically in Figure 2.12a. More accurate values can be obtained from books of standard mathematical functions. Note that since $\text{erf}(0.5) \simeq 0.5$ the depth at which the carbon concentration is midway between C_s and C_0 is given by $(x/2\sqrt{(Dt)} \simeq 0.5$, that is

$$x \simeq \sqrt{(Dt)} \tag{2.24}$$

Thus the thickness of the carburized layer is $\sim\sqrt{(Dt)}$. Note also that the depth of any isoconcentration line is directly proportional to $\sqrt{(Dt)}$; that is, to obtain a twofold increase in penetration requires a fourfold increase in time.

For the case of carbon diffusion in austenite at 1000°C, D. 4×10^{-11} m^2 s^{-1}, which means that a carburized layer 0.2 mm thick requires a time of $(0.2\times10^{-3})^2/4 \times 10^{-11}$, i.e., 1000 s (17 min).

There are other situations in which the solution to the diffusion equation is very similar to Equation 2.23. For example during decarburization of steel the surface concentration is reduced to a very low value and carbon diffuses out of the specimen. The carbon profile is then given by

$$C = C_0\text{erf}\left(\frac{x}{2\sqrt{(Dt)}}\right) \tag{2.25}$$

Another situation arises if two semi-infinite specimens of different compositions C_1 and C_2 are joined together and annealed. The profiles in this case are shown in Figure 2.12b and the relevant solution is

$$C = \left(\frac{C_1 + C_2}{2}\right) - \left(\frac{C_1 - C_2}{2}\right)\text{erf}\left(\frac{x}{2\sqrt{(Dt)}}\right) \tag{2.26}$$

2.3 Substitutional Diffusion

Diffusion in dilute interstitial alloys was relatively simple because the diffusing atoms are always surrounded by vacant sites to which they can jump whenever they have enough to overcome the energy barrier for migration. In substitutional diffusion, however, an atom can only jump if there happens to be vacant site at one of the adjacent lattice positions as shown in Figure 2.2. The simplest case of substitutional diffusion is the self-diffusion of atoms in a pure metal. This is amenable to a simple atomic model similar to the case of interstitial diffusion and will be treated first. Substitutional diffusion in binary alloys is more complex and will be dealt with separately.

2.3.1 Self-Diffusion

The rate of self-diffusion can be measured experimentally by introducing a few radioactive A atoms (A*) into pure A and measuring the rate at which penetration occurs at various temperatures. Since A* and A atoms are chemically identical their jump frequencies are also almost identical. Thus the diffusion coefficient can be related to the jump frequency by Equation 2.3, that is

$$D_A^* = D_A = \frac{1}{6}\alpha^2\Gamma \tag{2.27}$$

where Γ is the jump frequency of both the A* and A atoms. Strictly speaking, Equation 2.3 was derived on the assumption that each atomic jump is unrelated to the previous jump. This is a good assumption for interstitial diffusion, but it is less valid for substitutional diffusion. The difference is that once an atom has jumped into a vacancy the next jump is not equally probable in all directions, but is most likely to occur back into the same vacancy. Such jumps do not contribute to the diffusive flux and therefore Equation 2.27 should be replaced by $D_A^* = fD_A = f \cdot \alpha^2\Gamma/6$ (known as a correlation factor) is less than unity. However, the effect is small and f is close to unity. (See P.G. Shewmon, *Diffusion in Solids,* McGraw-Hill, New York, 1963, p. 100.)

Consider the atomic jump shown in Figure 2.2. An atom next to a vacancy can make a jump provided it has enough thermal energy to overcome the activation energy barrier to migration, ΔG_m. Therefore the probability that any attempt at jumping will be successful is given by $\exp(-\Delta G_m/RT)$ as in the case of interstitial migration. However, most of the time the adjacent site will not be vacant and the jump will not be possible. The probability that an adjacent site is vacant is given by zX_v where z is the number of nearest neighbors and X_v is the probability that any one site is vacant, which is just the mole fraction of vacancies in the metal. Combining all these probabilities

atoms are vibrating with a temperature-independent frequency v the number of successful jumps any given atom will make in 1 s is given by

$$\Gamma = vzX_{\text{v}} \exp\frac{-\Delta G_{\text{m}}}{RT} \tag{2.28}$$

But, if the vacancies are in thermodynamic equilibrium, $X_{\text{V}} = X_{\text{v}}^{\text{e}}$ as given by Equation 1.57, i.e.,

$$X_{\text{v}}^{\text{e}} = \exp\frac{-\Delta G_{\text{v}}}{RT} \tag{2.29}$$

Combining Equations 2.27 through 2.29 gives

$$D_{\text{A}} = \frac{1}{6}\alpha^2 zv \exp\frac{-(\Delta G_{\text{m}} + \Delta G_{\text{v}})}{RT} \tag{2.30}$$

Substituting $\Delta G = \Delta H - T\Delta S$ gives

$$D_{\text{A}} = \frac{1}{6}\alpha^2 zv \exp\frac{\Delta S_{\text{m}} + \Delta S_{\text{v}}}{R} \exp -\left(\frac{\Delta H_{\text{m}} + \Delta H_{\text{v}}}{RT}\right) \tag{2.31}$$

For most metals v is $\sim 10^{13}$. In fcc metals $z = 12$ and $\alpha = a/\surd$ the jump distance. This equation can be written more concisely as

$$D_{\text{A}} = D_0 \exp\frac{-Q_{\text{SD}}}{RT} \tag{2.32}$$

where

$$D_0 = \frac{1}{6}\alpha^2 zv \exp\frac{\Delta S_{\text{m}} + \Delta S_{\text{v}}}{R} \tag{2.33}$$

and

$$Q_{\text{SD}} = \Delta H_{\text{m}} + \Delta H_{\text{v}} \tag{2.34}$$

Equation 2.32 is the same as was obtained for interstitial diffusion except that the activation energy for self-diffusion has an extra term (ΔH_{v}). This is because self-diffusion requires the presence of vacancies whose concentration depends on ΔH_{v}.

Some of the experimental data on substitutional self-diffusion are summarized in Table 2.2. It can be seen that for a given crystal structure and bond type Q/RT_{m} is roughly constant; that is, the activation enthalpy for self-diffusion, Q, is roughly proportional to the equilibrium melting temperature, T_{m}. Also, within each class, the diffusivity at the melting temperature, $D(T_{\text{m}})$, and D_0 are approximately constants. For example, for most close-packed

TABLE 2.2

Experimental Data for Substitutional Self-Diffusion in Pure Metals at Atmospheric Pressure

Class	Metal	T_m (K)	D_0 ($mm^2 s^{-1}$)	Q ($kJ mol^{-1}$)	$\frac{Q}{RT_m}$	$Q(T_m)$ ($\mu m^2 s^{-1}$)
bcc (rare earths)	ε-Pu	914	0.3	65.7	8.7	53
	δ-Ce	1071	1.2	90.0	10.1	49
	γ-La	1193	1.3	102.6	10.4	42
	γ-Yb	1796	1.2	121.0	8.1	3600
bcc (alkali metals)	Rb	312	23	39.4	15.2	5.8
	K	337	31	40.8	14.6	15
	Na	371	24.2	43.8	14.2	16
	Li	454	23	55.3	14.7	9.9
bcc (transition metals)	β-T1	577	40	94.6	19.7	0.11
	Eu	1095	100	143.5	15.8	14
	Er	1795	451	302.4	20.3	0.71
	α-Fe[a]	1811	200	239.7	15.9	26
	δ-Fe[a]	1811	190	238.5	15.8	26
	β–Ti	1933	109	251.2	15.6	18
	β–Zρ	2125	134	273.5	15.5	25
	Cr	2130	20	308.6	17.4	0.54
	V	2163	28.8	309.2	17.2	0.97
	Nb	2741	1240	439.6	19.3	5.2
	Mo	2890	180	460.6	19.2	0.84
	Ta	3269	124	413.3	15.2	31
	W	3683	4280	641.0	20.9	3.4
hcp[a]	Cd	594	‖c 5	76.2	15.4	0.99
			⊥c 10	79.9	16.2	0.94
	Zn	692	‖c 13	91.6	15.9	1.6
			⊥c 18	96.2	16.7	0.98
	Mg	922	‖c 100	134.7	17.6	2.3
			⊥c 150	136.0	17.8	2.9
fcc	Pb	601	137	109.1	21.8	0.045
	Al	933	170	142.0	18.3	1.9
	Ag	1234	40	184.6	18.0	0.61
	Au	1336	10.7	176.9	15.9	1.3
	Cu	1356	31	200.3	17.8	0.59
	Ni	1726	190	279.7	19.5	0.65
	β-Co	1768	83	283.4	19.3	0.35
	γ-Fe[b]	1805	49	284.1	18.9	0.29
	Pd	1825	20.5	266.3	17.6	0.49
	Th	2023	120	319.7	19.0	6.6
	Pt	2046	22	278.4	16.4	0.17

(continued)

TABLE 2.2 (continued)

Experimental Data for Substitutional Self-Diffusion in Pure Metals at Atmospheric Pressure

Class	Metal	T_m (K)	D_0 ($mm^2\ s^{-1}$)	Q ($kJ\ mol^{-1}$)	$\frac{Q}{RT_m}$	$Q(T_m)$ ($\mu m^2\ s^{-1}$)
Tet.[a]	β-Sn	505 ‖c	770	107.1	25.5	0.0064
		⊥ c	1070	105.0	25.0	0.015
Diamond	Ge	1211	440	324.5	32.3	4.4×10^{-5}
Cubic	Si	1683	0.9×10^6	496.0	35.5	3.6×10^{-4}

Source: Data selected mainly from Brown, A.M. and Ashby, M.F., *Acta Metall.*, 28, 1085, 1980.

[a] Data selected from N.L. Peterson, in D. Turnbull and H. Ehrenreich (Eds.), *Solid State Physics*, Vol. 22, Academic Press, New York, 1968. The symbols for the hcp structures indicate directions parallel to and perpendicular to the crystallographic c axis.

[b] T_m for γ-Fe is the temperature at which γ-Fe would melt if δ-Fe did not intervene.

metals (fcc and hexagonal close-packed, hcp) Q/RT_m ~18 and $D(T_m)$ 1 $\mu m^{-2}\ s^{-1}$ ($10^{-12}\ m^{-2}\ s^{-1}$). The Q/RT_m and $D(T_m)$ data are also plotted in Figure 2.13 along with data for other materials for comparison. An immediate consequence of these correlations is that the diffusion coefficients of all materials with a given crystal structure and bond type will be approximately the same at the same fraction of their melting temperature, i.e., $D(T/T_m) =$ constant. (T/T_m is known as the homologous temperature.)

The correlations in Table 2.2 have been evaluated for atmospheric pressure. There are, however, limited experimental data that suggest the same correlations hold independently of pressure, provided of course the effect of pressure on T_m is taken into account. Since volume usually increases on melting, raising the pressure increases T_m and thereby lowers the diffusivity at a given temperature.

That a rough correlation exists between Q and T_m is not surprising: increasing the interatomic bond strength makes the process of melting more difficult; that is, T_m is raised. It also makes diffusion more difficult by increasing ΔH_v and ΔH_m.

Consider the effect of temperature on self-diffusion in Cu as an example. At 800°C (1073 K) the data in Table 2.2 give $D_{Cu} = 5 \times 10^{-9}\ mm^2\ s^{-1}$. The jump distance α in Cu is 0.25 nm and Equation 2.3 therefore gives $\Gamma_{Cu} = 5 \times 10^5$ jumps s^{-1}. After an hour at this temperature, $\sqrt{(Dt)} \sim 4\ \mu m$. Extrapolating the data to 20°C, however, gives $D_{Cu} \sim 10^{-34}\ mm^2\ s^{-1}$, i.e., $\Gamma \sim 10^{-20}$ jumps s^{-1}. Alternatively, each atom would make one jump every 10^{12} years!

Experimentally the usual method for determining the self-diffusion coefficient is to deposit a known quantity (M) of a radioactive isotope A* onto the ends of two bars of A which are then joined as shown in Figure 2.14a. After annealing for a known time at a fixed temperature. A* will have diffused into A and the concentration profile can be determined by machining away thin layers of the bar and measuring the radioactivity as a function of position. Since A and A* are chemically identical the diffusion of A* into A will occur

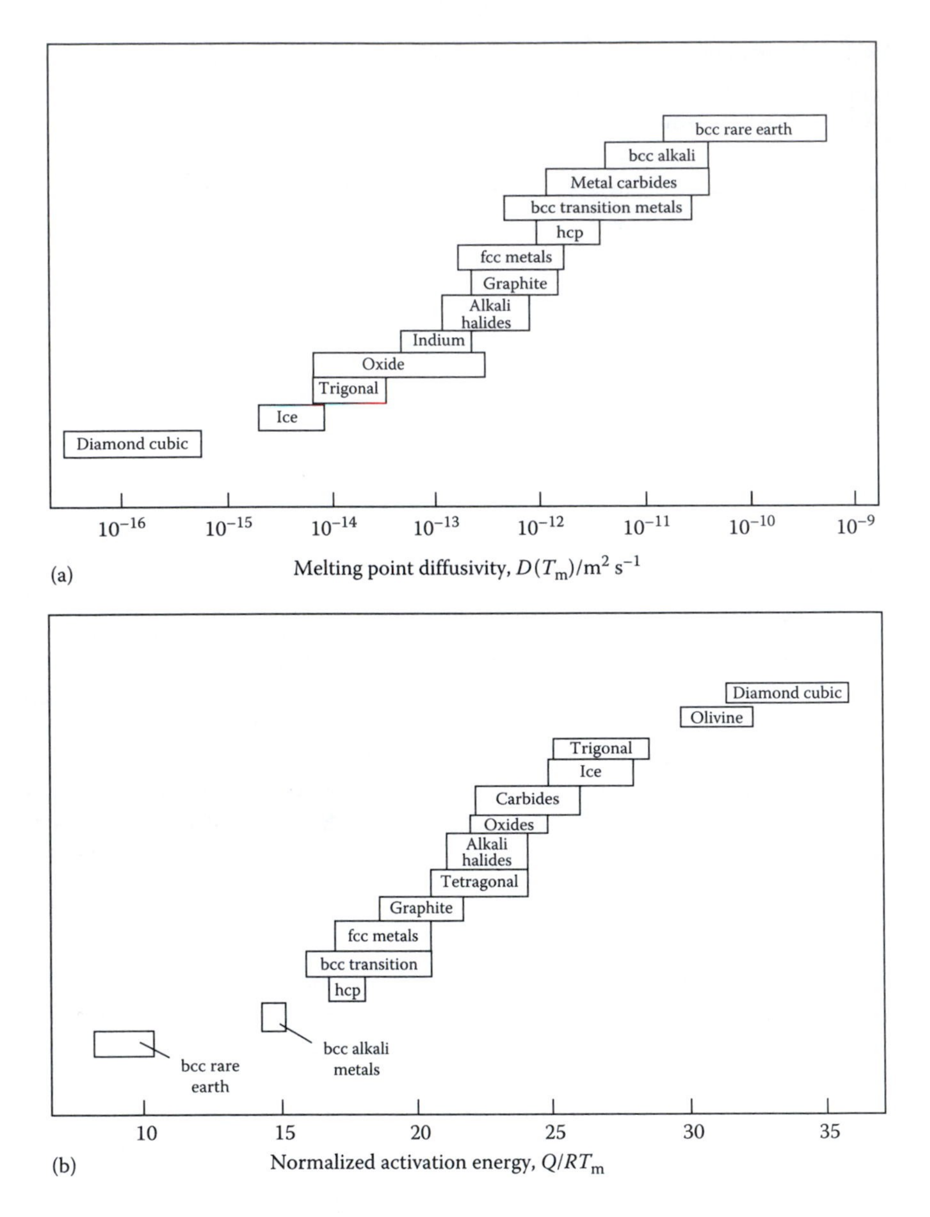

FIGURE 2.13
(a) Melting point diffusivities and (b) normalized activation energies for various classes of materials. (After Brown, A.M. and Ashby, M.F., *Acta Metall.*, 28, 1085, 1980.)

according to Equation 2.18. The solution of this equation for the present boundary conditions is

$$C = \frac{M}{2_{\mathrm{v}}(\pi Dt)} \exp\left(-\frac{x^2}{4Dt}\right) \tag{2.35}$$

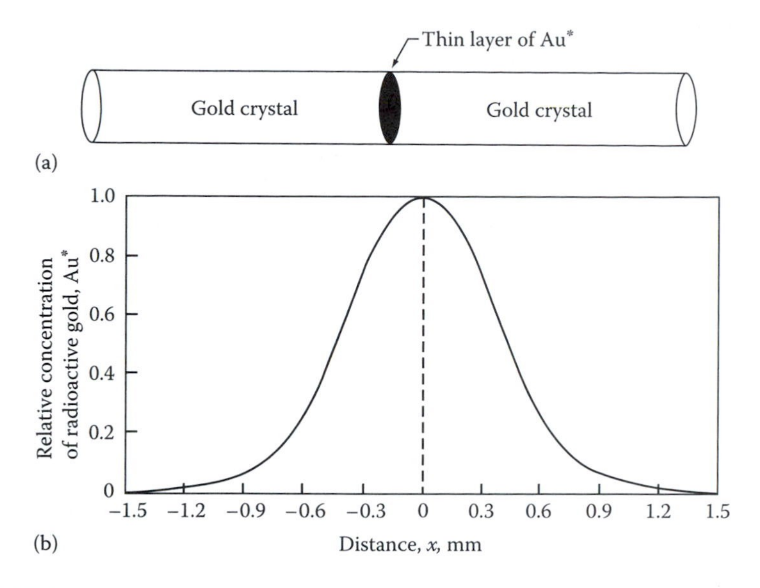

FIGURE 2.14
Illustration of the principle of tracer diffusion and of the planar source method for determining the self-diffusion coefficient of gold. (a) Initial diffusion couple with planar source of radioactive gold Au*. (b) Distribution of Au* after diffusion for 100 h at 920°C. (After Guy, A.G., *Introduction to Materials Science*, McGraw-Hill, New York, 1971.)

M has units [quantity m^{-2}] and C [quantity m^{-3}]. Figure 2.14b shows the form of this equation fitted to experimental points for self-diffusion in gold.

2.3.2 Vacancy Diffusion

The jumping of atoms into vacant sites can equally well be considered as the jumping of vacancies onto atom sites. If excess vacancies are introduced into the lattice they will diffuse at a rate which depends on the jump frequency. However, a vacancy is always surrounded by sites to which it can jump and it is thus analogous to an interstitial atom (see Section 2.2.2). Therefore a vacancy can be considered to have its own diffusion coefficient given by

$$D_v = \frac{1}{6}\alpha^2\Gamma_v \tag{2.36}$$

By analogy with Equation 2.8

$$D_v = \frac{1}{6}\alpha^2 z\nu \exp\frac{\Delta S_m}{R} \exp\frac{-\Delta H_m}{RT} \tag{2.37}$$

In this case ΔH_m and ΔS_m apply to the migration of a vacancy, and are therefore the same as for the migration of a substitutional atom. Comparing Equations 2.37 and 2.31 it can be seen that

$$D_v = D_A/X_v^e \quad (2.38)$$

This shows in fact that D_v is many orders of magnitude greater than D_A the diffusivity of substitutional atoms.

2.3.3 Diffusion in Substitutional Alloys

During self-diffusion all atoms are chemically identical. Thus the probability of finding a vacancy adjacent to any atom and the probability that the atom will make a jump into the vacancy is equal for all atoms. This leads to a simple relationship between jump frequency and diffusion coefficient. In binary substitutional alloys, however, the situation is more complex. In general, the rate at which solvent (A) and solute (B) atoms can move into a vacant site is not equal and each atomic species must be given its own intrinsic diffusion coefficient D_A or D_B.

The fact that the A and B atoms occupy the same sites has important consequences on the form that Fick's first and second laws assume for substitutional alloys. It will be seen later that when the A and B atoms jump at different rates the presence of concentration gradients induces a movement of the lattice through which the A and B atoms are diffusing.

D_A and D_B are defined such that Fick's first law applies to diffusion relative to the lattice, that is

$$J_A = -D_A \frac{\partial C_A}{\partial x} \quad (2.39)$$

$$J_B = -D_B \frac{\partial C_B}{\partial x} \quad (2.40)$$

where J_A and J_B are the fluxes of A and B atoms across a given lattice plane. This point did not need emphasizing in the case of interstitial diffusion because the lattice planes of the parent atoms were unaffected by the diffusion process. It will be seen, however, that the situation is different in the case of substitutional diffusion.

In order to derive Fick's second law let us consider the interdiffusion of A and B atoms in a diffusion couple that is made by welding together blocks of pure A and B as shown in Figure 2.15a. If the couple is annealed at a high enough temperature, a concentration profile will develop as shown.

If we make the simplifying assumption that the total number of atoms per unit volume is a constant, C_0, independent of composition, then

$$C_0 = C_A + C_B \quad (2.41)$$

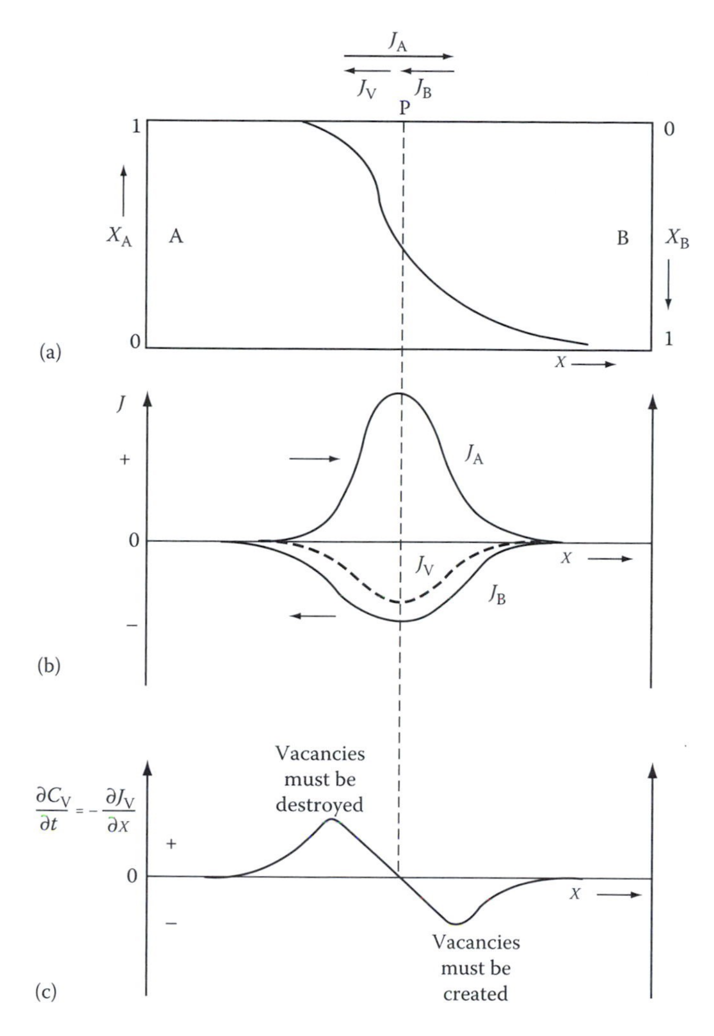

FIGURE 2.15
Interdiffusion and vacancy flow. (a) Composition profile after interdiffusion of A and B. (b) The corresponding fluxes of atoms and vacancies as a function of position x. (c) The rate at which the vacancy concentration would increase or decrease if vacancies were not created or destroyed by dislocation climb.

and

$$\frac{\partial C_A}{\partial x} = -\frac{\partial C_B}{\partial x} \tag{2.42}$$

Hence at a given position the concentration gradients driving the diffusion of A and B atoms are equal but opposite, and the fluxes of A and B relative to the lattice can be written as

$$J_A = -D_A \frac{\partial C_A}{\partial x}$$
$$J_A = -D_B \frac{\partial C_A}{\partial x} \quad (2.43)$$

These fluxes are shown schematically in Figure 2.15 for the case $D_A > D_B$, i.e., $|J_A| > |J_B|$.

When atoms migrate by the vacancy process the jumping of an atom into a vacant site can equally well be regarded as the jumping of the vacancy onto the atom, as illustrated in Figure 2.16. In other words, if there is a net flux of atoms in one direction there is an equal flux of vacancies in the opposite direction. Thus in Figure 2.15a there is a flux of vacancies $-J_A$ due to the migration of A atoms plus of vacancies $-J_B$ due to the diffusion of B atoms. As $J_A > J_B$ there will be a net flux of vacancies

$$J_v = -J_A - J_B \quad (2.44)$$

This is indicated in vector notation in Figure 2.15a. In terms of D_A and D_B, therefore

$$J_v = (D_A - D_B)\frac{\partial C_A}{\partial x} \quad (2.45)$$

This leads to a variation in J_v across the diffusion couple as illustrated in Figure 2.15b.

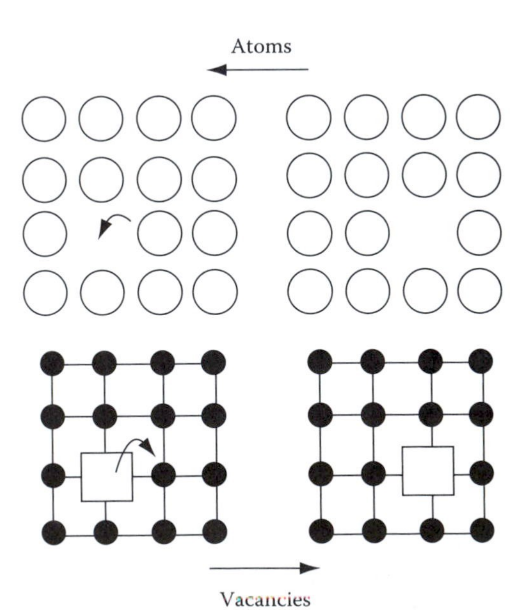

FIGURE 2.16
The jumping of atoms in one direction can be considered as the jumping of vacancies in the other direction.

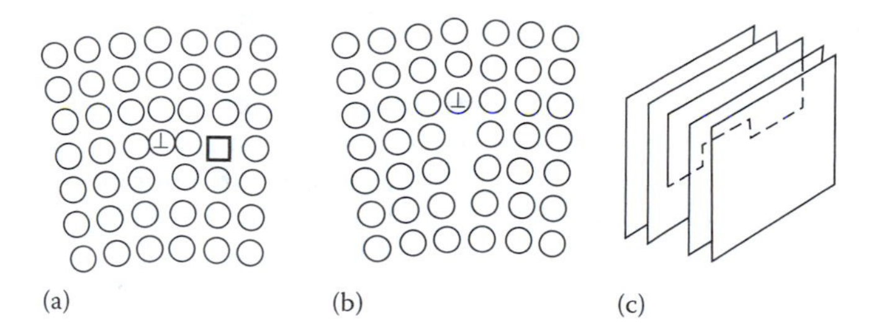

FIGURE 2.17
(a) Before, (b) after: a vacancy is absorbed at a jog on an edge dislocation (positive climb). (b) Before, (a) after: a vacancy is created by negative climb of an edge dislocation. (c) Perspective drawing of a jogged edge dislocation.

In order to maintain the vacancy concentration everywhere near equilibrium vacancies must be created on the B-rich side and destroyed on the A-rich side. The rate at which vacancies are created or destroyed at any point is given by $\partial C_v/\partial t = -\partial J_v/\partial x$ (Equation 2.16) and this varies across the diffusion couple as shown in Figure 2.15c.

It is the net flux of vacancies across the middle of the diffusion couple that gives rise to movement of the lattice. Jogged edge dislocations can provide a convenient source or sink for vacancies as shown in Figure 2.17. Vacancies can be absorbed by the extra half-plane of the edge dislocation shrinking while growth of the plane can occur by the emission of vacancies. If this or a similar mechanism operates on each side of the diffusion couple then the required flux of vacancies can be generated as illustrated in Figure 2.18. This means that extra atomic planes will be introduced on the B-rich side while whole planes of atoms will be eaten away on the A-rich

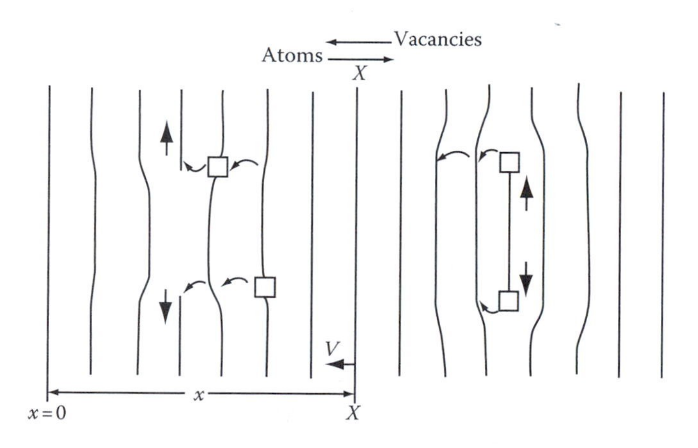

FIGURE 2.18
A flux of vacancies causes the atomic planes to move through the specimen.

side. Consequently the lattice planes in the middle of the couple will be shifted to the left.

The velocity at which any given lattice plane moves, v, can be related to the flux of vacancies crossing it. If the plane has an area A, during a small time interval δt, the plane will sweep out a volume of $Av \cdot \delta t$ containing $Av \cdot \delta t \cdot C_0$ atoms. This number of atoms is removed by the total number of vacancies crossing the plane in the same time interval, i.e., $J_v A \cdot \delta t$, giving

$$J_v = C_0 v \tag{2.46}$$

Thus the velocity of the lattice planes will vary across the couple in the same way as J_v, see Figure 2.15b. Substituting Equation 2.45 gives

$$v = (D_A - D_B)\frac{\partial X_A}{\partial x} \tag{2.47}$$

where the mole fraction of A, $X_A = C_A/C_0$.

In practice, of course, internal movements of lattice planes are usually not directly of interest. More practical questions concern how long homogenization of an alloy takes, or how rapidly the composition will change at a fixed position relative to the ends of a specimen. To answer these questions we can derive Fick's second law for substitutional alloys.

Consider a thin slice of material δx thick at a fixed distance x from one end of the couple which is outside the diffusion zone as shown in Figure 2.19. If the total flux of A atoms entering the slice across plane 1 is J_A' and the total flux leaving is $J_A' + (\partial J_A'/\partial x)\delta x$ the same arguments as were used to derive Equation 2.16 can be used to show that

$$\frac{\partial C_A}{\partial t} = -\frac{\partial J_A'}{\partial x} \tag{2.48}$$

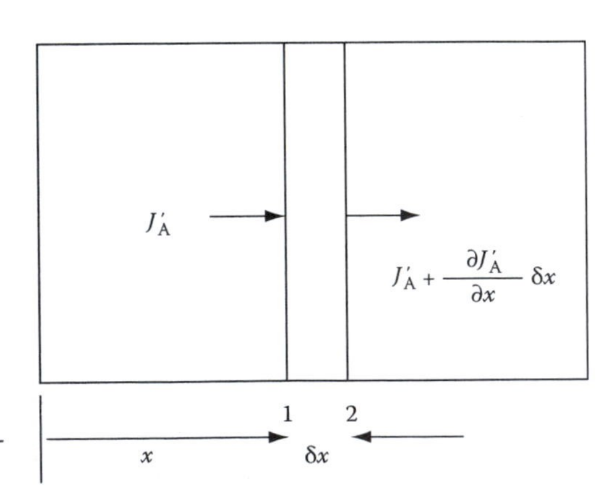

FIGURE 2.19
Derivation of Fick's second law for interdiffusion. (See text for details.)

The total flux of A atoms across a stationary plane with respect to the specimen is the sum of two contributions: (1) a diffusive flux $J_A = -D_A \partial C_A/\partial x$ due to diffusion relative to the lattice and (2) a flux $v \cdot C_A$ due to the velocity of the lattice in which diffusion is occurring. Therefore

$$J'_A = -D_A \frac{\partial C_A}{\partial x} + vC_A \tag{2.49}$$

By combining this equation with Equation 2.47 we obtain the equivalent of Fick's first law for the flux relative to the specimen ends:

$$J'_A = -(X_B D_A + X_A D_B) \frac{\partial C_A}{\partial x} \tag{2.50}$$

where $X_A = C_A/C_0$ and $X_B = C_B/C_0$ are the mole fractions of A and B, respectively. This can be simplified by defining an interdiffusion coefficient $\widetilde{D}$ as

$$\widetilde{D} = X_B D_A + X_A D_B \tag{2.51}$$

so that Fick's first law becomes

$$J'_A = -\widetilde{D} \frac{\partial C_A}{\partial x} \tag{2.52}$$

Likewise,

$$J'_B = -\widetilde{D} \frac{\partial C_B}{\partial x} = \widetilde{D} \frac{\partial C_A}{\partial x}$$

i.e.,

$$J'_B = -J'_A$$

Substitution of Equation 2.52 into Equation 2.48 gives

$$\frac{\partial C_A}{\partial t} = \frac{\partial}{\partial x}\left(\widetilde{D} \frac{\partial C_A}{\partial x}\right) \tag{2.53}$$

Equation 2.53 is Fick's second law for diffusion in substitutional alloys. The only difference between Equations 2.53 and 2.18 (for interstitial diffusion) is that the interdiffusion coefficient D for substitutional alloys depends on D_A and D_B whereas in interstitial diffusion D_B alone is needed. Equations 2.47 and 2.51 were first derived by Darken[5] and are usually known as Darken's equations.

By solving Equation 2.53 with appropriate boundary conditions it is possible to obtain $C_A(x, t)$ and $C_B(x, t)$, i.e., the concentration of A and B at any position (x) after any given annealing time (t). The solutions that were given in Section 2.2.5 will be applicable to substitutional alloys provided the range of compositions is small enough that any effect of composition of $\tilde{D}$ can be ignored. For example, if $\tilde{D}$ is known the characteristic relaxation time for an homogenization anneal would be given by Equation 2.21 using $\tilde{D}$ in place of D_B, i.e.,

$$\tau = \frac{l^2}{\pi^2 \tilde{D}} \tag{2.54}$$

If the initial composition differences are so great that changes in $\tilde{D}$ become important then more complex solutions to Equation 2.53 must be used. These will not be dealt with here, however, as they only add mathematical complexities without increasing our understanding of the basic principles.[6]

Experimentally it is possible to measure $\tilde{D}$ by determining the variation of X_A or X_B after annealing a diffusion couple for a given time such as that shown in Figure 2.15a. In cases where $\tilde{D}$ can be assumed constant a comparison of Equation 2.26 and the measured concentration profile would give $\tilde{D}$. When $\tilde{D}$ is not constant there are graphical solutions of Fick's second law that enable $\tilde{D}$ to be determined at any composition. In order to determine D_A and D_B separately it is also necessary to measure the velocity of the lattice at a given point in the couple. This can be achieved in practice by inserting insoluble wires at the interface before welding the two blocks together. These wires remain in effect fixed to the lattice planes and their displacement after a given annealing time can be used to calculate v. When v and $\tilde{D}$ are known, Equations 2.47 and 2.51 can be used to calculate D_A and D_B for the composition at the markers.

The displacement of inert wires during diffusion was first observed by Smigelskas and Kirkendall in 1947[7] and is usually known as the Kirkendall effect. In this experiment a block of α-brass (Cu-30 wt% Zn) was wound with molybdenum wire and encapsuled in a block or pure Cu, as shown in Figure 2.20. After annealing at a high temperature it was found that the separation of the markers (w) had decreased. This is because $D_{Zn} > D_{Cu}$ and the zinc atoms diffuse out of the central block faster than they are replaced by copper atoms diffusing in the opposite direction. Similar effects have been demonstrated in many alloy systems. In general it is found that in any given couple, atoms with the lower melting point posses a higher D. The exact value of D, however, varies with the composition of the alloy. Thus in Cu–Ni alloys, D_{Cu} and $\tilde{D}$ are all composition dependent, increasing as X_{Cu}, increasing, Figure 2.21.

In Figure 2.17 it was assumed that the extra half-planes of atoms that grew or shrank due to the addition or loss of atoms, were parallel to the original weld interface so that there were no constraints on the resultant local expansion or contraction of the lattice. In practice, however, these planes can be

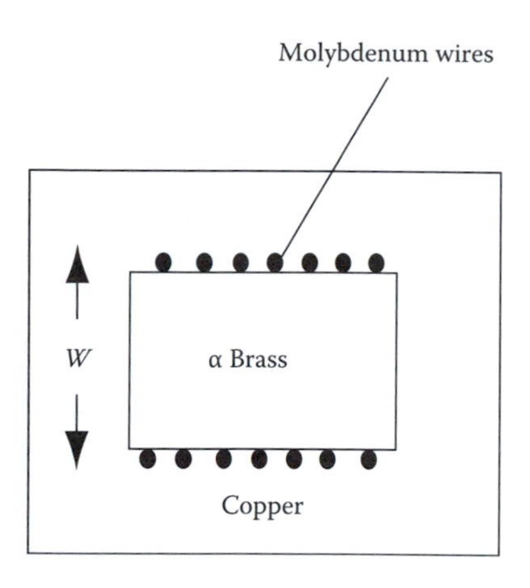

FIGURE 2.20
An experimental arrangement to show the Kirkendall effect.

oriented in many directions and the lattice will also try to expand or contract parallel to the weld interface. Such volume changes are restricted by the surrounding material with the result that two-dimensional compressive stresses develop in regions where vacancies are created, while tensile stresses arise in regions where vacancies are destroyed. These stress fields can even induce plastic deformation resulting in microstructures characteristic of hot deformation.

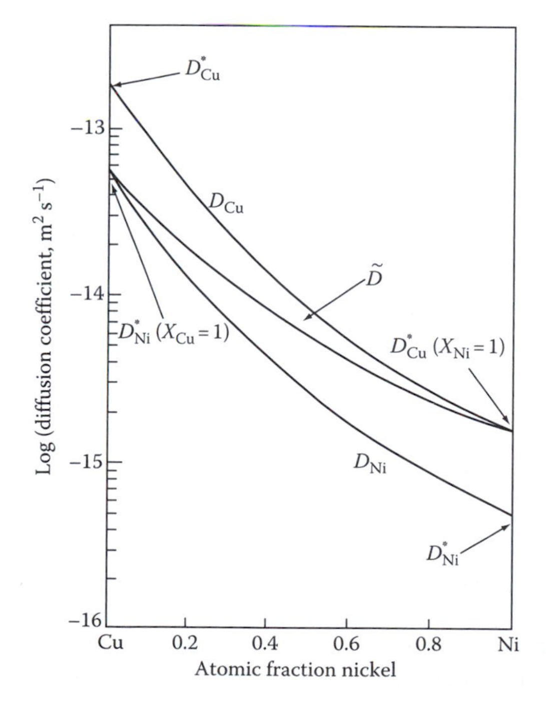

FIGURE 2.21
The relationship between the various diffusion coefficients in the Cu–Ni system at 1000°C. (After Guy, A.G., *Introduction to Materials Sciences*, McGraw-Hill, New York, 1971.)

Vacancies are not necessarily all annihilated at dislocations, but can also be absorbed by internal boundaries and free surfaces. However, those not absorbed at dislocations mainly agglomerate to form holes or voids in the lattice. Void nucleation is difficult because it requires the creation of a new surface and it is generally believed that voids are heterogeneously nucleated at impurity particles. The tensile stresses that arise in conjunction with vacancy destruction can also play a role in the nucleation of voids. When voids are formed the equations derived above cannot be used without modification.

In concentrated alloys the experimentally determined values of $\tilde{D}$, D_A, and D_B are also found to show the same form of temperature dependence as all other diffusivities, so that

$$\tilde{D} = \tilde{D}_0 \exp\frac{-Q}{RT} \tag{2.55}$$

$$D_A = D_{A0} \exp\frac{-Q_A}{RT} \tag{2.56}$$

$$D_B = D_{B0} \exp\frac{-Q_B}{RT} \tag{2.57}$$

However the factors that determine D_0 and Q in these cases are uncertain and there is no simple atomistic model for concentrated solutions.

The variation of $\tilde{D}$ with composition can be estimated in cases where it has not been measured, by utilizing two experimental observations:[8]

1. For a given crystal structure, $\tilde{D}$ at the melting point is roughly constant. Therefore if adding B to A decreases the melting point, $\tilde{D}$ will increase, at a given temperature, and vice versa.
2. For a given solvent and temperature, both interstitial and substitutional diffusion are more rapid in a bcc lattice than a close-packed lattice. For example, for the diffusion of carbon in Fe at 910°C, $D_c^{\alpha}/D_c^{\gamma} \sim 100$. At 850°C the self-diffusion coefficients for Fe are such that $D_{Fe}^{\alpha}/D_{Fe}^{\gamma} \sim 100$. The reason for this difference lies in the fact that the bcc structure is more open and the diffusion processes require less lattice distortion.

2.3.4 Diffusion in Dilute Substitutional Alloys

Another special situation arises with diffusion in dilute alloys. When $X_B \sim 0$ and $X_A \sim 1$, Equation 2.51 becomes

$$\tilde{D} = D_B \tag{2.58}$$

This is reasonable since it means that the rate of homogenization in dilute alloys is controlled by how fast the solute (B) atoms can diffuse. Indeed the only way homogenization can be achieved is by the migration of the B atoms into the solute-depleted regions. D_B for a dilute solution of B in A is called the impurity diffusion coefficient. Such data are more readily available than interdiffusion data in concentrated alloys. One way in which impurity diffusion coefficients can be measured is by using radioactive tracers.

It is often found that D_B in a dilute solution of B in A is greater than D_A. The reason for this is that the solute atoms can attract vacancies so that there is more than a random probability of finding a vacancy next to a solute atom with the result that they can diffuse faster than the solvent. An attraction between a solute atom and a vacancy can arise if the solute atom is larger than the solvent atoms or if it has higher valency. If the binding energy is very large the vacancy will be unable to escape from the solute atom. In this case the solute–vacancy pair can diffuse through the lattice together.

2.4 Atomic Mobility

Fick's first law is based on the assumption that diffusion eventually stops, that is equilibrium is reached, when the concentration is the same everywhere. Strictly speaking this situation is never true in practice because real materials always contain lattice defects such as grain boundaries, phase boundaries, and dislocations. Some atoms can lower their free energies if they migrate to such defects and at equilibrium their concentrations will be higher in the vicinity of the defect than in the matrix. Diffusion in the vicinity of these defects is therefore affected by both the concentration gradient and the gradient of the interaction energy. Fick's law alone is insufficient to describe how the concentration will vary with distance and time.

As an example consider the case of a solute atom that is too big or too small in comparison to the space available in the solvent lattice. The potential energy of the atom will then be relatively high due to the strain in the surrounding matrix. However, this strain energy can be reduced if the atom is located in a position where it better matches the space available, e.g., near dislocations and in boundaries, where the matrix is already distorted.

Segregation of atoms to grain boundaries, interfaces, and dislocations is of great technological importance. For example the diffusion of carbon or nitrogen to dislocations in mild steel is responsible for strain aging and blue brittleness. The segregation of impurities such as Sb, Sn, P, and As to grain boundaries in low-alloy steels produces temper embrittlement. Segregation to grain boundaries affects the mobility of the boundary and has pronounced effects on recrystallization, texture, and grain growth. Similarly the rate at which phase transformations occur is sensitive to segregation at dislocations and interfaces.

The problem of atom migration can be solved by considering the thermodynamic condition for equilibrium; namely that the chemical potential of an atom must be the same everywhere. Diffusion continues in fact until this condition is satisfied. Therefore it seems reasonable to suppose that in general the flux of atoms at any point in the lattice is proportional to the chemical potential gradient. Fick's first law is merely a special case of this more general approach.

An alternative way to describe a flux of atoms is to consider a net drift velocity (v) superimposed on the random jumping motion of each diffusing atom. The drift velocity is simply related to the diffusive flux via the equation

$$J_B = v_B C_B \tag{2.59}$$

Since atoms always migrate so as to remove differences in chemical potential it is reasonable to suppose that the velocity is proportional to the local chemical potential gradient, i.e.,

$$v_B = -M_B \frac{\partial \mu_B}{\partial x} \tag{2.60}$$

where M_B is a constant of proportionality known as the atomic mobility. Since μ_B has units of energy, the derivative of μ_B with respect to distance ($\delta\mu_B/\delta x$) is effectively the chemical force causing the atom to migrate.

Combining Equations 2.59 and 2.60 gives

$$J_B = -M_B C_B \frac{\partial \mu_B}{\partial x} \tag{2.61}$$

Intuitively it seems that the mobility of an atom and its diffusion coefficient must be closely related. The relationship can be obtained by relating $\partial\mu/\partial x$ to $\partial C/\partial x$ for a stress-free solid solution. Using Equation 1.70 and $C_B = X_B/V_m$, Equation 2.61 becomes

$$J_B = -M_B \frac{X_B}{V_m} \cdot \frac{RT}{X_B} \left\{ 1 + \frac{d \ln \gamma_B}{d \ln X_B} \right\} \frac{\partial X_B}{\partial x} \tag{2.62}$$

i.e.,

$$J_B = -M_B RT \left\{ 1 + \frac{d \ln \gamma_B}{d \ln X_B} \right\} \frac{\partial C_B}{\partial x} \tag{2.63}$$

Comparison with Fick's first law gives the required relationship:

$$D_B = M_B RT \left\{ 1 + \frac{d \ln \gamma_B}{d \ln X_B} \right\} \tag{2.64}$$

Similarly

$$D_A = M_A RT\left\{1 + \frac{d \ln \gamma_A}{d \ln X_A}\right\} \tag{2.65}$$

For ideal or dilute solutions ($X_B \to 0$)γ_B is a constant and the term in brackets is unity, i.e.,

$$D_B = M_B RT \tag{2.66}$$

For nonideal concentrated solutions the terms in brackets, the so-called thermodynamic factor, must be included. As shown by Equation 1.71 this factor is the same for both A and B and is simply related to the curvature of the molar free energy–composition curve.

When diffusion occurs in the presence of a strain energy gradient, for example, the expression for the chemical potential can be modified to include the effect of an elastic strain energy term E which depends on the position (x) relative to a dislocation, say

$$\mu_B = G_B + RT \ln \gamma_B X_B + E \tag{2.67}$$

Following the above procedure, this gives

$$J_B = -D_B \cdot \frac{\partial C_B}{\partial x} - \frac{D_B C_B}{RT} \cdot \frac{\partial E}{\partial x} \tag{2.68}$$

It can thus be seen that in addition to the effect of the concentration gradient the diffusive flux is also affected by the gradient of strain energy, $\partial E/\partial x$.

Other examples of atoms diffusing toward regions of high concentration can be found when diffusion occurs in the presence of an electric field or a temperature gradient. These are known as electromigration and thermomigration, respectively.[9] Cases encountered in phase transformations can be found where atoms migrate phase boundaries, or, as mentioned in the introduction, when the free energy curve has a negative curvature. The latter is known as spinodal decomposition.

2.5 Tracer Diffusion in Binary Alloys

The use of radioactive tracers was described in connection with self-diffusion in pure metals. It is, however, possible to use radioactive tracers to determine the intrinsic diffusion coefficients of the components in an alloy. The method is similar to that shown in Figure 2.14 except that a small quantity of a suitable radioactive tracer, e.g., B*, is allowed to diffuse into a homogeneous bar of A/B solution. The value obtained for D from Equation 2.35 is the tracer diffusion coefficient D_B^*.

Such experiments have been carried out on a whole series of gold–nickel alloys at 900°C.[10] At this temperature gold and nickel are completely soluble in each other, Figure 2.22a. The results are shown in Figure 2.22c. Since

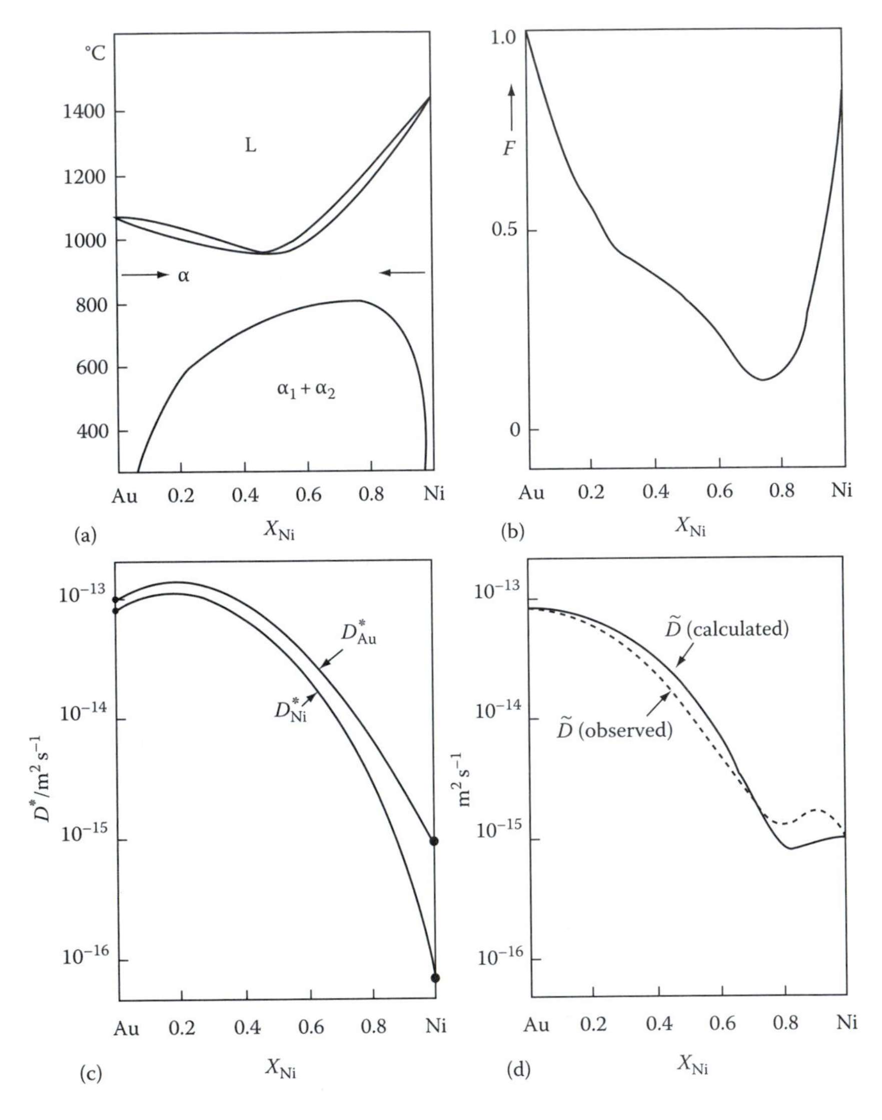

FIGURE 2.22
Interdiffusion in Au–Ni alloys at 900°C (a) Au–Ni phase diagram, (b) the thermodynamic factor, F, at 900°C, (c) experimentally measured tracer diffusivities at 900°C, and (d) experimentally measured interdiffusion coefficients compared with values calculated from (b) and (c). (From Reynolds, J.E., Averbach, B.L., and Cohen, M., *Acta Metall.*, 5, 29, 1957. With permission.)

radioactive isotopes are chemically identical it might appear at first sight that the tracer diffusivities (D^*_{Au} and D^*_{Ni}) should be identical to the intrinsic diffusivities (D_{Au} and D_{Ni}) determined by marker movement in a diffusion couple. This would be convenient as the intrinsic diffusivities are of more practical value whereas it is much easier to determine tracer diffusivities. However, it can be demonstrated that this is not the case. D^*_{Au} gives the rate at which Au* (or Au) atoms diffuse in a chemically homogeneous alloy, whereas D_{Au} gives the diffusion rate of Au when a concentration gradient is present.

The Au–Ni phase diagram contains a miscibility gap at low temperatures implying that $\Delta H_{mix} > 0$ (the gold and nickel atoms "dislike" each other). Therefore, whereas the jumps made by Au atoms in a chemically homogeneous alloy will be equally probable in all directions, in a concentration gradient they will be biased away from the Ni-rich regions. The rate of homogenization will therefore be slower in the second case, i.e., $D_{Au} < D^*_{Au}$ and $D_{Ni} < D^*_{Ni}$. On the other hand since the chemical potential gradient is the driving force for diffusion in both types of experiment it is reasonable to suppose that the atomic mobilities are not affected by the concentration gradient. If this is true the intrinsic chemical diffusivities and tracer diffusivities can be related as follows.

In the tracer diffusion experiment the tracer essentially forms a dilute solution in the alloy. Therefore from Equation 2.66

$$D^*_B = M^*_B RT = M_B RT \tag{2.69}$$

The second equality has been obtained by assuming M^*_B in the tracer experiment equals M_B in the chemical diffusion case. Substitution into Equations 2.64 and 2.51 therefore leads to the following relationships

$$\begin{aligned} D_A &= F D^*_A \\ D_B &= F D^*_B \end{aligned} \tag{2.70}$$

and

$$\tilde{D} = F\left(X_B D^*_A + X_A D^*_B\right) \tag{2.71}$$

where F is the thermodynamic factor, i.e.,

$$F = \left\{1 + \frac{d \ln \gamma_A}{d \ln X_A}\right\} = \left\{1 + \frac{d \ln \gamma_B}{d \ln X_B}\right\} = \frac{X_A X_B}{RT} \frac{d^2 G}{dX^2} \tag{2.72}$$

The last equality follows from Equation 1.71.

In the case of the Au–Ni system, diffusion couple experiments have also been carried out so that data are available for the interdiffusion coefficient $\widetilde{D}$, the full line in Figure 2.22d. In addition there is also enough thermodynamic data on this system for the thermodynamic factor F to be evaluated, Figure 2.22b. It is therefore possible to check the assumption leading to the second equality in Equation 2.69 by combining the data in Figure 2.22b and c using Equation 2.71. This produces the solid line in Figure 2.22d. The agreement is within experimental error.

Before leaving Figure 2.22 it is interesting to note how the diffusion coefficients are strongly composition dependent. There is a difference of about three orders of magnitude across the composition range. This can be explained by the lower liquidus temperature of the Au-rich compositions. Also in agreement with the rules of thumb given earlier, Au, with the lower melting temperature, diffuses faster than Ni at all compositions.

2.6 Diffusion in Ternary Alloys

The addition of a third diffusing species to a solid solution produces mathematical complexities which will not be considered here. Instead let us consider an illustrative example of some of the additional effects that can arise. Fe–Si–C alloys are particularly instructive for two reasons. First silicon raises the chemical potential (or activity) of carbon in solution, i.e., carbon will not only diffuse from regions or high carbon concentration but also from regions rich in silicon. Second the mobilities of carbon and silicon are widely different. Carbon, being and interstitial solute, is able to diffuse far more rapidly than the substitutionally dissolved silicon.

Consider two pieces of steel, one containing 3.8% silicon and 0.48% carbon and 0.48% carbon and the 0.44% carbon but no silicon. If the two pieces are welded together austenitized at 1050°C, the carbon concentration profile shown in Figure 2.23b is produced. The initial concentrations of silicon and carbon in the couple are shown in Figure 2.23a and the resultant chemical potential of carbon by the dotted line in Figure 2.23c. Therefore carbon atoms on the silicon-rich side will jump over to the silicon-free side until the difference in concentration at the interface is sufficient to equalize the activity, or chemical potential, of carbon on both sides. The carbon atoms at the interface are therefore in local equilibrium and the interfacial compositions remain constant as long as the silicon atoms do not migrate. Within each half of the couple the silicon concentration is initially uniform and the carbon atoms diffuse down the concentration gradients as shown in Figure 2.23b. The resultant chemical potential varies smoothly across the whole specimen (Figure 2.23c). If the total length of the diffusion couple is sufficiently small the carbon concentration in each block will eventually equal the interfacial compositions and the chemical potential of carbon will be the same everywhere. The alloys are now in a state of partial equilibrium. It is only partial because the chemical potential of the silicon is not uniform. Given sufficient

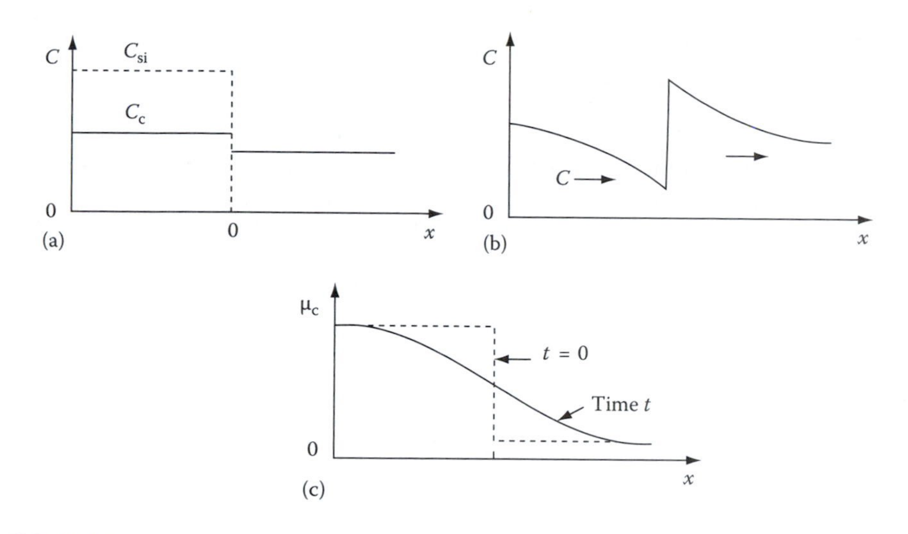

FIGURE 2.23
(a) Carbon and silicon distribution at $t = 0$. (b) Carbon distribution after high-temperature anneal. (c) Chemical potential of carbon versus distance.

time the silicon atoms will also diffuse over significant distances and the carbon atoms will continually redistribute themselves to maintain a constant chemical potential. In the final equilibrium state the concentrations of carbon and silicon are uniform everywhere. The change in composition of two points on opposite sides of the weld will be as illustrated on the ternary diagram of Figure 2.24.

The redistribution of carbon in Fe–Si–C system is particularly interesting since the mobilities of carbon and silicon are so different. Similarly, though less striking effects can arise in ternary systems where all three components diffuse substitutionally if the diffusivities (or mobilities) are unequal.

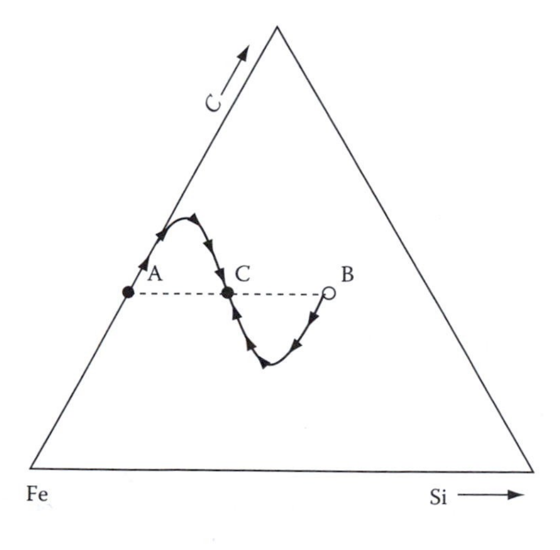

FIGURE 2.24
Schematic diagram showing the change in composition of two points (A and B) on opposite sides of the diffusion couple in Figure 2.23. C is the final equilibrium composition of the whole bar. (After Darken, L.S., *Trans. AIME*, 180, 430, 1949.)

2.7 High-Diffusivity Paths

In Section 2.4 the diffusion of atoms toward or away from dislocations, interfaces, grain boundaries, and free surfaces was considered. In this section diffusion along these defects will be discussed. All of these defects are associated with a more open structure and it has been shown experimentally that the jump frequency for atoms migrating along these defects is higher than that for diffusion in the lattice. It will become apparent that under certain circumstances diffusion along these defects can be the dominant diffusion path.

2.7.1 Diffusion along Grain Boundaries and Free Surfaces

It is found experimentally that diffusion along grain boundaries and free surfaces can be described by

$$D_b = D_{b0} \exp \frac{-Q_b}{RT} \quad (2.73)$$

or

$$D_s = D_{s0} \exp \frac{-Q_s}{RT} \quad (2.74)$$

where

D_b and D_s are the grain boundary and surface diffusivities
D_{b0} and D_{s0} are the frequency factors
Q_b and Q_s are the experimentally determined values of the activation energies for diffusion

In general, at any temperature the magnitudes of D_b and D_s relative to the diffusivity through defect-free lattice D_1 are such that

$$D_s > D_b > D_1 \quad (2.75)$$

This mainly reflects the relative ease with which atoms can migrate along free surfaces, interior boundaries, and through the lattice. Surface diffusion can play an important role in many metallurgical phenomena, but in an average metallic specimen the total grain boundary area is much greater than the surface area so that grain boundary diffusion is usually most important.

The effect of grain boundary diffusion can be illustrated by considering a diffusion couple made by welding together two metals A and B, as shown in Figure 2.25. A atoms diffusing along the boundary will be able to penetrate much deeper than atoms which only diffuse through the lattice. In addition, as the concentration of solute builds up in the boundaries atoms will also

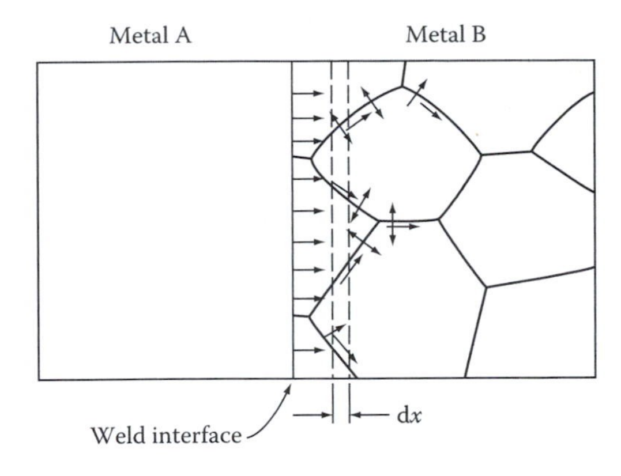

FIGURE 2.25
The effect of grain boundary diffusion combined with volume diffusion. (After Reed-Hill, R.E., *Physical Metallurgy Principles*, 2nd edn., Van Nostrand, New York, 1973.)

diffuse from the boundary into the lattice. The process can be compared to the conduction of heat through a plastic in which a continuous network of aluminum sheets is embedded. The temperature at any point in such a specimen would be analogous to the concentration of solute in the diffusion couple. Points in the lattice close to grain boundaries can receive solute via the high-conductivity path much more rapidly than if the boundaries were absent. Rapid diffusion along the grain boundaries increases the mean concentration in a slice such as dx in Figure 2.25 and thereby produces an increase in the apparent diffusivity in the material as a whole. Consider now under what conditions grain boundary diffusion is important.

For simplicity let us take a case of steady-state diffusion through a sheet of material in which the grain boundaries are perpendicular to the sheet as shown in Figure 2.26. Assuming that the concentration gradients in the lattice and along the boundary are identical, the fluxes of solute through the lattice J_1 and along the boundary J_b will be given by

$$J_1 = -D_1 \frac{dC}{dx} \quad J_b = -D_b \frac{dC}{dx} \tag{2.76}$$

However the contribution of grain boundary diffusion to the total flux through the sheet will depend on the relative cross-sectional areas through which the solute is conducted.

If the grain boundary has an effective thickness δ and the grain size is d the total flux will be given by

$$J = (J_b\delta + J_1 d)/d = -\left(\frac{D_b\delta + D_1 d}{d}\right)\frac{dC}{dx} \tag{2.77}$$

Thus the apparent diffusion coefficient in this case,

$$D_{app} = D_1 + D_b\delta/d \tag{2.78}$$

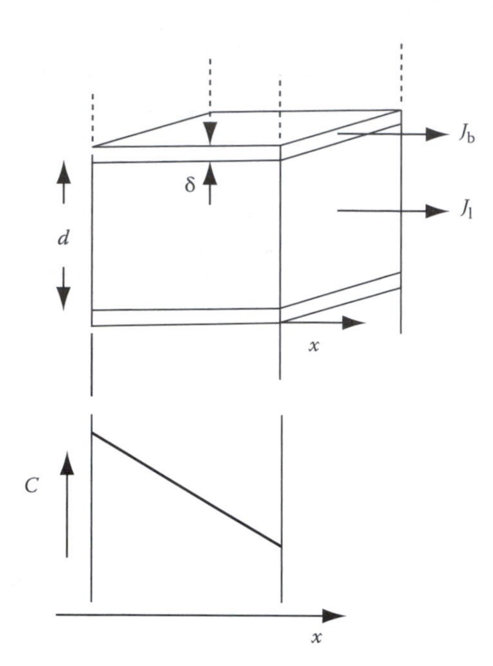

FIGURE 2.26
Combined lattice and boundary fluxes during steady-state diffusion through a thin slab of material.

or

$$\frac{D_{\text{app}}}{D_1} = 1 + \frac{D_b\delta}{D_1 d} \tag{2.79}$$

It can be seen that the relative importance of lattice and grain boundary diffusion depends on the ratio $D_b\delta/D_1 d$. When $D_b\delta \gg D_1 d$ diffusion through the lattice can be ignored in comparison to grain boundary diffusion. Thus grain boundary diffusion makes a significant contribution to the total flux when

$$D_b\delta > D_1 d \tag{2.80}$$

The effective width of a grain boundary is ~0.5 nm. Grain sizes on the other hand can vary from ~1 to 1000 μm and the effectiveness of the grain boundaries will vary accordingly. The relative magnitudes of $D_b\delta$ and $D_1 d$ are most sensitive to temperature. This is illustrated in Figure 2.27 which shows the effect or temperature on both D_1 and D_b. Note that although $D_b > D_1$ at all temperatures the difference increases as temperatures decreases. This is because the activation energy for diffusion along grain boundaries (Q_b) is lower than that for lattice diffusion (Q_1). For example, in fcc metals it is generally found that $Q_b \sim 0.5Q_1$. This means that when the grain boundary diffusivity is scaled by the factor δ/d (Equation 2.78) the grain boundary contribution to the total, or apparent, diffusion coefficient is negligible in comparison to the lattice diffusivity at high temperatures, but

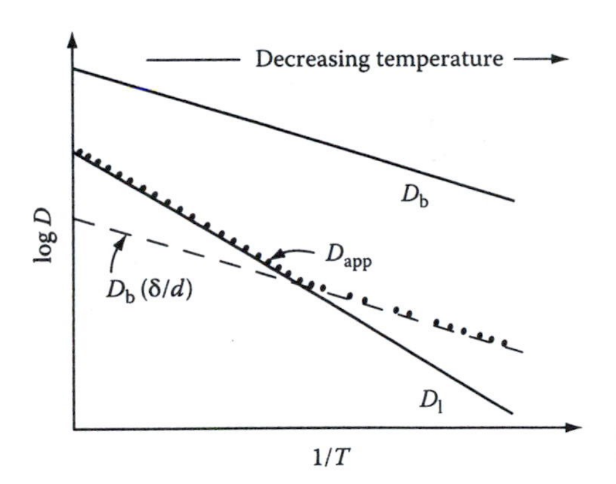

FIGURE 2.27
Diffusion in a polycrystalline metal.

dominates at low temperatures. In general it is found that grain boundary diffusion becomes important below about 0.75–0.8T_m, is the equilibrium melting temperature in degrees Kelvin.

The rate at which atoms diffuse along different boundaries is not the same, but depends on the atomic structure of the individual boundary. This in turn depends on the orientation of the adjoining crystals and the plane of the boundary. Also the diffusion coefficient can vary with direction within a given boundary plane. The reasons for these differences will become apparent in Chapter 3.

2.7.2 Diffusion along Dislocations

If grain boundary diffusion is compared to the conduction of heat through a material made of sheets of aluminum in a plastic matrix, the analogy for diffusion along dislocations would be aluminum wires in a plastic matrix. The dislocations effectively act as pipes along which atoms can diffuse with a diffusion coefficient D_p. The contribution of dislocations to the total diffusive flux through a metal will of course depend on the relative cross-sectional areas of pipe and matrix. Using the simple model illustrated in Figure 2.28 it can easily be shown that the apparent diffusivity through a single crystal containing dislocations, D_{app}, is related to the lattice diffusion coefficient by

$$\frac{D_{app}}{D_1} = 1 + g \cdot \frac{D_p}{D_1} \tag{2.81}$$

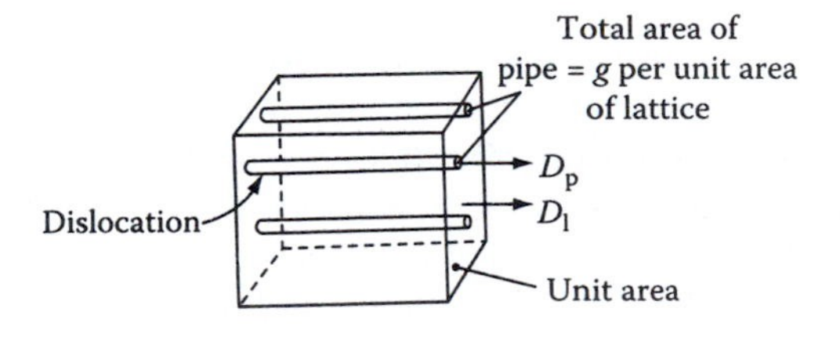

FIGURE 2.28
Dislocations act as a high conductivity path through the lattice.

where g is the cross-sectional area of pipe per unit area of matrix. In a well-annealed material there are roughly 10^5 dislocations mm^{-2}. Assuming that the cross section of a single pipe accommodates about 10 atoms while the matrix contains about 10^{13} atoms mm^{-2}, makes $g \,.\, 10^{-7}$.

At high temperatures diffusion through the lattice is rapid and gD_p/D_l is very small so that the dislocation contribution to the total flux of atoms is negligible. However, since the activation energy for pipe diffusion is less than for lattice diffusion, D_l decreases much more rapidly than D_p with decreasing temperature, and a low temperature gD_p/D_l can become so large that the apparent diffusivity is entirely due to diffusion along dislocations.

2.8 Diffusion in Multiphase Binary Systems

So far only diffusion in single-phase systems has been considered. In most practical cases, however, diffusion occurs in the presence of more than one phase. For example diffusion is involved in solidification transformations and diffusional transformations in solids (Chapters 4 and 5). Another example of multiphase diffusion arises when diffusion couples are made by welding together two metals that are not completely miscible in each other. This situation arises in practice with galvanized iron and hot-dipped tin plate for example. In order to understand what happens in these cases consider the hypothetical phase diagram in Figure 2.29a. A diffusion couple made by welding together pure A and pure B will result in a layered structure containing α, β, and γ. Annealing at temperature T_1 will produce a phase distribution and composition profile as shown in Figure 2.29b. Usually X_B varies as shown from 0 to a in the α-phase, from b to c in the β-phase, and from d to 1 in the γ-phase, where a, b, c, and d are the solubility limits of the phases at T_1. The compositions a and b are seen to be the equilibrium compositions of the α- and β-phases are therefore in local equilibrium across the α/β interface. Similarly β and γ are in local equilibrium across the β/γ interface. A sketch of the free energy–composition diagram for this system at T_1 will show that the chemical potentials (or activities) of A and B will vary continuously across the diffusion couple. Figure 2.29c shows how the activity of B varies across the couple (see Excercise 2.8). Clearly the equilibrium condition $a_B^\alpha = a_B^\beta$ is satisfied at the α/β interface (point p in Figure 2.29c). Similar considerations apply for A and for the β/γ interface.

The α/β and β/γ interfaces are not stationary but move as diffusion progresses. For example if the overall composition of the diffusion couple lies between b and c the final equilibrium state will be a single block of β.

A complete solution of the diffusion equations for this type of diffusion couple is complex. However an expression for the rate at which the boundaries move can be obtained as follows. Consider the planar α/β interface as shown in Figure 2.30. If unit area of the interface moves a distance dx,

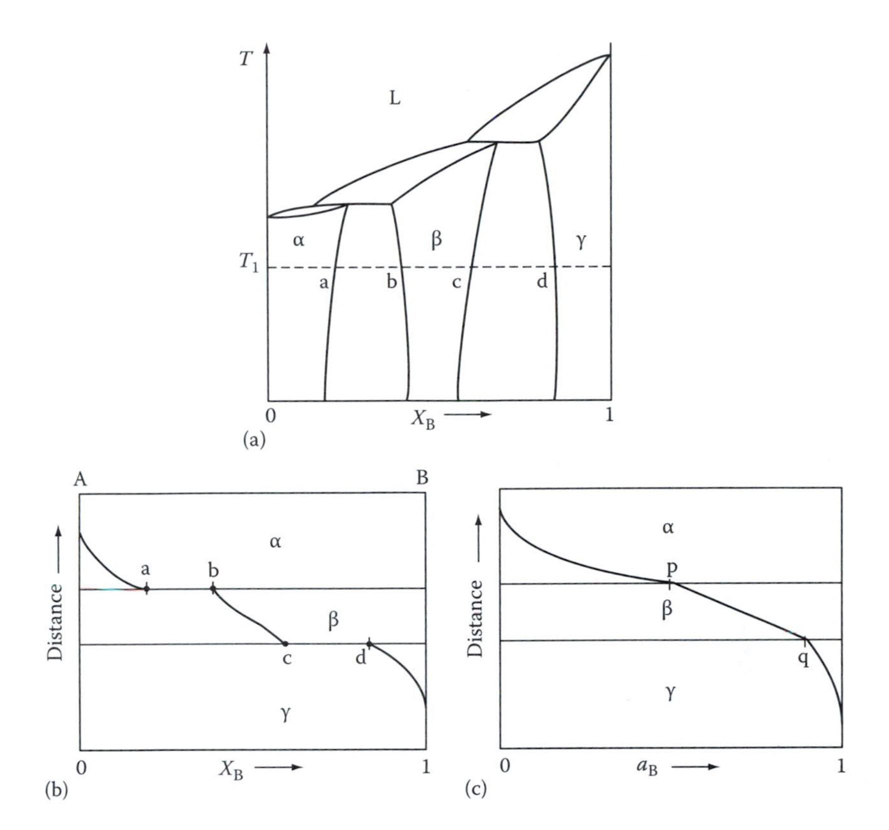

FIGURE 2.29
(a) A hypothetical phase diagram. (b) A possible diffusion layer structure for pure A and B welded together and annealed at T_1. (c) A possible variation of the activity of B (a_B) across the diffusion couple.

a volume $(dx \cdot 1)$ will be converted from α containing C_B^a B-atoms m^{-3} to β containing C_B^b B-atoms m^{-3}. This means that a total of

$$\left(C_B^b - C_B^a\right)dx$$

B atoms must accumulate at the α/β interface (the shaded area in Figure 2.30). There is a flux of B toward the interface from the β-phase equal to $-\tilde{D}(\beta)\partial C_B^b/\partial x$ and a similar flux away from the interface into the α-phase equal to $-\tilde{D}(\alpha)\partial C_B^a/\partial x$. In a time dt, therefore, there will be an accumulation of B atoms given by

$$\left\{\left(-\tilde{D}(\beta)\frac{\partial C_B^b}{\partial x}\right) - \left(-\tilde{D}(\alpha)\frac{\partial C_B^a}{\partial x}\right)\right\}dt$$

Equating the above expressions gives the instantaneous velocity of the α/β interface v as

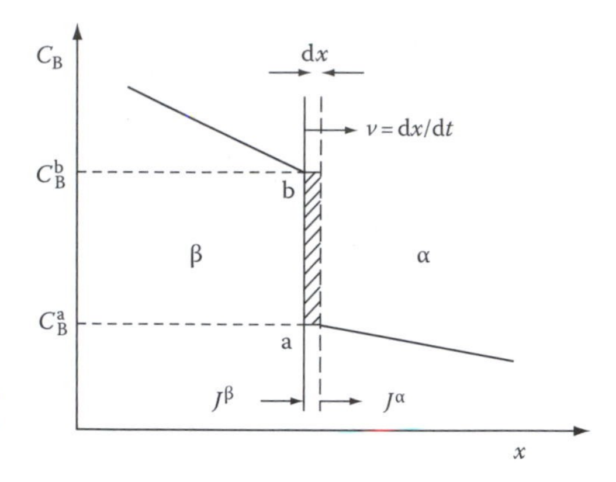

FIGURE 2.30
Concentration profile across the α/β interface and its associated movement assuming diffusion control.

$$v = \frac{dx}{dt} = \frac{1}{(C_B^b - C_B^a)}\left\{\tilde{D}(\alpha)\frac{\partial C_B^a}{\partial x} - \tilde{D}(\beta)\frac{\partial C_B^b}{\partial x}\right\} \tag{2.82}$$

In the above treatment it has been assumed that the α/β interface moves as fast as allowed by the diffusive fluxes in the two adjacent phases. This is quite correct when the two phases are in local equilibrium, and is usually true in diffusion-couple experiments. However, it is not true for all moving interphase interfaces. By assuming local equilibrium at the interface it has also been assumed that atoms can be transferred across the interface as easily as they can diffuse through the matrix. Under these circumstances μ_B and a_B are continuous across the interface. However, in general this need not be true. If, for some reason, the interface has a low mobility the concentration difference across the boundary $(C_B^b - C_B^a)$ will increase, thereby creating a discontinuity of chemical potential across the boundary. The problem of evaluating the boundary velocity in this case is more complex. Not only must the flux of atoms to the interface balance the rate of accumulation due to the boundary migration and the rate of diffusion away into the other phase, but it must also balance with the rate of transfer across the interface. In extreme cases the interface reaction, as it is sometimes called, can be so slow that there are virtually no concentration gradients in the two phases. Under these circumstances the interface migration is said to be interface controlled. The subject of interface migration is treated further in Section 3.5.

Exercises

2.1 A thin sheet of iron is in contact with a carburizing gas on one side and a decarburizing gas on the other at temperature of 1000°C.
   a. Sketch the resultant carbon concentration profile when a steady state has been reached assuming the surface concentrations are maintained at 0.15 and 1.4 wt% C.

b. If D_c increases from 2.5×10^{-11} m^2 s^{-1} at 0.15% C to 7.7×10^{-11} m^2 s^{-1} at 1.4% C what will be the quantitative relationship between the concentration gradients at the surfaces?

c. Estimate an approximate value for the flux of carbon through the sheet if the thickness is 2 mm (0.8 wt% C = 60 kg m^{-3} at 1000°C).

2.2 It was stated in Section 2.2.1 that $D = \Gamma\alpha^2/6$ applies to any diffusing species in any cubic lattice. Show that this is true for vacancy diffusion in a pure fcc metal. (Hint: consider two adjacent {111} planes and determine what fraction of all possible jumps results in the transfer of a vacancy between the two planes. Is the same result obtained by considering adjacent {100} planes?)

2.3 A small quantity of radioactive gold was deposited on the end of a gold cylinder. After holding for 24 h at a high temperature the specimen was sectioned and the radioactivity of each slice was as follows:

Distance from end of bar to center of slice (μm)	10	20	30	40	50
Activity	83.8	66.4	42.0	23.6	8.74

2.4 Prove by differentiation that Equation 2.20 is a solution of Fick's second law.

2.5 Fourier analysis is a powerful tool for the solution of diffusion problems when the initial concentration profile is not sinusoidal. Consider for example the diffusion of hydrogen from an initially uniform sheet of iron. If the concentration outside the sheet is maintained at zero the resultant concentration profile is initially a top-hat function. Fourier analysis of this function shows that it can be considered as an infinite series of sine terms:

$$C(x) = \frac{4C_0}{\pi}\sum_{i=0}^{\infty}\frac{1}{2i+1}\sin\frac{(2i+1)\pi x}{l}$$

where

l is the thickness of the sheet

C_0 is the initial concentration

a. Plot the first two terms of this series. If during diffusion the surface concentration is maintained close to zero each Fourier component can be considered to decrease exponentially with time with a time constant $\tau_i = l^2/(2i+1)^2\pi^2 D$. The solution to the diffusion equation therefore becomes

$$C(x,t) = \frac{4C_0}{\pi}\sum_{i=0}^{\infty}\frac{1}{2i+1}\sin\left\{\frac{2i+1\pi x}{l}\right\}\exp\left(-\frac{t}{\tau_i}\right)$$

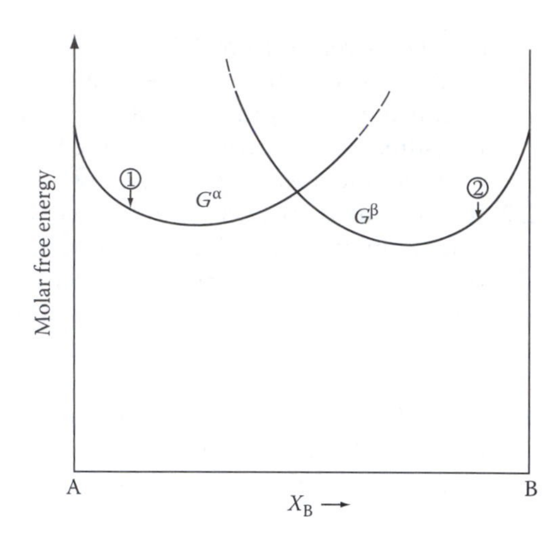

FIGURE 2.31

b. Derive an equation for the time at which the amplitude of the second term is less than 5% of the first term.

c. Approximately how long will it take to remove 95% of all the hydrogen from an initially uniform plate of α-iron at 20°C if (1) the plate is 10 mm thick and if (2) it is 100 mm thick, assuming the surface concentration is maintained constant at zero? (Use data in Table 2.1.)

2.6 Figure 2.31 shows the molar free energy–composition diagram for the A–B system at temperature T_1. Imagine that a block of α with composition (1) is welded to a block of β-phase composition (2). By considering the chemical potentials of the A and B atoms in both the α- and β-phases predict which way the atoms will move during a diffusion anneal at T_1. Show that this leads to a reduction of the molar free energy of the couple. Indicate the compositions of the two phases when equilibrium is reached.

2.7 A diffusion couple including inert wires was made by plating pure copper on to a block of α-brass with a composition Cu–30 wt% Zn, Figure 2.20. After 56 days at 785°C the marker velocity was determined as 2.6×10^{-8} mm s^{-1}. Microanalysis showed that the composition at the markers was $X_{Zn} = 0.22$, $X_{Cu} = 0.78$, and that $\partial X_{Zn}/\partial x$ was 0.089 mm^{-1}. From an analysis of the complete penetration curve $\tilde{D}^\alpha$ at the markers was calculated as 4.5×10^{-13} m^2 s^{-1}. Use this data to calculate D^α_{Zn} and D^α_{Cu} in brass at 22 atomic % Zn. How would you expect D^α_{Zn}, D^α_{Cu}, and $\tilde{D}^\alpha$ to vary as a function of composition?

2.8 Draw possible free energy–composition curves for the system in Figure 2.29 at T_1. Derive from this a μ_B–X_B and an a_B–X_B diagram (similar to

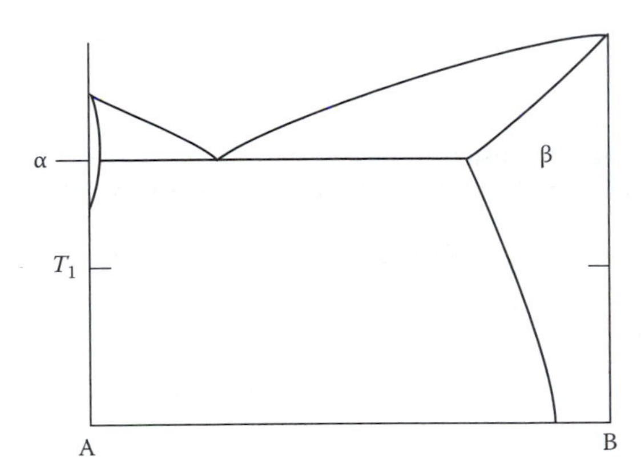

FIGURE 2.32

Figure 1.28). Mark the points corresponding to p and q in Figure 2.29c. Sketch diagrams similar to Figure 2.29c to show a_A, μ_A, and μ_B across the diffusion couple. What will be the final composition profile when the couple reaches equilibrium if the overall composition lies (1) between a and b, (2) below a?

2.9 Figure 2.32 is a hypothetical phase diagram for the A–B system. At a temperature T_1 B is practically insoluble in A, whereas B can dissolve 10 atomic % A. diffusion couple made by welding together pure A and pure B is annealed at T_1. Show by a series of sketches how the concentration profiles and α/β interface position will vary with time. If the overall composition of the couple is 50 atomic % B what will be the maximum displacement of the α/β interface? (Assume α and β have equal molar volumes.)

References

1. P.G. Shewmon, *Diffusion in Solids*, McGraw-Hill, New York, 1963, p. 47.
2. C. Wert, *Phys. Rev.*, 79: 601, 1950.
3. C. Wert, *J. Appl. Phys.*, 21: 1196, 1950.
4. E. Johnson and M. Hill, *Trans. AIME*, 218: 1104, 1960.
5. L.S. Darken, *Trans. Met. Soc. AIME*, 175: 184, 1948.
6. J. Crank, *The Mathematics of Diffusion*, Oxford University Press, Oxford, 1956.
7. A.D. Smigelskas and E.O. Kirkendall, *Trans. Met. Soc. AIME*, 171: 130, 1947.
8. P.G. Shewmon, *Diffusion in Solids*, McGraw-Hill, New York, 1963, p. 134.
9. A.G. Guy, *Introduction to Materials Science*, McGraw-Hill, New York, 1971, p. 284.
10. J.E. Reynolds, B.L. Averbach, and M. Cohen, *Acta Metall.*, 5: 29, 1957.

Further Reading

J.L. Bocquet, G. Brébec, and Y. Limoge, Diffusion in metals and alloys, in *Physical Metallurgy*, R.W. Cahn and P. Haasen (Eds.), North-Holland, Amsterdam, 1983, Chapter 8.

A.M. Brown and M.F. Ashby, Correlations for diffusion constants, *Acta Metallurgica*, 28: 1085, 1980.

C.P. Flynn, *Point Defects and Diffusion*, Oxford University Press, Oxford, 1972.

S. Mrowec, *Defects and Diffusion in Solids–An Introduction*, Elsevier, Amsterdam, 1980.

P.G. Shewmon, *Diffusion in Solids*, 2nd edn., McGraw-Hill, New York, 1989.

3

Crystal Interfaces and Microstructure

Basically three different types of interface are important in metallic systems:

1. The free surfaces of a crystal (solid/vapor interface)
2. Grain boundaries (α/α interfaces)
3. Interphase interfaces (α/β interfaces)

All crystals possess the first type of interface. The second type separates crystals with essentially the same composition and crystal structure, but a different orientation in space. The third interface separates two different phases that can have different crystal structures or compositions and therefore also includes solid/liquid interfaces.

The great majority of phase transformations in metals occur by the growth of a new phase (β) from a few nucleation sites within the parent phase (α)—a nucleation and growth process. The α/β interface therefore plays an important role in determining the kinetics of phase transformations and is the most important class of interface listed. It is, however, also the most complex and least understood and this chapter thus begins by first considering the simpler interfaces, (1) and (2).

The solid/vapor interface is of course itself important in vaporization and condensation transformations, whereas grain boundaries are important in recrystallization, i.e., the transformation of a highly deformed grain structure into new undeformed grains. Although no new phase is involved in recrystallization it does have many features in common with phase transformations.

The importance of interfaces is not restricted to what can be called the primary transformation. Since interfaces are in almost essential feature of the transformed microstructure, a second (slower) stage of most transformations is the microstructural coarsening that occurs with time.[1] This is precisely analogous to the grain coarsening or grain growth that follows a recrystallization transformation.

3.1 Interfacial Free Energy

It is common practice to talk of interfacial energy. In reality, however, what is usually meant and measured by experiment is the interfacial free energy, γ. The free energy of a system containing an interface of area A and free energy, γ per unit area is given by

$$G = G_0 + A\gamma \tag{3.1}$$

where G_0 is the free energy of the system assuming that all material in the system has the properties of the bulk—γ is therefore the excess free energy arising from the fact that some material lies in or close to the interface. It is also the work that must be done at constant T and P to create unit area of interface.

Consider for simplicity a wire frame suspending a liquid film, Figure 3.1. If one bar of the frame is movable it is found that a force F per unit length must be applied to maintain the bar in position. If this force moves a small distance so that the total area of the film is increased by dA the work done by the force is $F\mathrm{d}A$. This work is used to increase the free energy of the system by dG. From Equation 3.1

$$\mathrm{d}G = \gamma\,\mathrm{d}A + A\,\mathrm{d}\gamma$$

Equating this with $F\mathrm{d}A$ gives

$$F = \gamma + A\frac{\mathrm{d}\gamma}{\mathrm{d}A} \tag{3.2}$$

In the case of a liquid film the surface energy is independent of the area of the interface and $\mathrm{d}\gamma/\mathrm{d}A = 0$. This leads to the well-known result

$$F = \gamma \tag{3.3}$$

i.e., a surface with a free energy γ J m^{-2} exerts a surface tension of γ N m^{-1}.

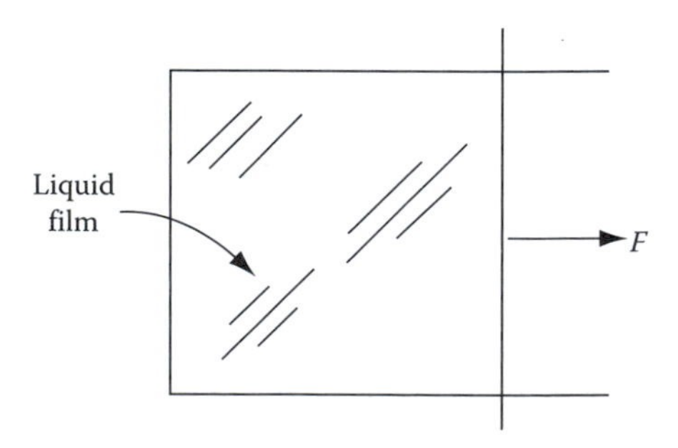

FIGURE 3.1
A liquid film on a wire frame.

In the case of interfaces involving solids, however, it is not immediately obvious that γ is independent of area. Since a liquid is unable to support shear stresses, the atoms within the liquid can rearrange during the stretching process and thereby maintain a constant surface structure. Solids, however, are much more viscous and the transfer of atoms from the bulk to the surface, which is necessary to maintain an unchanged surface structure and energy, will take much longer. If this time is long in comparison to the time of the experiment then $d\gamma/dA \neq 0$ and surface free energy and surface tension will not be identical. Nevertheless, at temperatures near the melting point the atomic mobility is usually high enough for Equation 3.3 to be applicable.

3.2 Solid/Vapor Interfaces

To a first approximation the structure of solid surfaces can be discussed in terms of a hard sphere model. If the surface is parallel to a low-index crystal plane the atomic arrangement will be the same as in the bulk, apart from perhaps a small change in lattice parameter. (This assumes that the surface is uncontaminated: in real system surfaces will reduce their free energies by the adsorption of impurities.) Figure 3.2 for example shows the {111} {200} {220} atom planes in the face-centered cubic (fcc) metals. Note how the density of atoms in these planes decreases as $(h^2 + k^2 + l^2)$ increases. (The notation {200} and {220} has been used instead of {100} and {110} because the spacing of equivalent atom planes is then given by $a/\hat{\mathrm{I}}\,(h^2 + k^2 + l^2)$ where a is the lattice parameter.)

The origin of the surface free energy is that atoms in the layers nearest the surface are without some of their neighbors. Considering only the nearest neighbors it can be seen that the atoms on a {111} surface, for example, are deprived of 3 of their 12 neighbors. If the bond strength of the metal is ε each bond can be considered as lowering the internal energy of each atom by $\varepsilon/2$. Therefore every surface atom with three "broken bonds" has an excess internal energy of $3\varepsilon/2$ over that of the atoms in the bulk. For a pure metal ε can be estimated from the heat of sublimation L_S. (The latent heat of

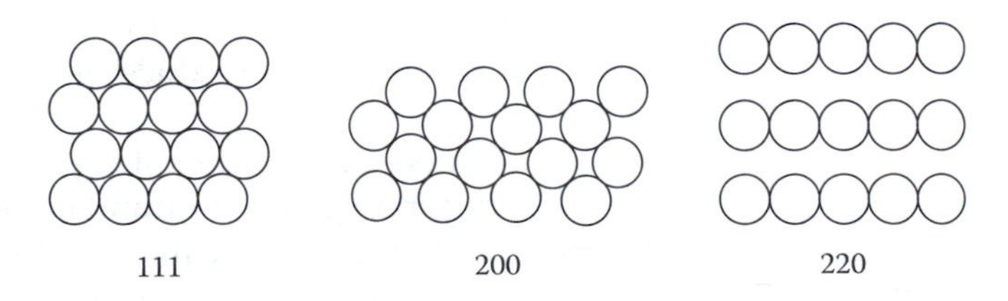

FIGURE 3.2
Atomic configurations on the three closest-packed planes in fcc crystals: (111), (200), and (220).

sublimation is equal to the sum of the latent heat of melting (or fusion) and the latent heat of vaporization.) If 1 mol of solid is vaporized 12 N_a broken bonds are formed. Therefore $L_s = 12N_a\varepsilon/2$. Consequently the energy of a [111] surface should be given by

$$E_{SV} = 0.25L_S/N_a \text{ J/surface atom} \tag{3.4}$$

This result will only be approximate since second nearest neighbors have been ignored and it has also been assumed that the strengths of the remaining bonds in the surface are unchanged from the bulk values.

From the definition of Gibbs free energy the surface free energy will be given by

$$\gamma = E + PV - TS \tag{3.5}$$

Thus even if the *PV* term is ignored surface entropy effects must be taken into account. It might be expected that the surface atoms will have more freedom of movement and therefore a higher thermal entropy compared to atoms in the bulk. Extra configurational entropy can also be introduced into the surface by the formation of surface vacancies for example. The surface of a crystal should therefore be associated with a positive excess entropy which will partly compensate for the high internal energy of Equation 3.4.

Experimental determination of γ_{SV} is difficult[2] but the measured values for pure metals indicate that near the melting temperature the surface free energy averaged over many surface planes is given by

$$\gamma_{SV} = 0.15L_S/N_a \text{ J/surface atom} \tag{3.6}$$

As a result of entropy effects γ_{SV} is slightly dependent on temperature. From Equation 1.10

$$\left(\frac{\partial\gamma}{\partial T}\right)_p = -S \tag{3.7}$$

Measured values of S are positive and vary between 0 and 3 mJ m^{-2} K^{-1}. Some selected values of γ_{SV} at the melting point are listed in Table 3.1. Note that metals with high melting temperatures have high values for L_s and high surface energies.

It can be seen from the above simple model that different crystal surfaces should have different values for E_{SV} depending on the number of broken bonds (see Exercise 3.1). A little consideration will show that for the surfaces shown in Figure 3.2 the number of broken bonds at the surface will increase through the series {111} {200} {220}. Therefore ignoring possible differences in the entropy terms γ_{SV} should also increase along the same series.

TABLE 3.1

Average Surface Free Energies of Selected Metals

Crystal	T_m (°C)	γ_{sv} (mJ m^{-2})
Sn	232	680
Al	660	1080
Ag	961	1120
Au	1063	1390
Cu	1084	1720
δ-Fe	1536	2080
Pt	1769	2280
W	3407	2650

Source: Values selected from Jones, H., *Metal Sci. J.*, 5, 15, 1971.

Note: Experimental errors are generally about 10%. The values have been extrapolated to the melting temperature, T_m.

When the macroscopic surface plane has a high or irrational {*hkl*} index the surface will appear as a stepped layer structure where each layer is a close-packed plane. This is illustrated for a simple cubic crystal in Figure 3.3.

A crystal plane at an angle θ to the close-packed plane will contain broken bonds in excess of the close-packed plane due to the atoms at the steps. For unit length of interface in the plane of the diagram and unit length out of the paper (parallel to the steps) there will be $(\cos\theta/a)(1/a)$ broken bonds out of the close-packed plane and $(\sin|\theta|/a(1/a)$ additional broken bonds from the atoms on the steps. Again attributing $\varepsilon/2$ energy to each broken bond, then

$$E_{sv} = (\cos\theta + \sin|\theta|)\varepsilon/2a^2 \tag{3.8}$$

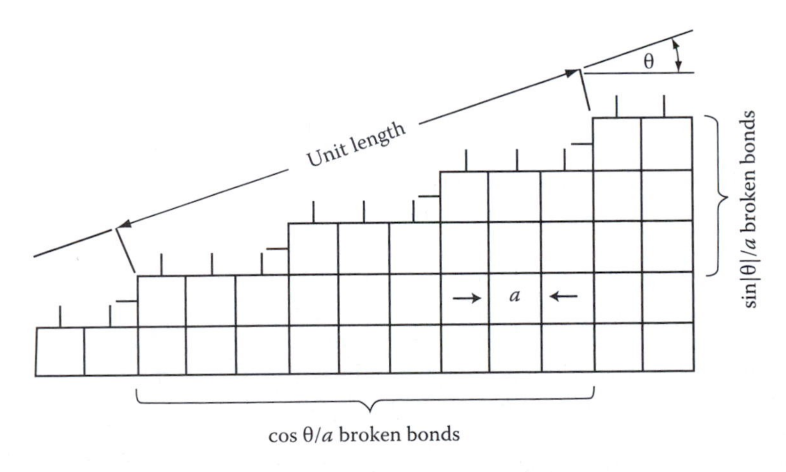

FIGURE 3.3
The "broken-bond" model for surface energy.

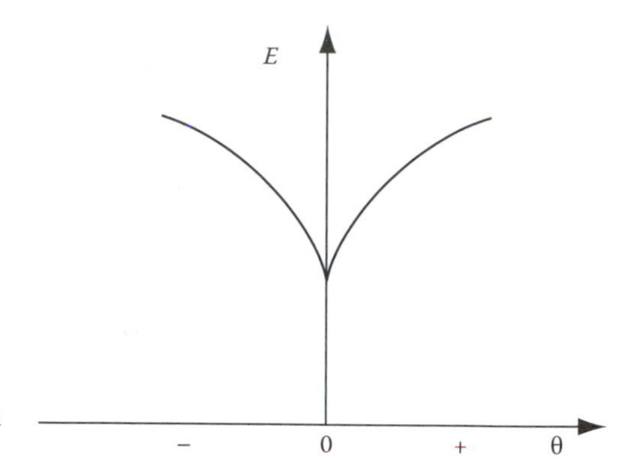

FIGURE 3.4
Valuation of surface energy as a function of θ in Figure 3.3.

This is plotted as a function of θ in Figure 3.4. Note that the close-packed orientation ($\theta = 0$) lies at a cusped minimum in the energy plot. Similar arguments can be applied to any crystal structure for rotations about any axis from any reasonably close-packed plane. All low-index planes should therefore be located at low-energy cusps.

If γ is plotted versus θ similar cusps are found, but as a result of entropy effects they are less prominent than in the $E-\theta$ plot, and for the higher index planes they can even disappear.

A convenient method for plotting the variation of γ with surface orientation in three dimensions is to construct a surface about an origin such that the free energy of any plane is equal to the distance between the surface and the origin when measured along the normal to the plane in question. A section through such a surface is shown in Figure 3.5a. This type of polar representation of γ is known as a γ-plot and has the useful property of being able to predict the equilibrium shape of an isolated single crystal.

For an isolated crystal bounded by several planes A_1, A_2, etc., with energies γ_1, γ_2, etc., the total surface energy will be given by the sum $A_1\gamma_1 + A_2\gamma_2 + \cdots$ The equilibrium shape has the property that $\Sigma A_i\gamma_i$ is a minimum and the shape that satisfies this condition is given by the following, the so-called Wulff construction.[3] For every point on the γ surface, such as A in Figure 3.5a, a plane is drawn through the point and normal to the radius vector OA. The equilibrium shape is then simply the inner envelope of all such planes. Therefore when the γ-plot contains sharp cusps the equilibrium shape is a polyhedron with the largest facets having the lowest interfacial free energy.

Equilibrium shapes can be determined experimentally by annealing small single crystals at high temperatures in an inert atmosphere, or by annealing small voids inside a crystal.[4] For example, fcc crystals usually assume a form showing {100} and {111} facets as shown in Figure 3.5b. Of course when γ is isotropic, as for liquid droplets, both the γ-plots and equilibrium shapes are spheres.

When the equilibrium shape is known it is possible to use the Wulff theorem in reverse to give the relative interfacial free energies of the observed

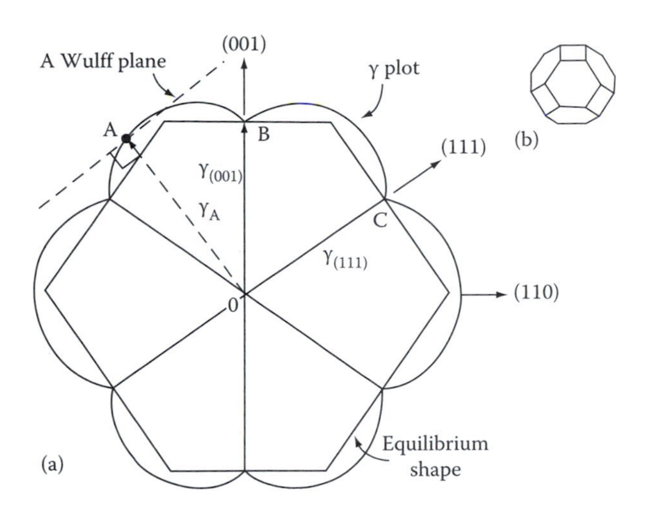

FIGURE 3.5
(a) A possible (110) section through the γ-plot of an fcc crystal. The length OA represents the free energy of a surface plane whose normal lies in the direction OA. Thus OB = $\gamma_{(001)}$, OC = $\gamma_{(111)}$, etc. Wulff planes are those such as that which lies normal to the vector OA. In this case the Wulff planes at the cusps (B,C., etc.) give the inner envelope of all Wulff planes and thus the equilibrium shape. (b) The equilibrium shape in three dimensions showing {100} (square faces) and {111} (hexagonal faces).

facet planes. In Figure 3.5 for example the widths of the crystal in the (111) and (100) directions will be in the ratio of γ(111):γ(100). {110} facets are usually missing from the equilibrium shape of fcc metals, but do however appear for base-centered cubic (bcc) metals.[5]

The aim of the section has been to show, using the simplest type of interface, the origin of interfacial free energy, and to show some of the methods available for estimating this energy. Let us now consider the second type of interface, grain boundaries.

3.3 Boundaries in Single-Phase Solids

The grains in a single-phase polycrystalline specimen are generally in many different orientations and many different types of grain boundary are therefore possible. The nature of any given boundary depends on the misorientation of the two adjoining grains and the orientation of the boundary plane relative to them. The lattices of any two grains can be made to coincide by rotating one of them through a suitable angle about a single axis. In general the axis of rotation will not be simply oriented with respect to either grain or the grain-boundary plane, but there are two special types of boundary that are relatively simple. These are pure tilt boundaries and pure twist

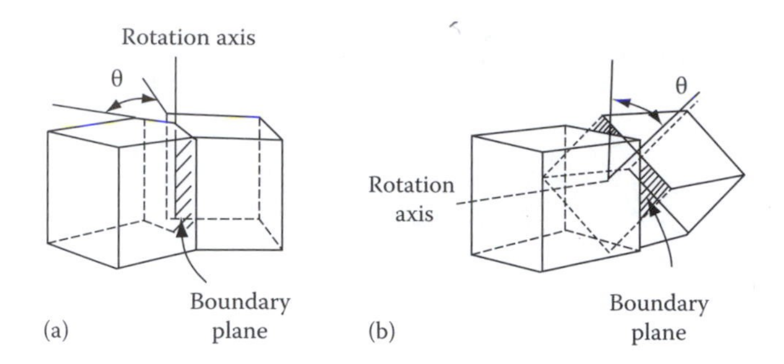

FIGURE 3.6
The relative orientations of the crystals and the boundary forming (a) a tilt boundary and (b) a twist boundary.

boundaries, as illustrated in Figure 3.6. A tilt boundary occurs when the axis of rotation is parallel to the plane of the boundary, Figure 3.6a, whereas a twist boundary is formed when the rotation axis is perpendicular to the boundary, Figure 3.6b.

3.3.1 Low-Angle and High-Angle Boundaries

It is simplest to first consider what happens when the misorientation between two grains is small. This type of boundary can be simply considered as an array of dislocation. Two idealized boundaries are illustrated in Figure 3.7. These are symmetrical low-angle tilt and low-angle twist boundaries.

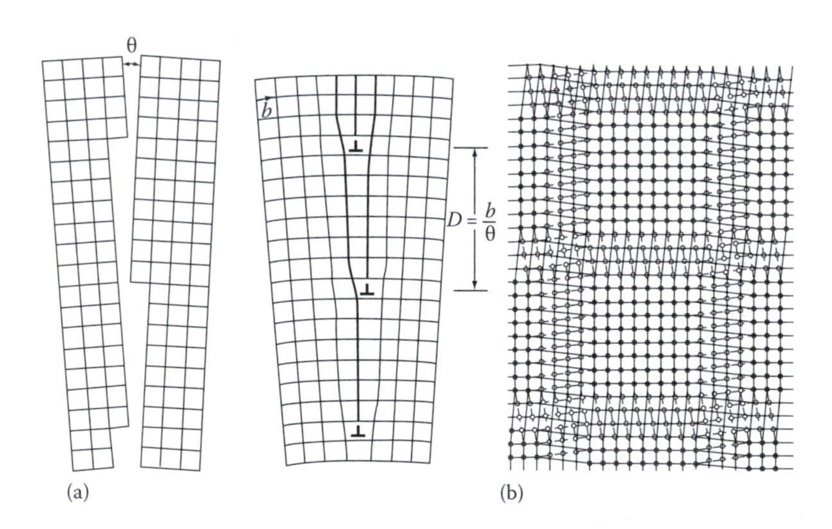

FIGURE 3.7
(a) Low-angle tilt boundary and (b) low-angle twist boundary: **s** atoms in crystal below boundary, **d** atoms in crystal above boundary. (After Read, W.T., Jr., *Dislocations in Crystals*, McGraw-Hill, New York, 1953.)

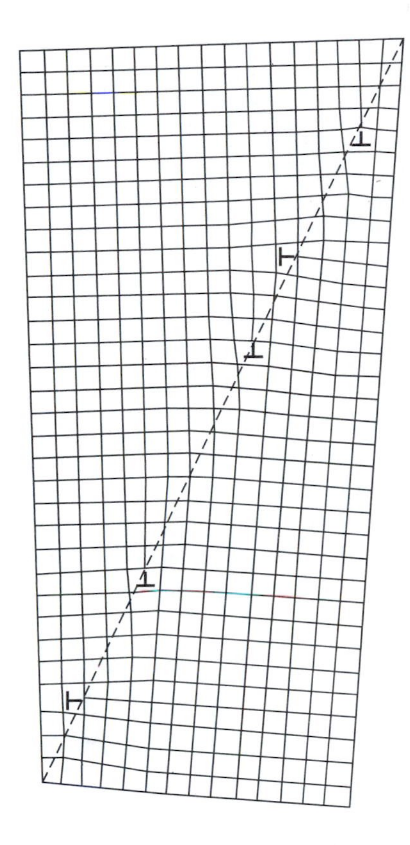

FIGURE 3.8
An unsymmetric tilt boundary. Dislocations with two different Burgers vectors are present. (After Read, W.T., Jr., *Dislocations in Crystals*, McGraw-Hill, New York, 1953.)

The low-angle tilt boundary is an array of parallel edge dislocations, whereas the twist boundary is a cross-grid of two sets of screw dislocations. In each case the atoms in the regions between the dislocations fit almost perfectly into both adjoining crystals whereas the dislocation cores are regions of poor fit in which the crystal structure is highly distorted.

The tilt boundary need not be symmetrical with respect to the two adjoining crystals. However, if the boundary is unsymmetrical dislocations with different Burgers vectors are required to accommodate the misfit, as illustrated in Figure 3.8. In general boundaries can be a mixture of the tilt and twist type in which case they must contain several sets of different edge and screw dislocation.

The energy of a low-angle grain boundary is simply the total energy of the dislocations within unit area of boundary. (For brevity the distinction between internal energy and free energy will usually not be made from now on except where essential to understanding.) This depends on the spacing of the dislocations which, for the simple arrays in Figure 3.7, is given by

$$D = \frac{b}{\sin\theta} \cdot \frac{b}{\theta} \tag{3.9}$$

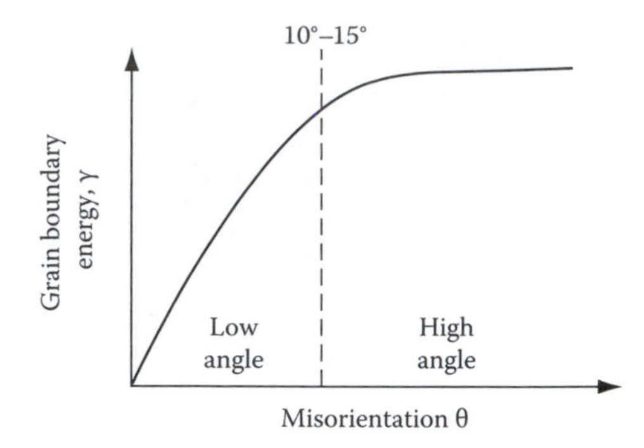

FIGURE 3.9
Variation of grain-boundary energy with misorientation (schematic).

where
- b is the Burgers vector of the dislocations
- θ is the angular misorientation across the boundary

At very small values of θ the dislocation spacing is very large and the grain-boundary energy γ is approximately proportional to the density of dislocations in the boundary $(1/D)$, i.e.,

$$\gamma \propto \theta \tag{3.10}$$

However as θ increases the strain fields of the dislocations progressively cancel out so that γ increases at a decreasing rate as shown in Figure 3.9. In general when θ exceeds 10°–15° the dislocation spacing is so small that the dislocation cores overlap and it is then impossible to physically identify the individual dislocations (Figure 3.10). At this stage the grain-boundary energy is almost independent of misorientation, Figure 3.9.

When $\theta > 10°–15°$ the boundary is known as a random high-angle grain boundary. The difference in structure between low-angle and high-angle grain boundaries is lucidly illustrated by the bubble-raft model in Figure 3.11. High-angle boundaries contain large areas of poor fit and have a relatively open structure. The bonds between the atoms are broken or highly distorted and consequently the boundary is associated with a relatively high

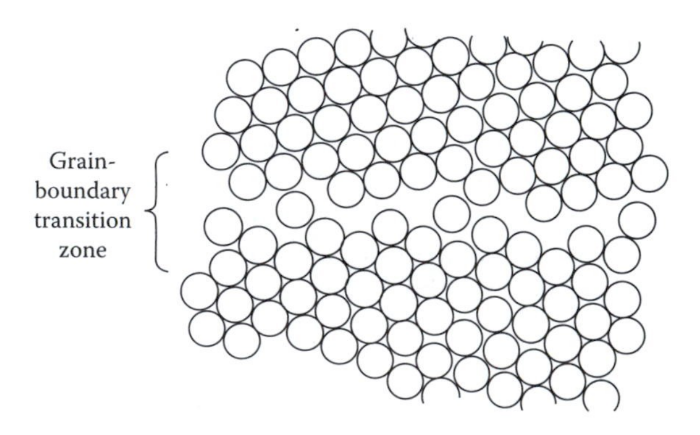

FIGURE 3.10
Disordered grain-boundary structure (schematic).

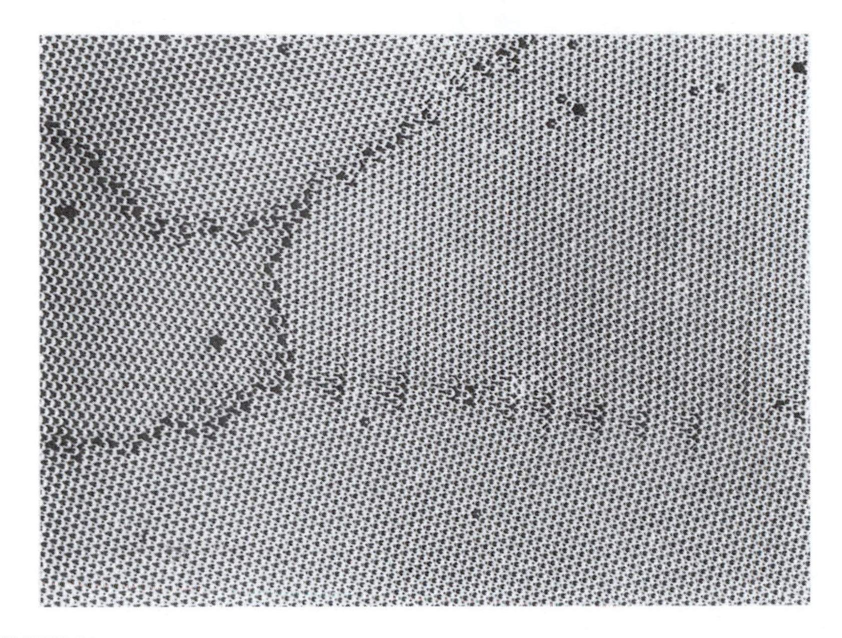

FIGURE 3.11
Rafts of soap bubbles showing several grains of varying misorientation. Note that the boundary with the smallest misorientation is made up of a row of dislocations, whereas the high-angle boundaries have a disordered structure in which individual dislocations cannot be identified. (After Shewmon, P.G., *Transformations in Metals*, McGraw-Hill, New York, 1969, from C.S. Smith.)

energy. In low-angle boundaries, however, most of the atoms fit very well into both lattices so that there is very little free volume and the interatomic bonds are only slightly distorted. The regions of poor fit are restricted to the dislocation cores which are associated with a higher energy similar to that of the random high-angle boundary.

Measured high-angle grain-boundary energies γ_b are often found to be roughly given by

$$\gamma_b \cdot \frac{1}{3}\gamma_{sv} \tag{3.11}$$

Some selected values for γ_b and γ_b/γ_{sv} are listed in Table 3.2. As for surface energies γ_b is temperature dependent decreasing somewhat with increasing temperature.

3.3.2 Special High-Angle Grain Boundaries

Not all high-angle boundaries have an open disordered structure. There are some special high-angle boundaries which have significantly lower energies than the random boundaries. These boundaries only occur at particular misorientations and boundary planes which allow the two adjoining lattices to fit together with relatively little distortion of the interatomic bonds.

TABLE 3.2

Measured Grain-Boundary Free Energies

Crystal	γ (mJ m^{-2})	T (°C)	γ_b/γ_{sv}
Sn	164	223	0.24
Al	324	450	0.30
Ag	375	950	0.33
Au	378	1000	0.27
Cu	625	925	0.36
γ-Fe	756	1350	0.40
δ-Fe	468	1450	0.23
Pt	660	1300	0.29
W	1080	2000	0.41

Source: Values selected from compilation given in Murr, L.E., *Interfacial Phenomena in Metals and Alloys*, Addison-Wesley, London, 1975.

The simplest special high-angle grain boundary is the boundary between two grains. If the twin boundary is parallel to the twinning plane the atoms in the boundary fit perfectly into both grains. The result is a coherent twin boundary as illustrated in Figure 3.12a. In fcc metals this is a {111} close-packed plane. Because the atoms in the boundary are essentially in

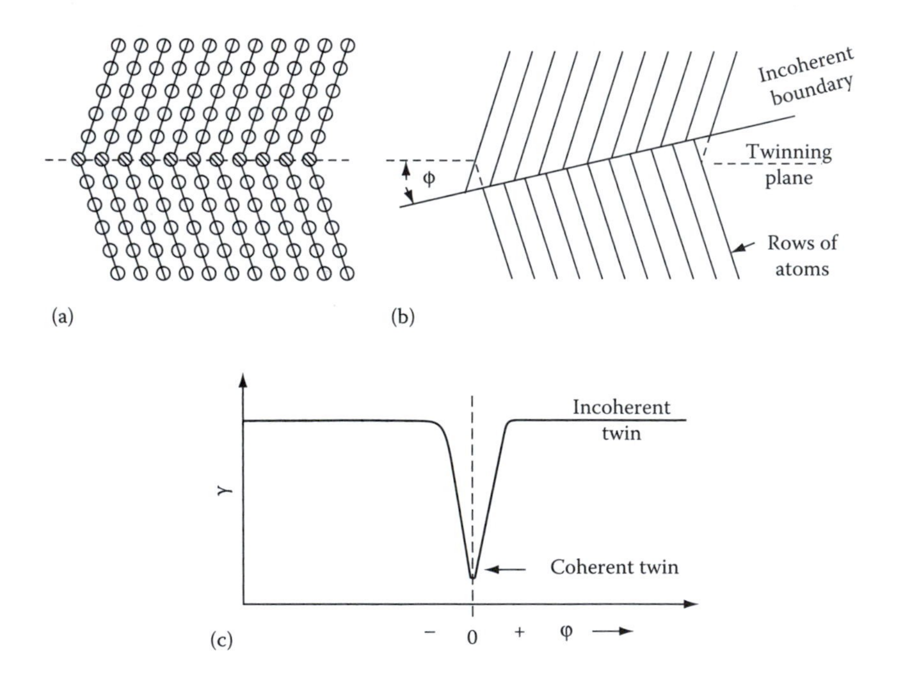

FIGURE 3.12
(a) A coherent twin boundary. (b) An incoherent twin boundary. (c) Twin-boundary energy as a function of the grain-boundary orientation.

TABLE 3.3

Measured Boundary Free Energies for Crystals in Twin Relationships (Units mJ m^{-2})

Crystal	Coherent Twin-Boundary Energy	Incoherent Twin-Boundary Energy	Grain-Boundary Energy
Cu	21	498	623
Ag	8	126	377
Fe-Cr-Ni (stainless steel type 304)	19	209	835

Source: Values selected from compilation given in Murr, L.E., *Interfacial Phenomena in Metals and Alloys*, Addison-Wesley, London, 1975.

undistorted positions the energy of a coherent twin boundary is extremely low in comparison to the energy of a random high-angle boundary.

If the twin boundary does not lie exactly parallel to the twinning plane, Figure 3.12b, the atoms do not fit perfectly into each grain and the boundary energy is much higher. This is known as an incoherent twin boundary. The energy of a twin boundary is therefore very sensitive to the orientation of the boundary plane. If γ is plotted as a function of the boundary orientation a sharp cusped minimum is obtained at the coherent boundary position as shown in Figure 3.12c. Table 3.3 lists some experimentally measured values of coherent and incoherent twins along with high-angle grain-boundary energies for comparison.

Twin orientations in fcc metals correspond to a misorientation of 70.5° about a ⟨110⟩ axis. Therefore a twin boundary is a special high-angle grain boundary, and a coherent twin boundary is a symmetrical tilt boundary between the two twin-related crystals. Figure 3.13 shows measured grain-boundary energies for various symmetric tilt boundaries in aluminum. When

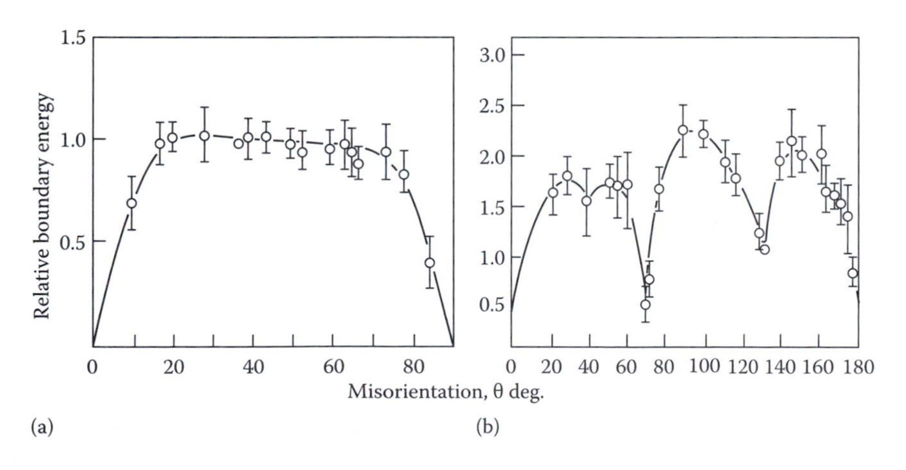

FIGURE 3.13
Measured grain-boundary energies for symmetric tilt boundaries in Al (a) when the rotation axis is parallel to ⟨100⟩ and (b) when the rotation axis is parallel to ⟨110⟩. (After Hasson, G. and Goux, C., *Scripta Metall.*, 5, 889, 1971.)

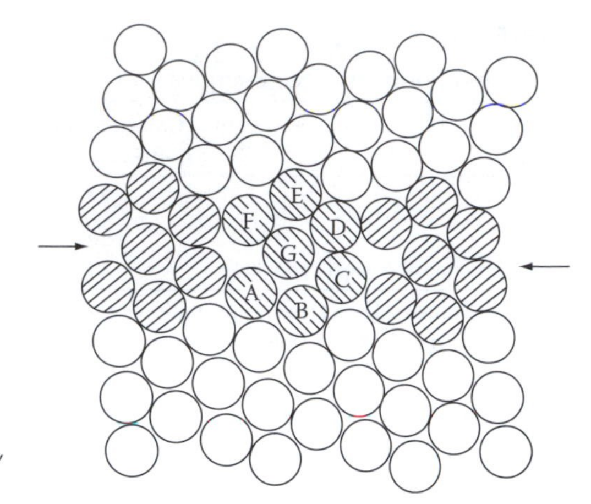

FIGURE 3.14
Special grain boundary. (After Gleiter, H., *Phys. Status Solidi* (b), 45, 9, 1971.)

the two grains are related by a rotation about a (100) axis, Figure 3.13a, it can be seen that most high-angle boundaries have about the same energy and should therefore have a relatively disordered structure characteristic of random boundaries. However, when the two grains are related by a rotation about a (110) axis there are several large-angle orientations which have significantly lower energies than the random boundaries (Figure 3.13b). $\theta = 70.5°$ corresponds to the coherent twin boundary discussed above, but low-energy boundaries are also found for several other values of θ. The reasons for these other special grain boundaries are not well understood. However, it seems reasonable to suppose that the atomic structure of these boundaries is such that they contain extensive areas of good fit. A two-dimensional example is shown in Figure 3.14. This is a symmetrical tilt boundary between grains with a misorientation of 38.2°. The boundary atoms fit rather well into both grains leaving relatively little free volume. Moreover, a small group of atoms (shaded) are repeated at regular intervals along the boundary.

3.3.3 Equilibrium in Polycrystalline Materials

Let us now examine how the possibility of different grain-boundary energies affects the microstructure of a polycrystalline material. Figure 3.15 shows the microstructure of an annealed austenitic stainless steel (fcc). The material contains high- and low-angle grain boundaries as well as coherent and incoherent twin boundaries. This microstructure is determined by how the different grain boundaries join together in space. When looking at two-dimensional microstructures like this it is important to remember that in reality the grains fill three dimensions, and only one section of the three-dimensional network of internal boundaries is apparent. Note that two grains meet in a plane (a grain boundary), three grains meet in a line (a grain edge), and four grains meet at a point (a grain corner). Let us now consider the factors that control the grain shapes in a recrystallized polycrystal.

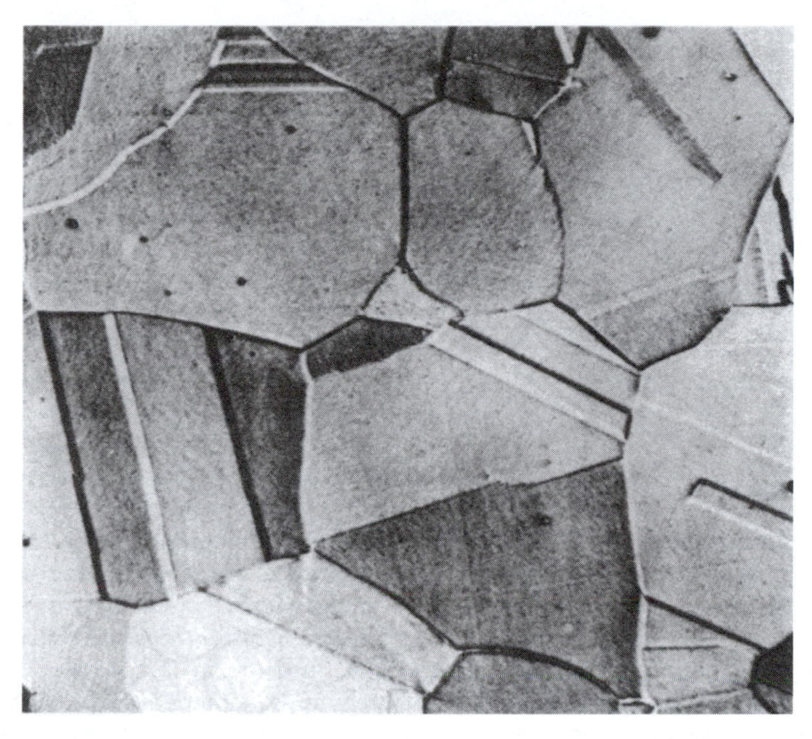

FIGURE 3.15
Microstructure of an annealed crystal of austenitic stainless steel. (After Shewmon, P.G., *Transformations in Metals*, McGraw-Hill, New York, 1969.)

The first problem to be solved is why grain boundaries exist at all in annealed materials. The boundaries are all high-energy regions that increase the free energy of a polycrystal relative to a single crystal. Therefore a polycrystalline material is never a true equilibrium structure. However the grain boundaries in a polycrystal can adjust themselves during annealing to produce a metastable equilibrium at the grain-boundary intersections.

The conditions for equilibrium at a grain-boundary junction can be obtained either by considering the total grain-boundary energy associated with a particular configuration or, more simply, by considering the forces that each boundary exerts on the junction. Let us first consider a grain-boundary segment of unit width and length OP as shown in Figure 3.16. If the boundary is mobile then forces F_x and F_y must act at O and P to maintain the boundary in equilibrium. From Equation 3.3, $F_x = \gamma \cdot F_y$ can be calculated as follows: if P is moved a small distance δy. This must balance the increase in boundary energy caused by the change in orientation $\delta\theta$, i.e.,

$$F_y \delta y = l \frac{d\gamma}{d\theta} \delta\theta$$

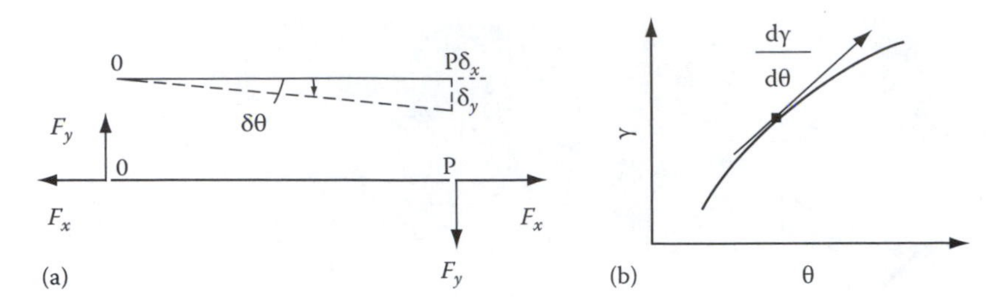

FIGURE 3.16
(a) Equilibrium forces F_x and F_y supporting a length l of boundary OP and (b) the origin of F_y.

Since $\delta y = l\delta\theta$

$$F_y = \frac{d\gamma}{d\theta} \tag{3.12}$$

This means that if the grain-boundary energy is dependent on the orientation of the boundary (Figure 3.16b) a force $d\gamma/d\theta$ must be applied to the ends of the boundary to prevent it rotating into a lower energy orientation. $d\gamma/d\theta$ is therefore known as a *torque term*. Since the segment OP must be supported by forces F_x and F_y the boundary exerts equal but opposite forces $-F_x$ and $-F_y$ on the ends of the segment which can be junctions with other grain boundaries.

If the boundary happens to be at the orientation of a cusp in the free energy, e.g., as shown in Figure 3.12c, there will be no torque acting on the boundary since the energy is a minimum in that orientation. However, the boundary will be able to resist a pulling force F_y of up to $(d\gamma/d\theta)_{cusp}$ without rotating.

If the boundary energy is independent of orientation the torque term is zero and the grain boundary behaves like a soap film. Under these conditions the requirement for metastable equilibrium at a junction between three grains (Figure 3.17) is that the boundary tensions γ_1, γ_2, and γ_3 must balance. In mathematical terms

$$\frac{\gamma_{23}}{\sin\theta_1} = \frac{\gamma_{13}}{\sin\theta_2} = \frac{\gamma_{12}}{\sin\theta_3} \tag{3.13}$$

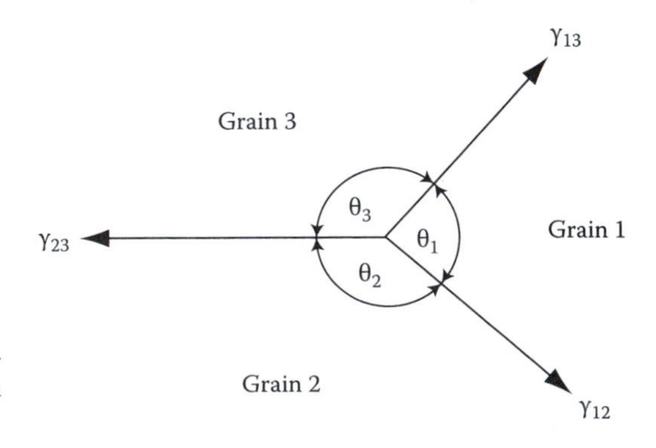

FIGURE 3.17
The balance of grain-boundary tensions for a grain-boundary intersection in metastable equilibrium.

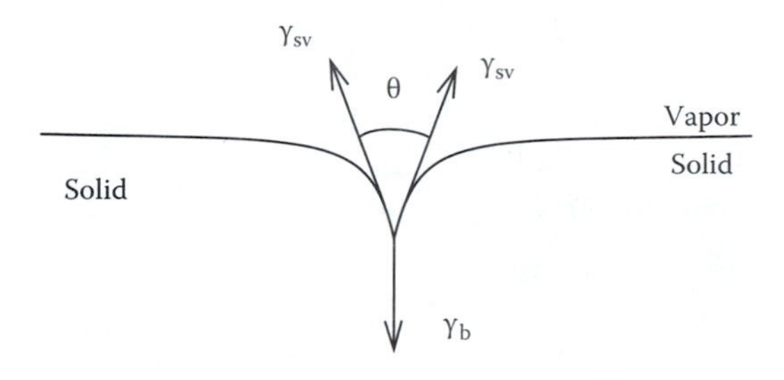

FIGURE 3.18
The balance of surface and grain-boundary tensions at the intersection of a grain boundary with a free surface.

Equation 3.13 applies to any three boundaries so that grain 1 for example could be a different phase to grains 2 and 3. Alternatively grain 1 could be a vapor phase in which case γ_{13} and γ_{12} would be the surface energies of the solid. The relationship is therefore useful for determining relative boundary energies.

One method of measuring grain-boundary energy is to anneal a specimen at a high temperature and then measure the angle at the intersection of the surface with the boundary, see Figure 3.18. If the solid–vapor energy (γ_{sv}) is the same for both grains, balancing the interfacial tensions gives

$$2\gamma_{sv}\cos\frac{\theta}{2} = \gamma_b \tag{3.14}$$

Therefore if γ_{sv} is known γ_b can be calculated.

When using Equation 3.14 it must be remembered that the presence of any torque terms has been neglected and such an approximation may introduce large errors. To illustrate the importance of such effects let us consider the junction between coherent and incoherent twin boundary segments, Figure 3.19. As a result of the orientation dependence of twin boundary energy, Figure 3.12c, it is energetically favorable for twin boundaries to align themselves parallel to the twining plane. If, however, the boundary is constrained to follow a macroscopic plane that is near but not exactly parallel to the twinning plane the boundary will usually develop a stepped appearance with large coherent low-energy facets and small incoherent high-energy risers as shown in Figure 3.19. Although this configuration does not minimize the total twin boundary area it does minimize the total free energy.

It is clear that at the coherent/incoherent twin junction the incoherent twin boundary tension γ_i must be balanced by a torque term. Since the maximum value of the resisting force is $d\gamma_c/d\theta$, the condition that the configuration shown in Figure 3.19 is stable is

$$\gamma_i \leq \frac{d\gamma_c}{d\theta} \tag{3.15}$$

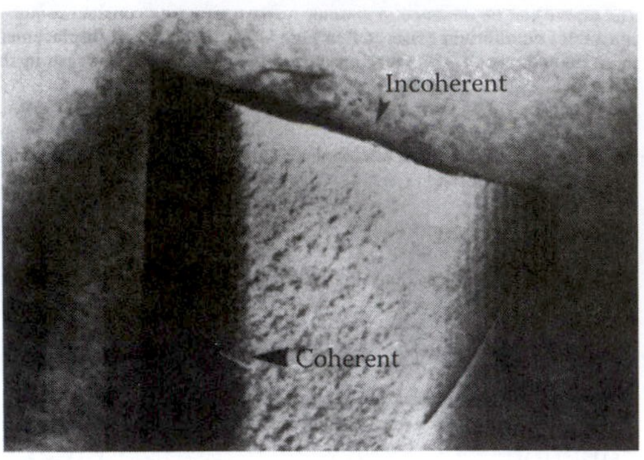

(a)

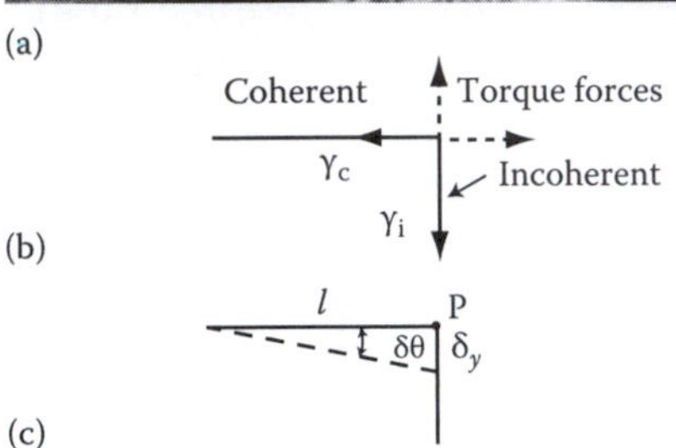

FIGURE 3.19
(a) A twin boundary in a thin foil specimen as imaged in the transmission electron microscope. (After Thompson, M.N. and Chen, C.W., *Philips Electron Optics Bulletin*, EM 112-1979/1 Eindhoven, 1979.) (b) and (c) the coherent and incoherent segments of the twin boundary.

Likewise the "incoherent" facet must also be a special boundary showing rather good fit in order to provide a force resisting γ_c. That is

$$\gamma_c \leq \frac{d\gamma_i}{d\theta} \tag{3.16}$$

However, since γ_c is usually very small the incoherent interface need only lie in a rather shallow energy cusp.

The above can be obtained in another way from energy considerations. If (metastable) equilibrium exists at P in Figure 3.19c, then a small displacement such as that shown should either produce no change, or an increase in the total free energy of the system, i.e.,

$$dG \geq 0$$

Considering unit depth a small displacement δy at P will increase the total free energy by an amount

$$dG = l\frac{d\gamma_c}{d\theta}\delta\theta - \gamma_i\delta y \geq 0$$

Since $l\delta\theta = \delta y$ this leads to the same result as given by Equation 3.15.

3.3.4 Thermally Activated Migration of Grain Boundaries

In Section 3.3.3 it was shown that metastable equilibrium at the grain-boundary junctions requires certain conditions to be satisfied for the angles at which three boundaries intersect. For simplicity, if all grain boundaries in a polycrystal are assumed to have the same grain-boundary energy independent of boundary orientation, Equation 3.13 predicts that $\theta_1 = \theta_2 = \theta_3 = 120°$. It can be similarly shown that the grain-boundary edges meeting at a corner formed by four grains will make an angle of 109° 28′. If these, or similar, angular conditions are satisfied then metastable equilibrium can be established at all grain-boundary junctions. However, for a grain structure to be in complete metastable equilibrium the surface tensions must also balance over all the boundary faces between the junctions. If a boundary is curved in the shape of a cylinder, Figure 3.20a, it is acted on by a force of magnitude γ/t toward its center of curvature. Therefore the only way the boundary tension forces can balance in three dimensions is if the boundary is planar ($r = \infty$) or if it is curved with equal radii in opposite directions, Figure 3.20b and c. It is theoretically possible to construct a three-dimensional polycrystal in which the boundary tension forces balance at all faces and junctions, but in a random polycrystalline aggregate, typical of real metallurgical specimens, there are always boundaries with a net curvature in one direction. Consequently a random grain structure is inherently unstable and, on annealing at high temperatures, the unbalanced forces will cause the boundaries to migrate toward their centers of curvature.

The effect of different boundary curvatures in two dimensions is shown in Figure 3.21. Again for simplicity it has been assumed that equilibrium at each boundary junction results in angles of 120°. Therefore if a grain has six boundaries they will be planar and the structure metastable. However, if the total number of boundaries around a grain is less than six each boundary must be concave inwards, Figure 3.21. These grains will therefore shrink and eventually disappear during annealing. Larger grains, on the other hand, will have more than six boundaries and will grow. The overall result of such boundary migration is to reduce the number of grains, thereby increasing the

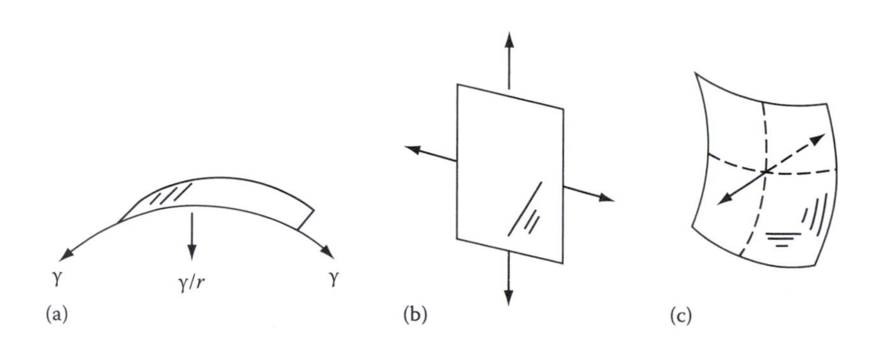

FIGURE 3.20
(a) A cylindrical boundary with a radius of curvature r is acted on by a force γ/r. (b) A planar boundary with no net force, and (c) A doubly curved boundary with no net force.

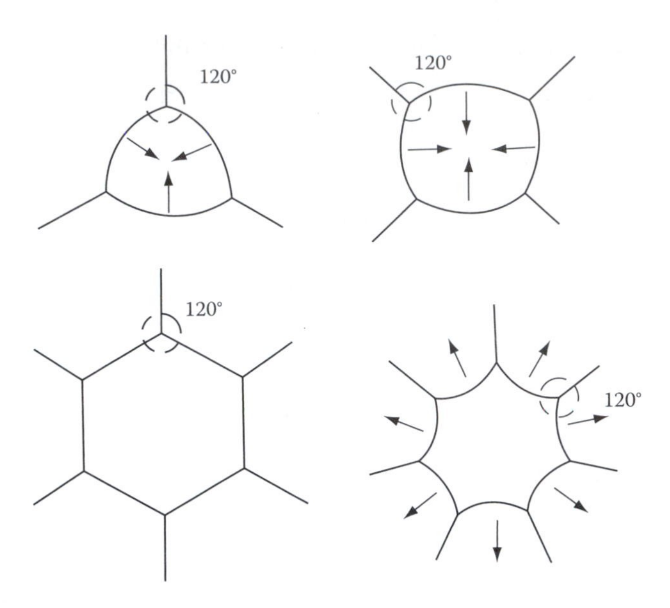

FIGURE 3.21
Two-dimensional grain-boundary configurations. The arrows indicate the directions boundaries will migrate during grain growth.

mean grain size and reducing the total grain-boundary energy. This phenomenon is known as grain growth or grain coarsening. It occurs in metals at temperatures about $0.5T_m$ where the boundaries have significant mobility. A soap froth serves as a convenient analog to demonstrate grain growth as shown in Figure 3.22.

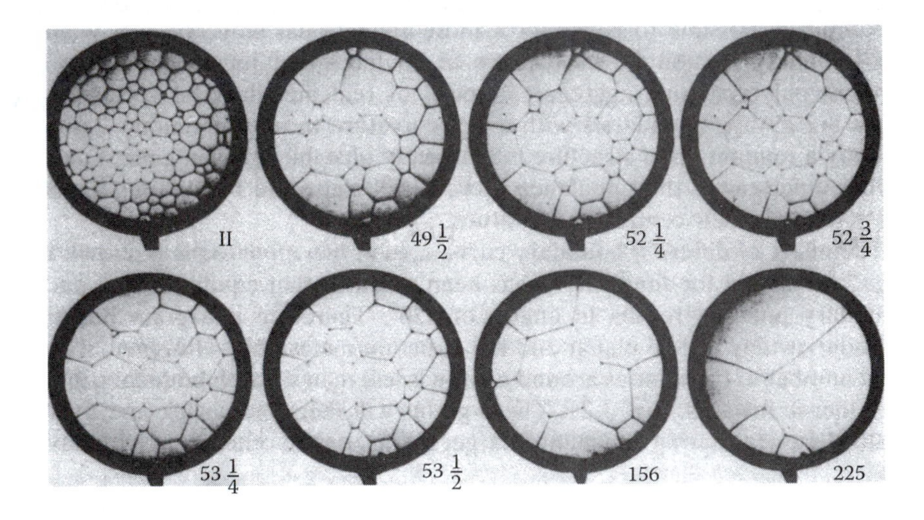

FIGURE 3.22
Two-dimensional cells of a soap solution illustrating the process of grain growth. Numbers are time in minutes. (After Smith, C.S., *Metal Interfaces*, American Society for Metals, Cleveland, 1952, 81.)

In the case of the cells in a soap froth the higher pressure on the concave side of the films induces the air molecules in the smaller cells to diffuse through the film into the larger cells, so that the small cells eventually disappear. A similar effect occurs in metal grains. In this case the atoms in the shrinking grain detach themselves from the lattice on the high pressure side of the boundary and relocate themselves on a lattice site of the growing grain. For example in Figure 3.23a if atom C jumps from grain 1 to grain 2 the boundary locally advances a small distance.

The effect of the pressure difference caused by a curved boundary is to create a difference in free energy (ΔG) or chemical potential ($\Delta\mu$) that drives the atoms across the boundary, see Figure 3.24. In a pure metal ΔG that $\Delta\mu$ are identical and are given by Equation 1.58 as

$$\Delta G = \frac{2\gamma V_{\rm m}}{r} = \Delta\mu \tag{3.17}$$

This free energy difference can be thought of as a force pulling the grain boundary toward the grain with the higher free energy. As shown in Figure 3.25, if unit area of grain boundary advances a distance δx the number of moles of material that enter grain B is $\delta x \cdot 1/V_{\rm m}$ and the free energy released is given by

$$\Delta G \cdot \delta x / V_{\rm m}$$

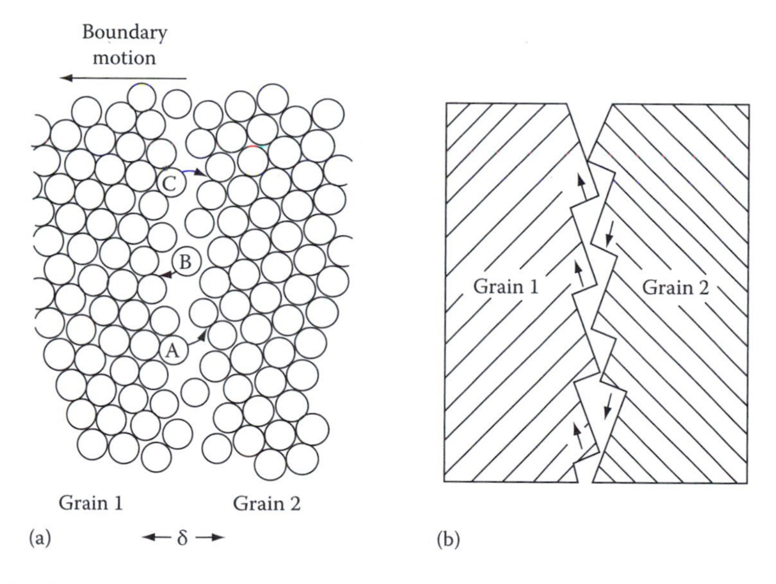

FIGURE 3.23
(a) The atomic mechanism of boundary migration. The boundary migrates to the left if the jump rate from grain 1 → 2 is greater than 2 → 1. Note that the free volume within the boundary has been exaggerated for clarity. (b) Steplike structure where close-packed planes protrude into the boundary.

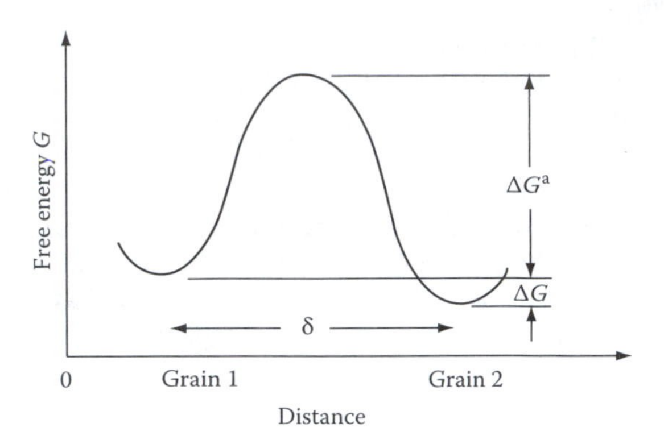

FIGURE 3.24
The free energy of an atom during the process of jumping from one grain to the other.

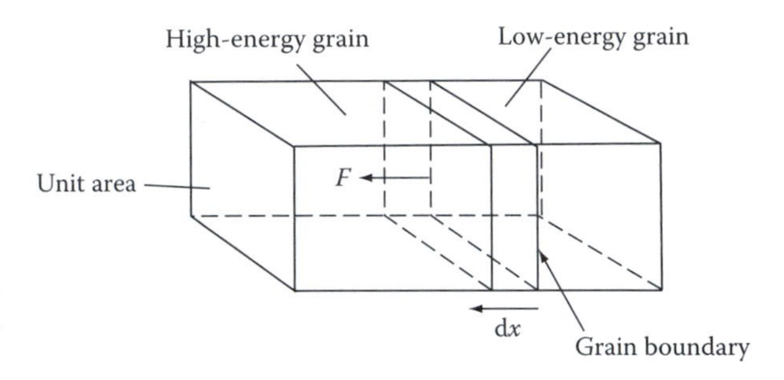

FIGURE 3.25
A boundary separating grains with different free energies is subjected to a pulling force F.

This can be equated to the work done by the pulling force $F\delta x$. Thus the pulling force per unit area of boundary is given by

$$F = \frac{\Delta G}{V_m} \text{ N m}^{-2} \tag{3.18}$$

In other words the force on the boundary is simply the free energy difference per unit volume of material.

In the case of grain growth ΔG arises from the boundary curvature, but Equation 3.18 applies equally to any boundary whose migration causes a decrease in free energy. During recrystallization, for example, the boundaries between the new strain-free grains and the original deformed grains are acted on by a force $\Delta G/V_m$ where, in the case, ΔG is due to the difference in dislocation strain energy between the two grains. Figure 3.26 shows a dislocation-free recrystallized grain expanding into the heavily deformed surroundings. In this case the total grain-boundary area is increasing; therefore the driving force for recrystallization must be greater than the opposing boundary tension forces. Such forces are greatest when the new grain is smallest, and the effect is therefore important in the early stages of recrystallization.

Let us now consider the effect of the driving force on the kinetics of boundary migration. In order for an atom to be able to break away from

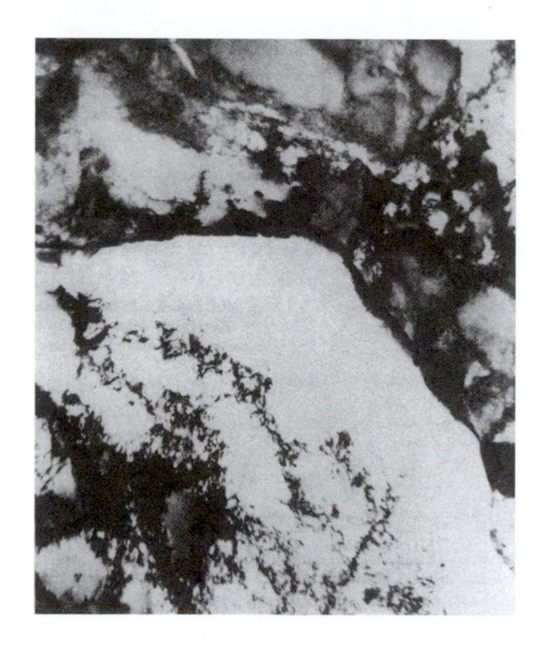

FIGURE 3.26
Grain-boundary migration in nickel pulled 10% and annealed 10 min at 425°C. The region behind the advancing boundary is dislocation free. (After Bailey, J. and Hirsch, P., *Proc. R. Soc. Lond.*, A267, 11, 1962.)

grain 1 it must acquire, by thermal activation, an activation energy ΔG^{a}, Figure 3.24. If the atoms vibrate with a frequency v_1 the number of times per second that an atom has this energy is $v_1 \exp(-\Delta G^{\mathrm{a}}/RT)$ jumps $\mathrm{m}^{-2}\,\mathrm{s}^{-1}$ away from grain 1. It is possible that not all these atoms will find a suitable site and "stick" to grain 2. If the probability of being accommodated ingrain 2 is A_2 the effective flux of atoms from grain 1 to 2 will be

$$A_2 n_1 v_1 \exp(-\Delta G^{\mathrm{a}}/RT)\ \mathrm{m}^{-2}\,\mathrm{s}^{-1}$$

There will also be a similar flux in the reverse direction, but if the atoms in grain 2 have a lower free energy than the atoms in grain 1 by ΔG (mol^{-1}) the flux from 2 to 1 will be

$$A_1 n_2 v_2 \exp -(\Delta G^{\mathrm{a}} + \Delta G)/RT\ \mathrm{m}^{-2}\,\mathrm{s}^{-1}$$

When $\Delta G = 0$ the twin grains are in equilibrium and there should therefore be no net boundary movement, i.e., the rates at which atoms cross the boundary in opposite directions must be equal. Equating the above expressions then gives

$$A_1 n_2 v_2 = A_2 n_1 v_1$$

For a high-angle grain boundary it seems reasonable to expect that there will not be great problems with accommodation so that $A_1 \simeq A_2 \simeq 1$. Assuming the above equality also holds for small nonzero driving forces, with $\Delta G > 0$ there will be a net flux from grain 1 to 2 given by

$$J_{\text{net}} = A_2 n_1 v_1 \exp\left(\frac{\Delta G^{\text{a}}}{RT}\right)\left\{1 - \exp\left(-\frac{\Delta G}{RT}\right)\right\} \tag{3.19}$$

If the boundary is moving with a velocity v the above flux must also be equal to $v/(V_{\text{m}}/N_{\text{a}})$, where $(V_{\text{m}}/N_{\text{a}})$ is the atomic volume. Therefore expanding exp $(-\Delta G/RT)$ for the usual case of $\Delta G \ll RT$ gives

$$v = \frac{A_2 n_1 v_1 V_{\text{m}}^2}{N_{\text{a}} RT} \exp\left(-\frac{\Delta G^{\text{a}}}{RT}\right)\frac{\Delta G}{V_{\text{m}}} \tag{3.20}$$

In other words v should be proportional to the driving force $\Delta G/V_{\text{m}}$ (N m^{-2}). Equation 3.20 can be written more simply as

$$v = M \cdot \Delta G/V_{\text{m}} \tag{3.21}$$

where M is the mobility of the boundary, i.e., the velocity under unit driving force. Substituting for ΔG^{a} gives

$$M = \left\{\frac{A_2 n_1 v_1 V_{\text{m}}^2}{N_{\text{a}} RT} \exp\left(\frac{\Delta S^{\text{a}}}{R}\right)\right\} \exp\left(\frac{-\Delta H^{\text{a}}}{RT}\right) \tag{3.22}$$

Note how this simple model predicts an exponential increase in mobility with temperature. This result should of course be intuitively obvious since the boundary migration is a thermally activated process like diffusion. Indeed boundary migration and boundary diffusion are closely related processes. The only difference is that diffusion involves transport along the boundary whereas migration requires atomic movement across the boundary.

The model used to derive Equations 3.20 and 3.22 particularly simple and gross assumptions are involved. In real grain boundaries it is likely that not all atoms in the boundary are equivalent and some will jump more easily than others. For example, atoms may jump preferably to and from atomic steps or ledges, like atoms A, B, and C in Figure 3.23a. In fcc metals such ledges should exist where the close-packed {111} planes protrude into the boundary. Boundary migration could then be effected by the growth of the ledges in one grain combined with the shrinking of corresponding ledges in the other grain as shown in Figure 3.23b.

From our discussion of grain-boundary structure it might be argued that the relatively open structure of a random high-angle boundary should lead to a high mobility whereas the denser packing of the special boundaries should be associated with a low mobility. Indeed, the coherent twin boundary, in which the atoms fit perfectly into both grains, has been found to be

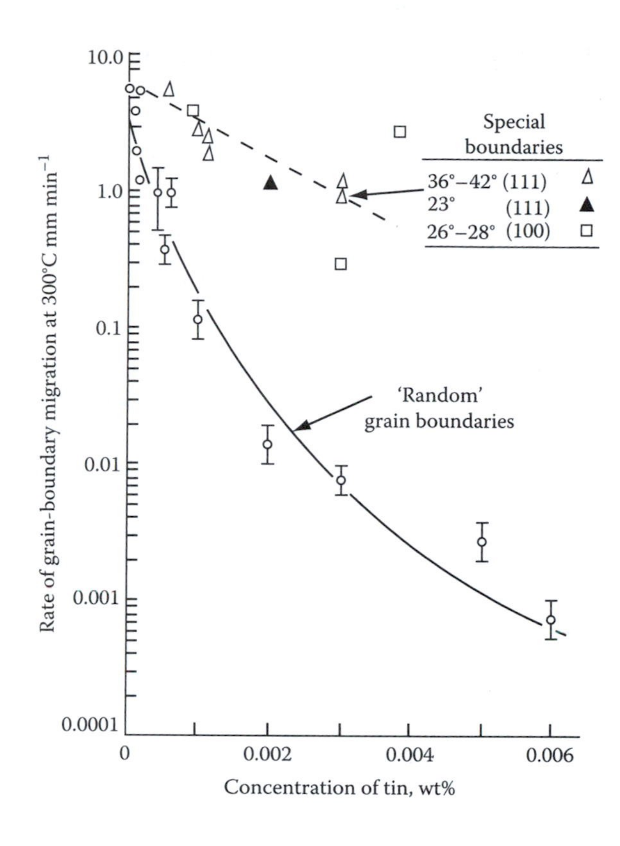

FIGURE 3.27
Migration rates of special and random boundaries at 300°C in zone-refined lead alloyed with tin under equal driving forces. (After Aust, K. and Rutter, J.W., *Trans. Metall. Soc. AIME*, 215, 119, 1959.)

almost entirely immobile.[6] However, experiments have shown that the other special boundaries are usually more mobile than random high-angle boundaries. The reason for this is associated with the presence of impurity or alloying elements in the metal. Figure 3.27 shows data for the migration of various boundaries in zone-refined lead alloyed with different concentrations of tin. For a given driving force the velocity of the random boundaries decreases rapidly with increasing alloy content. Note that only very low concentrations of impurity are required to change the boundary mobility by orders of magnitude. The special grain boundaries on the other hand are less sensitive to impurities. It is possible that if the metal were "perfectly" pure the random boundaries would have the higher mobility. The reason for this type of behavior arises from differences in the interactions of alloy elements or impurities with different boundaries.

Generally the grain-boundary energy of a pure metal changes on alloying. Often (though not always) it is reduced. Under these circumstances the concentration of alloying element in the boundary is higher than that in the matrix. In grain-boundary segregation theory, grain-boundary solute concentrations (X_b) are expressed as fractions of a monolayer. One monolayer ($X_b = 1$) means that the solute atoms in the boundary could be arranged to form a single close-packed layer of atoms. Approximately, for low mole fractions of solute in the matrix (X_0), the boundary solute concentrations X_b is given by

$$X_b \cdot X_0 \exp \frac{\Delta G_b}{RT} \tag{3.23}$$

ΔG_b is the free energy released per mole when a solute atom is moved from the matrix to the boundary ΔG_b is usually positive and roughly increases as the size misfit between the solute and matrix increases and as the solute–solute bond strength decreases.

Equation 3.23 shows how grain-boundary segregation decreases as temperature increases, i.e., the solute "evaporates" into the matrix. For sufficiently low temperatures or high values of ΔG_b, X_b increases toward unity and Equation 3.23 breaks down as X_b approaches a maximum saturation value.

The variation boundary mobility with alloy concentration varies markedly from one element to another. It is a general rule that ΔG_b, which measures the tendency for segregation, increases as the matrix solubility decreases. This is illustrated by the experimental data in Figure 3.28.

When the boundary moves the solute atoms migrate along with the boundary and exert a drag that reduces the boundary velocity. The magnitude of the drag will depend on the binding energy and the concentration in the boundary. The higher mobility of special boundaries can, therefore, possibly be attributed to a low solute drag on account of the relatively more close-packed structure of the special boundaries.

The variation of boundary mobility with alloy concentration varies markedly from one element to another. It is a general rule that Q_B, which measures the tendency for segregation, increases as the matrix solubility decreases. This is illustrated by the experimental data in Figure 3.28.

It is possible that the higher mobility of special grain boundaries plays a role in the development of recrystallization textures. If a polycrystalline metal is heavily deformed, by say rolling to a 90% reduction, a deformation texture develops such that the rolled material resembles a deformed single crystal. On heating to a sufficiently high temperature new grains nucleate and begin to grow. However, not all grains will grow at the same rate: those grains which are specially oriented with respect to the matrix should have higher mobility boundaries and should overgrow the boundaries of the randomly oriented grains. Consequently the recrystallized structure should have a special orientation with respect to the original "single crystal." Thus a new texture results which is called the recrystallization texture. Recrystallization is, however, incompletely understood and the above explanation of recrystallization texture may be an oversimplification. It is possible for example that the nuclei for recrystallization are themselves specially oriented with respect to the deformed matrix.

A recrystallization texture is sometimes an advantage. For example the proper texture in Fe–3 wt% Si alloys makes them much better soft magnets

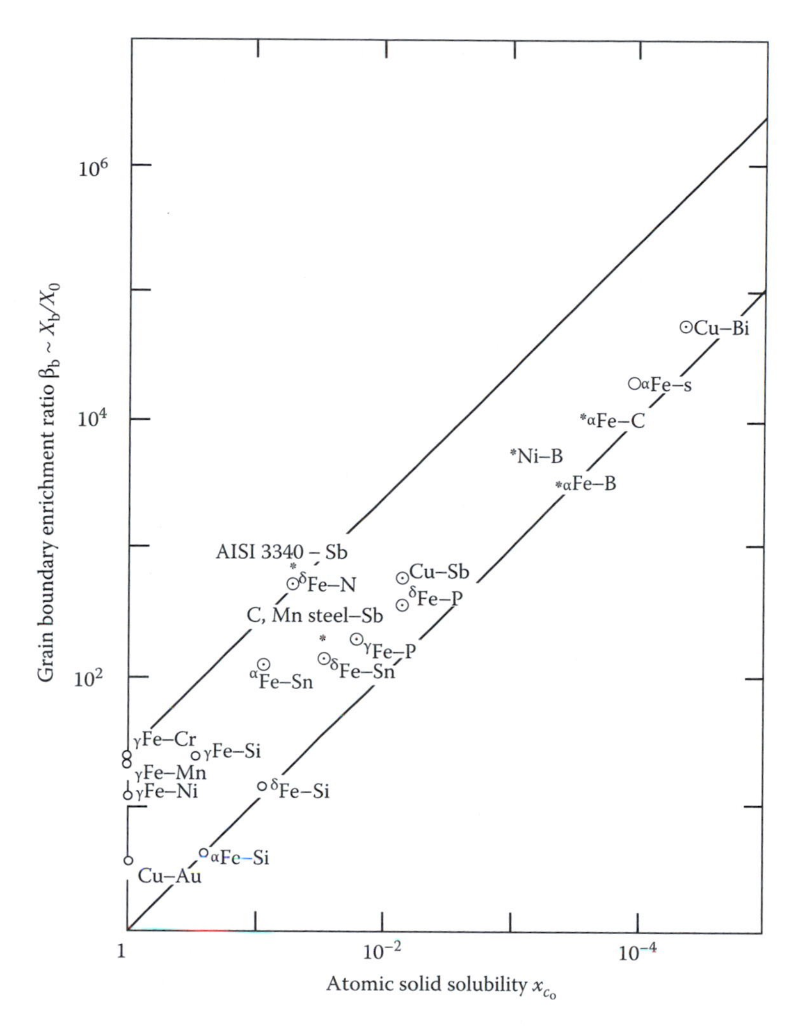

FIGURE 3.28
Increasing grain boundary enrichment with decreasing solid solubility in a range of systems. (After Hondras, E.D. and Seah, M.P., *Int. Metall. Rev.*, December 1977, Review 222.)

for use in transformers. Another application is in the production of textured sheet for the deep drawing of such materials as low-carbon steel. The only way to avoid a recrystallization texture is to give an intermediate anneal before a deformation texture has been produced.

3.3.5 Kinetics of Grain Growth

It was shown in Section 3.3.4 that at sufficiently high temperatures the grain boundaries in a recrystallized specimen will migrate so as to reduce the total number of grains and thereby increase the mean grain diameter. In a single-phase metal the rate at which the mean grain diameter is $\overline{D}$, then mean driving force for grain growth will be proportional to $2\gamma/\overline{D}$ (Equation 3.17). Therefore

$$\bar{v} = \alpha M \frac{2\gamma}{\overline{D}} \cdot \frac{d\overline{D}}{dt} \tag{3.24}$$

where α is a proportionality constant of the order of unity.

Note that this equation implies that the rate of grain growth is inversely proportional to $\overline{D}$ and increases rapidly with increasing temperature due to increased boundary mobility, M. Integration of Equation 3.24 taking $\overline{D} = D_0$ when $t = 0$ gives

$$\overline{D}^2 = D_0^2 + Kt \tag{3.25}$$

where $K = 4\alpha M\gamma$.

Experimentally it is found that grain growth in single-phase metals follows a relationship of the form

$$\overline{D} = K' t^n \tag{3.26}$$

where K' is a proportionality constant which increases with temperature. This is equivalent to Equation 3.25 with $n = 0.5$ if $D > D_0$. However, the experimentally determined values of n are usually much less than 0.5 and only approach 0.5 in very pure metals or at very high temperatures. The reasons for this are not fully understood, but the most likely explanation is that the velocity of grain-boundary migration, v, is not a linear function of the driving force. ΔG, i.e., the mobility in Equation 3.21 is not a constant but varies with ΔG and therefore also with D. It has been suggested that such a variation of M could arise from solute drag effects.[7]

The above type of grain growth is referred to as normal. Occasionally so-called abnormal grain growth to very large diameters. These grains then expand consuming the surrounding grains, until the fine grains are entirely replaced by a coarse-grained array. The effect is illustrated in Figure 3.29 and is also known as discontinuous grain growth, coarsening, or secondary recrystallization. It can occur when normal grain growth ceases due to the presence of a fine precipitate array.

The nature of normal grain growth in the presence of a second phase deserves special consideration. The moving boundaries will be attached to the particles as shown in Figure 3.30a, so that the particles exert a pulling force on the boundary restricting its motion. The boundary shown in Figure 3.30b will be attached to the particle along a length $2\pi r \cos\theta$. Therefore if the boundary intersects the particle surface at 90° the particle will feel a pull of $(2\pi r \cdot \cos\theta \cdot \gamma) \sin\theta$. This will be counterbalanced by an equal and opposite force acting on the boundary. As the boundary moves over the particle

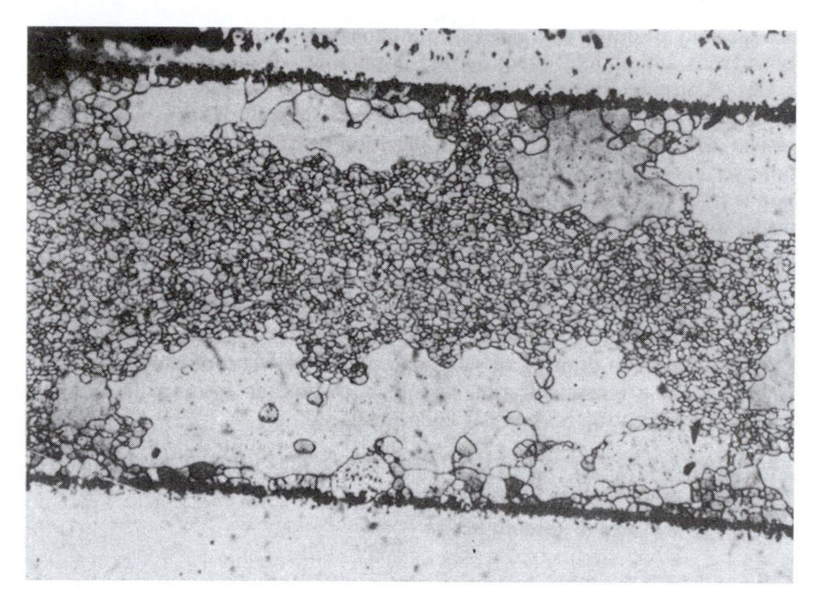

FIGURE 3.29
Optical micrograph (×130) showing abnormal grain growth in a fine-grained steel containing 0.4 wt% carbon. The matrix grains are prevented from growing by a fine dispersion of unresolved carbide particles. (After Gawne, D.T. and Higins, G.T., *J. Iron Steel Inst.*, 209, 562, 1971.)

surface θ changes and the drag reaches a maximum value when sin θ · cos θ is a maximum, i.e., at θ = 45°. The maximum force exerted by a single particle is therefore given by $\pi r \gamma$.

If there is a volume fraction f of particles all with a radius r the mean number of particles intersecting unit area of a random plane is $3f/2\pi r^2$ so that the restraining force per unit area of boundary is approximately

$$P = \frac{3f}{2\pi r^2} \cdot \pi r \gamma = \frac{3f\gamma}{2r} \tag{3.27}$$

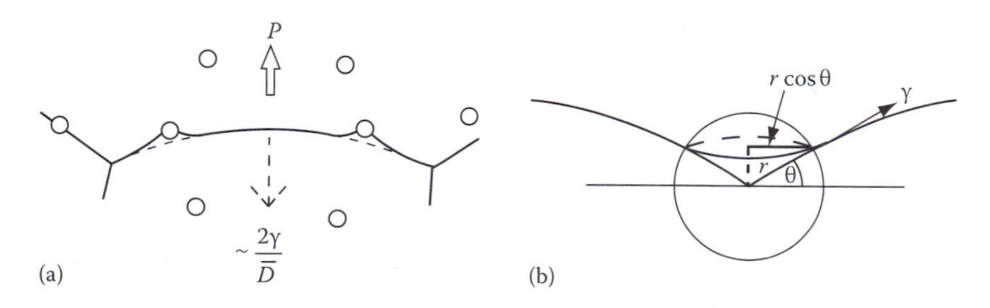

FIGURE 3.30
The effect of spherical particles on grain-boundary migration.

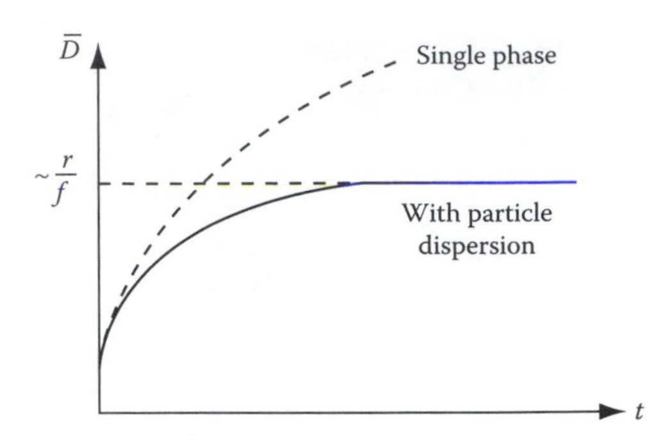

FIGURE 3.31
Effect of second-phase particles on grain growth.

This force will oppose the driving force for the grain growth, namely $\sim 2\gamma/\overline{D}$ as shown in Figure 3.30a. When $\overline{D}$ is small P will be relatively insignificant, but as $\overline{D}$ increases the driving force $2\gamma/\overline{D}$ decreases and when

$$\frac{2\gamma}{\overline{D}} = \frac{3f\gamma}{2r}$$

the driving force will be insufficient to overcome the drag of the particles and grain growth stagnates. A maximum grain size will be given by

$$\overline{D}_{\max} = \frac{4r}{3f} \tag{3.28}$$

The effect of a particle dispersion on grain growth is illustrated in Figure 3.31. It can be seen that the stabilization of a fine grain size during heating at high temperatures requires a large volume fraction of very small particles. Unfortunately, if the temperature is too high, the particles tend to coarsen or dissolve. When this occurs some boundaries can break away before the others and abnormal grain growth occurs, transforming the fine-grain array into a very coarse-grain structure. For example, aluminum-killed steels contain aluminum nitride precipitates which stabilize the austenite grain size during heating. However, their effectiveness disappears above about 1000°C when the aluminum nitride precipitates start to dissolve.

3.4 Interphase Interfaces in Solids

Section 3.3 dealt in some detail with the structure and properties of boundaries between crystals of the same solid phase. In this section we will be dealing with boundaries between different solid phases, i.e., where the two adjoining crystals can have different crystal structures or compositions. Interphase boundaries in solids can be divided on the basis of their atomic structure into three classes: coherent, semicoherent, and incoherent.

3.4.1 Interface Coherence

3.4.1.1 Fully Coherent Interfaces

A coherent interface arises when the two crystals match perfectly at the interface plane so that the two lattices are continuous across the interface, Figure 3.32. This can only be achieved if, disregarding chemical species, the interfacial plane has the same atomic configuration in both phases, and this requires the two crystals to be oriented relative to each other in a special way. For example, such an interface is formed between the hexagonal close-packed (hcp) silicon-rich κ-phase and the fcc copper-rich α-matrix in Cu–Si alloys. The lattice parameters of these two phases are such that the $(111)_{fcc}$ plane is identical to the $(0001)_{hcp}$ plane. Both planes are hexagonally close-packed (Figure 3.33) and in this particular case the interatomic distances are also identical. Therefore when the two crystals are joined along their close-packed planes with the close-packed directions parallel the resultant interface is completely coherent. The requirement that the close-packed planes and directions are parallel produces an orientation relationship between the two phases such that

$$(111)_{\alpha}//(0001)_{\kappa}$$
$$(\bar{1}10]_{\alpha}//[11\bar{2}0]_{\kappa}$$

Note that the relative orientation of two crystals can always be specified by giving two parallel planes (*hkl*) and two parallel directions [*uvw*] that lie in those planes.

Within the bulk of each phase every atom has an optimum arrangement of nearest neighbors that reduces a low energy. At the interface, however, there

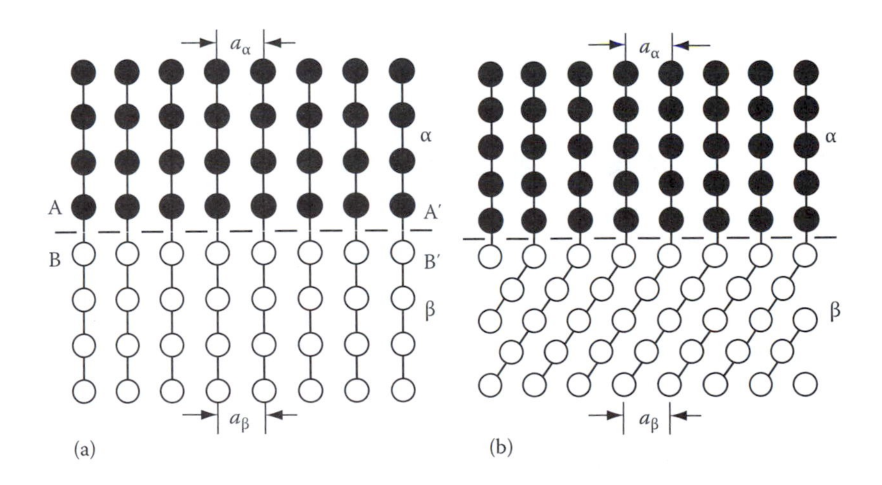

FIGURE 3.32
Strain-free coherent interfaces. (a) Each crystal has a different chemical composition but the same crystal structure. (b) The two phases have different lattices.

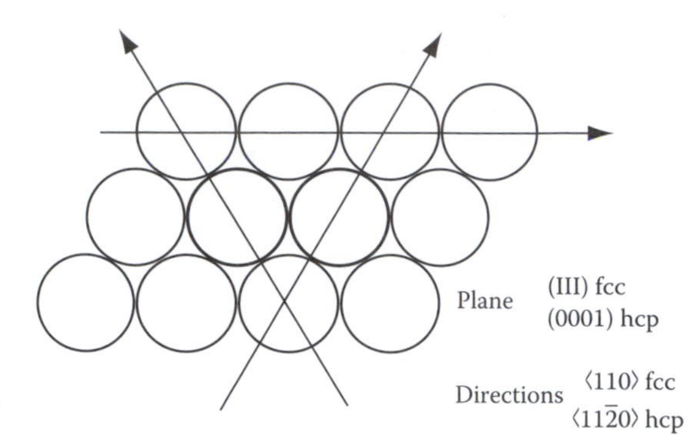

FIGURE 3.33
The close-packed plane and directions in fcc and hcp structures.

is usually a change in composition so that each atom is partly bonded to wrong neighbors across the interface. This increases the energy of the interfacial atoms and leads to a chemical contribution to the interfacial energy (γ_{ch}). For a coherent interface this is the only contribution, i.e.,

$$\gamma \text{ (coherent)} = \gamma_{ch} \tag{3.29}$$

In the case of the α–κ interface in Cu–Si alloys the interfacial energy has been estimated to be as low as 1 mJ m^{-2}. In general coherent interfacial energies range up to about 200 mJ m^{-2}.

In the case of an hcp/fcc interface there is only one plane that can form a coherent interface: no other plane is identical in both crystal lattices. If, however, the two adjoining phases have the same crystal structure and lattice parameter then, apart from differences in composition, all lattice planes are identical.

When the distance between the atoms in the interface is not identical it is still possible to maintain coherency by straining one or both of the two lattices as illustrated in Figure 3.34. The resultant lattice distortions are known as coherency strains.

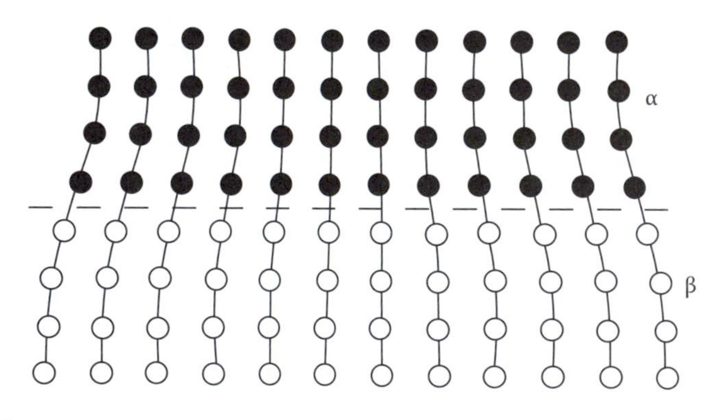

FIGURE 3.34
A coherent interface with slight mismatch leads to coherency strains in the adjoining lattices.

3.4.1.2 Semicoherent Interfaces

The strains associated with a coherent interface raise the total energy of the system, and for sufficiently large atomic misfit, or interfacial area, it becomes energetically more favorable to replace the coherent interface with a semicoherent interface in which the disregistry is periodically taken up by misfit dislocations, Figure 3.35.

If d_α and d_β are the unstressed interplanar spacings of matching planes in the α- and β-phases, respectively, the disregistry, or misfit between the two lattices (δ) is defined by

$$\delta = \frac{d_\beta - d_\alpha}{d_\alpha} \tag{3.30}$$

It can be shown that in one dimension the lattice misfit can be completely accommodated without any long-range strain fields by a set of edge dislocations with a spacing D given by

$$D = \frac{d_\beta}{\delta} \tag{3.31}$$

or approximately, for small δ

$$D \cdot \frac{b}{\delta} \tag{3.32}$$

where $b = (d_\alpha + d_\beta)/2$ is the Burgers vector of the dislocations. The matching in the interface is now almost perfect except around the dislocation cores where the structure is highly distorted and the lattice planes are discontinuous.

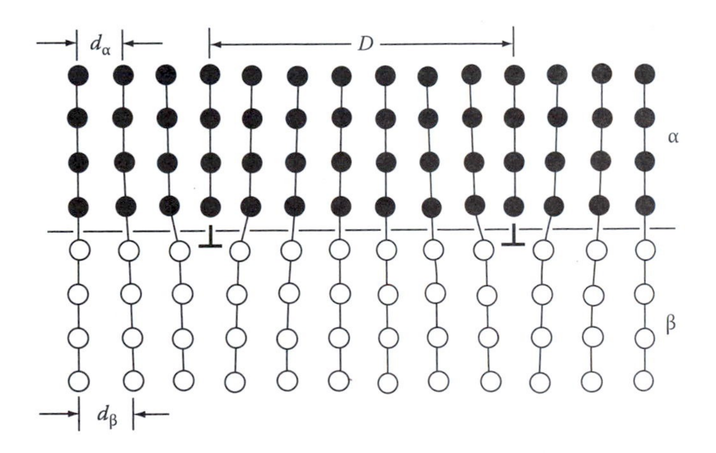

FIGURE 3.35
A semicoherent interface. The misfit parallel to the interface is accommodated by a series of edge dislocations.

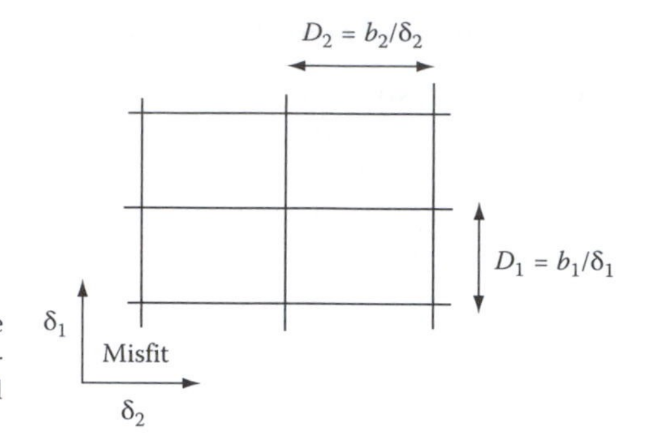

FIGURE 3.36
Misfit in two directions (δ_1 and δ_2) can be accommodated by a cross-grid of edge dislocations with spacings $D_1 = b_1/\delta_1$ and $D_2 = b_2/\delta_2$.

In practice misfit usually exists in two dimensions and in this case the coherency strain fields can be completely relieved if the interface contains two nonparallel sets of dislocations with spacings $D_1 = b_1/\delta_1$ and $D_2 = b_2/\delta_2$, as shown in Figure 3.36. If, for some reason, the dislocation spacing is greater than given by Equation 3.32, the coherency strains will have been only partially relieved by the misfit dislocations and residual long-range strain fields will still be present.

The interfacial energy of a semicoherent interface can be approximately considered as the sum of two parts: (a) a chemical contribution, γ_{st}, which is the extra energy due to the structural distortions caused by the misfit dislocations, i.e.,

$$\gamma \text{ (semicoherent)} = \gamma_{ch} + \gamma_{st} \tag{3.33}$$

Equation 3.32 shows that as the misfit δ increases the dislocation spacing diminishes. For small values of δ the structural contribution to the interfacial energy is approximately proportional to the density of dislocations in the interface, i.e.,

$$\gamma_{st} \propto \delta \text{ (for small } \delta) \tag{3.34}$$

However γ_{st} increases less rapidly as δ becomes larger and it levels out when $\delta \cdot 0.25$ in a similar way to the variation of grain-boundary energy with θ shown in Figure 3.9. The reason for such behavior is that as the misfit dislocation spacing decreases the associated strain fields increasingly overlap and annul each other. The energies of semicoherent interfaces are generally in the range 200–500 mJ m^{-2}.

When $\delta > 0.25$, i.e., one dislocation for every four interplanar spacings, the regions of poor fit around the dislocation cores overlap and the interface cannot be considered as coherent, i.e., it is incoherent.

3.4.1.3 *Incoherent Interfaces*

When the interfacial plane has a very different atomic configuration in the two adjoining phases there is no possibility of good matching across the

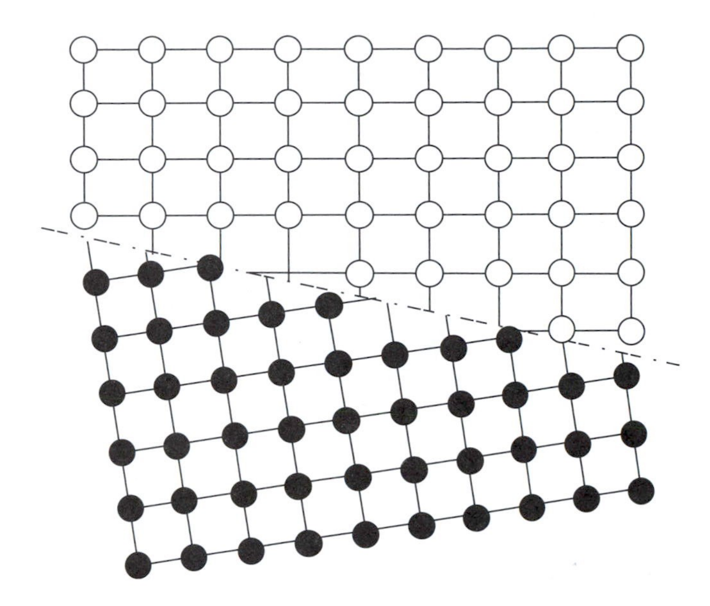

FIGURE 3.37
An incoherent interface.

interface. The pattern of atoms may either be very different in the two phases or, if it is similar, the interatomic distances may differ by more than 25%. In both cases the interface is said to be incoherent. In general, incoherent interfaces result when two randomly oriented crystals are joined across any interfacial plane as shown in Figure 3.37. They may, however, also exist between crystals with an orientation relationship if the interface has a different structure in the two crystals.

Very little is known about the detailed atomic structure of incoherent interfaces, but they have many features in common with high-angle grain boundaries. For example they are characterized by a high energy (~500–1000 mJ m^{-2}) which is relatively insensitive to the orientation of the interfacial plane. They probably have a disordered atomic structure in that the interface lacks the long-range periodicity of coherent and semicoherent interfaces; although, like high-angle grain boundaries, they may have a steplike structure caused by low-index planes protruding into the interface, as in Figure 3.23b.

3.4.1.4 Complex Semicoherent Interfaces

The semicoherent interfaces considered above have been observed at boundaries formed by low-index whose atom patterns and spacings are clearly almost the same. However, semicoherent interfaces, i.e., interfaces containing misfit dislocations, can also form between phases when good lattice matching is not initially obvious. For example, fcc and bcc crystals often appear with the closest-packed planes in each phase, $(111)_{fcc}$ and $(110)_{bcc}$, almost parallel to each other. Two variants of this relationship are found: the so-called Nishiyama–Wasserman (N–W) relationship:

$$(110)_{bcc}//(111)_{fcc}, \quad [001]_{bcc}//[\bar{1}101]_{fcc}$$

and the so-called Kurdjumov–Sachs (K–S) relationship:

$$(110)_{bcc}//(111)_{fcc}, \quad [1\bar{1}1]_{bcc}//[0\bar{1}1]_{fcc}$$

(The only difference between these two is a rotation in the closest-packed planes of 5.26°.) Figure 3.38 shows that the matching between a $\{111\}_{fcc}$ and $\{110\}_{bcc}$ plane bearing the N–W relationship is very poor. Good fit is restricted to small diamond-shaped areas that only contain ~8% of the interfacial atoms. A similar situation can be shown to exist for the K–S orientation relationship. Thus it can be seen that a coherent or semicoherent interface between the two phases is impossible for large interfaces parallel to $\{111\}_{fcc}$ and $\{110\}_{bcc}$. Such interfaces would be incoherent.

The degree of coherency can, however, be greatly increased if a macroscopically irrational interface is formed (i.e., the indices of the interfacial plane in either crystal structure are not small integers). The detailed structure of such interfaces is, however, uncertain due to their complex nature.[8,9]

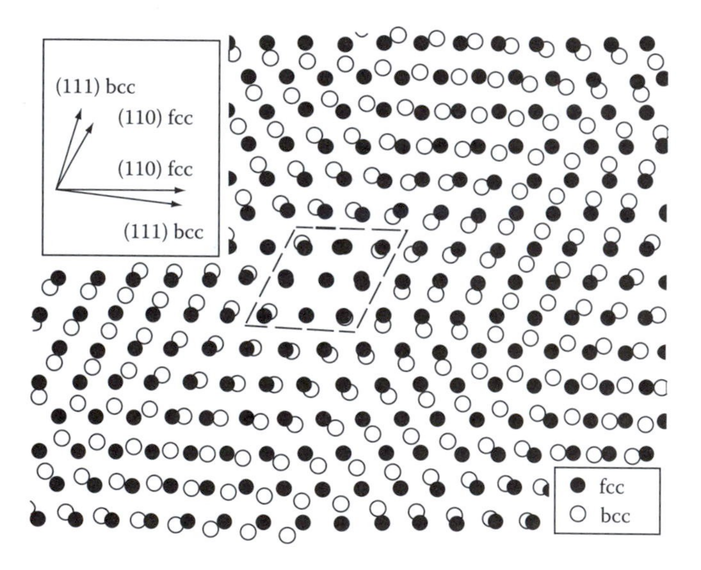

FIGURE 3.38
Atomic matching across a $(111)_{fcc}/(110)_{bcc}$ interface bearing the N–W Orientation relationship for lattice parameters closely corresponding to the case of fcc and bcc iron. (From Hall, M.G. et al., *Surf. Sci.*, 31, 257, 1972. With permission.)

3.4.2 Second-Phase Shape: Interfacial Energy Effects

In a two-phase microstructure one of the phases is often dispersed within the other, for example β-precipitates in an α-matrix. Consider for simplicity a system containing one β-precipitate embedded in a single α-crystal, and assume for the moment that both the precipitate and matrix are strain free. Such a system will have a minimum free energy when the shape of the precipitate and its orientation relationship to the matrix are optimized to give the lowest total interfacial free energy ($\Sigma A\gamma_1$). Let us see how this can be achieved for different types of precipitate.

3.4.2.1 Fully Coherent Precipitates

If the precipitate (β) has the same crystal structure and a similar lattice parameter to the parent α-phase the two phases can form low-energy coherent interfaces on all sides—provided the two lattices are in a parallel orientation relationship—as shown in Figure 3.39a. This situation arises during the early stages of many precipitation-hardening heat treatments, and the β-phase is then termed a fully coherent precipitate or a GP zone. (GP for Guinier and Preston who first discovered their existence. This discovery was made independently by Preston in the United States and Guinier in France,

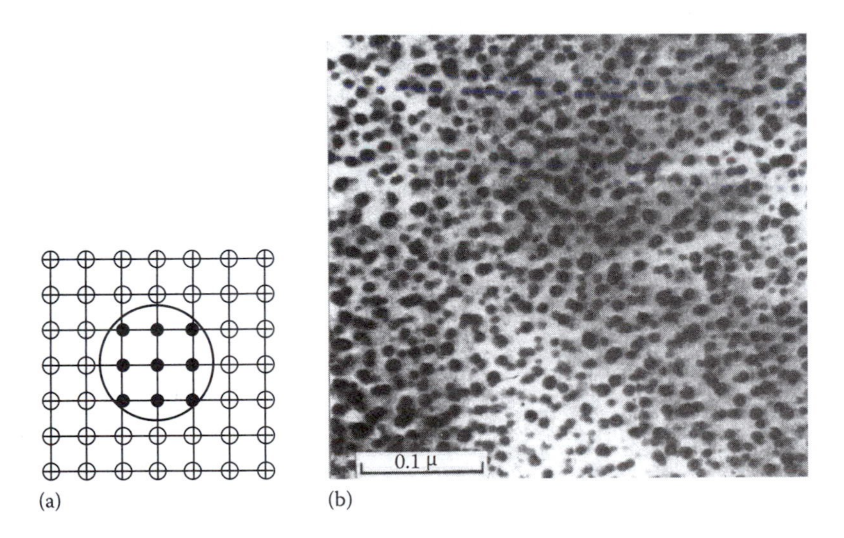

FIGURE 3.39
(a) A zone with no misfit (s Al, d Ag, for example), (b) Electron micrograph of Ag-rich zones in an Al–4 atomic % Ag alloy (×300,000). (After Nicholson, R.B., Thomas, G., and Nutting, J., *J. Inst. Metals*, 87, 431, 1958–1959.)

both employing x-ray diffraction techniques. Their work was later confirmed by transmission electron microscopy.) Since the two crystal structures match more or less perfectly across all interfacial planes the zone can be any shape and remain fully coherent. Thus a γ-plot of the α/β interfacial energy would be largely spherical and, ignoring coherency strains, the equilibrium shape of a zone should be a sphere. Figure 3.39b shows an example of GP zones, ~10 nm in diameter, in an Al–4 atomic% Ag alloy. The zones are a silver-rich fcc region within the aluminum-rich fcc matrix. Since the atomic diameters of aluminum and silver differ by only 0.7% the coherency strains make a negligible contribution to the total free energy of the alloy. In other systems such as Al–Cu where the atomic size difference is much larger strain energy is found to be more important than interfacial energy in determining the equilibrium shape of the zone. This point is discussed further in Section 3.4.3.

3.4.2.2 Partially Coherent Precipitates

From an interfacial energy standpoint it is favorable for a precipitate to be surrounded by low-energy coherent interfaces. However, when the precipitate and matrix have different crystal structures it is usually difficult to find a lattice plane that is common to both phases. Nevertheless, for certain phase combinations there may be one plane that is more or less identical in each crystal, and by choosing the correct orientation relationship it is then possible for a low-energy coherent or semicoherent interface to be formed. There are, however, usually no other planes of good matching and the precipitate must consequently also be bounded by high-energy incoherent interfaces.

A γ-plot of the interfacial energy in this case could look like that in Figure 3.40, i.e., roughly a sphere with two deep cusps normal to the coherent interface. The Wulff theorem would then predict the equilibrium shape to be a disk with a thickness/diameter ratio of γ_c/γ_i, where γ_c and γ_i are the energies of the (semi-)coherent and incoherent interfaces. Triangular, square, or hexagonal plate shapes would be predicted if the γ-plot also contained smaller cusps at symmetrically disposed positions in the plane of the plate.

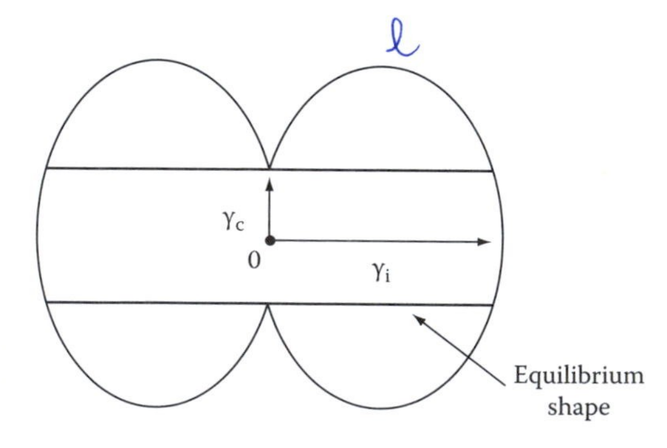

FIGURE 3.40
A section through a γ-plot for a precipitate showing one coherent or semicoherent interface, together with the equilibrium shape (a disk).

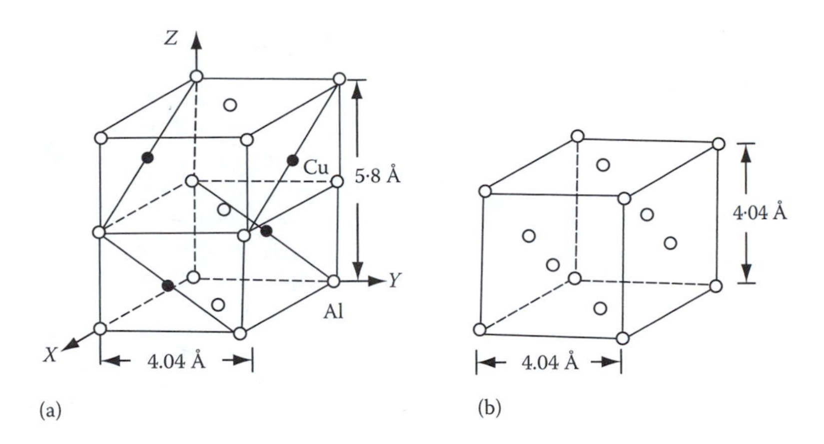

FIGURE 3.41
(a) The unit cell of the θ′ precipitate in Al–Cu alloys. (b) The unit cell of the matrix. (After Silcock, J.M., Heal, T.J., and Hardy, H.K., *J. Inst. Metals*, 82, 239, 1953–1954.)

The precipitate shapes observed in practice may deviate from this shape for two main reasons. First the above construction only predicts the equilibrium shape if misfit strain energy effects can be ignored. Second the precipitate may not be able to achieve an equilibrium shape due to constrains on how it can grow. For example disk-shaped precipitates may be much wider than the equilibrium shape if the incoherent edges grow faster than the broad faces.

Platelike precipitates occur in many systems. For example the hcp γ′-phase in aged Al–4 atomic% Ag alloys forms as plates with semicoherent broad faces parallel to the $\{111\}_\alpha$ matrix planes and with the usual hcp/fcc orientation relationship. The tetragonal θ′-phase in aged Al–4 wt% Cu alloys are also plate-shaped, but in this case the broad faces of the plate (known as the habit plane) are parallel to $\{100\}_\alpha$ matrix planes. Figure 3.41 shows that the $\{100\}_\alpha$ planes are almost identical to the $(100)_{\theta'}$ plane so that the orientation relationship between the θ′ and the aluminum-rich matrix (α) is

$$(001)_{\theta'}//(001)_\alpha$$
$$[100]_{\theta'}//[001]_\alpha$$

Examples of the precipitate shapes that are formed in these two systems are shown in Figures 3.42 and 3.43. Note that as a result of the cubic symmetry of the aluminum-rich matrix there are many possible orientations for the precipitate plates within any given grain. This leads to a very characteristic crystallographic microstructure known after its discoverer as a Widmanstätten morphology.

Besides platelike habits precipitates have also been observed to be lath-shaped (a plate elongated in one direction) and needlelike. For example the S-phase in Al–Cu–Mg alloys forms as laths and the β′-phase in Al–Mg–Si

FIGURE 3.42
Electron micrograph showing the Widmanstätten morphology of γ' precipitates in an Al–4 atomic% Ag alloy. GP zones can be seen between the γ', e.g., at H (×700). (From Nicholson, R.B. and Nutting, J., *Acta Metall.*, 9, 332, 1961. With permission.)

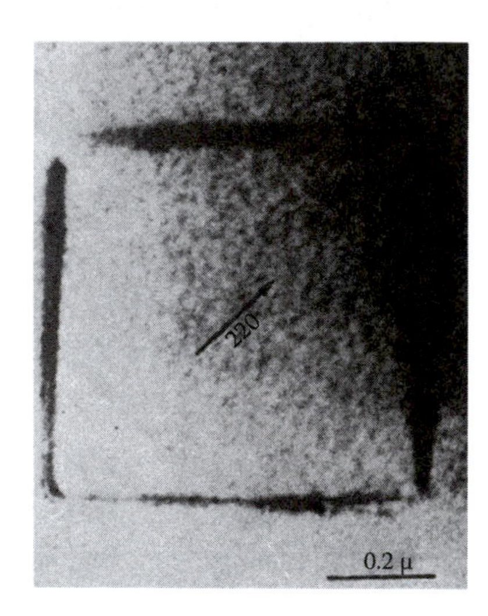

FIGURE 3.43
Electron micrograph of a single coherent θ' plate in an Al–3.9 wt% Cu alloy aged 24 h at 200°C. (×80,000) (From Sankaran, R. and Laird, C., *Acta Metall.*, 22, 957, 1974. With permission.)

alloys as needles.[10] In both cases the precipitates are also crystallographically related to the matrix and produce a Widmanstätten structure.

3.4.2.3 Incoherent Precipitates

When the two phases have completely different crystal structures, or when the two lattices are in a random orientation, it is unlikely that any coherent or

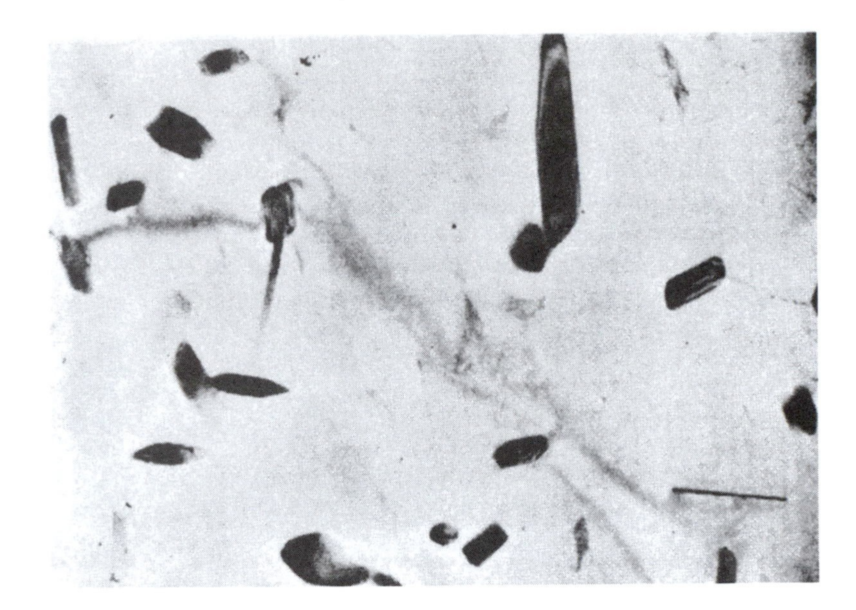

FIGURE 3.44
Electron micrograph showing incoherent particles of θ in an Al–Cu alloy. (After Chadwick, G.A., *Metallography of Phase Transformations*, Butterworths, London, 1972, from C. Laird.)

semicoherent interfaces form and the precipitate is said to be incoherent. Since the interfacial energy should be high for all interfacial planes, the γ-plot and the equilibrium inclusion shape will be roughly spherical. It is possible that certain crystallographic planes of the inclusion lie at cusps in the γ-plot so that polyhedral shapes are also possible. Such faceting, however, need not imply the existence of coherent or semicoherent interfaces.

The θ ($CuAl_2$ precipitate) in Al–Cu alloys is an example of an incoherent precipitate, Figure 3.44. It is found that there is an orientation relationship between the θ and aluminum matrix but this is probably because θ forms from the θ′-phase and does not imply that θ is semicoherent with the matrix.

3.4.2.4 Precipitates on Grain Boundaries

Rather special situations arise when a second-phase article is located on a grain boundary as it is necessary to consider the formation of interfaces with two differently oriented grains. Three possibilities now arise (Figure 3.45):

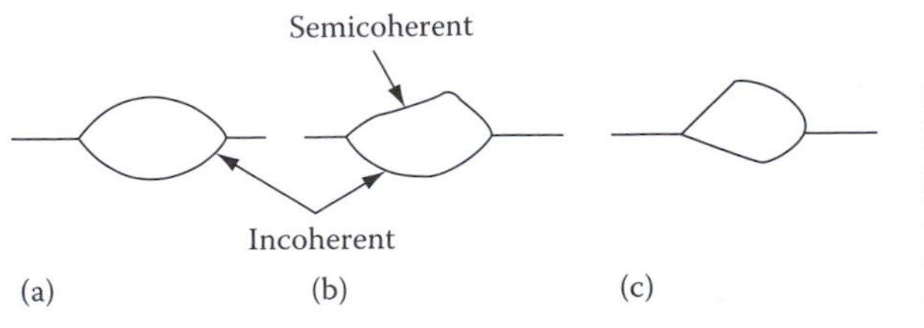

FIGURE 3.45
Possible morphologies for grain-boundary precipitates. Incoherent interfaces smoothly curved: coherent or semi-coherent interfaces planar.

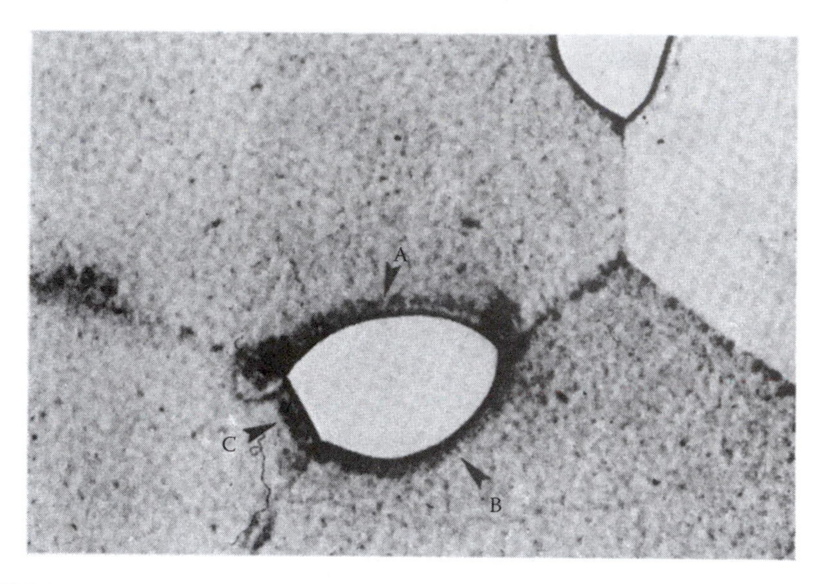

FIGURE 3.46
An α precipitate at a grain boundary triple point in an α–β Cu–In alloy. Interfaces A and B are incoherent, value C is semicoherent (×310). (After Chadwick, G.A., *Metallography of Phase Transformations*, Butterworths, London, 1972.)

the precipitate can have (1) incoherent interfaces with both grains, (2) a coherent or semicoherent interface with one grain and an incoherent interface with the other, or (3) it can have a coherent or semicoherent interface with both grains. The first two cases are commonly encountered but the third possibility is unlikely since the very restrictive crystallographic conditions imposed by coherency with one grain are unlikely to yield a favorable orientation relationship toward the other grain.

The minimization of interfacial energy in these cases also leads to planar semicoherent (or coherent) interfaces and smoothly curved incoherent interfaces as before, but now the interfacial tensions and torques must also balance at the intersection between the precipitate and the boundary. (The shape that produces the minimum free energy can in fact be obtained by superimposing the γ-plots for both grains in a certain way.)[11] An example of a grain-boundary precipitate is shown in Figure 3.46.

3.4.3 Second-Phase Shape: Misfit Strain Effects

3.4.3.1 Fully Coherent Precipitates

It was pointed out in Section 3.4.2 that the equilibrium shape of a coherent precipitate or zone can only be predicted from the γ-plot when the misfit between the precipitate and matrix is small. When misfit is present the formation of coherent interfaces raises the free energy of the system on

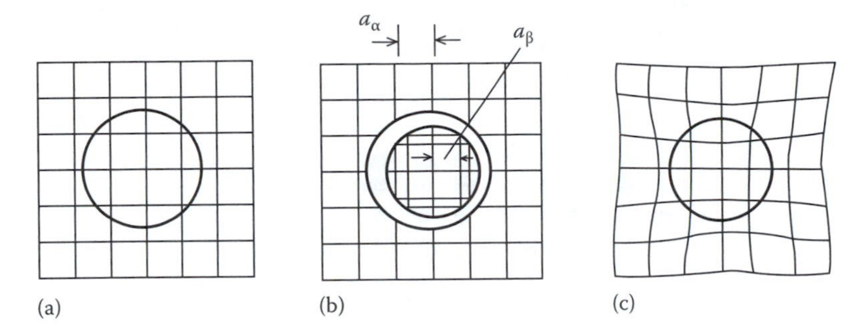

FIGURE 3.47
The origin of coherency strains. The number of lattice points in the hole is conserved.

accounted of the elastic strain fields that arise. If this elastic strain energy is denoted by ΔG_s the condition for equilibrium becomes

$$\Sigma A_i \gamma_i + \Delta G_s = \text{minimum} \tag{3.35}$$

The origin of the coherency strains for a misfitting precipitate is demonstrated in Figure 3.47. If the volume of matrix encircled in Figure 3.47a is cut out and the atoms are replaced by smaller atoms the cutout volume will undergo a uniform negative dilatational strain to an inclusion with a smaller lattice parameter, Figure 3.47b. In order to produce a fully coherent precipitate the matrix and inclusion must be strained by equal and opposite forces as shown in Figure 3.47c.[12]

If the lattice parameters of the unstrained precipitate and matrix are a_β and a_α, respectively, the unconstrained misfit δ is defined by

$$\delta = \frac{\alpha_\beta - a_\alpha}{a_\alpha} \tag{3.36}$$

However, the stresses maintaining coherency at the interfaces distort the precipitate lattice, and in the case of a spherical inclusion the distortion is purely hydrostatic, i.e., it is uniform in all directions, giving a new lattice parameters a'_β. The in situ or constrained misfit ε is defined by

$$\varepsilon = \frac{a'_\beta - a'_\alpha}{a'_\alpha} \tag{3.37}$$

If the elastic moduli of the matrix and inclusion are equal and Poisson's ratio is 1/3, ε and δ are simply related by

$$\varepsilon = \frac{2}{3}\delta \tag{3.38}$$

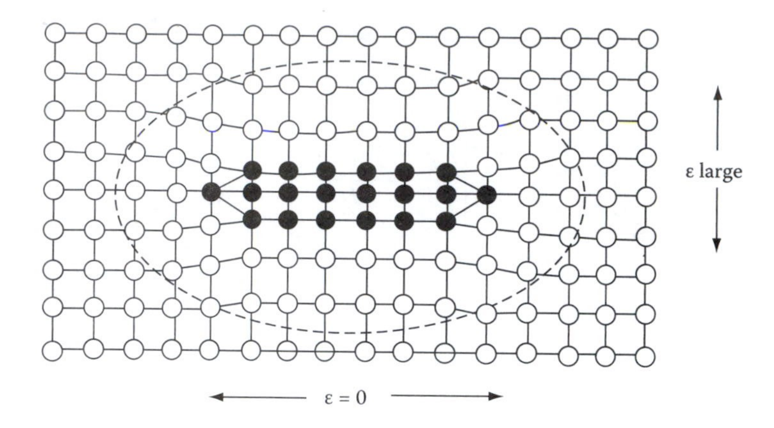

FIGURE 3.48
For a coherent thin disc there is little misfit parallel to the plane of the disk. Maximum misfit is perpendicular to the disc.

In practice the inclusion has different elastic constants to the matrix, nevertheless ε still usually lies in the range

$$0.5\delta < \varepsilon < \delta.$$

When the precipitate is a thin disk the in situ misfit is no longer equal in all directions, but instead it is large perpendicular to the disk and almost zero in the plane of the broad faces, as shown in Figure 3.48.

In general the total elastic energy depends on the shape and elastic properties of both matrix and inclusion. However, if the matrix is elastically isotropic and both precipitate and matrix have equal elastic moduli, the total elastic strain energy ΔG_s is independent of the shape of the shape of the precipitate, and assuming Poisson's ratio $(v) = 1/3$ it is given by

$$\Delta G_s \cdot 4\mu\delta^2 \cdot V \tag{3.39}$$

where
- μ is the shear modulus of the matrix
- V is the volume of the unconstrained hole in the matrix

Therefore coherency strains produce an elastic strain energy which is proportional to the volume of the precipitate and increases as the square of the lattice misfit (δ^2). If the precipitate and inclusion have different elastic moduli the elastic strain energy is no longer shape-independent but is a minimum for a sphere if the inclusion is hard and a disk if the inclusion is soft.

The above comments applied to isotropic matrices. In general, however, most metals are elastically anisotropic. For example, most cubic metals (except molybdenum) are soft in (100) directions and hard in <111>. The shape with a minimum strain energy under these conditions is a disk parallel to {100} since most of the misfit is then accommodated in the soft directions perpendicular to the disk.

The influence of strain energy on the equilibrium shape of coherent precipitates can be illustrated by reference to zones in various aluminum-rich precipitation hardening alloys: Al–Ag, Al–Zn, and Al–Cu. In each case zones containing 50%–100% solute can be produced. Assuming the zone is pure solute the misfit can be calculated directly from the atomic radii as shown below.

Atom Radius (Å)	**Al: 1.43**	**Ag: 1.44**	**Zn: 1.38**	**Cu: 1.28**
Zone misfit (δ)	—	+0.7%	−3.5%	−10.5%
Zone shape	—	Sphere	Sphere	Disk

When $\delta < 5\%$ strain energy effects are less important than interfacial energy effects and spherical zones minimize the total free energy. For $\delta \gtrsim 5\%$, as in the case of zones in Al–Cu, the small increase in interfacial energy caused by choosing a disk shape is more than compensated by the reduction in coherency strain energy.

3.4.3.2 Incoherent Inclusions

When the inclusion is incoherent with the matrix, there is an attempt at matching the two lattices and lattice sites are not conserved across the interface. Under these circumstances there are no coherency strains. Misfit strains can, however, still arise if the inclusion is the wrong size for the hole it is located in, Figure 3.49. In this case the lattice misfit δ has no significance and it is better to consider the volume misfit Δ as defined by

$$\Delta = \frac{\Delta V}{V} \tag{3.40}$$

where

V is the volume of the unconstrained hole in the matrix

$(V - \Delta V)$ is the volume of the unconstrained inclusion

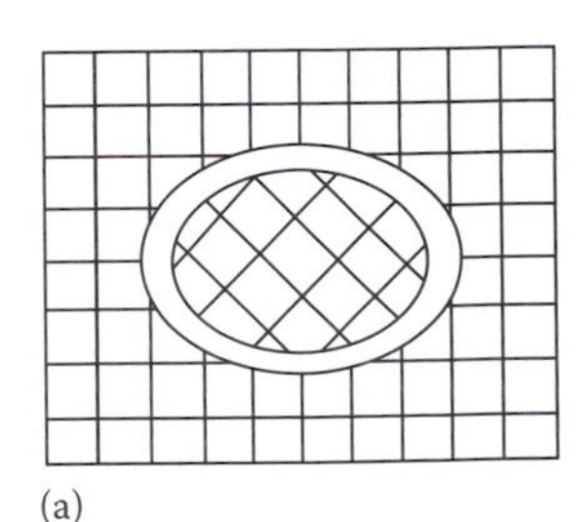

(a)

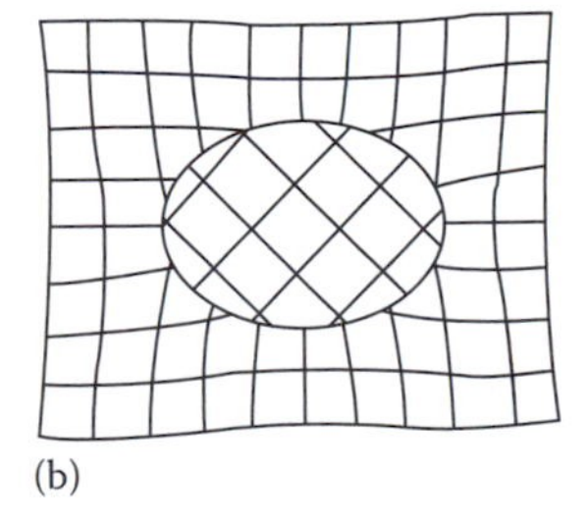

(b)

FIGURE 3.49
The origin of misfit strain for an incoherent inclusion (no lattice matching).

(For a coherent spherical inclusion the volume misfit and the linear lattice misfit are related by $\Delta = 3\delta$. But for a noncoherent sphere the number of lattice sites within the hole is not preserved (Figure 3.49) and in this case $\Delta \neq 3\delta$.) When the matrix hole and inclusion are constrained to occupy the same volume the elastic strain fields again result as shown in Figure 3.49b. The elasticity problem in this case has been solved for spheroidal inclusions which are described by the equation

$$\frac{x}{a^2} + \frac{y^2}{a^2} + \frac{z^2}{c^2} = 1 \tag{3.41}$$

Nabarro[13] gives the elastic strain energy for a homogeneous incompressible inclusion in an isotropic matrix as

$$\Delta G_s = \frac{2}{3}\mu\Delta^2 \cdot V \cdot f(c/a) \tag{3.42}$$

where μ is the shear modulus of the matrix. Thus the elastic strain energy is proportional to the square of the volume misfit Δ^2. The function $f(c/a)$ is a factor that takes into account the shape effects and as shown in Figure 3.50. Notice that, for a given volume, a sphere ($c/a = 1$) has the highest strain energy while a thin, oblate spheroid ($c/a \to 0$) has a very low strain energy, and a needle shape ($c/a = \infty$) lies between the two. If elastic anisotropy is included[14] it is found that the same general form for $f(c/a)$ is preserved and only small changes in the exact values are required. Therefore the equilibrium shape of an incoherent inclusion will be an oblate spheroid with c/a value that balances the opposing effects of interfacial energy and strain energy. When Δ is small interfacial energy effects should dominate and the inclusion should be roughly spherical.

3.4.3.3 Platelike Precipitates

Consider a platelike precipitate with coherent broad faces and incoherent or semicoherent edges, Figure 3.51. (The criterion for whether these interfaces

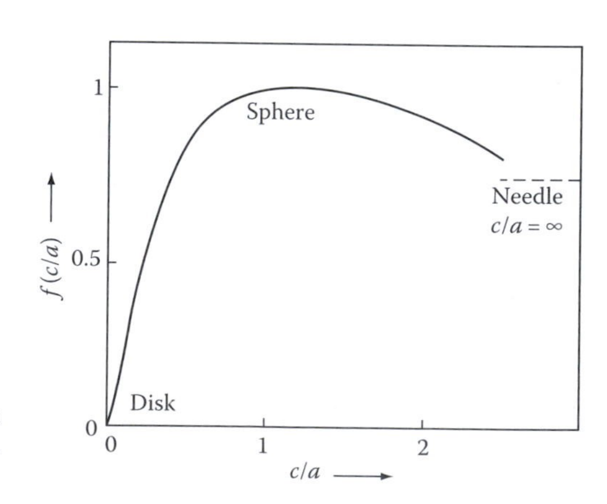

FIGURE 3.50
The variation of misfit strain energy with ellipsoid shape, $f(c/a)$. (After Nabarro, F.R.N., *Proc. R. Soc. A*, 175, 519, 1940.)

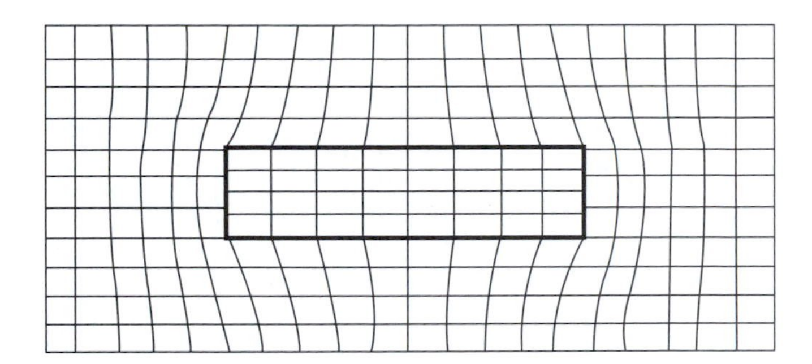

FIGURE 3.51
Coherency strains caused by the coherent broad faces of θ' precipitates.

are coherent or semicoherent is discussed in the following section.) Misfit across the broad faces results in large coherency strains parallel to the plate, but no coherency strains will exist across the edges. The in situ misfit across the broad faces increases with increasing plate thickness which leads to greater strains in the matrix and higher shear stresses at the corners of the plates.[15] Eventually it becomes energetically favorable for the broad faces to become semicoherent. Thereafter the precipitate behaves as an incoherent inclusion with comparatively little misfit strain energy. An example of a precipitate that can be either coherent or semicoherent in this way is θ' in Al–Cu alloys (see Section 5.5.1).

3.4.4 Coherency Loss

Precipitates with coherent interfaces have a low interfacial energy, but in the presence of misfit, they are associated with a coherency strain energy. On the other hand, if the same precipitate has noncoherent interfaces it will have a higher interfacial energy but the coherency strain energy will be absent. Let us now consider which state produces the lowest total energy for a spherical precipitate with a misfit δ and a radius r.

The free energy of a crystal containing a fully coherent spherical precipitate has contributions from (1) the coherency strain energy given by Equation 3.39 and (2) the chemical interfacial energy γ_{ch}. The sum of these two terms is given by

$$\Delta G\ (\text{coherent}) = 4\mu\delta^2 \cdot \frac{4}{3}\pi r^2 + 4\pi r^3 \cdot \gamma_{ch} \tag{3.43}$$

If the same precipitate has incoherent or semicoherent interfaces that completely relieve the unconstrained misfit there will be no misfit energy, but there will be an extra structural contribution to the interfacial energy γ_{st}. The total energy in this case is given by

$$\Delta G\ (\text{noncoherent}) = 0 + 4\pi r^2(\gamma_{ch} + \gamma_{st}) \tag{3.44}$$

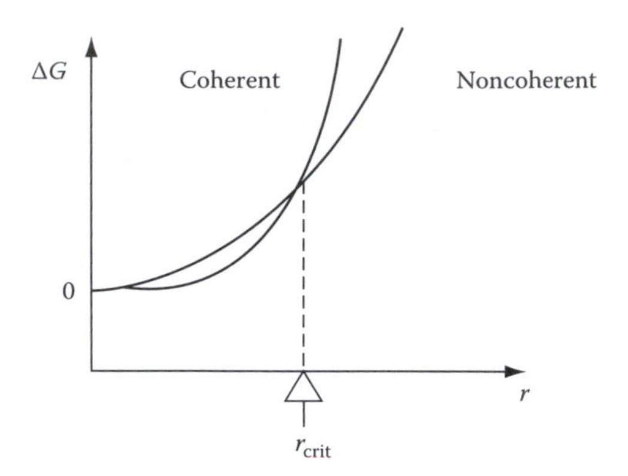

FIGURE 3.52
The total energy of matrix + precipitate versus precipitate radius for spherical coherent and noncoherent (semicoherent or incoherent) precipitates.

For a given δ, ΔG (coherent) and ΔG (noncoherent) vary with r as shown in Figure 3.52. When small, therefore, the coherent state give the lowest total energy, while it is more favorable for large precipitates to be semicoherent or incoherent (depending on the magnitude of δ). At the critical radius (r_{crit}) δG (coherent) $= \Delta G$ (noncoherent) giving

$$r_{crit} = \frac{3\gamma_{st}}{4\mu\delta^2} \tag{3.45}$$

If we assume that δ is small, a semicoherent interface will be formed with a structural energy $\gamma_{st} \propto \delta$. In which case

$$r_{crit} \propto \frac{1}{\delta} \tag{3.46}$$

If a coherent precipitate grows, during aging for example, it should lose coherency when it exceeds r_{crit}. However, as shown in Figure 3.53 loss of coherency requires the introduction of dislocation loops around the precipitate and in practice this can be rather difficult to achieve. Consequently coherent precipitates are often found with sizes much larger than r_{crit}.

There are several ways in which coherency may be lost and some of them are illustrated in Figure 3.54. The most straightforward way is for a dislocation loop to be punched out at the interface as shown in Figure 3.54a. This requires the stresses at the interface to exceed the theoretical strength of the matrix. However, it can be shown that the punching stress p_s is independent of the precipitate size and depends only on the constrained misfit ε. If the shear modulus of the matrix is μ

$$p_s = 3\mu\varepsilon \tag{3.47}$$

It has been estimated that the critical value of ε that can cause the theoretical strength of the matrix to be exceeded is approximately given by

$$\varepsilon_{crit} = 0.05 \tag{3.48}$$

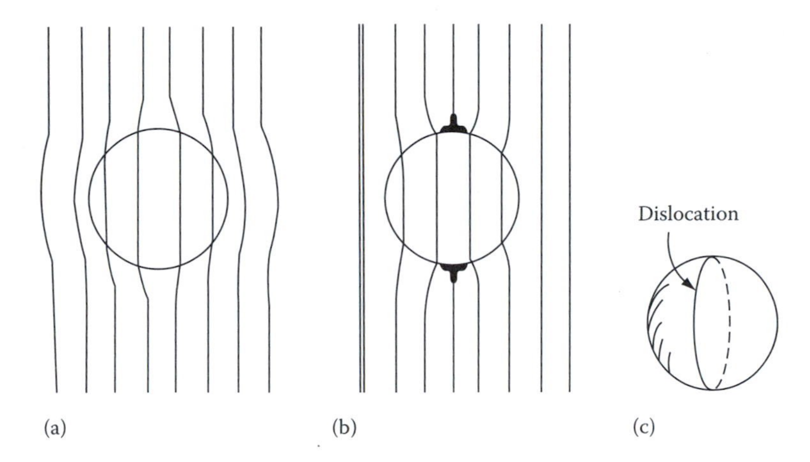

FIGURE 3.53
Coherency loss for a spherical precipitate. (a) Coherent. (b) Coherency strains replaced by dislocation loop. (c) In perspective.

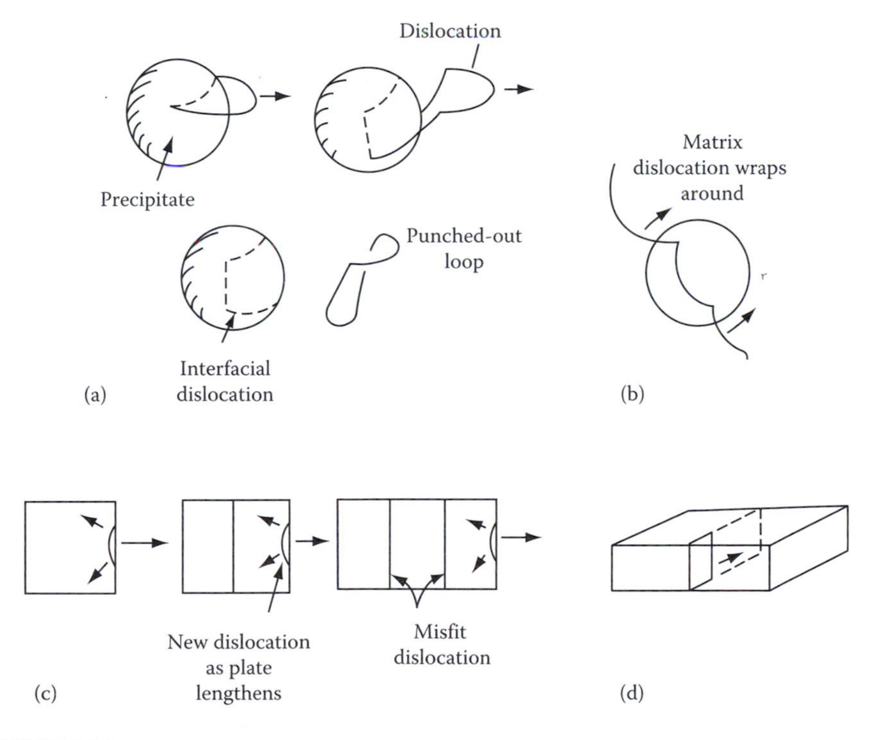

FIGURE 3.54
Mechanisms for coherency loss. (a) Dislocation punching from interface. (b) Capture of matrix dislocation. (c) Nucleation at the edge of the plate repeated as plate lengthens. (d) Loop expansion by vacancy condensation in the precipitate.

Consequently precipitates with a smaller value of ε cannot lose coherency by this mechanism, no matter how large.

There are several alternative mechanism, but all require the precipitate to reach a larger size than r_{crit}.[16] For example, the precipitate can attract a matrix dislocation with a suitable Burgers vector, and cause it to wrap itself around the precipitate, Figure 3.54b. This mechanisms is difficult in annealed specimens but is assisted by mechanical deformation.

In the case of platelike precipitates the situation is different and it is now possible for the high stresses at the edges of the plates to nucleate dislocations by exceeding the theoretical strength of the matrix. The process can be repeated as the plate lengthens so as to maintain a roughly constant interdislocation spacing, Figure 3.54c. Another mechanism that has been observed for platelike precipitates is the nucleation of dislocation loops within the precipitate.[17] Vacancies can be attracted to coherent interfaces[18] and "condense" to form a prismatic dislocation loop which can expand across the precipitate, as shown in Figure 3.54d.

3.4.5 Glissile Interfaces

In the treatment of semicoherent interfaces that has been presented in Section 3.4.4 it has been assumed that the misfit dislocations have Burgers vectors parallel to the interfacial plane. This type of interface is referred to as expitaxial. Glide of the interfacial dislocations cannot cause the interface to advance and the interface is therefore nonglissile. It is however possible, under certain circumstances, to have glissile semicoherent interfaces which can advance by the coordinated glide of the interfacial dislocations. This is possible if the dislocations have a Burgers vector that can glide on matching planes in the adjacent lattices as illustrated in Figure 3.55. The slip planes must be continuous across the interface, but not necessarily parallel. Any gliding dislocation shears the lattice above the slip plane relative to that below by the Burgers vector of the dislocation. In the same way the gliding of the dislocations in a glissile interface causes the receding lattice, α say, to be sheared into the β-structure.

As an aid to understanding the nature of glissile boundaries consider two simple cases. The first is the low-angle symmetric tilt boundary, shown in

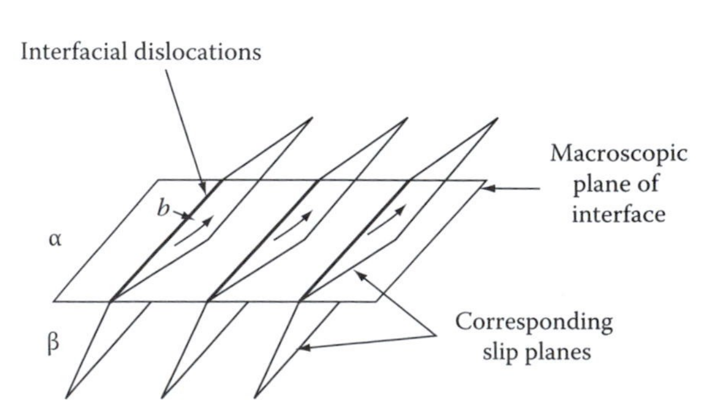

FIGURE 3.55
The nature of a glissile interface.

Figures 3.7a and 3.11. In this case the Burgers vectors are all pure edge in nature and as they glide one grain is rotated into the other grain. Strictly speaking this is not an interphase interface as there is no change in crystal structure, just a rotation of the lattice. A slightly more complex example of a glissile interface between two different lattices is that which can arise between the cubic and hcp lattices. To understand the structure of this interface requires a slight digression to consider the nature of Shockley partial dislocations.

Both fcc and hcp lattices can be formed by staking close-packed layers of atoms one above the other. If the centers of the atoms in the first layer are denoted as A-positions, the second layer of atoms can be formed either by filling the B-positions, or C-positions as shown in Figure 3.56. Either position produces the same atomic configuration at this stage. Let us assume therefore that the atoms in the second layer occupy B-sites. There are now two none-quivalent ways of stacking the third layer. If the third layer is placed directly above the first layer the resulting stacking sequence is ABA and the addition of further layer in the same sequence ABABABABAB... has hexagonal symmetry and is known as a hcp arrangement. The unit cell and stacking sequence of this structure are shown in Figure 3.57. The close-packed plane can therefore be indexed as (0001) and the close-packed directions are of the type $\langle 11\bar{2}0 \rangle$.

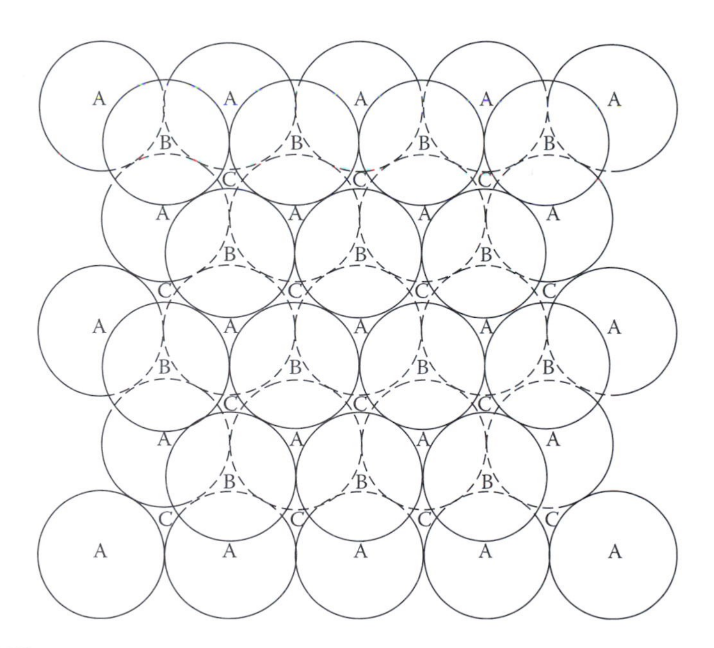

FIGURE 3.56
The location of A, B, and C sites in a close-packed layer of atoms. See also Figures 3.57 and 3.58. (After Martin, J.W. and Doherty, R.D., *Stability of Microstructure in Metallic Systems*, Cambridge University Press, Cambridge, 1976.)

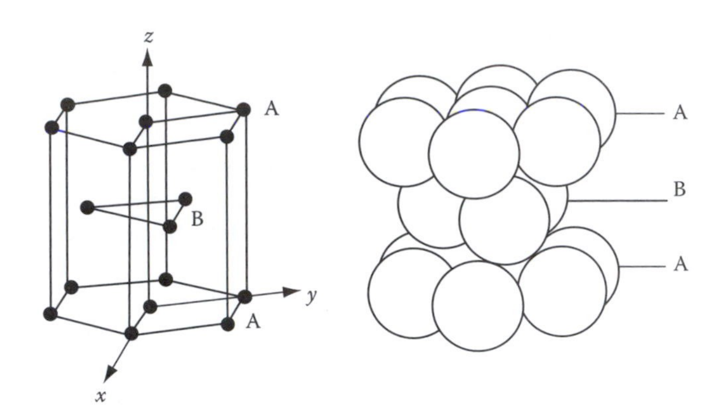

FIGURE 3.57
A hexagonal close-packed unit cell and stacking sequence.

If the atoms in the third layer are placed on the C-sites to form ABC and the same sequence is then repeated, the stacking sequence becomes ABCABCAB... which produces a cubic close-packed arrangement with a fcc unit cell as shown in Figure 3.58. The close-packed atomic planes in this case become the {111}-type and the close-packed directions the $\langle 110 \rangle$-type.

In terms of the fcc unit cell that distance between the B- and C-sites measured parallel to the close-packed planes corresponds to vectors of the type $\frac{a}{6}\langle 11\bar{2} \rangle$. Therefore if a dislocation with a Burgers vector $\frac{a}{6}[11\bar{2}]$ glides between two (111) layers of an fcc lattice, say layers 4 and 5 in Figure 3.59, all layers above the glide plane (5, 6, 7, ...) will be shifted relative to those below the glide plane by a vector $\frac{a}{6}[11\bar{2}]$. Therefore all atoms above the glide plane in B-sites are moved to C-sites, atoms in C-sites move to A-sites, and atoms in A-sites move to B-sites, as shown in Figure 3.59. This type of dislocation with $b = \frac{a}{6}\langle 11\bar{2} \rangle$ is known as Shockley partial dislocation. They are called partial dislocations because vectors of the type $\frac{a}{6}\langle 11\bar{2} \rangle$ do not connect lattice points in the fcc structure. The gliding of Shockley partial dislocations therefore

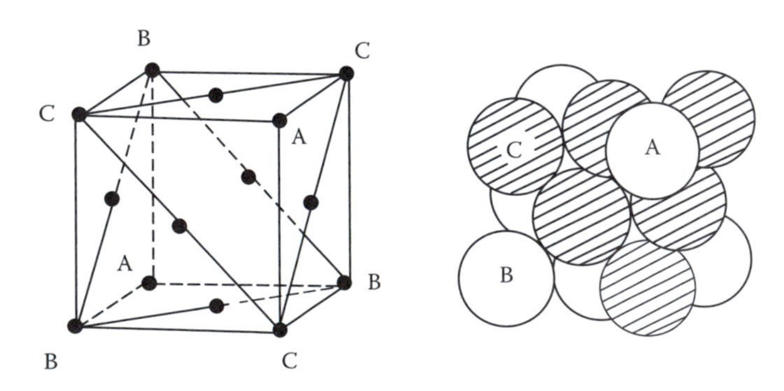

FIGURE 3.58
A cubic close-packed structure showing fcc unit cell and stacking sequence.

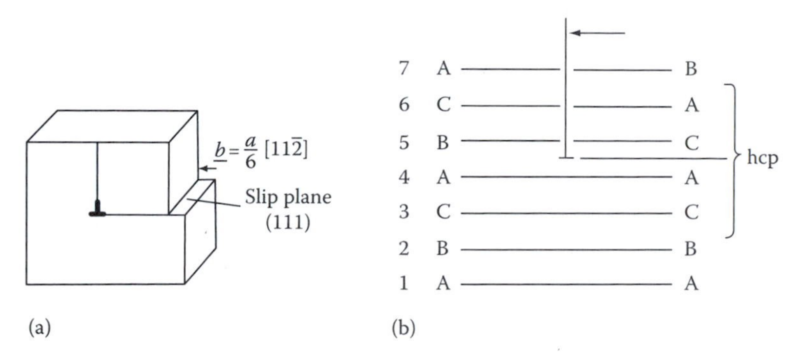

FIGURE 3.59
(a) An edge dislocation with a Burgers vector $b = \frac{a}{6}[11\bar{2}]$ on (111). (Shockley partial dislocation.) (b) The same dislocation locally changes the stacking sequence from fcc to hcp.

disrupts the crystal lattice and causes a stacking fault over the area of glide plane swept by the dislocation. Figure 3.59 shows that the nature of this fault is such that four layers of material are converted into a hcp sequence CACA. Therefore in thermodynamically stable fcc lattices the stacking fault is a region of high free energy. On the other hand if the fcc lattice is only metastable with respect to the hcp structure the stacking fault energy will be effectively negative and the gliding of Shockley partial dislocations will decrease the free energy of the system.

Consider now the effect of passing another $\frac{a}{6}[11\bar{2}]$ dislocation between layers 6 and 7 as shown in Figure 3.60. It can be seen that the region of hcp stacking is now extended by a further two layers. Therefore a sequence of Shockley partial dislocations between every other (111) plane will create a glissile interface separating fcc and hcp crystals, Figure 3.61.

The glide planes of the interfacial dislocations are continuous from the fcc to the hcp lattice and the Burgers vectors of the dislocations, which necessarily lie in the glide plane, are at an angle to the macroscopic interfacial

Layer	fcc		
10	A	B	C
9	C	A	B
8	B	C	A (hcp)
7	A	B	C (hcp)
6	C	A	A (hcp)
5	B	C	C (hcp)
4	A	A	A (hcp)
3	C	C	C (hcp)
2	B	B	B
1	A	A	A

FIGURE 3.60
Two Shockley partial dislocation on alternate (111) planes create six layers of hcp stacking.

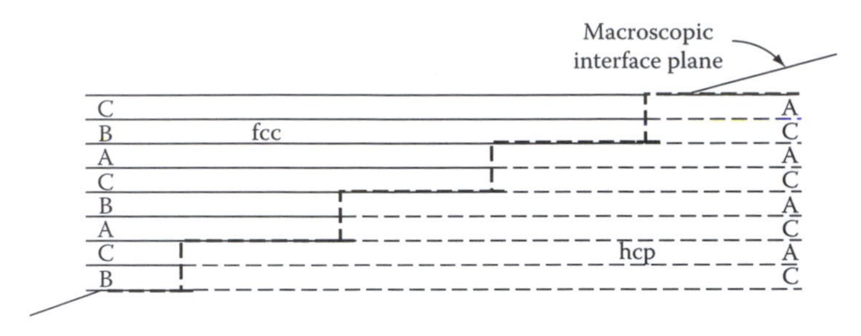

FIGURE 3.61
An array of Shockley partial dislocations forming a glissile interface between fcc and hcp crystals.

plane. If the dislocation network glides into the fcc crystal it results in a transformation of fcc → hcp, whereas a hcp → fcc transformation can be brought about by the reverse motion. Macroscopically the interfacial plane lies at an angle to the (111) or (0001) planes and need not be parallel to any low-index plane, i.e., it can be irrational. Microscopically, however, the interface is steeped into planner coherent facets parallel to $(111)_{fcc}$ and $(0001)_{hcp}$ with a step height the thickness of two closed-packed layers.

An important characteristic of glissile dislocation interfaces is that they can produce a macroscopic shape change in the crystal. This is illustrated for the fcc → hcp transformation in Figure 3.62a. If a single fcc crystal is transformed into an hcp crystal by the passage of the same Shockley partial over every

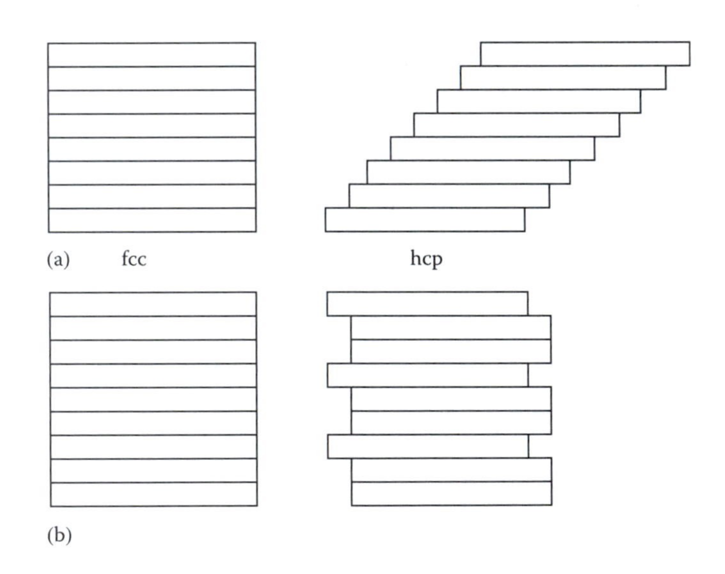

FIGURE 3.62
Schematic representation of the different ways of shearing cubic close-packed planes into hexagonal close-packed (a) using only one Shockley partial and (b) using equal numbers of all three Shockley partials.

(111) plane then there is a macroscopic shape change, in this case a simple shear, as shown. There are, however, two other Shockley partials which can also be used to transform fcc → hcp stacking, and if the transformation is achieved using all three partials in equal numbers there will be no overall shape change, Figure 3.62b.

The formation of martensite in steel and other alloy systems occurs by the motion of glissile–dislocation interfaces. These transformations are characterized by a macroscopic shape change and no change in composition. Usually, however, the interface must be more complex than the fcc/hcp case discussed above, although the same principles will still apply. Martensitic transformations are dealt with further in Chapter 6.

3.4.6 Solid/Liquid Interfaces[19]

Many of the ideas that were discussed with regard to solid/vapor interfaces can be carried over to solid/liquid interfaces, only now the low-density vapor phase is replaced by a high-density liquid, and this has important consequences for the structure and energy of the interface.

There are basically two types of atomic structure for solid/liquid interfaces. One is essentially the same as the solid/vapor interfaces described in Section 3.1, i.e., an atomically flat close-packed interfaces, Figure 3.63a. In this case the transition from liquid to solid occurs over a rather narrow transition zone approximately one atom layer thick. Such interfaces can also be described as smooth, faceted, or sharp. The other type is an atomically diffuse interface, Figure 3.63b, in which the transition from liquid to solid occurs over several atom layers. Thus there is a gradual weakening of the interatomic bonds and an increasing disorder across the interface into the bulk liquid phase; or in thermodynamic terms, enthalpy and entropy gradually change from bulk solid to bulk liquid values across the interface as shown in Figure 3.64. When the solid and liquid are in equilibrium (at T_m) the high enthalpy of the liquid is balanced by a high entropy so that both phases

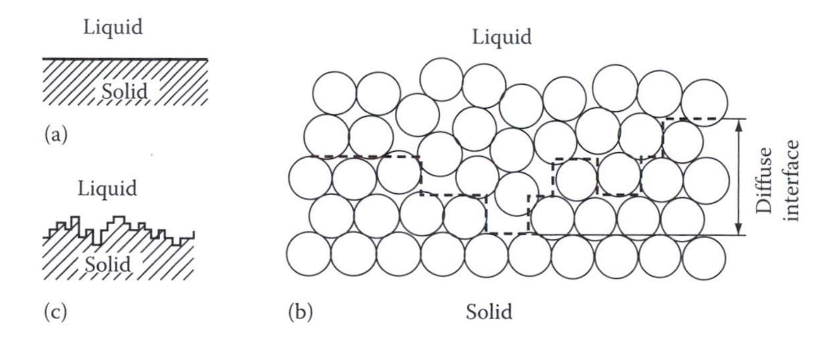

FIGURE 3.63
Solid/liquid interfaces: (a) atomically smooth, (b) and (c) atomically rough, or diffuse interfaces. (After Flemings, M.C., *Solidification Processing*, McGraw-Hill, New York, 1974.)

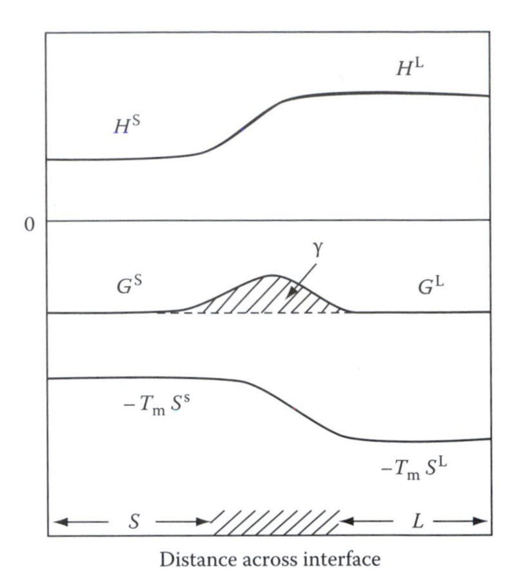

FIGURE 3.64
The variation of H, $-T_m S$, and G across the solid/liquid interface at the equilibrium melting temperature T_m, showing the origin of the solid/liquid interfacial energy γ.

have the same free energy. In the interface, however, the balance is disturbed thereby giving rise to an excess free energy, γ_{SL}.

Diffuse interfaces are also known as rough or nonfaceted. The dotted line in Figure 3.63b did an attempt to show the rough nature of the interface by dividing the atoms into the "solid" and "liquid." If this is done the schematic representation of Figure 3.63c can be used.

The type of structure chosen by a particular system will be that which minimizes the interfacial free energy. According to a simple theory developed by Jackson[20] the optimum atomic arrangement depends mainly on the latent heat of fusion (L_f) relative to the melting temperature (T_m). This theory predicts that there is a critical value of L_f/T_m. $4R$ above which the interface should be flat and below which it should be diffuse. Most metals have L_f/T_m. R and are therefore predicted to have rough interfaces. On the other hand some intermetallic compounds and elements such as Si, Ge, Sb as well as most nonmetals have high value of L_f/T_m and generally have flat close-packed interfaces. If the model is applied to solid/vapor interfaces L_s (the heat of sublimation) should be used instead of L_f and then flat surfaces are predicted even for metals, in agreement with observations.

If the broken-bond model is applied to the calculation of the energy of a solid/liquid interface it can be argued that the atoms in the interface are roughly half bonded to the solid and half to the liquid so that the interfacial enthalpy should be $\sim 0.5 L_f/N_a$ per atom. This appears to compare rather favorably with experimentally measured value of γ_{SL} which are $\sim 0.45 L_f/N_a$ per atom for most metals. However, the agreement is probably only fortuitous since entropy effects should also be taken into account, Figure 3.64. Some experimentally determined values of γ_{SL} are listed in Table 3.4.

TABLE 3.4

Experimentally Determined Solid/Liquid Interfacial Free Energies

Material	T_m (K)	γ_{SL} (mJ m^{-2})
Sn	505.7	54.5
Pb	600.7	33.3
Al	931.7	93
Ag	1233.7	126
Au	1336	132
Cu	1356	177
Mn	1493	206
Ni	1725	255
Co	1763	234
Fe	1803	204
Pd	1828	209
Pt	2043	240

Source: Values selected from Turnbull, D., *J. Appl. Phys.*, 21, 1022, 1950.

These values were determined by indirect means from homogeneous nucleation experiments (see Chapter 4) and many contain systematic errors. Comparison of Tables 3.2 and 3.3 indicates $\gamma_{SL} = 0.30\gamma_b$ (for a grain boundary). More direct experiments[21] imply that $\gamma_{SL} \cdot 0.45\gamma_b$ (.0.15γ_{SV}). Another useful empirical relationship is that

$$\gamma_{SV} > \gamma_{SL} + \gamma_{LV}$$

which means that for a solid metal close to T_m it is energetically favorable for the surface to melt and replace the solid/vapor interface with solid/liquid and liquid/vapor interfaces.

It is found experimentally that the free energies of diffuse interfaces do not vary with crystallographic orientation, i.e., γ-plots are spherical.[22] Materials with atomically flat interfaces, however, show strong crystallographic effects and solidify with low-index close-packed facets, Figure 3.65.

3.5 Interface Migration

The great majority of phase transformations in metals and alloys occur by a process known as nucleation and growth, i.e., the new phase (β) first appears at certain sites within the metastable (α) phase (nucleation) and this is subsequently followed by the growth of these nuclei into the surrounding matrix. In other words, an interface is created during the nucleation stage and then migrates into the surrounding parent phase during the growth

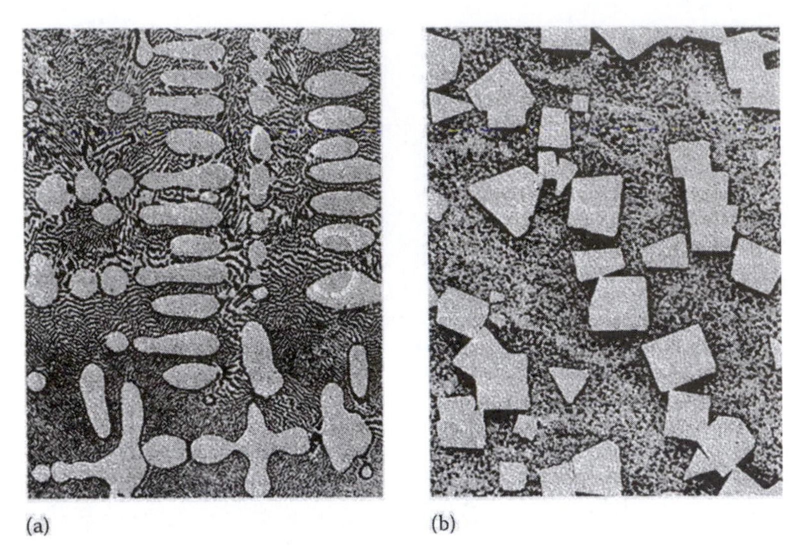

FIGURE 3.65
Example of solid/liquid interface structure in metallic systems. (a) Nonfaceted dendrites of silver in a copper–silver eutectic matrix (×330) and (b) faceted cuboids of β′-SnSb compound in a matrix of Sn-rich material (×110). (After Chadwick, G.A., *Metallography of Phase Transformations*, Butterworths, London, 1972.)

stage. This type of transformation is therefore essentially heterogeneous, i.e., at any time during the transformation the system can be divided into parent and product phases. The nucleation stage is very important and determines many features of the transformation. However, most of the transformation product is formed during the growth stage by the transfer of atoms across the moving parent/product interface.

There are basically two different types of interface: glissile and nonglissile. Glissile interfaces migrate by dislocation glide that results in the shearing of the parent lattice into the product. The motion of glissile interfaces is relatively insensitive to temperature and is therefore known as athermal migration. Most interfaces are nonglissile and migrate by the more or less random jumps of individual atoms across the interface in a similar way to the migration of a random high-angle grain boundary. The extra energy that the atom needs to break free of one phase and attach itself to the other is supplied by thermal activation. The migration of nonglissile interfaces is therefore extremely sensitive to temperature.

A convenient way of classifying nucleation and growth transformation is to divide them according to the way in which the product grows. Therefore two major groupings can be made by dividing the transformation according the whether growth involved glissile or nonglissile interfaces. Transformations produced by the migration of a glissile interface are referred to as military transformations. This emphasizes the analogy between the coordinated motion of atoms crossing the interface and that of soldiers moving in rank on the parade ground. In contrast the uncoordinated transfer of atoms across a nonglissile interface results in what is known as a civilian transformation.

During a military transformation the nearest neighbors of any tom are essentially unchanged. Therefore the parent and product phases must have the same composition and no diffusion is involved in the transformation. Martensitic transformations belong to this group. Glissile interfaces are also involved in the formation of mechanical twins and twinning therefore has much in common with martensitic transformations.

During civilian transformation the parent and product may or may not have the same composition. If there is no change in composition, e.g., the $\alpha \rightarrow \gamma$ transformation in pure ion, the new phase will be able to grow as fast as the atoms can cross the interface. Such transformations are said to be interface controlled. When the parent and product phases have different compositions, growth of the new phase will require long-range diffusion. For example, the growth of the B-rich β-phase into the A-rich α-phase shown in Figure 3.66 can only occur if diffusion is able to transport A away from,

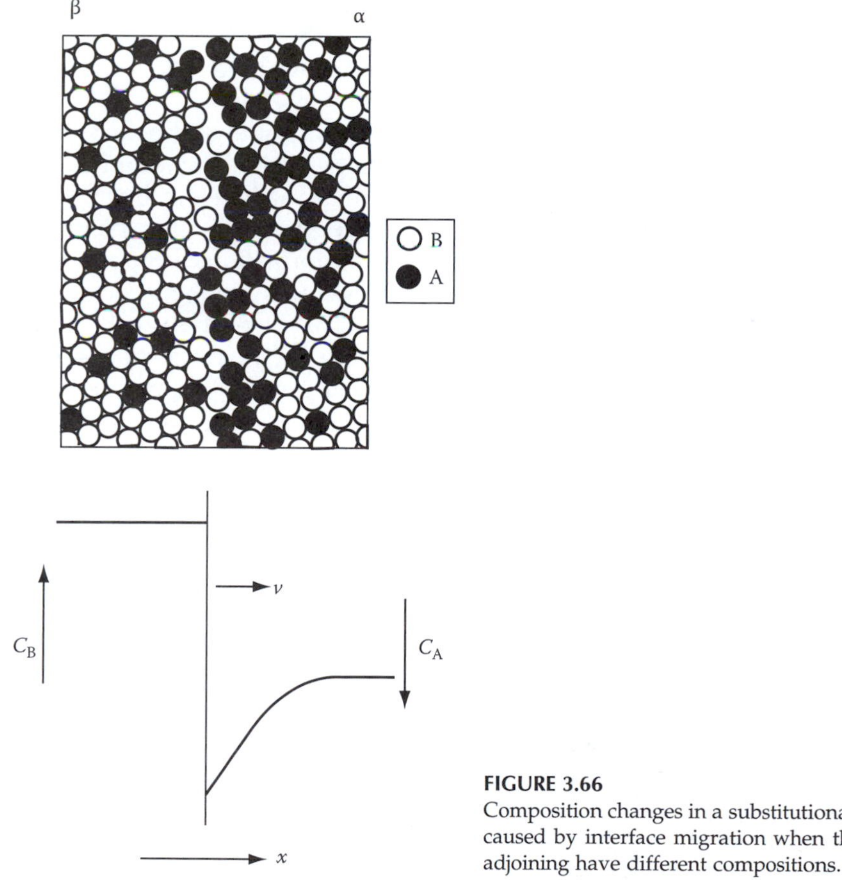

FIGURE 3.66
Composition changes in a substitutional alloy caused by interface migration when the two adjoining have different compositions.

and B toward the advancing interface. If the interfacial reaction is fast, i.e., the transfer of atoms across the interface is an easy process, the rate at which the β-phase can grow will be governed by the rate at which lattice diffusion can remove the excess atoms from ahead of the interface. This is therefore known as diffusion-controlled growth. However, if for some reason the interfacial reaction is slow, the growth rate will be governed by the interface kinetics. Under these circumstances growth is said to be interface controlled and a very small concentration gradient in the matrix is sufficient to provide the necessary flux of atoms to and from the interface. It is also possible that the interface reaction and diffusion process occur at similar rates in which case the interface is said to migrate under mixed control.

The above discussion of interface migration and classification of nucleation and growth transformations (also known as heterogeneous transformations) is summarized in Table 3.5, together with some examples of each class. This classification is adapted from that first proposed by Christian.[23,24] Nonglissile interfaces can be considered to include solid/liquid and solid/vapor interfaces as well as solid/solid (coherent, semicoherent, and incoherent) interfaces. Therefore solidification and melting can be included in the classification of civilian transformations under diffusion control (although the concept of diffusion may sometimes need to be extended to include the diffusion of heat). Condensation and evaporation at a free solid surface are also included although they will not be treated in any depth.[25]

While many transformations can be easily classified into the above system, there are other transformations where difficulties arise. For example, the bainite transformation takes place by the thermally activated growth, but it also produces a shape change similar to that product by the motion of a glissile interface. At present the exact nature of such transformations is unresolved.

There is a small class of transformation, known as homogeneous transformations that are not covered by Table 3.5. This is because they do not occur by the creation and migration of an interface, i.e., no nucleation stage is involved. Instead the transformation occurs homogeneously throughout the parent phase. Spinodal decomposition and certain ordering transformations are examples of this category and they will be discussed in Chapter 5.

3.5.1 Diffusion-Controlled and Interface-Controlled Growth[26]

Let us now look more closely at the migration of an interface separating two phases of different composition. Consider for simplicity a β precipitate of almost pure B growing behind a planar interface into A-rich α with an initial composition X_0 as illustrated in Figure 3.67. As the precipitate grows, the α adjacent to the interface becomes depleted of B so that the concentration of B in the α-phase adjacent to the interface X_i decreases below the bulk concentration, Figure 3.67a. Since growth of the precipitate requires a net flux of B atoms from the α- to the β-phase there must be a positive driving force across the interface $\Delta\mu_B^i$ as shown in Figure 3.67b. The origin of this chemical potential difference can be seen in Figure 3.67c. Clearly for growth to occur the interface

TABLE 3.5

Classification of Nucleation and Growth Transformations

Type	Military	Civilian			
Effect of temperature change	Athermal	Thermally activated			
Interface type	Glissile (coherent or semicoherent)	Nonglissile (coherent, semicoherent. Incoherent, solid/liquid, or solid/vapor)			
Composition of parent and product phase	Same composition	Same composition	Different compositions		
Nature of diffusion process	No diffusion	Short-range diffusion (across interface)	Long-range diffusion (through lattice)		
Interface, diffusion or mixed control?	Interface control	Interface control	Mainly interface control	Mainly diffusion control	Mixed control
Examples	Martensite twining	Massive ordering	Precipitation dissolution	Precipitation dissolution	Precipitation dissolution
	Symmetric tilt boundary	Polymorphic recrystallization	Bainite condensation	Soldification and melting	Eutectoid
		Grain growth	Evaporation		Cellular precipitation
		Condensation			
		Evaporation			

Source: Adapted from Christian, J.W., in *Phase Transformations*, Vol. 1, Institute of Metallurgists, 1979, p. 1.

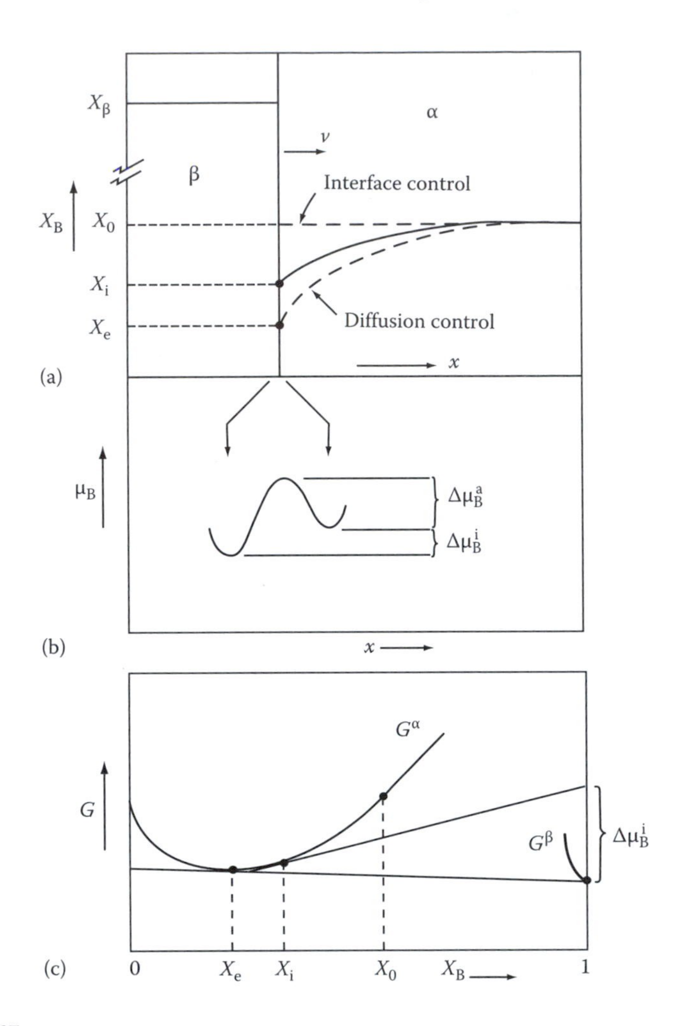

FIGURE 3.67
Interface migration with long-range diffusion. (a) Composition profiles across the interface. (b) The origin of the driving force for boundary migration into the α-phase. (c) A schematic molar free energy diagram showing the relationship between $\Delta\mu_B^i$, X_i, and X_c. (Note that the solubility of A in the β-phase is so low that the true shape of the free energy curve cannot be drawn on this scale.)

composition must be greater than the equilibrium concentration X_e. By analogy with the migration of a high-angle grain boundary (Section 3.3.4) the net flux of B across the interface will produce an interface velocity v given by

$$v = M\Delta\mu_B^i/V_m \tag{3.49}$$

where

M is the interface mobility

V_m is the molar volume of the β-phase

The corresponding flux across the interface will be given by

$$J_{\mathrm{B}}^{\mathrm{i}} = -M\Delta\mu_{\mathrm{B}}^{\mathrm{i}}/V_{\mathrm{m}}^{2} \text{ moles of B m}^{-2}\,\text{s}^{-1} \tag{3.50}$$

(The negative sign indicates that the flux is in the negative direction along the x-axis.) As a result of the concentration gradient in the α-phase there will also be a flux of B atoms toward the interface J_{B}^{α} given by

$$J_{\mathrm{B}}^{\alpha} = -D\left(\frac{\partial C_{\mathrm{B}}}{\partial x}\right)_{\text{interface}}$$

If a steady state exists at the interface these two fluxes must balance, i.e.,

$$J_{\mathrm{B}}^{\mathrm{i}} = J_{\mathrm{B}}^{\alpha} \tag{3.52}$$

If the interface mobility is very high, e.g., an incoherent interface, $\Delta\mu_{\mathrm{B}}^{\mathrm{i}}$ can be very small and $X_{\mathrm{i}} \cdot X_{\mathrm{e}}$. Under these circumstances there is effectively local equilibrium at the interface. The interface will then move as fast as diffusion allows, and growth takes place under diffusion control. The growth rate can then be evaluated as a function of time, say, by solving the diffusion equation with the boundary conditions $X_{\mathrm{i}} = X_{\mathrm{e}}$ and $X_{\mathrm{B}}(\infty) = X_0$. Simple examples of this problem will be given in subsequent chapters in connection with solidification and diffusive transformations in solids (Chapters 4 and 5).

When the interface has a lower mobility a greater chemical potential difference ($\Delta\mu_{\mathrm{B}}^{\mathrm{i}}$) is required to drive the interface reaction and there will be a departure from local equilibrium at the interface. The value of X_{i} and that is chosen will be the one which enables Equation 3.52 to be satisfied and the interface will then be migrating under mixed control. In the limit of a very low mobility it is possible that $X_{\mathrm{i}} \cdot X_0$ and $(\partial C/\partial x)_{\text{interface}}$ is almost zero. Under these conditions growth is said to be interface controlled and there is a maximum possible driving force $\Delta\mu_{\mathrm{B}}^{\mathrm{i}}$ across the interface.

It can easily be shown that for a dilute or ideal solution, the driving force $\Delta\mu_{\mathrm{B}}^{\mathrm{i}}$ is given by

$$\Delta\mu_{\mathrm{B}}^{\mathrm{i}} = RT\,\ln\frac{X_{\mathrm{i}}}{X_{\mathrm{e}}} \cdot \frac{RT}{X_{\mathrm{e}}}(X_{\mathrm{i}} - X_{\mathrm{e}}) \tag{3.53}$$

provided $(X_{\mathrm{i}} - X_{\mathrm{e}}) \ll X_{\mathrm{e}}$ (see Exercise 3.20). Thus the rate at which the interface moves under interface control should be proportional to the deviation of the interface concentration from equilibrium $(X_{\mathrm{i}} - X_{\mathrm{e}})$.

Let us now consider the question of why interface control should occur at all when the two phases have a different composition. At first sight it may appear that interface control should be very unlikely in practice. After all, the necessary long-range diffusion involves a great many atom jumps while the interface reaction essentially involves only one jump. Furthermore the activation energy for diffusion across the interface is not likely to be greater than for diffusion through the lattice—quite the contrary. On the basis, therefore, all interface reactions should be very rapid in comparison to lattice diffusion, i.e., all growth should be diffusion controlled. In many cases the above arguments are quite valid, but under certain conditions they are insufficient and may even be misleading.

Consider again the expression that was derived for the mobility of a high-angle grain boundary, Equation 3.22. A similar expression can be derived for the case of an interphase interface with $\Delta\mu_B^i$ replacing ΔG (see Exercise 3.19). It can be seen, therefore, that the above arguments neglect the effect of the accommodation factor (A), i.e., the probability that an atom crossing the boundary will be accommodated on arrival at the new phase. It is likely that incoherent interfaces and diffuse solid/liquid interfaces, as high-angle grain boundaries, will have value of A close to unity. These interfaces should therefore migrate under diffusion control. However, as will be demonstrated later, it is possible for certain types of coherent or semicoherent interfaces, as well as smooth solid/liquid interfaces to have such low values of A that some degree of interface control is easily possible.

If two phases with different compositions, but the same crystal structure are separated by a coherent interface as shown in Figure 3.22a, the interface can advance by the replacement of the α atoms in phase AA′ with β atoms by normal lattice diffusion involving vacancies. There is no need for a separate interface reaction and the migration of this type of interface is therefore diffusion controlled. This situation arises during the growth of GP zones for example. The same arguments will apply if the interface is semicoherent provided the misfit dislocations can climb by vacancy creation or annihilation.

Quite a different situation arises when the two phases forming a coherent or semicoherent interface have different crystal structures. Consider for example the coherent close-packed interface between fcc and hcp crystals, Figure 3.68a. If growth of the hcp phase is to occur by individual atomic jumps (i.e., so-called continuous growth) then an atom on a C site in the fcc phase must change into a B position as shown in Figure 3.68b. It can be seen,

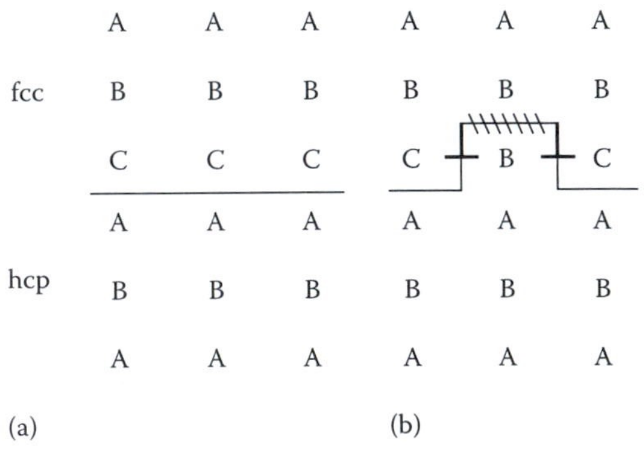

FIGURE 3.68
Problems associated with the continuous growth of coherent interfaces between phases with different crystal structures. (After Martin, J.W. and Doherty, R.D., *Stability of Microstructure in Metallic Systems*, Cambridge University Press, Cambridge, 1976.)

however, that this results in a very high energy, unstable configuration with two atoms directly above each other on B sites. In addition a loop of Shockley partial dislocation is effectively created around the atom. An atom attempting such a jump will, therefore, by unstable and be forced back to its original position. The same situation will be encountered over the coherent regions of semicoherent interfaces separating phases with different crystal structures. Solid/vapor as well as smooth solid/liquid interfaces should behave in a similar manner, through perhaps to a lesser extent. If a single atom attaches itself to a flat close-packed interface it will raise the interfacial free energy and will therefore tend to detach itself again. It can thus be seen that continuous growth at the above type of interfaces will be very difficult, i.e., very low accommodation factors and low mobility are expected.

A way of avoiding the difficulties of continuous growth encountered in the above cases is provided by the "ledge" mechanism shown in Figure 3.69. If the interface contains a series of ledges BC, DE normal to the facets AB, CD, EF, atoms will be able to transfer more easily across the ledges than the immobile facets and interface migration is therefore affected by the transverse migration of the ledges as shown.

Growth ledges have in fact been seen with the aid of the electron microscope on the surfaces of growing precipitates. For example Figure 3.70 shows an electron micrograph and a schematic drawing of the growth ledge on an Mg_2Si plate in an Al–Mg–Si alloy.[27] Note that growth ledges are usually hundreds of atom layers high.

When existing ledges have grown across the interface there is a problem of generating new ones. In Figure 3.70 the source of new ledges is through to be heterogeneous nucleation at the point of contract with another precipitate. The same problem will not be encountered if the precipitate is dissolving, however, since the edges of the plate will provide a continual source of ledges.[28] It is though that once nucleated, the rate at which ledges migrate across the planar facets should be diffusion controlled, i.e., controlled by how fast diffusion can occur to and form the ledges. However, the problem of nucleating new ledges may often lead to a degree of interface control on the overall rate at which the coherent or semicoherent interface can advance perpendicular to itself.

Growth ledges are by no means restricted to solid/solid systems. The first evidence for the existence of growth ledges came from studies of solid/vapor interfaces. They are also found on faceted solid/liquid interfaces.

The mechanism of interface migration can have important effects on the shape of second-phase inclusions. It was shown in Section 3.4.2 that in

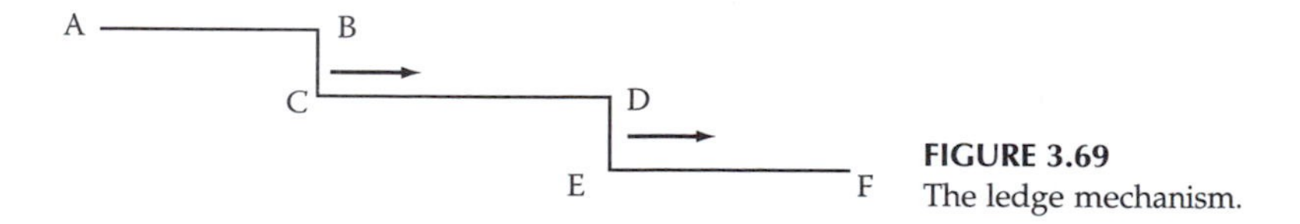

FIGURE 3.69
The ledge mechanism.

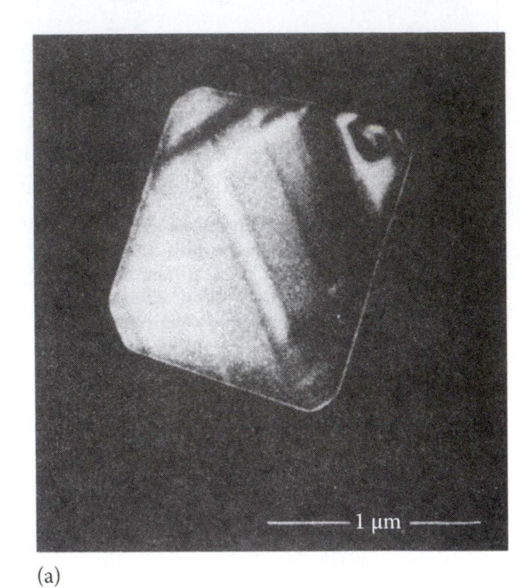

(a)

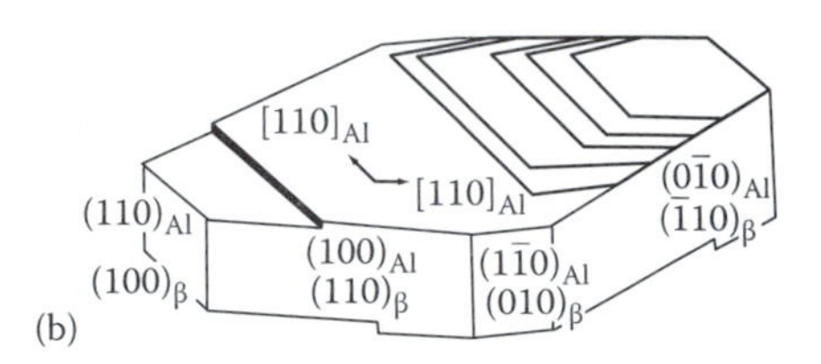

(b)

FIGURE 3.70
(a) Growth ledges at an Mg_2 Si plate in Al–1.5 wt% Mg_2Si, solution treated and aged 2 h at 350°C. Dark field micrograph. (b) Schematic diagram of (a) showing ledges on Mg_2Si plate. (After Weatherly, G.C., *Acta Metall.*, 19, 181, 1971.)

the absence of strain energy effects the equilibrium shape of a precipitate should be determined by the relative energies of the bounding interfaces. For example, a partially coherent precipitate should be disk- or plate-shaped with an aspect ratio γ_i/γ_c where γ_i is the energy of the incoherent edges and γ_c is the energy of the coherent or semicoherent broad faces. However, the precipitate shape observed in practice may be prevented from achieving this equilibrium shape by the relative rates at which the coherent and incoherent interface can migrate. For example if there are problems of ledge nucleation the easier growth of the incoherent plate edge may lead to a larger aspect ratio than the equilibrium.

Exercises

3.1 Use the method of Section 3.1 to estimate the surface energy of {111}, {200}, and {220} surface plane in an fcc crystal. Express your answer in J per surface atom and in $J\,m^{-2}$.

3.2 Differentiate Equation 3.8 to obtain the slope of the $E_{SV}-\theta$ curve at $\theta = 0$.

3.3 If a two-dimensional rectangular crystal is bounded by sides of lengths l_1 and l_2 show by differentiation that the equilibrium shape is given by

$$\frac{l_1}{l_2} = \frac{\gamma_2}{\gamma_1}$$

where γ_1 and γ_2 are the energies of the sides l_1 and l_2, respectively. (The area of the crystal $l_1 l_2$.)

3.4 (a) Measure θ for the low-angle tilt boundary in Figure 3.11.
(b) Determine the Burgers vector of the interface dislocations by making a Burgers circuit around one of the dislocations. Does the mean spacing of the dislocations agree with that predicted by Equation 3.9?

3.5 Explain why grain boundaries move toward their center of curvature during grain growth but away from their center of curvature during recrystallization.

3.6 (a) Suppose a recrystallized dislocation-free grain is growing into a deformed matrix containing a dislocation density of 10^{16} m^{-2} (i.e., 10^{16} mm^{-3}). If the dislocations have an energy of $\mu b^2/4$ J m^{-1} calculate the pulling force acting on the recrystallized grain boundary. (Assume a shear modulus $\mu = 10^{10}$ N m^{-2} and a Burgers vector $b = 0.28$ nm.)
(b) If the recrystallized grains grow from spherically shaped nuclei, what is the diameter of the smallest nucleus that can expand into the surrounding matrix? (Assume a grain-boundary energy of 0.5 J m^{-2}.)

3.7 Look up the equilibrium phase diagrams for the Al–Fe and Al–Mg systems. On the basis of these diagrams would you expect the grain boundary enrichment of Fe in dilute Al–Fe alloys to be greater or less than or Mg in dilute Al–Mg alloys at the same temperatures?

3.8 Derive Equation 3.31.

3.9 When a precipitate is surrounded by a spherical interface of radius r it is subjected to a pressure above that of the matrix by $2\gamma/r$. Consider a faceted precipitate with an equilibrium shape that of a square plate with a thickness of $2x_1$ and width $2x_2$. If the free energies of the broad faces and edges are, respectively, γ_1 and γ_2, show that broad faces exert a pressure on the precipitate (ΔP) given by

$$\Delta P = 2\gamma_2/x_2$$

(Hint: consider the total force acting on the periphery of the broad faces.) Show that the same result can be obtained by considering the pressure exerted by one of the edge faces of the plate.

3.10 Explain the structure and energies of coherent, semicoherent, and incoherent interfaces, with particular reference to the role of orientation relationships and misfit.

3.11 Fe-rich GP zones can form in dilute Al–Fe alloys. Given that the atomic radii are 1.43 Å Fe, would you expect the zones to be spherical- or disc-shaped?

3.12 Mg can dissolve in Al to form a substitutional solid solution. Mg atoms are, however, bigger than Al atoms and each Mg atom therefore distorts the surrounding Al lattice, i.e., a coherency strain field effectively exist around each Mg atom. Using Equation 3.39 estimate the misfit strain energy. Express the answer in kJ mol^{-1} and eV $atom^{-1}$. (The shear modulus of Al = 25 GPa, the radius of an Al atom = 1.43 Å, the radius of a Mg atom = 1.60 Å.) What assumptions are implicit in this calculation?

3.13 Explain why fully coherent precipitates tend to lose coherency as they grow.

3.14 Show that the passage of a Shockley partial dislocation over every one of a given set of close-packed planes in fcc crystals produces a twin of the original crystal.

3.15 If the ledges on the planar semicoherent interface in Figure 3.69 move with a transverse velocity u what will be the overall velocity of the interface perpendicular to CD. Assuming an infinite array of identical ledges of height (BX) = h and spacing (CD) = l.

3.16 Using arguments similar to those used in connection with Figure 3.68 show that a coherent twin boundary in an fcc metal will not migrate by the random jumping of atoms across the interface. Suggest an interfacial structure that would result in a highly mobile interface (see Exercise 3.15).

3.17 What are the most likely atomic processes involved in the migration of (1) solid/vapor interfaces, (2) solid/liquid interfaces in nonmetals, and (3) solid/liquid interfaces in metals.

3.18 By using a similar approach to the derivation of Equation 3.20 for a high-angle grain boundary, show that the net flux of B atoms across the α/β interface in Figure 3.67 is given by

$$J_B^i = \frac{A_B n_\alpha v_\alpha}{RT} \exp\left(-\frac{\Delta\mu^a}{RT}\right)\Delta\mu_B^i$$

3.19 Derive Equation 3.53 for an ideal or dilute solution.

3.20 If an alloy containing β precipitates in an α-matrix is given a solution treatment by heating to a temperature above the equilibrium β solvus the precipitates will dissolve. (See for example the phase diagram in Figure 1.36.) Show with diagrams how the composition will change in the vicinity of an α/β interface during dissolution if the dissolution is (1) diffusion controlled, (2) interface controlled, and (3) under mixed control. Indicate compositions by reference to a phase diagram where appropriate.

References

1. This subject is covered in detail in J.W. Martin and R.D. Doherty, *Stability of Microstructure in Metallic Systems*, Cambridge University Press, Cambridge, 1976, and in M. McLean, Microstructural instabilities in metallurgical systems—a review, *Metal Sci.*, March 1978, p. 113.
2. Methods are described in J.M Blakely, *Introduction to the Properties of Crystal Surfaces*, Pergamon, Oxford, 1973, p. 53.

3. For a more detailed discussion, see J.W. Christian, *The Theory of Transformations in Metals and Alloys*, 2nd edn., Part I. Pergamon, Oxford, 1975, p. 153.
4. R.S. Nelson, D.J. Mazey, and R.S. Barnes, *Philos. Mag.*, 11: 91, 1965.
5. B.E. Sundquist, *Acta Metall.*, 12: 37, 1964.
6. C.J. Simpson, K.T. Aust, and W.C. Winegard, *Scripta Metall.*, 3: 171, 1969.
7. See for example J.W. Martin and R.W. Doherty, *Stability of Microstructure in Metallic Systems*, Cambridge University Press, Cambridge, 1976, p. 221.
8. J.W. Christian, *Metall. Trans.*, 21A: 799, 1990.
9. J.W. Christian, *Mater. Sci. Eng.*, A127: 215, 1990.
10. G.C. Weatherly and R.B. Nicholson, *Philos. Mag.*, 17: 801, 1968.
11. The interested reader is referred to M.I. Aaronson and J.K. Lee, *Lectures on the Theory of Phase Transformations*, H.I. Aronson (Ed.), Met. Soc. AIME, New York, 1975.
12. A quite general method for calculating the stress and strain fields in and around misfitting inclusions has been elucidated by J.D. Eshelby, in *Proc. R. Soc. Lond.*, A241: 376, 1957, and in *Progress in Solid Mechanics*, Vol. II, Chapter III, North Holland, Amsterdam, 1961.
13. F.R.N. Nabarro, *Proc. R. Soc. A*, 175: 519, 1940.
14. J.K. Lee and W.C. Johnson, *Acta Metall.*, 26: 541–545, 1978.
15. R. Sankaran and C. Laird, *J. Mech. Phys. Solids*, 24: 251–262, 1976 and *Philos. Mag.*, 29: 179, 1974.
16. L.M. Brown and G.R. Woolhouse, *Philos. Mag.*, 21: 329, 1970.
17. R. Sankaran and C. Laird, *Philos. Mag.*, 21: 329, 1970.
18. R. Sankaran and C. Lard, *Philos. Mag.*, 29: 179, 1974.
19. A detailed treatment of solid/liquid interfaces can be found in D.P. Woodruff, *The Solid–Liquid Interface*, Cambridge University Press, Cambridge, 1973.
20. K.A. Jackson, *Liquid Metal and Solidification*, ASM, Cleveland, OH, 1958, p. 174.
21. W.A. Miller and G.A Chadwick, *Acta Metall.*, 15: 607, 1967.
22. W.A Miller and G.A. Chadwick, Proc. *R. Soc.*, A312: 251, 1969.
23. J.W. Christian, *The Theory of Transformations in Metal and Alloys*, 2nd edn., Part 1, Pergamon, Oxford, 1975, p. 9.
24. J.W. Christian, Phase transformations in metals and alloys—an introduction, in *Phase Transformations*, Vol. 1, Institute of Metallurgists, London, 1979, p. 1.
25. The interested reader is referred to E. Rutner et al. (Eds.), *Condensation and Evaporation of Solids*, Gordon and Breach, New York, 1964.
26. For an advanced discussion of this subject see M. Hillert, Diffusion and interface control of reactions in alloys, *Metall. Trans. A.*, 6A: 1975.
27. G.C. Weatherly, *Acta Metall.*, 19: 181, 1971.
28. J.W. Martin and R.D. Doherty, *Stability* of *Microstructure in Metallic Systems*, Cambridge University Press, Cambridge, 1976, p. 17.

Further Reading

J.M. Blakely, *Introduction to the Properties of Crystal Surface*, Pergamon Press, Oxford, 1973.

G.A. Chadwick and D.A. Smith (Eds.), *Grain Boundary Structure and Properties*, Academic Press, London, 1976.

H. Gleiter and B. Chalmers, High angle grain boundaries, in B. Chalmers, J.W. Christian, and T.B. Massalski (Eds.), *Progress in Materials Science*, Vol. 16, Pergamon, Oxford, 1972.

J.W. Martin and R.D. Doherty, *Stability of Microstructure in Metallic Systems,* Cambridge University Press, Cambridge, 1976.

L.E. Murr, *Interfacial Phenomena in Metals and Alloys,* Addison-Wesley, London, 1975.

D.P. Woodruff, *The Solid–Liquid Interface, Cambridge* University Press, Cambridge, 1973.

4

Solidification

Solidification and melting are transformations between crystallographic and noncrystallographic states of a metal or alloy. These transformations are of course basic to such technological applications as ingot casting, foundry casting, continuous casting, single-crystal growth for semiconductors, directionally solidified composite alloys, and more recently rapidly solidified alloys and glasses. Another important and complex solidification and melting process, often neglected in textbooks on solidification concerns the process of fusion welding. An understanding of the mechanism of solidification and how it is affected by such parameters as temperature distribution, cooling rate, and alloying is important in the control of mechanical properties of cast metals and fusion welds. It is the objective of this chapter to develop some of the basic concepts of solidification, and apply these to some of the more important practical processes such as ingot casting, continuous casting, and fusion welding. We then consider a few practical examples illustratig the casting or welding the engineering alloys in the light of the theoretical introduction.

4.1 Nucleation in Pure Metals

If a liquid is cooled below its equilibrium melting temperature (T_m), there is a driving force for solidification ($G_L - G_S$) and it might be expected that the liquid phase would spontaneously solidify. However, this is not always the case. For example, under suitable conditions liquid nicked can be undercooled (or supercooled) to 250 K below T_m (1453°C) and held there indefinitely without any transformation occurring. The reason for this behavior is that the transformation begins by the formation of very small solid particles or nuclei. Normally undercoolings as large as 250 K are not observed, since in practice the walls of the liquid container and solid impurity particles in the liquid catalyze the nucleation of solid at undercoolings of only ~1 K. This is known as heterogeneous nucleation. The large undercoolings mentioned above are only obtained when no heterogeneous nucleation sites are available, i.e., when solid must form homogeneously from the liquid. Experimentally this can be achieved by dividing the liquid into tiny droplets, many of which remain impurity-free and do not solidify until every large undercoolings are reached.[1]

4.1.1 Homogeneous Nucleation

Consider a given volume of liquid at a temperature ΔT below T_m with a free energy G_1 (Figure 4.1a). If some of the atoms of the liquid cluster together to form a small sphere of solid, Figure 4.1b, the free energy of the system will change to G_2 given by

$$G_2 = V_S G_v^S + V_L G_v^L + A_{SL}\gamma_{SL}$$

where

V_S is the volume of the solid sphere
V_L is the volume of the liquid
A_{SL} is the solid/liquid interfacial area
G_v^S and G_v^L are the free energies per unit volume of solid and liquid, respectively
γ_{SL} is the solid/liquid interfacial free energy

The free energy of the system without any solid present is given by

$$G_1 = (V_S + V_L)G_v^L$$

Here, the formation of solid results in a free energy change $\Delta G = G_2 - G_1$ where

$$\Delta G = -V_S \Delta G_v + A_{SL}\gamma_{SL} \tag{4.1}$$

and

$$\Delta G_v = G_v^L - G_v^S \tag{4.2}$$

For an undercooling ΔT, ΔG_v is given by Equation 1.17 as

$$\Delta G_v = \frac{L_v \Delta T}{T_m} \tag{4.3}$$

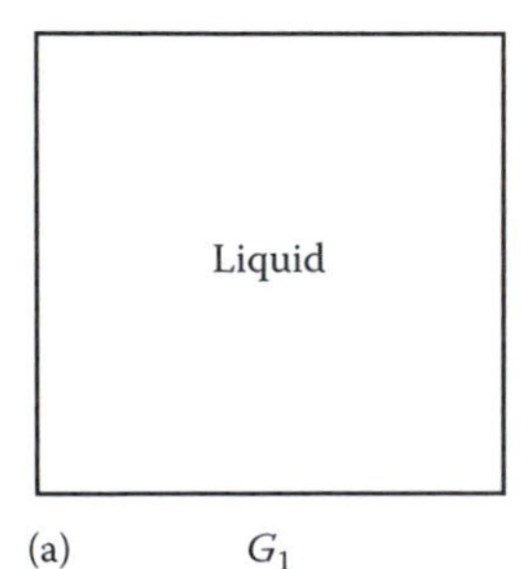

FIGURE 4.1
Homogeneous nucleation.

where L_v is the latent heat of fusion per unit volume. Below T_m, ΔG_v is positive so that the free energy change associated with the formation of a small volume of solid has a negative contribution due to the lower free energy of a bulk solid, but there is also a positive contribution due to the creation of a solid/liquid interface. The excess free energy associated with the solid particle can be minimized by the correct choice of particle shape. If γ_{SL} is isotropic this is a sphere of radius r. Equation 4.1 then becomes

$$\Delta G_r = -\frac{4}{3}\pi r^3 \Delta G_v + 4\pi r^2 \gamma_{SL} \tag{4.4}$$

This is illustrated in Figure 4.2. Since the interfacial term increases as r^2 whereas the volume free energy released only increase as r^3, the creation of small particles of solid always leads to a free energy increase. It is this increase that is able to maintain the liquid phase in a metastable state almost indefinitely at temperatures below T_m. It can be seen from Figure 4.2 that for a given undercooling there is a certain radius, r^*, which is associated with a maximum excess free energy. If $r < r^*$ the system can lower its free energy by dissolution of the solid, whereas when $r > r^*$ the free energy of the system decreases if the solid grows. Unstable solid particles with $r < r^*$ are known as clusters or embryos while stable particles with $r > r^*$ are referred to as nuclei—r^* is known as the critical nucleus size. Since $dG = 0$ when $r = r^*$ the critical nucleus is effectively in (unstable) equilibrium with the surrounding liquid.

It can easily be shown by differentiation of Equation 4.4 that

$$r^* = \frac{2\gamma_{SL}}{\Delta G_v} \tag{4.5}$$

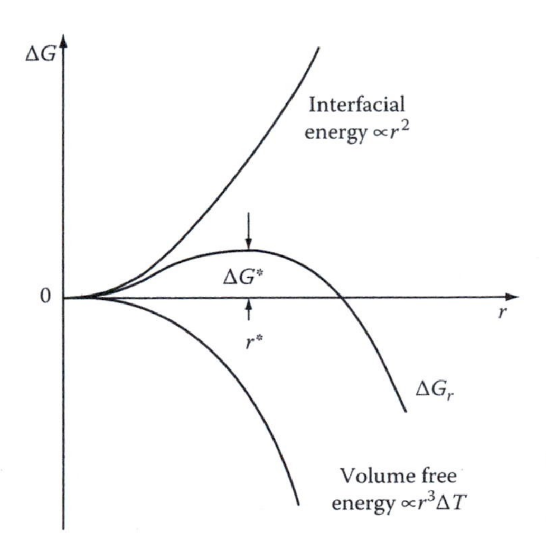

FIGURE 4.2
Free energy change associated with homogeneous nucleation of a sphere of radius r.

and

$$\Delta G^* = \frac{16\pi\gamma_{SL}^3}{3(\Delta G_v)^2} \tag{4.6}$$

Substituting Equation 4.3 for ΔG_v gives

$$r^* = \left(\frac{2\gamma_{SL}T_m}{L_v}\right)\frac{1}{\Delta T} \tag{4.7}$$

and

$$\Delta G^* = \left(\frac{16\pi\gamma_{SL}^3 T_m^2}{3L_v^2}\right)\frac{1}{(\Delta T)^2} \tag{4.8}$$

Note how r^* and ΔG^* decrease with increasing undercooling (ΔT).

Equation 4.5 could also have been obtained from the Gibbs–Thomson equilibrium with the surrounding liquid, the solidified sphere and liquid must then have the same free energy. From Equation 1.58 a solid sphere of radius r will have a free energy greater than that of bulk solid by $2\gamma V_m/r$ per mole or $2\gamma/r$ per unit volume. Therefore, it can be seen from Figure 4.3 that equality of free energy implies

$$\Delta G_v = 2\gamma_{SL}/r^* \tag{4.9}$$

which is identical to Equation 4.5.

To understand how it is possible for a stable solid nucleus to form homogeneously from the liquid, it is first necessary to examine the atomic structure of the liquid phase. From dilatometric measurements it is known that at the melting point the liquid phase has a volume 2%–4% greater than the solid. Therefore, there is a great deal more freedom of movement of atoms in the

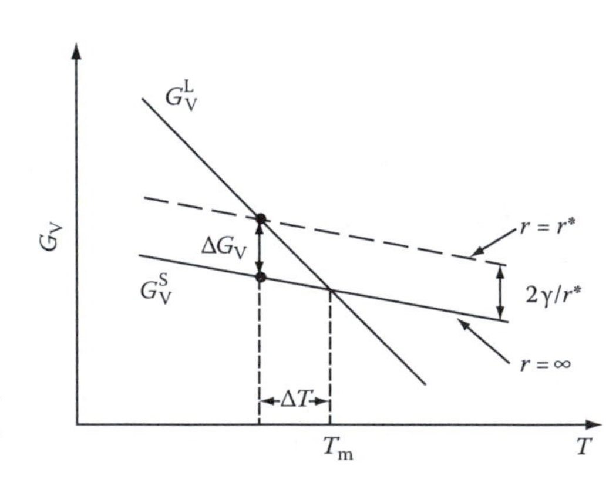

FIGURE 4.3
Volume free energy as a function of temperature for solid and liquid phases, showing the origin of ΔG_v and r^*.

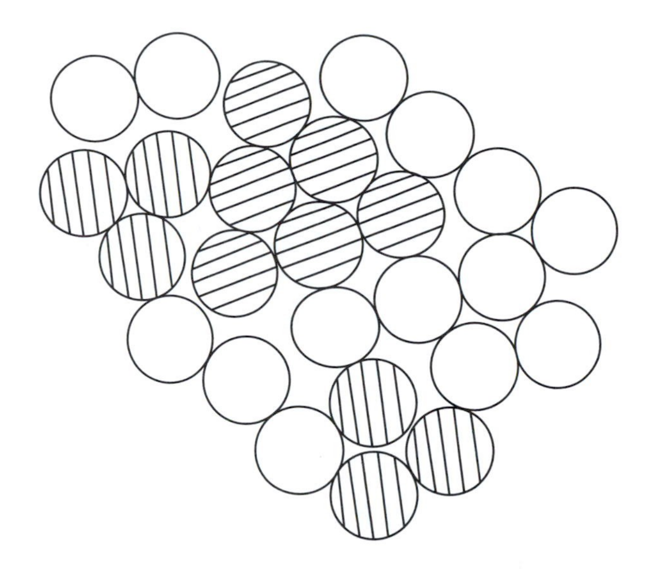

FIGURE 4.4
A two-dimensional representation of an instantaneous picture of the liquid structure. Many close-packed crystal-like clusters (shaded) are present.

liquid and when averaged over a period of time the atom positions appear completely random. However, an instantaneous picture of the liquid would reveal the presence of many small close-packed clusters of atoms which are temporarily in the same crystalline array as in the solid (Figure 4.4). On average the number of spherical clusters of radius r is given by

$$n_r = n_0 \exp\left(\frac{\Delta G_r}{kT}\right) \tag{4.10}$$

where

n_0 is the total number of atoms in the system
ΔG_r is the excess free energy associated with the cluster, Equation 4.4
k is Boltzmann's constant

For a liquid above T_m this relationship applies for all values of r. Below T_m it only applies for $r \leq r^*$ because clusters greater than the critical size are stable nuclei of solid and no longer part of the liquid. Since n_r decreases exponentially with ΔG_r (which itself increases rapidly with r) the probability of finding a given cluster decreases very rapidly as the cluster size increases. For example, by combining Equations 4.4 and 4.10 it can be shown (Exercise 4.2) that 1 mm^3 of copper at its melting point ($\sim 10^{20}$ atoms) should on average contain $\sim 10^{14}$ clusters of 0.3 nm radius (i.e., ~10 atoms) but only ~10 clusters with a radius of 0.6 nm (i.e., ~60 atoms). These numbers are of course only approximate. Such small clusters of atoms cannot be considered

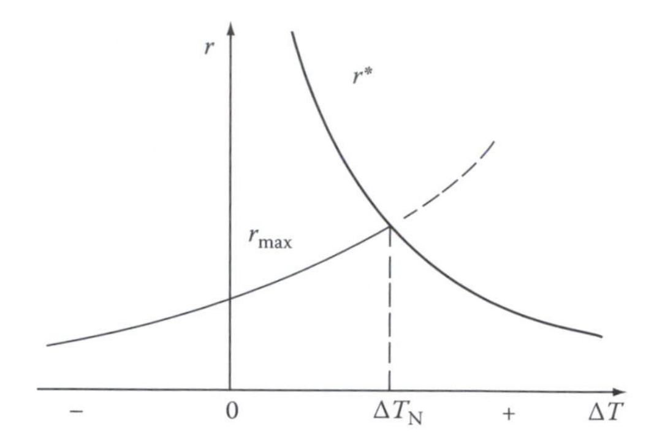

FIGURE 4.5
Variation of r^* and r_{max} with undercooling ΔT.

to be spherical, and even more important the effective value of γ used in calculating ΔG_r (Equation 4.4) is very probably a function of the cluster size. However, the above calculations do illustrate how sensitively cluster density depends on their size. Also, it can be seen that there is effectively a maximum cluster size, ~100 atoms, which has a reasonable probability of occurring in the liquid. The same sort of calculations can be made at temperatures other than T_m. Below T_m there is an increasing contribution from ΔG_v in Equation 4.4 as the solid becomes progressively more stable and this has the effect of increasing the maximum cluster size somewhat. Figure 4.5 shows schematically how r_{max} varies with ΔT. Of course larger cluster than r_{max} are possible in large enough systems or given sufficient time, but the probability of finding clusters only slightly larger than r_{max} is extremely small.

The critical nucleus size r^* is also shown in Figure 4.5. It can be seen that at small undercoolings, r^* is so large that there will be virtually no chance of forming a stable nucleus. But as ΔT increases r^* and ΔG^* decrease, and for supercoolings of ΔT_N or greater there is a very good chance of some clusters reaching r^* and growing into stable solid particles. In the small droplet experiment, therefore, homogeneous nucleation should occur when the liquid is undercooled by $\sim\Delta T_N$.

The same conclusion can also be reached by an energy approach. The creation of a critical nucleus can be considered to be a thermally activated process, i.e., a solid-like cluster must be able to cross the nucleation barrier ΔG^* before it becomes a stable nucleus. Since the probability of achieving this energy is proportional to exp $(-\Delta G^*/kT)$ nucleation will only become possible when ΔG^* is reduced below some critical value which can be shown to be $\sim 78kT$ (see below).

4.1.2 Homogeneous Nucleation Rate

Let us consider how fast solid nuclei will appear in the liquid at a given undercooling. If the liquid contains C_0 atoms per unit volume, the number of clusters that have reached the critical size (C^*) can be obtained from Equation 4.10 as

$$C^* = C_0 \exp\left(\frac{\Delta G_{\text{hom}}}{kT}\right) \text{clusters m}^{-3} \tag{4.11}$$

The addition of one more atom to each of these clusters will convert them into stable nuclei and, if this happens with a frequency f_0, the homogeneous nucleation rate will be given by

$$N_{\text{hom}} = f_0 C_0 \exp\left(-\frac{\Delta G^*_{\text{hom}}}{kT}\right) \text{nuclei m}^{-3}\ \text{s}^{-1} \tag{4.12}$$

where f_0 is a complex function that depends on the vibration frequency of the atoms, the activation energy for diffusion in the liquid, and the surface area of the critical nuclei. Its exact nature is not important here and it is sufficient to consider it a constant equal to $\sim 10^{11}$.* Since C_0 is typically $\sim 10^{29}$ atoms m^{-3} a reasonable nucleation rate ($1\ \text{cm}^{-3}\ \text{s}^{-1}$) is obtained when $\Delta G^* \sim 78kT$.

$$N_{\text{hour}} = f_0 C_0 \exp\left\{-\frac{A}{(\Delta T)^2}\right\} \tag{4.13}$$

where A is relatively insensitive to temperature and is given by

$$A = \frac{16\pi\gamma_{\text{SL}}^3 T_{\text{m}}^2}{3L_{\text{v}}^2 kT}$$

N_{hom} is plotted as a function of ΔT in Figure 4.6. As a result of the $(\Delta T)^2$ term, inside the exponential N_{hom} changes by orders of magnitude form essentially zero to very high values over a very narrow temperature range, i.e., there is effectively a critical undercooling for nucleation ΔT_{N}. This is the same as ΔT_{N} in Figure 4.5, but Figure 4.6 demonstrates more vividly how virtually no nuclei are formed until ΔT_{N} is reached after which there is an "explosion" of nuclei.

The small droplet experiments of Turnbull and Cech[1] have shown that ΔT_{N} is $\sim 0.2T_{\text{m}}$ for most metals (i.e., ~ 200 K). The measured values of ΔT_{N} have in fact been used along with Equation 4.13 to derive the values of interfacial free energy given in Table 3.4.

In practice homogeneous nucleation is rarely encountered in solidification. Instead heterogeneous nucleation occurs at crevices in mold walls, or at impurity particles suspended in the liquid.

---

* Since atomic jumps from the liquid on to the cluster are thermally activated, f_0 will in fact diminish with decreasing temperature. In some metallic systems the liquid can be rapidly cooled to temperatures below the so-called glass transition temperature without the formation of crystalline solid. F_0 is very small at these temperatures and the supercooled liquid is a relatively stable glass or amorphous metal. The variation of f_0 with temperature is very important with solid-state transformations, and it is covered in Chapter 5. For further details on alloys rapidly quenched from the melt see R.W. Chan and P. Haasen (Eds.), *Physical Metallurgy*, North-Holland, Amsterdam, 1983, Chapter 28.

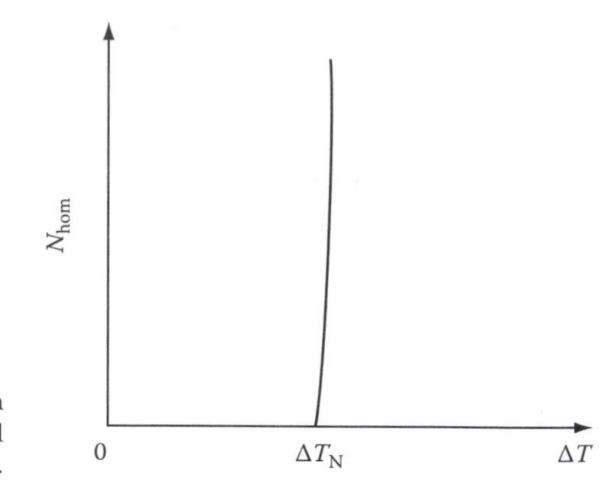

FIGURE 4.6
Homogeneous nucleation rate as a function of undercooling ΔT. ΔT_N is the critical undercooling for homogeneous nucleation.

4.1.3 Heterogeneous Nucleation

From the expression for ΔG^* (Equation 4.8) it can be seen that if nucleation is to be made easier at small undercoolings the interfacial energy term must be reduced. A simple way of effectively achieving this is if the nucleus forms in contact with the mold wall. Consider a solid embryo forming in contact with a perfectly flat mold wall as depicted in Figure 4.7. Assuming γ_{SL} is isotropic it can be shown that for a given volume of solid the total interfacial energy of the system is minimized if the embryo has the shape of a spherical cap with a "wetting" angle θ given by the condition that the interfacial tensions γ_{ML}, γ_{SM}, and γ_{SL} balance in the plane of the mold wall.

$$\gamma_{ML} = \gamma_{SM} + \gamma_{SL} \cos\theta$$

or

$$\cos\theta = (\gamma_{ML} - \gamma_{SM})/\gamma_{SL} \qquad (4.14)$$

Note that the vertical component of γ_{SL} remains unbalanced. Given time this force would pull the mold surface upward until the surface tension forces balance in all directions. Therefore, Equation 4.14 only gives the optimum embryo shape on the condition that the mold walls remain planar.

The formation of such an embryo will be associated with an excess free energy given by

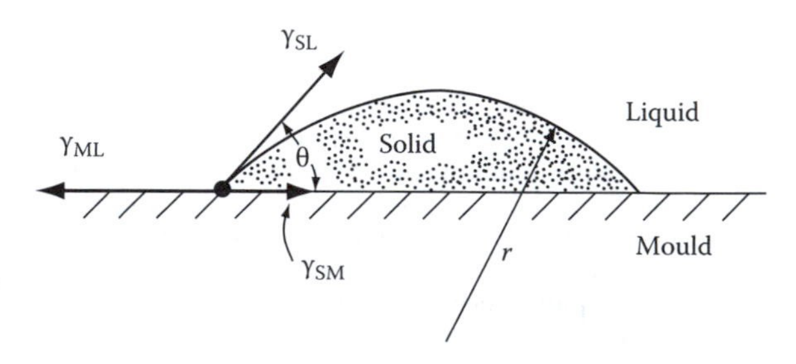

FIGURE 4.7
Heterogeneous nucleation of spherical cap on a flat mold wall.

$$\Delta G_{het} = -V_s \Delta G_v + A_{SL}\gamma_{SL} + A_{SM}\gamma_{SM} - A_{SM}\gamma_{ML} \tag{4.15}$$

where

V_S is the volume of the spherical cap

A_{SL} and A_{SM} are the areas of the solid/liquid and solid/mold interfaces

γ_{SL}, γ_{SM}, and γ_{ML} are the free energies of the solid/liquid, solid/mold, and mold/liquid interfaces, respectively

Note that there are now three interfacial energy contributions. The first two are positive as they arise from interfaces created during the nucleation process. The third, however, is due to the destruction of the mold/liquid interface under the spherical cap and results in a negative energy contribution.

It can be easily shown (see Exercise 4.6) that the above equation can be written in terms of the wetting angle (θ) and the cap radius (r) as

$$\Delta G_{het} = \left\{ -\frac{1}{3}\pi r^3 \Delta G_v + 4\pi r^2 \gamma_{SL} \right\} S(\theta) \tag{4.16}$$

where

$$S(\theta) = (2 + \cos\theta)(1 - \cos\theta)^2/4 \tag{4.17}$$

Note that except for factor $S(\theta)$ this expression is the same as that obtained for homogeneous nucleation, Equation 4.4. $S(\theta)$ has a numerical value ≤ 1 dependent only on θ, i.e., the shape of the nucleus. It is therefore referred to as a shape factor. ΔG_{het} is shown in Figure 4.8 along with ΔG_{hom} for comparison. By differentiation of Equation 4.16 it can be shown that

$$r^* = \frac{2\gamma_{SL}}{\Delta G_v} \tag{4.18}$$

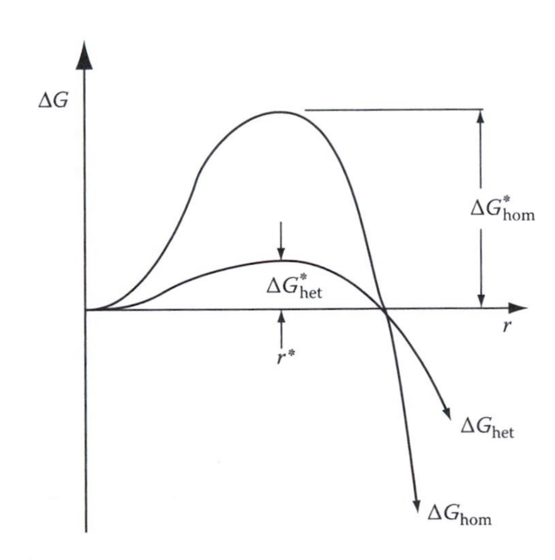

FIGURE 4.8
The excess free energy of solid clusters for homogeneous and heterogeneous nucleation. Note r^* is independent of the nucleation site.

and

$$\Delta G^* = \frac{16\pi\gamma_{SL}^3}{3\Delta G_v^2} \cdot S(\theta) \tag{4.19}$$

Therefore, the activation energy barrier against heterogeneous nucleation (ΔG^*_{het}) is smaller than ΔG^*_{hom} by the shape factor $S(\theta)$. In addition, the critical nucleus radius (r^*) is unaffected by the mold wall and only depends on the undercooling. This result was to be expected since equilibrium across the curved interface is unaffected by the presence of the mold wall.

Combining Equations 4.6 and 4.19 gives

$$\Delta G^*_{het} = S(\theta)\Delta G^*_{hom} \tag{4.20}$$

If for example $\theta = 10°$, $S(\theta) \sim 10^{-4}$, i.e., the energy barrier for heterogeneous nucleation can be very much smaller than for homogeneous nucleation. Significant reductions are also obtained for higher values of θ, e.g., when $\theta = 30°$, $S = 0.02$; even when $\theta = 90°$, $S = 0.5$. It should be noted that the above model breaks down for $\theta = 0$. In this case, the nucleus must be modeled in some other way, e.g., as shown in Figure 4.12.

The effect of undercooling on ΔG^*_{het} and ΔG^*_{hom} is shown schematically in Figure 4.9. If there are n_1 atoms in contact with the mold wall the number of nuclei should be given by

$$n^* = n_1 \exp\left(\frac{\Delta G^*_{het}}{kT}\right) \tag{4.21}$$

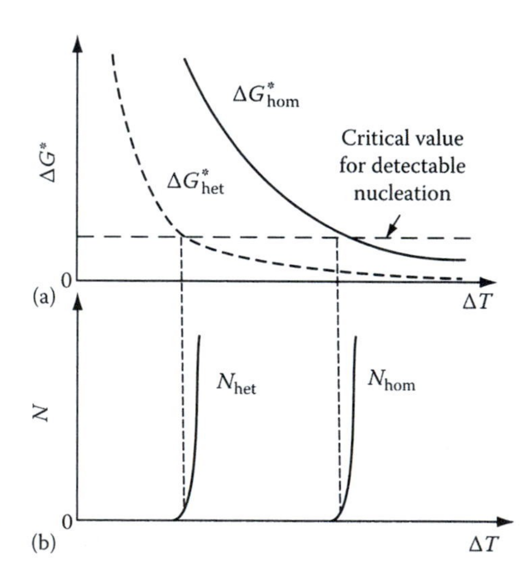

FIGURE 4.9
(a) Variation of ΔG^* with undercooling (ΔT) for homogeneous and heterogeneous nucleation. (b) The corresponding nucleation rates assuming the same critical value of ΔG^*.

Therefore, heterogeneous nucleation should become feasible when ΔG^*_{het} becomes sufficiently small. The critical value for ΔG^*_{het} should not be very different from the critical value for homogeneous nucleation. It will mainly depend on the magnitude of n_1 in the above equation. Assuming for the sake of simplicity that the critical value is again $\sim 78kT$ it can be seen from Figure 4.9 that heterogeneous nucleation will be possible at much lower undercoolings than are necessary for homogeneous nucleation.

To be more precise, the volume rate of heterogeneous nucleation ought to be given by an equation of the form

$$N_{het} = f_1 C_1 \exp\left(-\frac{\Delta G^*_{het}}{kT}\right) \tag{4.22}$$

where
- f_1 is a frequency factor similar to f_0 in Equation 4.12
- C_1 is the number of atoms in contact with heterogeneous nucleation sites per unit volume of liquid

So far it has been assumed that the mold wall is microscopically flat. In practice, however, it is likely to contain many microscopic cracks or crevices. It is possible to write down equations for the formation of a nucleus on such a surface but the result can be obtained more easily as follows. In both of the nucleation types considered so far it can be shown that

$$\Delta G^* = \frac{1}{2} V^* \, \Delta G_v \tag{4.23}$$

where V^* is the volume of the critical nucleus (sphere or cap). This equation, as well as Equation 4.7, is in fact quite generally true for any nucleation geometry. Thus, if a nucleus forms at the root of a crack the critical volume can be very small even if the wetting angle θ is quite large. Figure 4.10 shows an example where $\theta = 90°$. Therefore, nucleation from cracks or crevices should be able to occur at very small undercoolings even when the wetting angle θ is relatively large. Note, however, that for the crack to be effective the

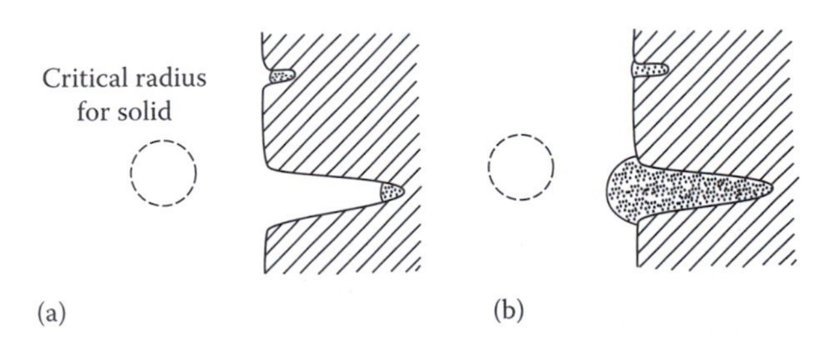

FIGURE 4.10
Heterogeneous nucleation in mold-wall cracks. (a) The critical nuclei. (b) The upper nucleus cannot grow out of the crack while the lower one can. (From Shewmon, P.G., *Transformations in Metals*, McGraw-Hill, New York, 1996. With permission.)

crack opening must be large enough to allow the solid to grow out without the radius of the solid/liquid interface decreasing below r^*.

In commercial practice heterogeneous nucleation is often enhanced by the addition of inoculants to the melt in order to refine the final grain size. The inoculating agent forms a solid compound with one of the components of the melt which then acts as a site for nucleation. According to the theory of heterogeneous nucleation outlined above the effectiveness of an inoculant should depend on the wetting angle and the surface roughness. Low values of θ are favored by a low-energy interface between the inoculant and solid nucleus, γ_{SM}, which should in turn be favored by good lattice matching between the particle and solid. However, lattice matching alone is unable to account for the effectiveness of nucleants. Other contributing factors include chemical effects, as well as surface segregation and roughness. It is thus difficult at present to predict the effectiveness of a given nucleant. In practice, the aim of inoculant additions is of course not to reduce undercooling but to achieve a fine grain size, and then other variables such as the concentration of nucleating particles also becomes important.

4.1.4 Nucleation of Melting

Although nucleation during solidification usually requires some undercooling, melting invariably occurs at the equilibrium melting temperature even at relatively high rates of heating. This is due to the relative free energies of the solid/vapor, solid/liquid, and liquid/vapor interfaces. It is always found that

$$\gamma_{SL} + \gamma_{LV} < \gamma_{SV} \tag{4.24}$$

Therefore, the wetting angle $\theta = 0$ and no superheating is required for nucleation of the liquid. In fact this interfacial energy relationship implies that a thin liquid layer should even be able to form below T_m (see Exercise 4.10). This phenomenon has not, however, been verified for metals as yet.

It is interesting to note that although T_m is a well-defined parameter in metallurgy, the actual atomic mechanism of melting is still not properly understood (for a good discussion of this phenomenon see, e.g., Cahn[2]). The solid → melt transformation in metals corresponds to an equivalent increase in vacancy concentration of as much as 10%, which is difficult to explain in the usual terms of defect structures. The melt, on this basis, might simply be considered to consist of an array of voids (condensed vacancies) surrounded by loose regions of disordered crystal (Frenkel's theory[3]). The sudden change from long-range crystallographic order to this loose, disordered structure may be associated with the creation of avalanches of dislocations which effectively break up the close-packed structure as melting occurs, as proposed by Cotterill et al.[4] on the basis of computer simulation experiments. There are, however, problems of quantifying this dislocation mechanism with dilatometric observations, and a more refined theory of melting is awaited.

4.2 Growth of a Pure Solid

It was shown in Section 3.4.6 that there are basically two different types of solid/liquid interface: an atomically rough or diffuse interface associated with metallic systems, and an atomically flat or sharply defined interface often associated with nonmetals. Because of the differences in atomic structure these two types of interface migrate in quite different ways. Rough interfaces migrate by a continuous growth process while flat interfaces migrate by a lateral growth process involving ledges.

4.2.1 Continuous Growth

The migration of a diffuse solid/liquid interface can be treated in a similar way to the migration of a random high-angle grain boundary. The free energy of an atom crossing the S/L interface will vary as shown in Figure 3.24 except one solid grain is replaced by the liquid phase. The activation energy barrier ΔG^a should be approximately the same as that for diffusion in the liquid phase and the driving force for solidification (ΔG) will then be given by

$$\Delta G = \frac{L}{T_m} \cdot \Delta T_i \tag{4.25}$$

where

L is the latent heat of melting

ΔT_i is the undercooling of the interface below the equilibrium melting temperature T_m

By analogy with Equation 3.21, therefore, the net rate of solidification should be given by an equation of the form

$$v = k_1 \Delta T_i \tag{4.26}$$

where k_1 has the properties of boundary mobility. A full theoretical treatment indicates that k_1 has such a high value that normal rates of solidification can be achieved with interfacial undercoolings (ΔT_i) of only a fraction of Kelvin. For most purposes, therefore, ΔT_i can be ignored and the solid/liquid interface is assumed to be at the equilibrium melting temperature. In other words, the solidification of metals is usually a diffusion-controlled process. For pure metals growth occurs at a rate controlled by heat conduction (diffusion) whereas alloy solidification is controlled by solute diffusion.

The above treatment is applicable to diffuse interface where it can be assumed that atoms can be received at any site on the solid surface, i.e., the accommodation factor A in Equation 3.22 is approximately unity. For this reason it is known as continuous growth. Such a mode of growth is reasonable because the interface is disordered and atoms arriving at random positions on the solid will not significantly disrupt the equilibrium configuration of the interface. The situation is, however, more complex when the equilibrium interface structure is atomically smooth as in the case of many nonmetals.

4.2.2 Lateral Growth

It will be recalled that materials with a high entropy of melting prefer to form atomically smooth, close-packed interfaces. For this type of interface the minimum free energy also corresponds to the minimum internal energy, i.e., a minimum number of broken "solid" bonds. If a single atom leaves the liquid and attaches itself to the flat solid surface (Figure 4.11a), it can be seen that the number of broken bonds associated with the interface, i.e., the interfacial energy, will be increased. There is, therefore, little probability of the atom remaining attached to the solid and it is likely to jump back into the liquid. In other words, atomically smooth interfaces have inherently low accommodation factors. However, if the interface contains ledges (Figure 4.11b), "liquid" atoms will be able to join the ledges with a much lower resulting increase in interfacial energy. If the ledge contains a jog, J, atoms from the liquid can join the solid without any increase in the number of broken bonds and the interfacial energy remains unchanged. Consequently the probability of an atom remaining attached to the solid at these positions is much greater than for an atom joining a facet. Smooth solid/liquid interfaces can, therefore, be expected to advance by the lateral growth of ledges similar to that described for coherent solid/solid interfaces in Section 3.5.1. Since the ledges and jogs are a nonequilibrium feature of the interface, growth will be very dependent on how the ledges and jogs can be supplied. It is thought that there are basically three different ways in which this can be achieved. These are (1) by repeated surface nucleation, (2) by spiral growth, and (3) from twin boundaries.

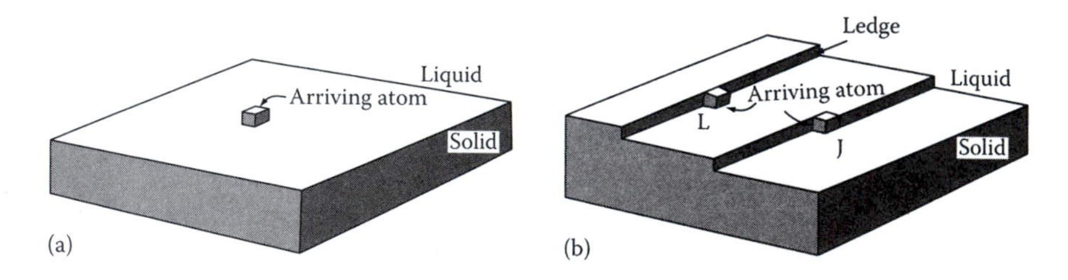

FIGURE 4.11
Atomically smooth solid/liquid interfaces with atoms represented by cubes. (a) Addition of a single atom onto a flat interface increases the number of broken bonds by four. (b) Addition to a ledge (L) only increases the number of broken bonds by two, whereas at a jog in a ledge (J) there is no increase.

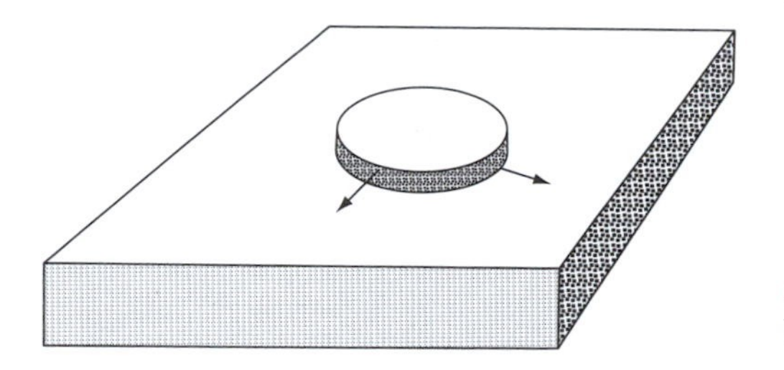

FIGURE 4.12
Ledge creation by surface nucleation.

Surface nucleation
It was pointed out above that a single atom "solidifying" on to a flat solid surface will be unstable and tend to rejoin the melt. However, if a sufficiently large number of atoms can come together to form a disk-shaped layer as shown in Figure 4.12, it is possible for the arrangement to become self-stabilized and continue to grow. The problem of disk creation is the two-dimensional analog of cluster formation during homogeneous nucleation. In this case, the edges of the disk contribute a positive energy which must be counterbalanced by the volume free energy released in the process. There will, therefore, be a critical radius (r^*) associated with the two-dimensional nucleus which will decrease with increasing interface undercooling. Once nucleated the disk will spread rapidly over the surface and the rate of growth normal to the interface will be governed by the surface nucleation rate. A full theoretical treatment shows that

$$v \propto \exp\left(-k_2/\Delta T_{\mathrm{i}}\right) \tag{4.27}$$

where k_2 is roughly a constant. This is shown schematically in Figure 4.14. Note that this mechanism is very ineffective at small undercoolings where r^* is very large.

Spiral growth
If the solid contains dislocations that intersect the S/L interface the problem of creating new interfacial steps can be circumvented.

Consider for simplicity the introduction of a screw dislocation into a block of perfect crystal. The effect will be to create a step or ledge in the surface of the crystal as shown in Figure 4.13a. The addition of atoms to the ledge will cause it to rotate about the point where the dislocation emerges, i.e., the ledge will never run out of the interface. If, on average, atoms add at an equal rate to all points along the step the angular velocity of the step will be initially greatest nearest to the dislocation core. Consequently as growth proceeds the ledge will develop into a growth spiral as shown in Figure 4.13b. The spiral tightens until it reaches a minimum radius of curvature r^* at which it is in equilibrium with the surrounding liquid and can decrease no more. Furthermore, the radius of curvature is less and the spiral can advance at a greater rate. Eventually a steady state is reached when the spiral appears to be rotating with a constant angular velocity. A complete theoretical treatment

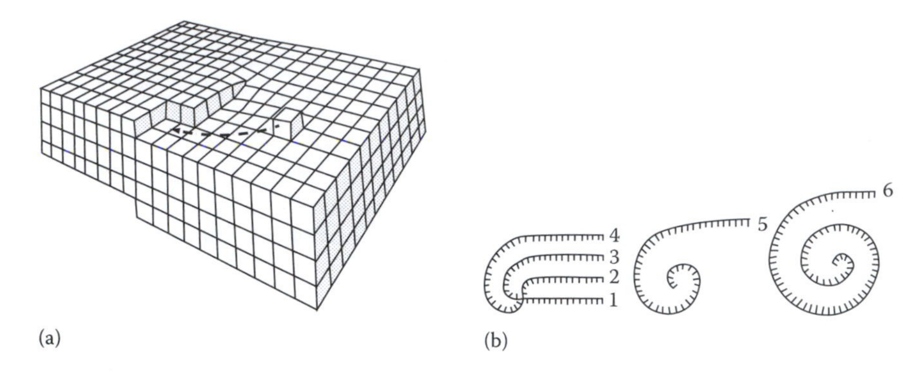

FIGURE 4.13
Spiral growth. (a) A screw dislocation terminating in the solid/liquid interface showing the associated ledge. (After Read, W.T., Jr., *Dislocations in Crystals*, McGraw-Hill, New York, 1953. With permission.) (b) Addition of atoms at the ledge causes it to rotate with an angular velocity decreasing away from the dislocation core so that a growth spiral develops. (After Christian, J.W., *The Theory of Phase Transformations in Metals and Alloys*, Pergamon Press, Oxford, 1965.)

of this situation shows that for spiral growth the normal growth rate v and the undercooling of the interface ΔT_i are related by an expression of the type

$$v = k_3(\Delta T_i)^2 \tag{4.28}$$

where k_3 is a materials constant. This variation is shown in Figure 4.14 along with the variations for continuous growth and two-dimensional nucleation. Note that for a given solid growth rate the necessary undercooling at the interface is least for the continuous growth of rough interfaces. For a given

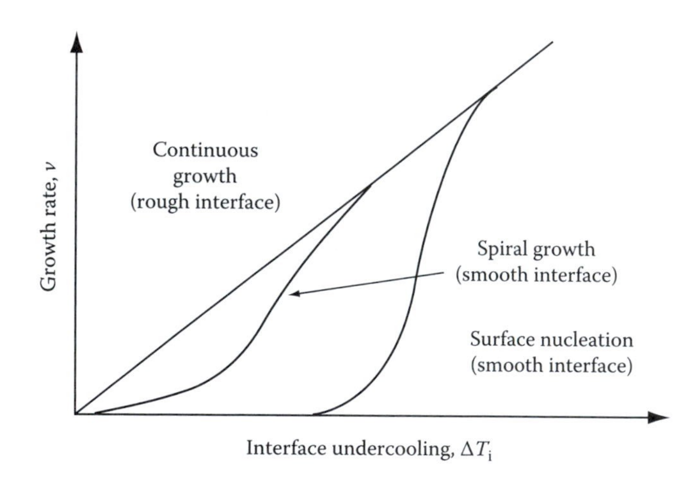

FIGURE 4.14
The influence of interface undercooling (ΔT_i) on growth rate for atomically rough and smooth interfaces.

undercooling, faceted interfaces are much less mobile and it is to be expected that the spiral growth mechanism will normally be more important than repeated nucleation.

Growth from twin intersections

Another permanent source of steps can arise where two crystals in different orientations are in contact. In solidification it is quite common for materials showing faceting to solidify as two crystals in twin orientations. Interfacial facets will therefore intersect at the twin boundary which can act as a permanent source of new steps thereby providing an easy growth mechanism similar to the growth spiral mechanism.

4.2.3 Heat Flow and Interface Stability

In pure metals solidification is controlled by the rate at which the latent heat of solidification can be conducted away from the solid/liquid interface. Conduction can take place either through the solid or the liquid depending on the temperature gradients at the interface. Consider for example solid growing at a velocity v with a planar interface into a superheated liquid (Figure 4.15a). The heat flow away from the interface through the solid must balance that from the liquid plus the latent heat generated at the interface, i.e.,

$$K_S T'_S = K_L T'_L + vL_v \tag{4.29}$$

where

K is the thermal conductivity

T' is the temperature gradient (dT/dx), the subscripts S and L stand for solid and liquid

v is the rate of growth of the solid

L_v is the latent heat of fusion per unit volume

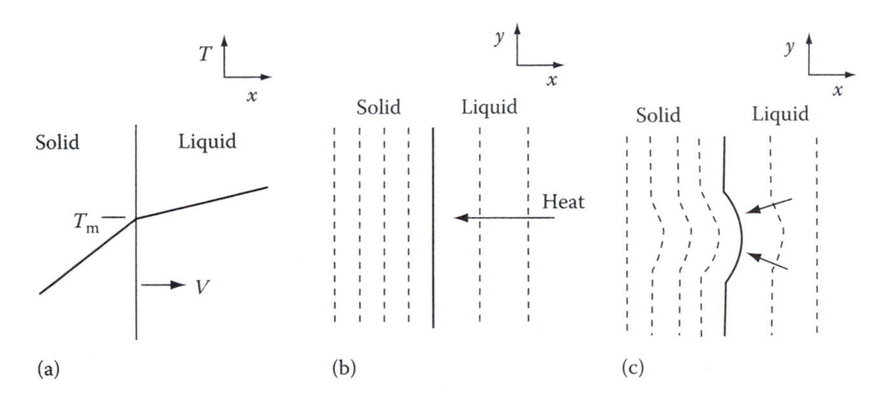

FIGURE 4.15
(a) Temperature distribution for solidification when heat is extracted through the solid. Isotherms (b) for a planar S/L interface and (c) for a protrusion.

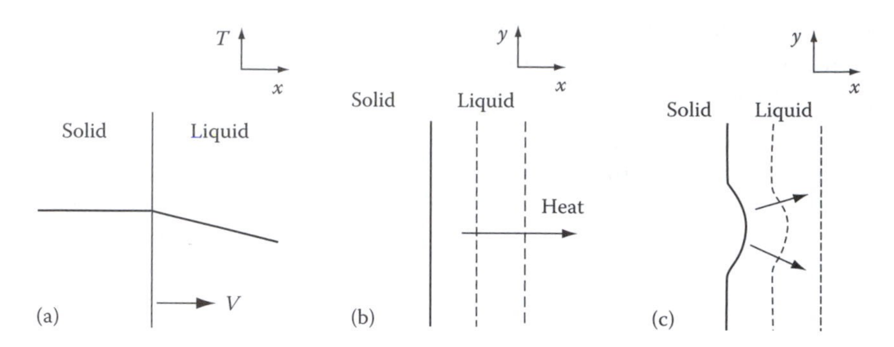

FIGURE 4.16
As Figure 4.15, but for heat conduction into the liquid.

This equation is quite general for a planar interface and even holds when heat is conducted into the liquid ($T'_1 < 0$) (Figure 4.16a).

When a solid grows into a superheated liquid, a planar solid/liquid interface is stable. This can be shown as follows. Suppose that as a result of a local increase in v a small protrusion forms at the interface (Figure 4.15c). If the radius of curvature of the protrusion is so large that the Gibbs–Thomson effect can be ignored the solid/liquid interface remains isothermal at essentially T_m. Therefore, the temperature gradient in the liquid ahead of the nodule will increase while that in the solid decreases. Consequently more heat will be conducted into the protruding solid and less away so that the growth rate will decrease below that of the planar regions and the protrusion will disappear.

The situation is, however, different for a solid growing into supercooled liquid (Figure 4.16). If a protrusion forms on the solid in this case the negative temperature gradient in the liquid becomes even more negative. Therefore heat is removed more effectively from the tip of the protrusion than from the surrounding regions allowing it to grow preferentially. A solid/liquid interface advancing into supercooled liquid is thus inherently unstable.

Heat conduction through the solid as depicted in Figure 4.15, arises when solidification takes place from mold walls which are cooler than the melt. Heat flow into the liquid, however, can only arise if the liquid is supercooled below T_m. Such a situation can arise at the beginning of solidification if nucleation occurs at impurity particles in the bulk of the liquid. Since a certain supercooling is required before nucleation can occur, the first solid particles will grow into supercooled liquid and the latent heat of solidification will be conducted away into the liquid. An originally spherical solid particle will therefore develop arms in many directions as shown in Figure 4.17. As the primary arms elongate their surfaces will also become unstable and break up into secondary and even tertiary arms. This shape of solid is known as a dendrite. Dendrite comes from the Greek for "tree." Dendrites in pure metals are usually called thermal dendrites to distinguish

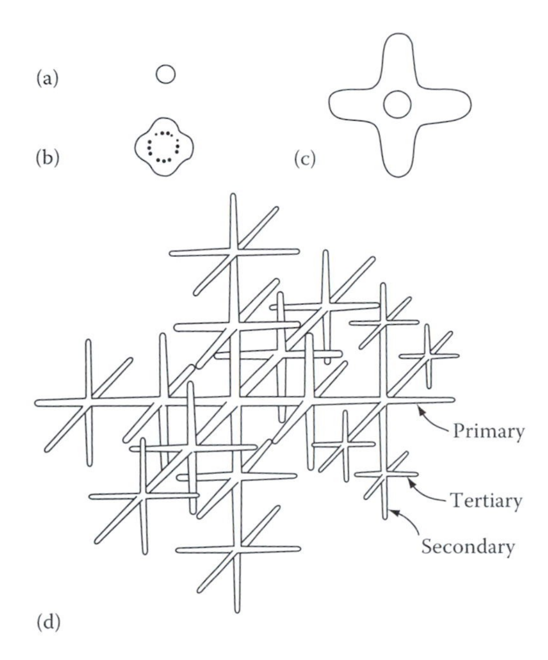

FIGURE 4.17
The development of thermal dendrites: (a) a spherical nucleus; (b) the interface becomes unstable; (c) primary arms develop in crystallographic directions (100) in cubic crystals; (d) secondary and tertiary arms develop. (After Reed-Hill, R.E., *Physical Metallurgy Principles*, 2nd edn., Van Nostrand, New York, 1973.)

them from dendrites in alloys (see below). It is found experimentally that the dendrite arms are always in certain crystallographic directions: e.g., <100> in cubic metals, and <1100> in hcp metals.[5]

Let us now take a closer look at the tip of a growing dendrite. The situation is different from that of a planar interface because that can be conducted away from the tip in three dimensions. If we assume that the solid is isothermal ($T'_S = 0$) the growth rate of the tip v will be given by a similar equation to Equation 4.29 provided T'_L is measured in the direction of v. A solution to the heat-flow equations for a hemispherical tip shows that the (negative) temperature gradient T'_L is approximately given by $\Delta T_c/r$ where ΔT_c is the difference between the interface temperature (T_i) and the temperature of the supercooled liquid far from the dendrite (T_∞) as shown in Figure 4.18. Equation 4.29 therefore gives

$$v = \frac{-K_L T'_L}{L_v} \simeq \frac{K_L}{L_v} \cdot \frac{\Delta T}{r} \tag{4.30}$$

Thus for a given ΔT, rapid growth will be favored by small values of r due to the increasing effectiveness of heat conduction as r diminishes. However ΔT is not independent of r. As a result of the Gibbs–Thomson effect equilibrium across a curved interface occurs at an undercooling ΔT_r below T_m given by

$$\Delta T_r = \frac{2\gamma T_m}{L_v r}$$

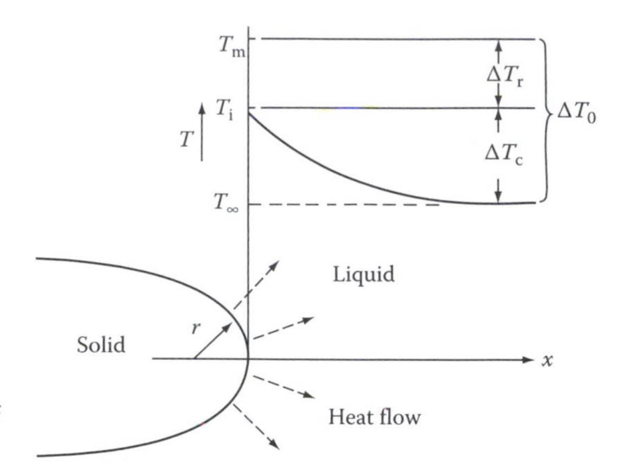

FIGURE 4.18
Temperature distribution at the tip of a growing thermal dendrite.

The minimum possible radius of curvature of the tip is when ΔT_r equals the total undercooling $\Delta T_0 = T_m - T_\infty$. This is just the critical nucleus radius r^* given by $(2\gamma T_m / L_v \Delta T_0)$. Therefore in general ΔT_r is given by $\Delta T_0 r^*/r$.

Finally since $\Delta T_0 = \Delta T_c + \Delta T_r$ Equation 4.30 becomes

$$v \simeq \frac{K_L}{L_v} \cdot \frac{1}{r}\left(1 - \frac{r^*}{r}\right) \tag{4.31}$$

It can thus be seen that the tip velocity tends to zero as $r \to r^*$ due to the Gibbs–Thomson effect and as $r \to \infty$ due to slower heat conduction. The maximum velocity is obtained when $r = 2r^*$.

4.3 Alloy Solidification

The solidification of pure metals is rarely encountered in practice. Even commercially pure metals contain sufficient impurities to change the characteristics of solidification from pure metal to alloy behavior. We now develop the theory a step further and examine the solidification of single-phase binary alloys. Following this we then consider the solidification of eutectic and peritectic alloys.

4.3.1 Solidification of Single-Phase Alloys

The alloys of interest in this section are those such as X_0 in Figure 4.19. This phase diagram has been idealized by assuming that the solidus and liquidus are straight lines. It is useful to define a partition coefficient k as

$$k = \frac{X_S}{X_L} \tag{4.32}$$

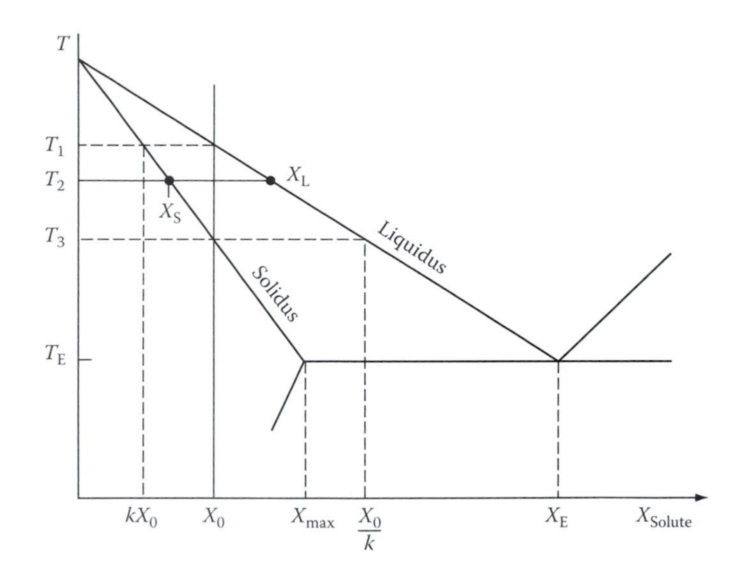

FIGURE 4.19
A hypothetical phase diagram. $k = X_S/X_L$ is constant.

where X_S and X_L are the mole fractions of solute in the solid and liquid in equilibrium at a given temperature. For the simple case shown in Figure 4.19, k is independent of temperature.

The way in which such alloys solidify in practice depends in rather a complex way on temperature gradients, cooling rates, and growth rates. Therefore, let us simplify matters by considering the movement of a planar solid/liquid interface along a bar of alloy as shown in Figure 4.20a. Such unidirectional solidification can be achieved in practice by passing the alloy in a crucible through a steep temperature gradient in a specially constructed furnace in which heat is confined to flow along the axis of the bar.

Let us examine three limiting cases:

1. Infinitely slow (equilibrium) solidification
2. Solidification with no diffusion in the solid but perfect mixing in the liquid
3. Solidification with no diffusion in the solid and only diffusional mixing in the liquid

Equilibrium solidification

Alloy X_0 in Figure 4.19 begins to solidify at T_1 with the formation of a small amount of solid with composition kX_0. As the temperature is lowered more solid forms and, provided cooling is slow enough to allow extensive solid-state diffusion, the solid and liquid will always be homogeneous with compositions following the solidus and liquidus lines (Figure 4.20b). The relative

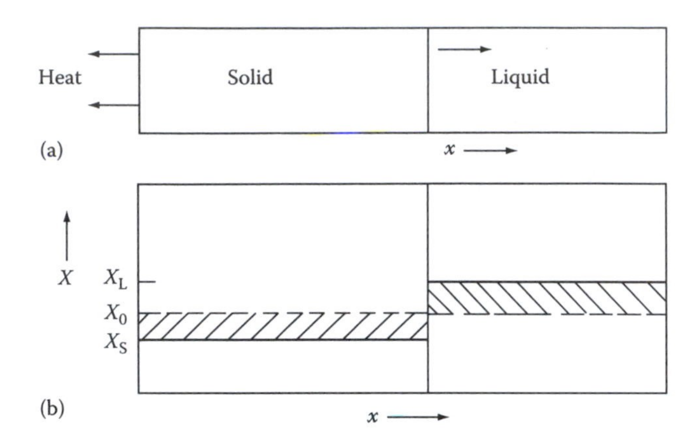

FIGURE 4.20
Unidirectional solidification of alloy X_0 in Figure 4.19. (a) A planar S/L interface and axial heat flow. (b) Corresponding composition profile at T_2 assuming complete equilibrium. Conservation of solute requires the two shaded areas to be equal.

amounts of solid and liquid at any temperature are simply given by the lever rule. Note that, since solidification is one-dimensional, conversation of solute requires the two shaded areas in Figure 4.20b to be equal (ignoring the differences in molar volume between the two phases). At T_3 the last drop of liquid will have a composition X_0/k and the bar of solid will have a composition X_0 along its entire length.

No diffusion in solid, perfect mixing in liquid
Very often the rate of cooling will be too rapid to allow substantial diffusion in the solid phase. Therefore, let us assume no diffusion takes place in the solid but that the liquid composition is kept homogeneous during solidification by efficient stirring. Again, assuming unidirectional solidification, the first solid will appear when the cooled end of the bar reaches T_1 in Figure 4.21a, at which stage solid containing kX_0 mole of solute forms. Since $kX_0 < X_0$, this first solid will be purer than the liquid from which it forms so that solute is rejected into the liquid and raises its concentration above X_0 (Figure 4.21b). The temperature of the interface must therefore decrease below T_1 before further solidification can occur, and the next layer of solid will be slightly richer in solute than the first. As this sequence of events continues the liquid becomes progressively richer in solute and solidification takes place at progressively lower temperatures (Figure 4.21c). At any stage during solidification local equilibrium can be assumed to exist at the solid/liquid interface, i.e., for a given interface temperature the compositions of the solid and liquid in contact with one another will be given by the equilibrium phase diagram. However, since there is no diffusion in the solid, the separate layers of solid retain their original compositions. Thus, the mean composition of the solid ($\overline{X}_S$) is always lower than the composition at the

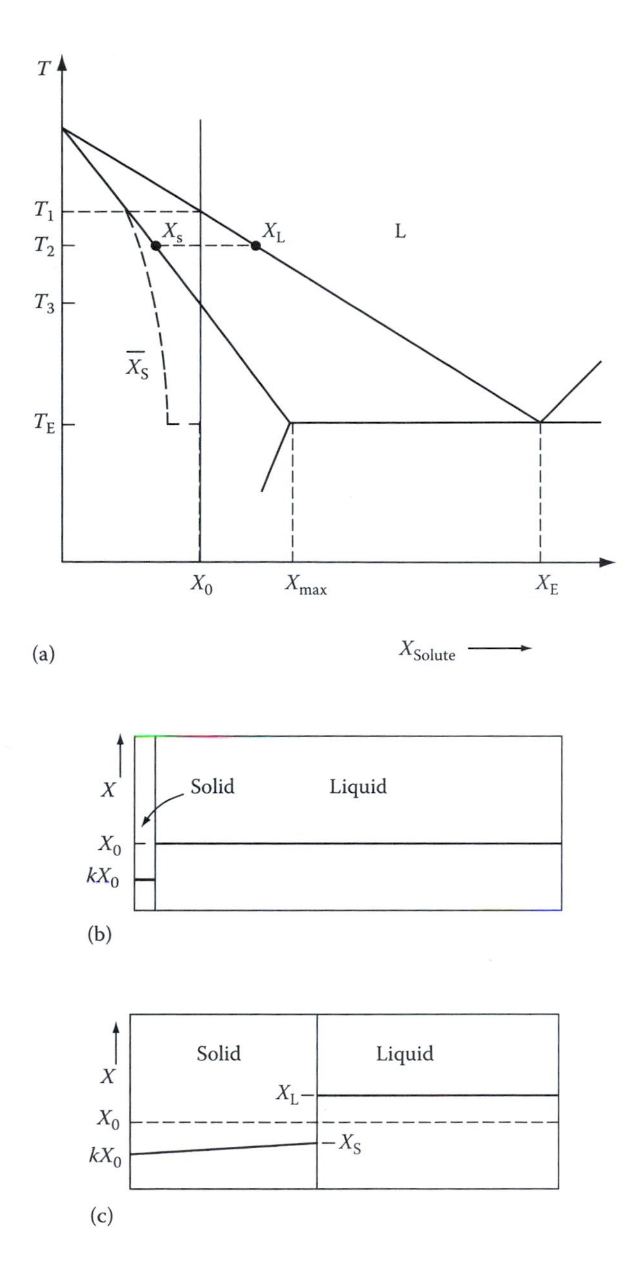

FIGURE 4.21
Planar-front solidification of alloy X_0 in Figure 4.19 assuming no diffusion in the solid, but complete mixing in the liquid. (a) As Figure 4.19, but including the mean composition of the solid. (b) Composition profile just under T_1. (c) Composition profile at T_2 (compare with the profile and fraction solidified in Figure 4.20b).

(continued)

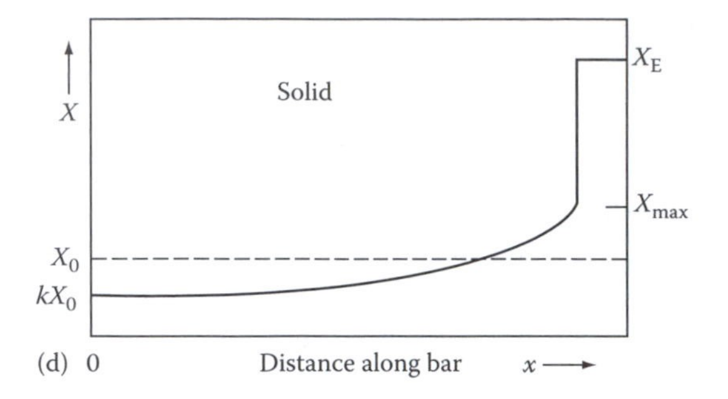

FIGURE 4.21 (continued)
(d) Composition profile at the eutectic temperature and below.

solid/liquid interface, as shown by the dashed line in Figure 4.21a. The relative amounts of solid and liquid for a given interface temperature are thus given by the lever rule using $\overline{X}_S$ and X_L. It follows that the liquid can become much richer in solute than X_0/k and it may even reach a eutectic composition, X_E, for example. Solidification will thus tend to terminate close to T_E with the formation of a eutectic structure of $\alpha + \beta$. The completely solidified bar will then have a solute distribution as shown in Figure 4.21d with $\overline{X}_S = X_0$.

The variation of X_S along the solidified bar can be obtained by equating the solute rejected into the liquid when a small amount of solid forms with the resulting solute increase in the liquid. Ignoring the difference in molar volume between the solid and liquid gives

$$(X_L - X_S)df_S = (1 - f_S)dX_L$$

where f_S is the volume fraction solidified. Integrating this equation using the boundary condition $X_S = kX_0$ when $f_S = 0$ gives

$$X_S = kX_0(1 - f_S)^{(k-1)}$$

and

$$X_L = X_0 f_L^{(k-1)} \tag{4.33}$$

These equations are known as the nonequilibrium lever rule or the Scheil equations.

Note that for $k < 1$ these equations predict that when there is no diffusion in the solid there will always be some eutectic in the last drop of liquid to solidify, no matter how little solute is present. Also the equation is quite generally applicable even for nonplanar solid/liquid interfaces provided the liquid composition is uniform and that the Gibbs–Thomson effect is negligible.

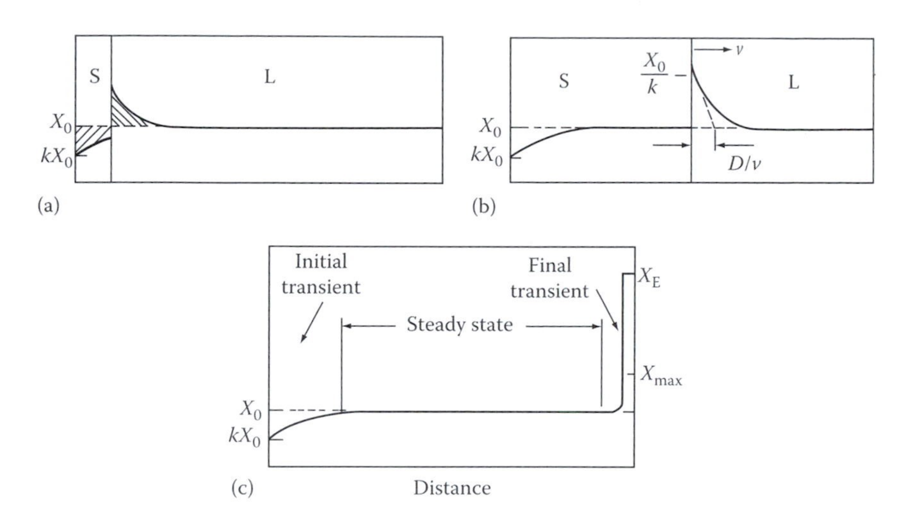

FIGURE 4.22
Planar-front solidification of alloy X_0 in Figure 4.19 assuming no diffusion in the solid and no stirring in the liquid. (a) Composition profile when S/L interface temperature is between T_2 and T_3 in Figure 4.19. (b) Steady-state solidification at T_3. The composition solidifying equals the composition of the liquid far ahead of the solid (X_0). (c) Composition profile at T_E and below, showing the final transient.

No diffusion in solid, diffusional mixing in liquid
If there is no stirring or convection in the liquid phase the solute rejected from the solid will only be transported away by diffusion. Hence there will be a rapid build up of solute ahead of the solid and a correspondingly rapid increase in the composition of the solid formed (Figure 4.22a). This is known as the initial transient. If solidification is made to occur at a constant rate, v, it can be shown that a steady state is finally obtained when the interface temperature reaches T_3 in Figure 4.19.[6] The liquid adjacent to the solid then has a composition X_0/k and the solid forms with the bulk composition X_0.

During steady-state growth the concentration profile in the liquid must be such that the rate at which solute diffuses down the concentration gradient away from the interface is balanced by the rate at which solute is rejected from the solidifying liquid, i.e.,

$$-DC'_L = v(C_L - C_S) \tag{4.34}$$

where
- D is the diffusivity in the liquid
- C'_L stands for dC_L/dx at the interface
- C_L and C_S are the solute concentrations of the liquid and solid in equilibrium at the interface (m^{-3})

Note the similarity of this equation to that describing the rate at which solidification occurs in pure metals (Equation 4.29).

If the diffusion equation is solved for steady-state solidification it can be shown that the concentration profile in the liquid is given by

$$X_L = X_0\left\{1 + \frac{1-k}{k}\exp\left[-\frac{x}{(D/v)}\right]\right\} \tag{4.35}$$

i.e., X_L decreases exponentially from X_0/k at $x = 0$, the interface, to X_0 at large distances from the interface. The concentration profile has a characteristic width of D/v.

When the solid/liquid interface is within $\sim D/v$ of the end of the bar the bow-wave of solute is compressed into a very small volume and the interface composition rises rapidly leading to a final transient and eutectic formation (Figure 4.22c).

In practice, alloy solidification will usually possess features from all three of the cases discussed above. There will usually be some stirring either due to liquid turbulence caused by pouring, or because of convection currents, or gravity effects. However, stirring will not usually be sufficiently effective to prevent the formation of a boundary layer and some liquid diffusion will therefore be involved. Partial stirring does, however, have the effect of reducing the boundary layer thickness. The concentration profiles found in practice may thus exhibit features between those shown in Figures 4.21d and 4.22c. In many cases diffusion in the solid must also be taken into account, e.g., when interstitial atoms or bcc metals are involved. In this case, solute can diffuse away from the solidifying interface back into the solid as well as into the liquid, with the result that after solidification the alloy is more homogeneous.

Even when solidification is not unidirectional the above ideas can still often be applied at a microscopic level as will be discussed below. Unidirectional solidification has commercial importance in, for example, the production of creep resistant aligned microstructures for gas turbine blades. It is also used in the production of extremely pure metals (zone refining).[7]

Cellular and dendritic solidification

So far we have considered solidification in which the growth front is planar. However, the diffusion of solute into the liquid during solidification of an alloy is analogous to the conduction of latent heat into the liquid during the solidification of a pure metal. At first sight therefore it would seem that the planar front should break up into dendrites. The problem is complicated, however, by the possibility of temperature gradients in the liquid.

Consider steady-state solidification at a planar interface as shown in Figure 4.23. As a result of the varying solute concentration ahead of the solidification front there is a corresponding variation of the equilibrium solidification temperature, i.e., the liquidus temperature, as given by the line T_e in Figure 4.23b. However, apart from the temperature of the interface,

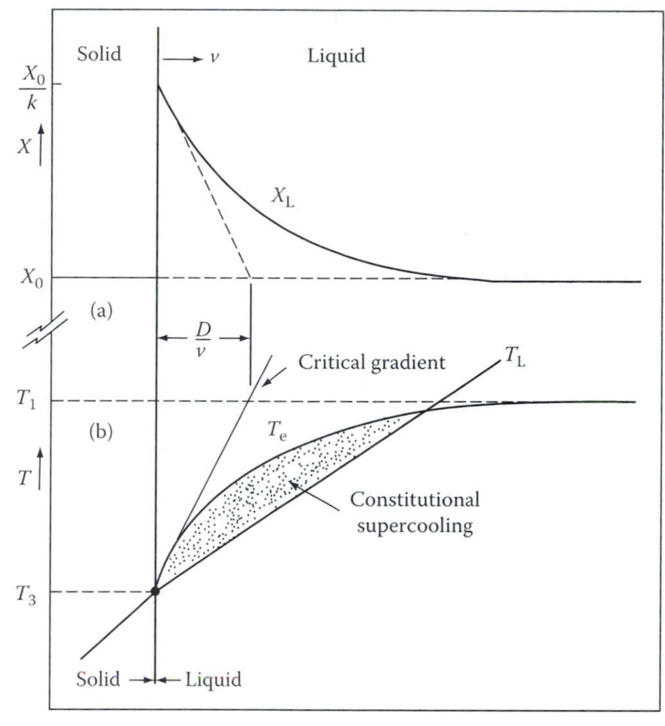

FIGURE 4.23
The origin of constitutional supercooling ahead of a planar solidification front. (a) Composition profile across the solid/liquid interface during steady-state solidification. The dashed line shows dX_L/dx at the S/L interface. (b) The temperature of the liquid ahead of the solidification front follows line T_L. The equilibrium liquidus temperature for the liquid adjacent to the interface varies as T_e. Constitutional supercooling arises when T_L lies under the critical gradient.

which is fixed by local equilibrium requirements, the actual temperature of the liquid can follow any line such as T_L. At the interface $T_L = T_e = T_3$ (defined in Figure 4.19). If the temperature gradient is less than the critical value shown in Figure 4.23b the liquid front of the solidification front exists below its equilibrium freezing temperature, i.e., it is supercooled. Since the supercooling arises from compositional or constitutional effects it is known as constitutional supercooling.

A necessary condition for the formation of stable protrusions on a planar interface is that there must exist a region of constitutional supercooling in the liquid. Assuming the T_L variation in Figure 4.23b the temperature at the tip of any protrusion that forms will be higher than that of the surrounding interface. (In contrast to pure metals the interface in alloys need not be isothermal.) However, provided the tip remains below the local liquidus temperature (T_c) solidification is still possible and the protrusion can develop. On the other hand, if the temperature gradient ahead of the interface is steeper than the critical gradient in Figure 4.23b the tip will be raised above the liquidus temperature and the protrusion will melt back.

Under steady-state growth the critical gradient can be seen from Figure 4.23 to be given by $(T_1 - T_3)/(D/v)$ where T_1 and T_3 are the liquidus and solidus temperatures for the bulk composition X_0 (Figure 4.19). The condition for a stable planar interface is therefore

$$T'_L > \frac{(T_1 - T_3)}{(D/v)}$$

where T'_L stands for (dT_L/dx) at the interface. Or, regrouping the experimentally adjustable parameters T'_L and v, the condition for no constitutional supercooling is

$$(T'_L/v) > (T_1 - T_3)/D \tag{4.36}$$

$(T_1 - T_3)$ is known as the equilibrium freezing range of the alloy. Clearly planar front solidification is most difficult for alloys with a large solidification range and high rates of solidification. Except under well-controlled experimental conditions alloys rarely solidify with planar solid/liquid interfaces. Normally the temperature gradients and growth rates are not individually controllable but are determined by the rate at which heat is conducted away from the solidifying alloy.

If the temperature gradient ahead of an initially planar interface is gradually reduced below the critical value the first stage in the breakdown of the interface is the formation of cellular structure (Figure 4.24). The formation of the first protrusion causes solute to be rejected laterally and pileup at the root of the protrusion (Figure 4.24b). This lowers the equilibrium solidification temperature causing recesses to form (Figure 4.24c), which in turn trigger the formation of other protrusions (Figure 4.24d). Eventually, the protrusions

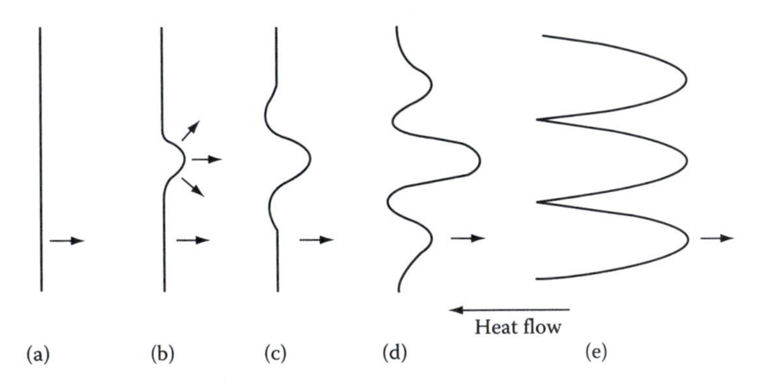

FIGURE 4.24
The breakdown of an initially planar solidification front into cells.

develop into long arms or cells growing parallel to the direction of heat flow (Figure 4.24e). The solute rejected from the solidifying liquid concentrates into the cell walls which solidify at the lowest temperatures. The tips of the cells, however, grow into the hottest liquid and therefore contain the least solute. Even if $X_0 \ll X_{max}$ (Figure 4.19) the liquid between the cells may reach the eutectic composition in which case the cell walls will contain a second phase. The interaction between temperature gradient, cell shape, and solute segregation is shown in Figure 4.25. Figure 4.26 shows the appearance of the cellular structure. Note that each cell has virtually the same orientation as its neighbors and together they form a single grain.

Cellular microstructures are only stable for a certain range of temperature gradients. At sufficiently low temperature gradients the cells, or primary

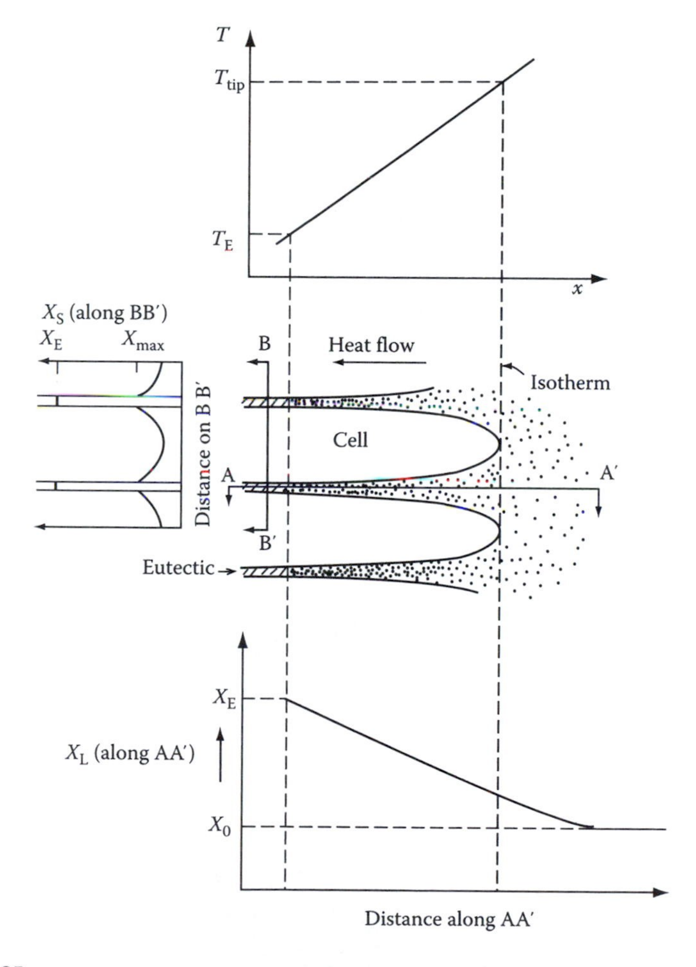

FIGURE 4.25
Temperature and solute distributions associated with cellular solidification. Note that solute enrichment in the liquid between the cells, and coring in the cells with eutectic in the cell walls.

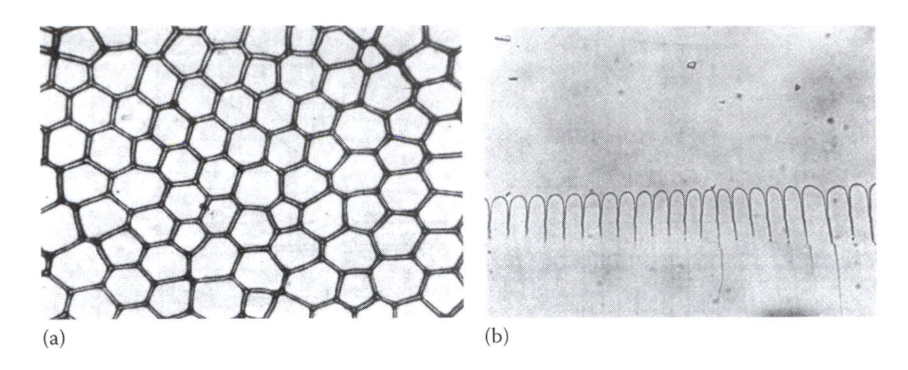

(a) (b)

FIGURE 4.26
Cellular microstructures. (a) A decanted interface of a cellularly solidified Pb–Sn alloy (×120). (After Rutter, J.W. in *Liquid Metals and Solidification*, American Society for Metals, Cleveland, OH 1958, p. 243.) (b) Longitudinal view of cells in carbon tetrabromide (×100). (After Jackson, K.A. and Hunt, J.D., *Acta Metall.*, 13 1212, 1965.)

arms of solid, are observed to develop secondary arms, and at still lower temperature gradients tertiary arms develop, i.e., dendrites form. Concomitant with this change in morphology there is a change in the direction of the primary arms away from the direction of heat flow into the crystallographically preferred directions such as ⟨100⟩ for cubic metals. The change in morphology from cells to dendrites can be seen in Figures 4.26b through 4.28. These pictures have been taken during in situ solidification of special

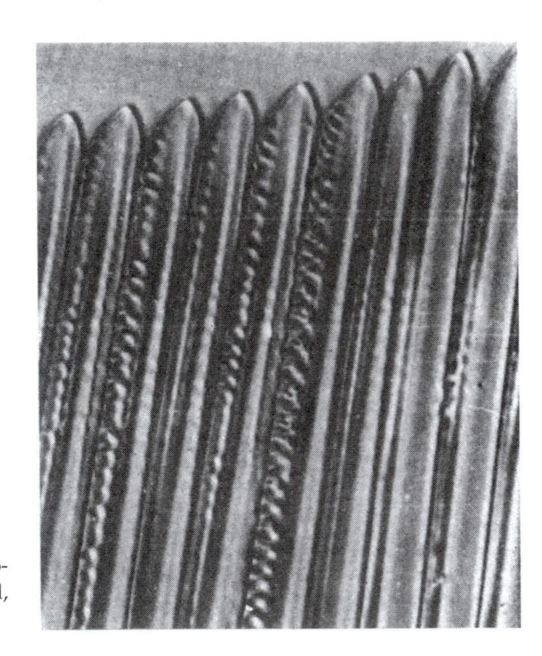

FIGURE 4.27
Cellular dendrites in carbon tetrabromide. (After Morris, L.R. and Winegard, W.C., *J. Cryst. Growth*, 6, 61, 1969.)

FIGURE 4.28
Columnar dendrites in a transparent organic alloy. (After Jackson, K.A. in *Solidification*, American Society for Metals, Cleveland, OH, 1971, p. 121.)

transparent organic compounds using a transmission light microscope.[8] The compounds used have low entropies of melting and solidify in the same way as metals. Alloys have been simulated by suitable combinations of such compounds.

In general, the tendency to form dendrites increases as the solidification range increases. Therefore, the effectiveness of different solutes can vary widely. For solutes with a very small partition coefficient (k) cellular or dendritic growth can be caused by the addition of a very small fraction of a percent solute.

The reason for the change from cells to dendrites is not fully understood. However, it is probably associated with the creation of constitutional supercooling in the liquid between the cells causing interface instabilities in the transverse direction. Note that for unidirectional solidification there is approximately no temperature gradient perpendicular to the growth direction. The cell or dendrite arm spacing developing is probably that which reduces the constitutional supercooling in the intervening liquid to a very low level. This would be consistent with the observation that cell and dendrite arm spacings both decrease with increasing cooling rate: higher cooling rates allow less time for lateral diffusion of the rejected solute and therefore require smaller cell or dendrite arm spacings to avoid constitutional supercooling.

Finally, it should be noted that although the discussion of alloy solidification has been limited to the case $k < 1$, similar arguments can be advanced for the case of $k > 1$ (see Exercise 4.13).

4.3.2 Eutectic Solidification[9]

In the solidification of a binary eutectic composition two solid phases form cooperatively from the liquid, i.e., L $\rightarrow \alpha + \beta$. Various different types of eutectic solidification are possible and these are usually classified as normal and anomalous. In normal structures the two phases appear either as alternate lamellae (Figure 4.29), or as rods of minor phase embedded in the other phase (Figure 4.30). During solidification both phases grow simultaneously behind an essentially planar solid/liquid interface. Normal structures occur when both phases have low entropies of fusion. Anomalous structures, on the other hand, occur in systems when one of the solid phases is capable of faceting, i.e., has a high entropy or melting. There are many variants of these structures the most important commercially being the flake structure of Al–Si alloys, This section will only be concerned with normal structures, and deal mainly with lamellar morphologies.

Growth of lamellar eutectics

Figure 4.31 shows how two phases can grow cooperatively behind an essentially planar solidification front. As the A-rich α-phase solidifies excess B diffuses a short distance laterally where it is incorporated in the B-rich β-phase. Similarly, the A atoms rejected ahead of the β diffuse to the tips of the adjacent α lamellae. The rate at which the eutectic grows will depend on how fast this diffusion can occur and this in turn will depend on the interlamellar spacing λ. Thus small interlamellar spacings should lead to rapid growth.

However, there is lower limit to λ determined by the need to supply the α/β interfacial energy, $\gamma_{\alpha\beta}$.

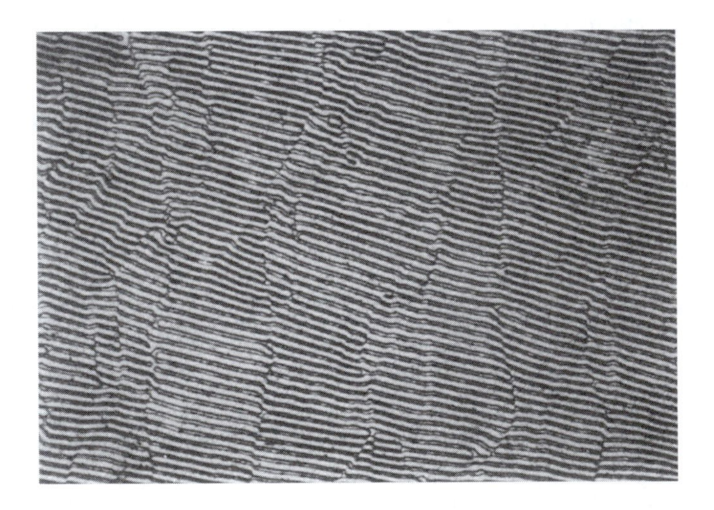

FIGURE 4.29
Al–Cu Al_2 lamellar eutectic normal to the growth direction ($\times$680). (Courtesy of J. Strid, University of Luleå, Sweden.)

FIGURE 4.30
Rodlike eutectic. Al_6Fe rods in Al matrix. Transverse section. Transmission electron micrograph (×70,000). (Courtesy of J. Strid, University of Luleå, Sweden.)

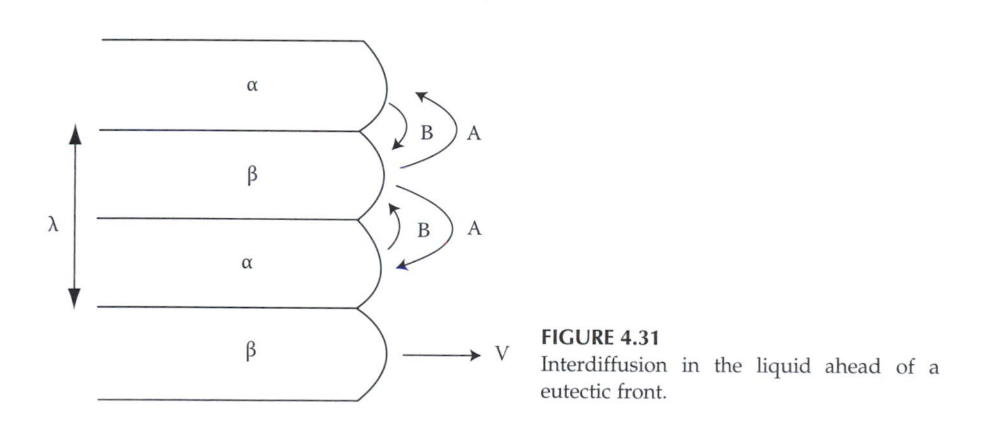

FIGURE 4.31
Interdiffusion in the liquid ahead of a eutectic front.

For an interlamellar spacing, λ, there is a total of $(2/\lambda)$ m^2 or α/β interface per m^3 of eutectic. Thus, the free energy change associated with the solidification of 1 mol of liquid is given by

$$\Delta G(\lambda) = -\Delta G(\infty) + \frac{2\gamma_{\alpha\beta} V_m}{\lambda} \quad (4.37)$$

where
V_m is the molar volume of the eutectic
$\Delta G(\infty)$ is the free energy decrease for very large values of λ

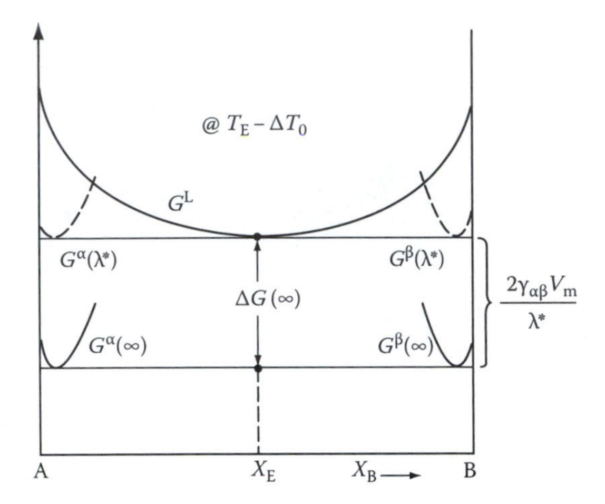

FIGURE 4.32
Molar free energy diagram at a temperature ΔT_0 below the eutectic temperature, for the case $\lambda = \lambda^*$.

Since solidification will not take place if $\Delta G(\lambda)$ is positive, $\Delta G(\infty)$ must be large enough to compensate for the interfacial energy term, i.e., the eutectic/liquid interface must be undercooled below the equilibrium eutectic temperature T_E (Figure 4.32). If the total undercooling is ΔT_0 it can be shown that $\Delta G(\infty)$ is then given approximately by

$$\Delta G(\infty) = \frac{\Delta H \cdot \Delta T_0}{T_E} \tag{4.38}$$

where ΔH is an enthalpy term. The minimum possible spacing (λ^*) is obtained by using the relation $\Delta G(\lambda^*) = 0$, hence

$$\lambda^* = \frac{2\gamma_{\alpha\beta}V_m T_E}{\Delta H \cdot \Delta T_0} \tag{4.39}$$

When the eutectic has this spacing the free energy of the liquid and eutectic is the same, i.e., all three phases are in equilibrium. This is because the α/β interface raises the free energies of the α and β from $G^\alpha(\infty)$ and $G^\beta(\infty)$ to $G^\alpha(\lambda^*)$ and $G^\beta(\lambda^*)$ as shown in Figure 4.32. The cause of increase is the curvature of the α/L and β/L interfaces arising from the need to balance the interfacial tensions at the $\alpha/\beta/L$ triple point, Figure 4.31 general, therefore, the increase will be different for the two phases, but for simple cases it can be shown to be $2\gamma_{\alpha\beta}V_m/\lambda$ for both (Figure 4.32).

Let us now turn to the mechanism of growth. If solidification is to occur at a finite rate there must be a flux of atoms between the tips of the α and β lamellae and this requires a finite composition difference. For example, the concentration of B must be higher ahead of the α-phase than ahead of the β-phase so that B rejected from the α can diffuse to the tips of the growing β. If $\lambda = \lambda^*$, growth will be infinitely slow because the liquid in contact with

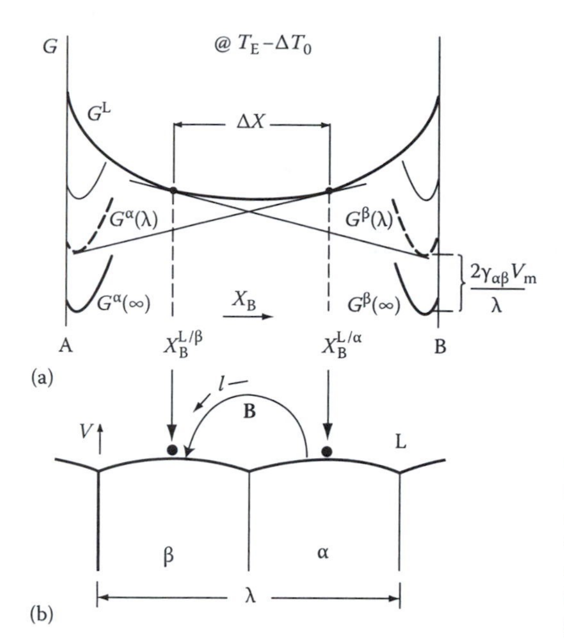

FIGURE 4.33
(a) Molar free energy diagram at ($T_E - \Delta T_0$) for the case $\lambda^* < \lambda < \infty$, showing the composition difference available to drive diffusion through the liquid (ΔX). (b) Model used to calculate the growth rate.

both phases has the same composition, X_E in Figure 4.32. However if the chosen spacing is greater than λ^*, less free energy is locked in the interfaces and G^α and G^β are correspondingly reduced (Figure 4.33a). Under these circumstances the liquid in local equilibrium with the α has a composition $X_B^{L/\alpha}$ which is richer in B than the composition in equilibrium with the β-phase $X_B^{L/\beta}$.

If the α/L and β/L interfaces are highly mobile, it is reasonable to assume that growth is diffusion controlled in which case the eutectic growth rate (v) should be proportional to the flux of solute through the liquid. This in turn will vary as $D\,dC/dl$ where D is the liquid diffusivity and dC/dl is the concentration gradient driving the diffusion. (l is measured along the direction of diffusion as shown in Figure 4.33b. In practice, dC/dl will not have a single value but will vary from place to place within the diffusion zone.) dC/dl should be roughly proportional to the maximum composition difference $\left(X_B^{L/\alpha} - X_B^{L/\beta}\right)$ and inversely proportional to the effective diffusion distance, which, in turn, will be linearly related to the interlamellar spacing (λ). Thus, we can write

$$v = k_1 D \frac{\Delta X}{\lambda} \tag{4.40}$$

where

k_1 is the proportionality constant
$\Delta X = X_B^{L/\alpha} - X_B^{L/\beta}$ as given in Figure 4.33

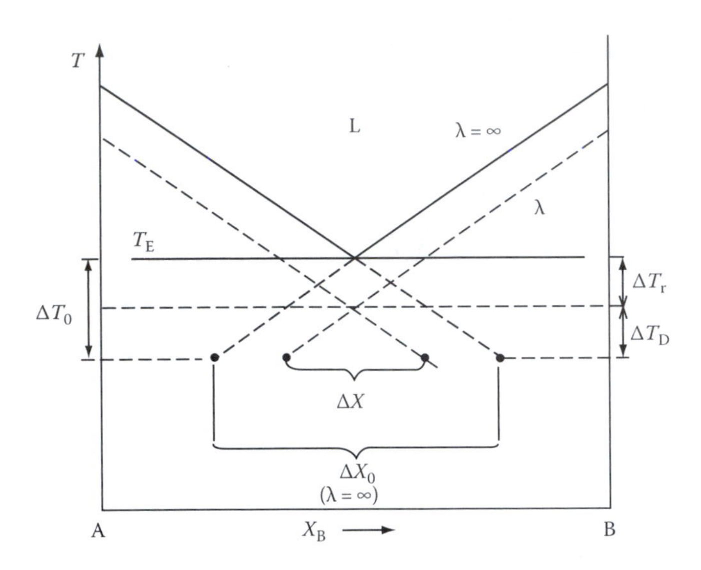

FIGURE 4.34
Eutectic phase diagram showing the relationship between ΔX and ΔX_0 (exaggerated for clarity).

ΔX will itself depend on λ for when $\lambda = \lambda^*$, $\lambda \Delta X = 0$, and as λ increases ΔX will tend to a maximum value, ΔX_0, say. Therefore, it is reasonable to write

$$\Delta X = \Delta X_0 \left(1 - \frac{\lambda^*}{\lambda}\right) \tag{4.41}$$

The magnitude of ΔX_0 can be obtained by extrapolating the equilibrium liquidus lines of the phase diagram ($\lambda = \infty$) as shown in Figure 4.34. For small undercoolings

$$\Delta X_0 \propto \Delta T_0 \tag{4.42}$$

The dashed lines in Figure 4.34 are the liquidus lines for $\lambda^* < \lambda < \infty$. ΔX is simply given by the extrapolation of these lines as shown. Combining the above equations gives

$$v = k_2 D \Delta T_0 \frac{1}{\lambda}\left(1 - \frac{\lambda^*}{\lambda}\right) \tag{4.43}$$

where k_2 is another proportionality constant. This equation shows that by varying the interface undercooling (ΔT_0) it is possible to vary the growth rate (v) and spacing (λ) independently. It is therefore impossible to predict the spacing that will be observed for a given growth rate. However, controlled growth experiments show that a specific value of λ is always associated with a given growth rate. Examination of Equation 4.43 shows that when $\lambda = 2\lambda^*$, the growth rate is a maximum for a given undercooling, or alternatively, the required undercooling is a minimum for a given growth rate. If it is assumed that growth occurs under these optimum conditions the observed spacing $\lambda_0 = 2\lambda^*$ and the observed growth rate is given by

$$v_0 = k_2 D \Delta T_0 / 2\lambda^*$$

However, from Equation 4.39, it is seen that $\Delta T_0 \propto 1/\lambda^*$ so that the following relationships are predicted:

$$v_0 \lambda_0^2 = k_3 \text{ (constant)} \tag{4.44}$$

and

$$\frac{v_0}{(\Delta T_0)^2} = k_4 \text{ (constant)} \tag{4.45}$$

There is in fact no physical basis for choosing $\lambda = 2\lambda^*$ and similar expressions can also be obtained using other assumptions concerning the spacing. Equations 4.44 and 4.45 are often found to be obeyed experimentally. For example, measurements on the lamellar eutectic in the Pb–Sn system[10] show that $k_3 \sim 33\ \mu m^3\ s^{-1}$ and $k_4 \sim 1\ \mu m\ s^{-1}\ K^{-2}$. Therefore for a solidification rate of $1\ \mu m\ s^{-1}$, $\lambda_0 \sim 5\ \mu m$ and $\Delta T_0 \sim 1$ K.

The total undercooling at the eutectic front (ΔT_0) has two contributions, i.e.,

$$\Delta T_0 = \Delta T_r + \Delta T_D \tag{4.46}$$

where

ΔT_r is the undercooling required to overcome the interfacial curvature effects

ΔT_D is the undercooling required to give a sufficient composition difference to drive the diffusion

Strictly speaking a ΔT_i term should also be added since a driving force is required to move the atoms across the interfaces, but this is negligible for high-mobility interfaces. A better theoretical treatment of eutectic solidification should take into account the fact that the composition of the liquid in contact with the interface and therefore ΔT_D vary continuously from the middle of the α to the middle of the β lamellae. Since the interface is essentially isothermal (ΔT_0 constant) the variation of ΔT_D must be compensated by a variation in ΔT_r, i.e., the interface curvature will change across the interface.[11]

A planar eutectic front is not always stable. If for example the binary eutectic alloy contains impurities, or if other allowing elements are present, the interface tends to break up to form a cellular morphology. The solidification direction thus changes as the cell walls are approached and the lamellar or rod structure fans out and may even change to an irregular

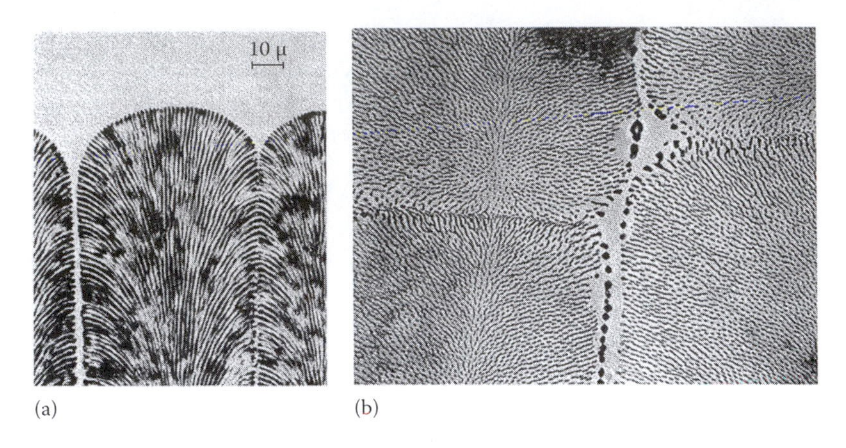

FIGURE 4.35
(a) Cellular eutectic solidification front in a transparent organic alloy. (After Hunt, J.D. and Jackson, K.A., *Trans. Metall. Soc. AIME*, 236, 843, 1996). (b) Transverse section through the cellular structure of an Al–Al_6Fe rod eutectic (×3500). (Courtesy of J. Strid, University of Luleå, Sweden.)

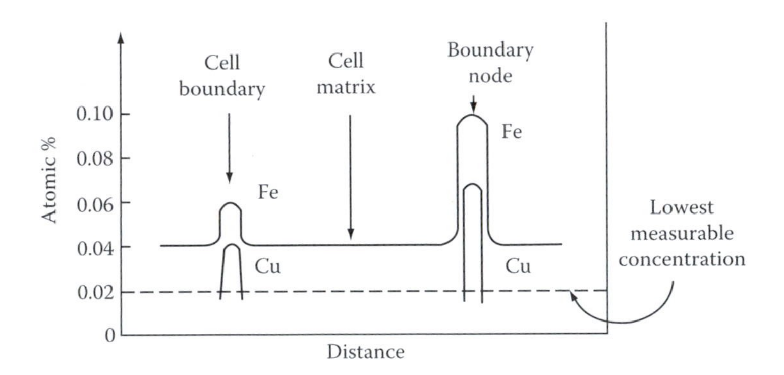

FIGURE 4.36
Composition profiles across the cells in Figure 4.35b.

structure (Figure 4.35). The impurity elements diffuse laterally and concentrate at the cell walls. In the case of the Al_6Fe–Al rodlike eutectic shown in Figure 4.35, the impurity causing the cellular structure is mainly copper. Figure 4.36 shows how the concentration of copper and iron in the aluminum matrix increases in the cell walls and boundary nodes.

Cell formation in eutectic structures is analogous to that in single-phase solidification, and under controlled conditions it is possible to stabilize a planar interface by solidifying in a sufficiently high temperature gradient.

4.3.3 Off-Eutectic Alloys

When the bulk alloy composition (X_0) deviates from the equilibrium eutectic composition (X_E) as shown in Figure 4.37, solidification usually begins close

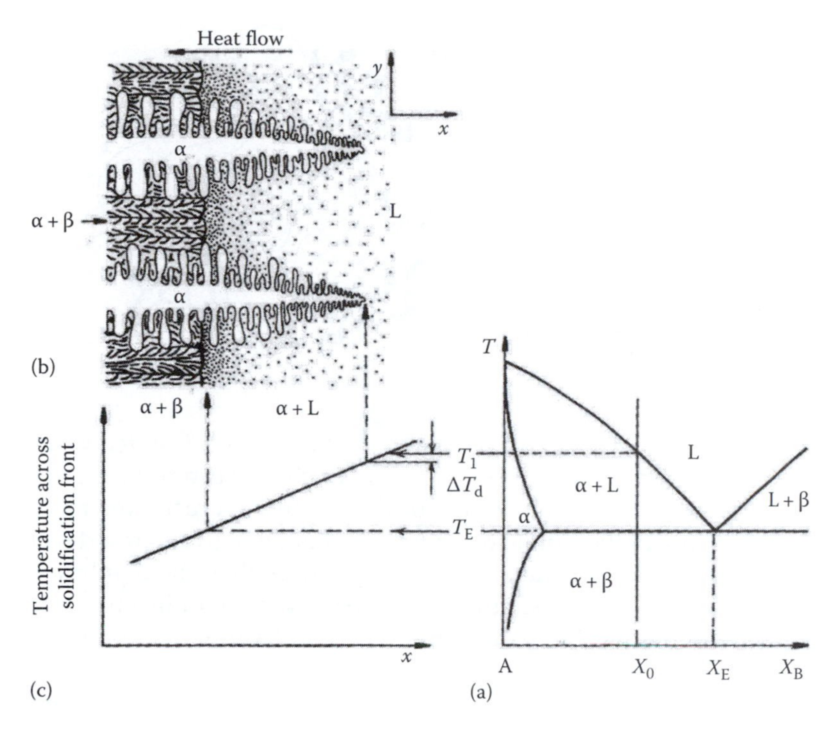

FIGURE 4.37
Solidification of an off-eutectic alloy in a temperature gradient. (a) Alloy composition in relation to the phase diagram. (b) Schematic solidification front. (c) Temperature variation across solidification front (isotherms normal to x).

to T_1 with the formation of primary (α) dendrites. As the dendrites thicken solute is rejected into the remaining liquid until its composition reaches X_E and the eutectic solidifies. Under steady-state unidirectional solidification conditions in the presence of a shallow temperature gradient the solidification front could appear as in Figure 4.37b. The tips of the dendrites are close to T_1 and the eutectic front, most probably cellular, close to T_E. Similar behavior is found during the solidification of castings and ingots. In the absence of solid-state diffusion the centers of the dendrites, which solidified close to T_1, will contain less solute than the outer layers that solidify at progressively lower temperatures. This leads to what is known as "coring" in the final microstructure (Figure 4.38). The eutectic does not always solidify as a two-phase mixture. When the volume fraction of one of the phases in the microstructure is very small it can form a so-called divorced eutectic. The minor phase then often appears as isolated islands and the other phase forms by the thickening of the dendrites.

Under controlled solidification conditions, e.g., in unidirectional solidification experiments, it is possible to solidify an off-eutectic alloy without permitting the formation of the primary dendritic phase. If the temperature gradient in the liquid is raised above a critical level the dendrite tips are

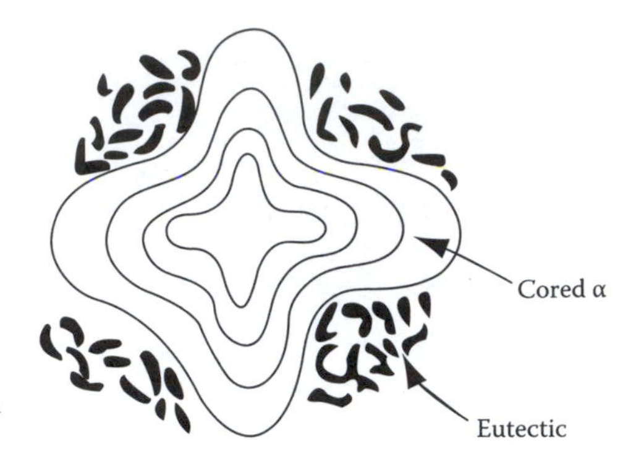

FIGURE 4.38
Transverse section through a dendrite in Figure 4.37.

overgrown by the eutectic and the alloy solidifies as 100% eutectic with an overall composition X_0 instead of X_E. A similar change can be brought about if the growth rate is raised above a critical level. In both cases, the reason for the disappearance of the primary dendrites is that for a given growth velocity the eutectic is able to grow at a higher temperature than the dendrite tips.[12] This phenomenon is of special interest in the production of in situ composite materials because the volume fraction of the two phases in the composite can be controlled by the choice of X_0.[13]

4.3.4 Peritectic Solidification[14]

A typical phase diagram containing a peritectic reaction, i.e., $L + \alpha \rightarrow \beta$ is shown in Figure 4.39a. During equilibrium solidification solid α with composition *a* and liquid with composition *c* react at the peritectic temperature T_P to give solid β of composition *b*. However, the transformation rarely goes to completion in practice.

Consider for example the solidification of an alloy X_0 at a finite velocity in a shallow temperature gradient (Figure 4.39b and c). As the temperature decreases, the first phase to appear is α with successive layers solidifying at composition in the dendrites is slow the liquid will eventually reach point c in Figure 4.39a and on further cooling it reacts with the α to produce a layer of β. However, the α dendrites are then often effectively isolated from further reaction and are retained to lower temperatures. Meanwhile the β-phase continues to precipitate from the liquid at compositions which follow the line bd. Again if there is no diffusion in the solid the liquid will finally reach point e and solidify as a $\beta + \gamma$ eutectic. The final solidified microstructure will then consist of cored α dendrites surrounded by a layer of β and islands of $\beta + \gamma$ eutectic, or divorced eutectic.

If alloy X_0 were directionally solidified at increasing values of (T'_L/v) the temperature of the dendrite tips would progressively fall from T_1 toward T_2 (Figure 4.39a) while the temperature at which the last liquid solidifies would increase toward T_2. Finally, solidification would take place behind a planar front at a temperature T_2, as discussed earlier in Section 4.3.1.

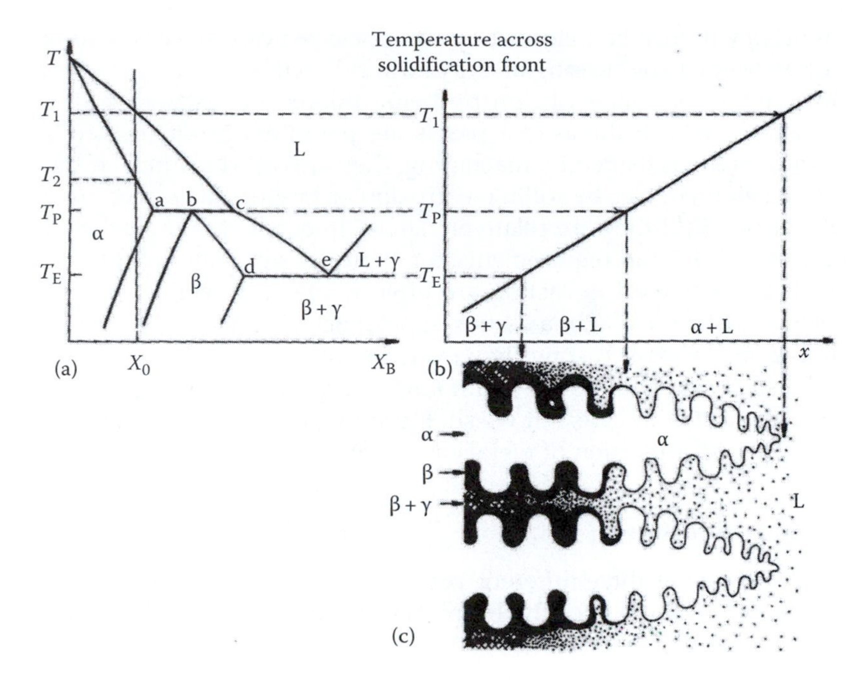

FIGURE 4.39
Peritectic solidification in a temperature gradient.

Planar-front solidification can also be obtained for alloys beyond *a* in used. Alloys between *a* and *b* then solidify with a eutectic-like α + β structure. (The structure is better described as composite to avoid confusion concerning the term eutectic.) Between b and d single-phase β forms, and beyond dβ + γ eutectic-like structures can be formed.

The Fe–C phase diagram also contains a peritectic (Figure 4.53a). However, due to the high diffusivity of carbon at these high temperatures the peritectic reaction is very rapid and is able to convert all of the primary (δ) dendrites into the more stable austenite.

4.4 Solidification of Ingots and Castings

This section is concerned with technological applications of the theory of solidification, as developed earlier. Two of the most important applications are casting and weld solidification and we shall first consider these. In modern constructions there is a tendency toward the use of stronger, heavier sections welded with higher energy techniques and faster speeds. It is thus important for the physical metallurgist to consider the effect of the various solidification parameters on the microstructure and properties of fusion welds. This will then be followed by some selected case studies of as-solidified or as-welded engineering alloys and weld metals.

Most engineering alloys begin by being poured or cast into a fireproof container or mold. If the as-cast pieces are permitted to retain their shape afterward, or are reshaped by machining, they are called castings. If they are later to be worked, e.g., by rolling, extrusion or forging, the pieces are called ingots, or blanks if they are relatively small. In either case the principles of solidification, and the requirements for high density and strength are the same. The molds used in casting are often made of a material that can be remolded or discarded after a casting series, such as sand. In the case of long casting series or ingot casting, however, the mold is of a more permanent material such as cast iron. The technological aspects of pouring and casting will not be dealt with here, but we shall confine our discussion simply to the mechanics of solidification of metals in a mold.

4.4.1 Ingot Structure

Generally speaking three different zones can be distinguished in solidified alloy ingots (Figure 4.40). These are (1) an outer chill zone of equiaxed

FIGURE 4.40
Schematic cast grain structure. (After Flemings, M.C., *Solidification Processing*, McGraw-Hill, New York, 1974.)

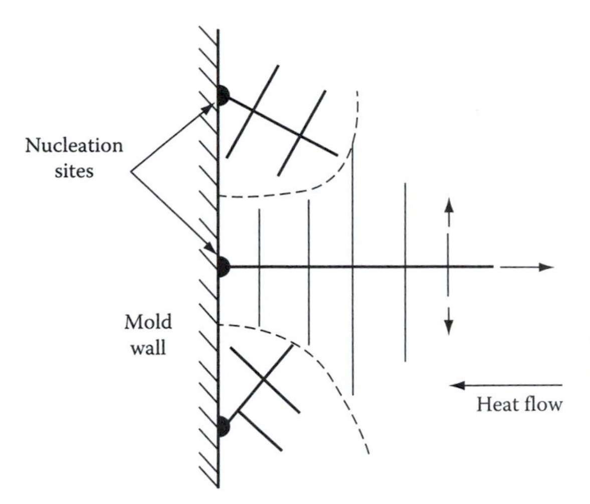

FIGURE 4.41
Competitive growth soon after pouring. Dendrites with primary arms normal to the mold wall, i.e., parallel to the maximum temperature gradient, outgrow less favorably oriented neighbors.

crystals, (2) a columnar zone of elongated or column-like grains, and (3) a central equiaxed zone.

Chill zone
During pouring the liquid in contact with the cold mold wall is rapidly cooled below the liquidus temperature. Many solid nuclei then form on the mold wall and begin to grow into the liquid (Figure 4.41). As the mold wall warms up it is possible for many of these solidified crystals to break away from the wall under the influence of the turbulent melt. If the pouring temperature is low the whole of the liquid will be rapidly cooled below the liquidus temperature and the crystals swept into the melt may be able to continue to grow. This is known as "big-bang" nucleation since the liquid is immediately filled with a myriad of crystals. This produces an entirely equiaxed ingot structure, i.e., no columnar zone forms. If the pouring temperature is high, on the other hand, the liquid in the center of the ingot will remain above the liquidus temperature for a long time and consequently the majority of crystals soon remelt after breaking away from the mold wall. Only those crystals remaining close to the wall will be able to grow to form the chill zone.

Columnar zone
Very soon after pouring the temperature gradient at the mold walls decreases and the crystals in the chill zone grow dendritically in certain crystallographic directions, e.g., <100> in the case of cubic metals. Those crystals with a <100> direction close to the direction of heat flow, i.e., perpendicular to the mold walls, grow fastest and are able to outgrow less favorably oriented neighbors (Figure 4.42). This leads to the formation of the columnar grains all with <100> almost parallel to the column axis. Note that each columnar crystal contains many primary dendrite arms. As the diameter of these grains increases additional primary dendrite arms appear by

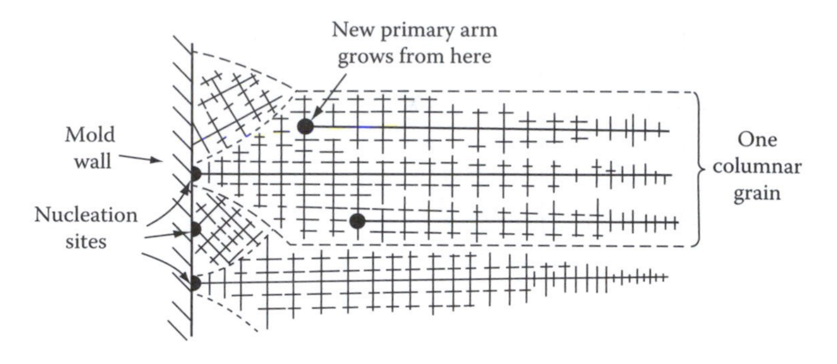

FIGURE 4.42
Favorably oriented dendrites develop into columnar grains. Each columnar grain originates from the same heterogeneous nucleation site, but can contain many primary dendrite arms.

a mechanism in which some tertiary arms grow ahead of their neighbors as shown in the figure.

The volume fraction of the melt solidified increases with increasing distance behind the tips of the dendrites and, when the structure is mainly single-phase, the secondary and tertiary arms of adjacent dendrites can linkup to form walls of solid containing many primary dendrite arms. The region between the tips of the dendrites and the point where the last drop of liquid is solidifying is known as the mushy or pasty zone. The length of this zone depends on the temperature gradient and the nonequilibrium freezing range of the alloy. In general, it is found that the secondary arms become coarser with distance behind the primary dendrite tips. This effect can be seen in Figure 4.28. The primary and secondary dendrite arm spacing is also often found to increase with increasing distance from the mold wall. This is simply due to a corresponding decrease in the cooling rate with time after pouring.

Equiaxed zone

The equiaxed zone consists of equiaxed grains randomly oriented in the center of the ingot. An important origin of these grains is thought to be melted-off dendrite side-arms. It can be seen from Figure 4.28 that the side-arms are narrowest at their roots. Therefore, if the temperature around the dendrite increases after it has formed, it will begin to melt and may become detached from the main stem. Provided the temperature falls again before the arm completely disappears it can then act as a seed for a new dendrite. An effective source of suitable temperature pulses is provided by the turbulent convection currents in the liquid brought about by the temperature differences across the remaining melt. Convection currents also provide a means of carrying the melted-off arms away to where they can develop uninhibited into equiaxed dendrites. If convection is reduced fewer seed crystals are created causing a larger final grain size and a greater preponderance of columnar grains. Convection also plays a dominant role in the formation of the outer chill zone. The mechanism whereby crystals are melted away from the mold walls is thought to be similar to the detachment of side-arms[15] and when convection is absent no chill zone is observed.

Shrinkage effects
Most metals shrink on solidification and this has important consequences for the final ingot structure. In alloys with a narrow freezing range the mushy zone is also narrow and as the outer shell of solid thickens the level of the remaining liquid continually decreases until finally when solidification is complete, the ingot contains a deep central cavity or pipe.

In alloys with a wide freezing range the mushy zone can occupy the whole of the ingot. In this case no central pipe is formed. Instead the liquid level gradually falls across the width of ingot as liquid flows down to compensate for the shrinkage of the dendrites. However, as the interdendritic channels close-up this liquid flow is inhibited so that the last pools of liquid to solidify leave small voids or pores.

4.4.2 Segregation in Ingots and Castings

Two types of segregation can be distinguished in solidified structures. There is macrosegregation, i.e., composition changes over distances comparable to the size of the specimen, and there is microsegregation that occurs on the scale of the secondary dendrite arm spacing.

It has already been shown that large differences in composition can arise across the dendrites due to coring and the formation of nonequilibrium phases in the last solidifying drops of liquid. Experimentally, it is found that while cooling rate affects the spacing of the dendrites it does not substantially alter the amplitude of the solute concentration profiles provided the dendrite morphology does not change and that diffusion in the solid is negligible. This result often applies to quite a wide range of practical cooling rates.

There are four important factors that can lead to macrosegregation in ingots. These are (1) shrinkage due to solidification and thermal contraction; (2) density differences in the interdendritic liquid; (3) density differences between the solid and liquid; and (4) convection currents driven by temperature-induced density differences in the liquid. All of these factors can induce macrosegregation by causing mass flow over large distances during solidification.

Shrinkage effects can give rise to what is known as inverse segregation. As the columnar dendrites thicken solute-rich liquid (assuming $k < 1$) must flow back between the dendrites to compensate for shrinkage and this raises the solute content of the outer parts of the ingot relative to the center. The effect is particularly marked in alloys with a wide freezing range, e.g., Al–Cu and Cu–Sn alloys.

Interdendritic liquid flow can also be induced by gravity effects. For example, during the solidification of Al–Cu alloys the copper rejected into

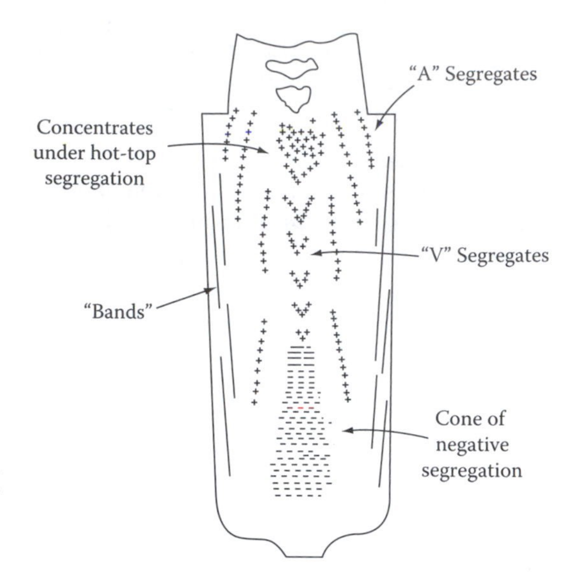

FIGURE 4.43
Segregation pattern in a large killed steel ingot. + positive, − negative segregation. (After Flemings, M.C., *Scand. J. Metall.*, 5, 1, 1976.)

the liquid raises its density and causes it to sink. The effect can be reinforced by convection currents driven by temperature differences in the ingot.

Gravity effects can also be observed when equiaxed crystals are forming. The solid is usually denser than the liquid and sinks carrying with it less solute than the bulk composition (assuming $k < 1$). This can, therefore, lead to a region of negative segregation near the bottom of the ingot.

The combination of all the above effects can lead to complex patterns of macrosegregation. Figure 4.43 for example illustrates the segregation patterns found in large steel ingots.[16]

In general, segregation is undesirable as it has marked deleterious effects on mechanical properties. The effects of microsegregation can be mitigated by subsequent homogenization heat treatment, but diffusion in the solid is far too slow to be able to remove macrosegregation which can only be combated by good control of the solidification process.

4.4.3 Continuous Casting

A number of industrial processes are nowadays employed in which casting is essentially a dynamic rather than a static process. In these cases, the molten metal is poured continuously into a water-cooled mold (e.g., copper) from which the solidified metal is continuously withdrawn in plate or rod form. This process is illustrated schematically in Figure 4.44.

It is seen that the speed of withdrawal is such that the solid–liquid interface is maintained in the shape and position illustrated. Ideally, the flow behavior of the liquid should be vertically downward, and if flow is maintained in this way the final composition across the ingot will be kept uniform. In practice, hydrodynamic effects do not allow this simple type of flow and there is a

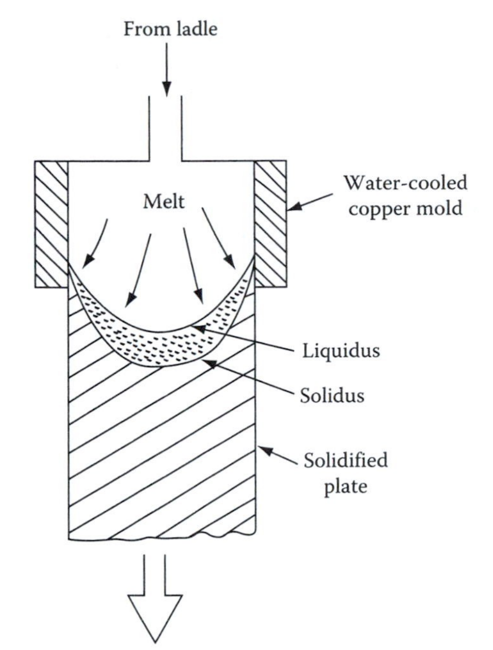

FIGURE 4.44
Schematic illustration of a continuous casting process.

tendency for the flow lines to fan outward (as shown by the arrows) producing negative segregation near the center. Solidification follows the maximum temperature gradient in the melt as given by the normals to the isotherms. In certain respects weld solidification has much in common with continuous casting in that it is also a dynamic process. As illustrated in Figure 4.45 the main difference is of course that in continuous casting the heat source (as defined by the mold) does not move, whereas in welding

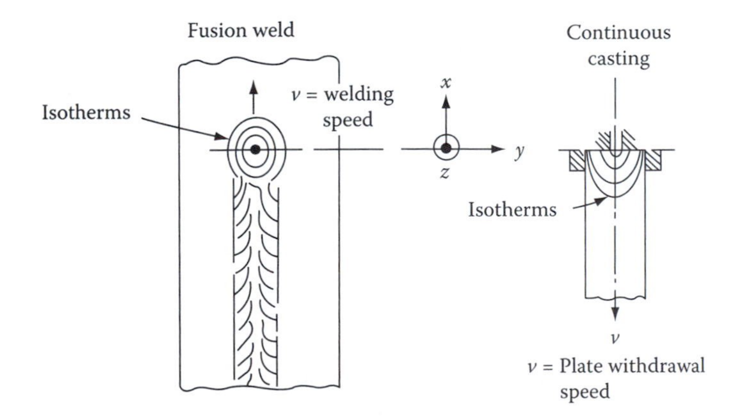

FIGURE 4.45
Illustrating the essential equivalence of isotherms about the heat sources in fusion welding and continuous casting.

the heat source (the electrode) is moving. We shall now consider the latter case in more detail, but it will be found that certain conclusions concerning weld solidification behavior can well be applied to both processes.

Heat flow in welding and continuous casting
As discussed earlier, there are many factors concerning heat distribution at the melt zone and the dynamics of the process, which are essentially fairly similar in both continuous casting and welding. As an example, we shall first consider the welding process and then discuss how the results may be applied to continuous casting. In contrast to continuous casting, weld solidification involves a mold that has approximately the same composition as the melt.

The most important variables in weld solidification or continuous casting are thus:

1. The rate of heat input, q (determined by type of weld process, weld size, etc.); in terms of continuous casting q is represented effectively by the volume and temperature of the melt
2. Speed of arc movement along join, v; in continuous casting, v is the velocity of plate withdrawal
3. Thermal conductivity of the metal being welded or cast, K_s
4. Thickness of plate being welded or cast, t

In the case of welding, assuming that the arc moves along the x coordinate, the resulting heat distribution in a three-dimensional solid plate is given by the solution to the heat-flow equation:[17]

$$\frac{\partial^2 T}{\partial x^2} + \frac{\partial^2 T}{\partial y^2} + \frac{\partial^2 T}{\partial z^2} = 2K_s v \frac{\partial T}{\partial (x - vt)} \tag{4.47}$$

where x, y, z are defined in Figure 4.45 and t is time.

Solving this equation gives the temperature distribution about the moving heat source in the form of isotherms in the solid metal, in which the distance between the isotherms in a given direction (x, y, z) is approximately given by

$$\lambda_{(x,y,z)} \propto \frac{q}{K_s v t} \tag{4.48}$$

Gray et al.[24] have solved Equation 4.47 and plotted isotherms for a number of different materials and welding speeds and some of their results are summarized in Figure 4.46.

Assuming a similar isotherm distribution in the melt, it is likely that the parameters K_s, v, t, and q will largely determine solidification morphology, in that dendrites always try to grow in directions as near normal to isotherms as their crystallography allows. It is seen in the above figure, for example, that

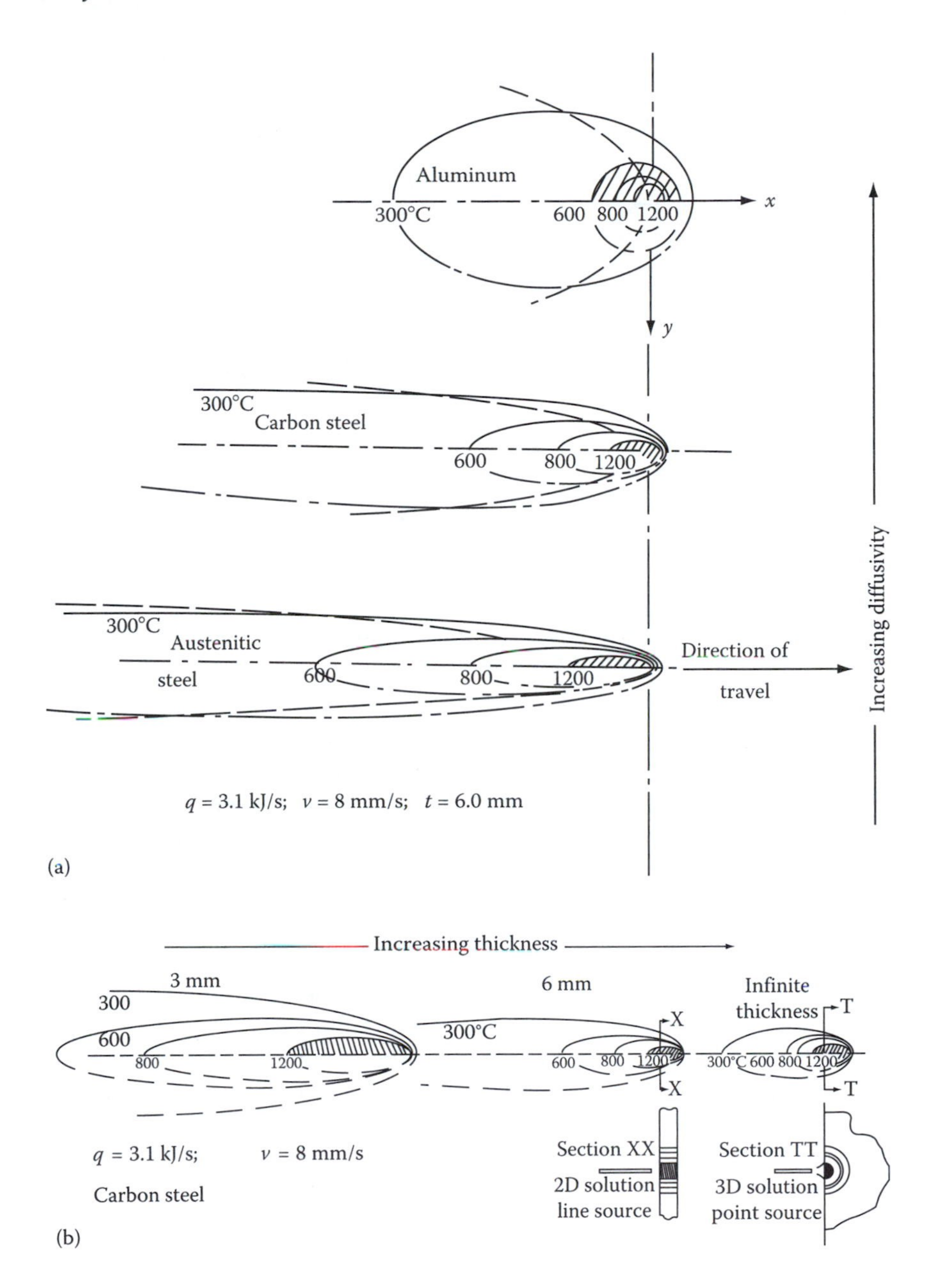

FIGURE 4.46
Effect of various parameters in Equation 5.36 on the isotherm distribution at a point heat source. (a) Effect of changes in thermal conductivity, K_S. (b) Effect of changes in plate thickness, t.

(continued)

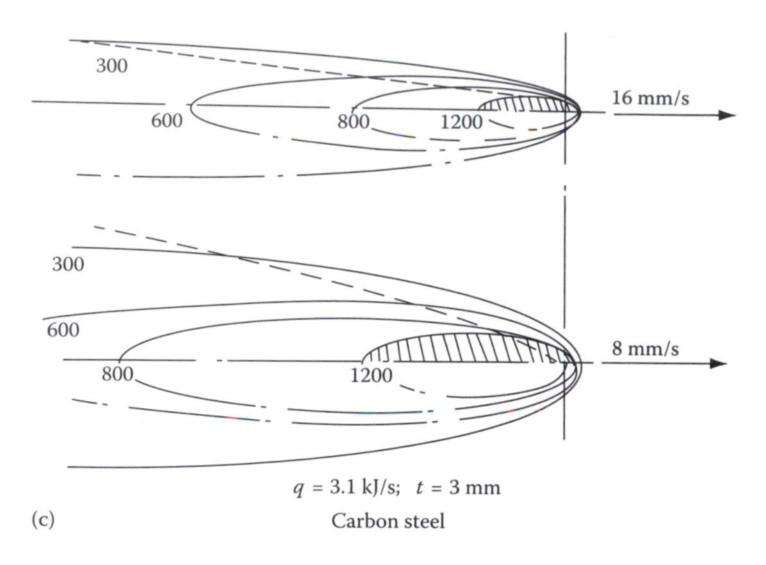

FIGURE 4.46 (continued)
(c) Effect of changes in movement of heat source, v. (After Gray, T.G., Spence, J., and North, T.H., *Rational Welding Design*, Newnes-Butterworth, London, 1975.)

holding q, v, and t constant (Figure 4.46a), the distance between isotherms, λ, increases substantially as a function of the change of heat conductivity, K_s, of the different materials: aluminum, carbon (ferritic) steel, and austenitic steel. If the plate thickness is increased (Figure 4.46b), or the weld speed is higher (Figure 4.46c), λ also decreases proportionally.

These results can also be applied to continuous casting, in that the isotherm distributions shown in Figure 4.46 are affected in a similar way by the conductivity of the solidifying metal and its speed of withdrawal from the mold. This means, for example, that the depth of the liquid pool in continuous casting is much greater for steel than for aluminum alloys under comparable conditions. This implies that in practice the maximum casting speed and billet cross section are less for steel than for aluminum or copper. Another practical difficulty resulting from a large depth of liquid is that the billet cannot be cut until it reaches a point well beyond the solidus line (see Figure 4.44), which requires in fact a very tall installation for high-speed casting.

4.5 Solidification of Fusion Welds

Contact between the weld melt and the base metal will initially cause melting back of the material and dilution of the filler metal as illustrated in Figure 4.47. The amount of dilution involved is not insignificant. Jesseman (1975)[18] reports for example that in microalloyed steel welds, the weld metal may

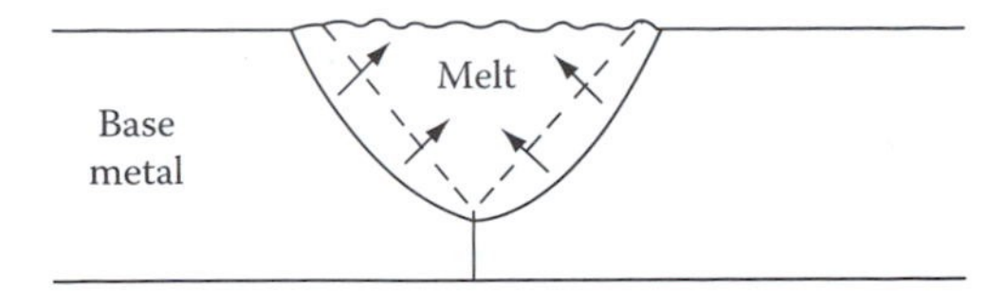

FIGURE 4.47
Illustrating the effect of dilution. In high-energy welds, the weld metal typically exhibits 50%–70% of the analysis of microalloying elements of the base metal through dilution.

contain 50%–70% of the amount of Nb, Ti, or V as analyzed in the base material. The effect of dilution is in fact threefold, and affects the weld metal as follows:

1. The composition of the melt is changed
2. The surface oxide layer of the base metal is removed (also into the melt)
3. It cools down the melt

Depending upon the type of material being welded as well as plate thickness, the base metal behaves as a very efficient heat sink, and already at $T \simeq T_e$ solidification nuclei form at the oxide-free surface of the melted-back base material. Since the melt has approximately the same composition as the base metal, wetting of the base metal is very efficient and $\theta \simeq 0$ (see Figure 4.7). This implies in turn that there is almost no nucleation barrier to solidification and hence very little undercooling occurs. Solidification is thus predicted to occur epitaxially, i.e., nuclei will have the same lattice structure and orientation as the grains at the solid–liquid surface of the base metal, and this is what is observed in practice.

Since the temperature of the melt beneath the arc is so high and the base material is such an efficient heat sink there is initially a steep temperature gradient in the liquid and consequently the degree of constitutional supercooling is low. The actual thermal gradient is of course dependent upon the welding process and the plate thickness (Equation 4.48). For example, TIG welding of thin plates will give steeper thermal gradients than submerged arc welding of thick plates, the latter process having the higher heat input. Since certain grains at the base metal are better oriented than others for <100> growth with respect to the isotherms of the melt, these quickly predominate and widen at the expense of the others. However, the general coarseness of the microstructure is largely determined at this stage by the grain size of the base metal. Unfortunately, the base metal at the transition zone receives the most severe thermal cycle and after high-energy welding in particular the grains in this zone tend to grow and become relatively coarse. The weld microstructure is thus inherently coarse grained.

Welding is essentially a dynamic process in which the heat source is continuously moving. This means that the maximum temperature gradients are constantly changing direction as the heat source moves away. The growing columnar crystals are thus faced with the necessity of trying to follow the

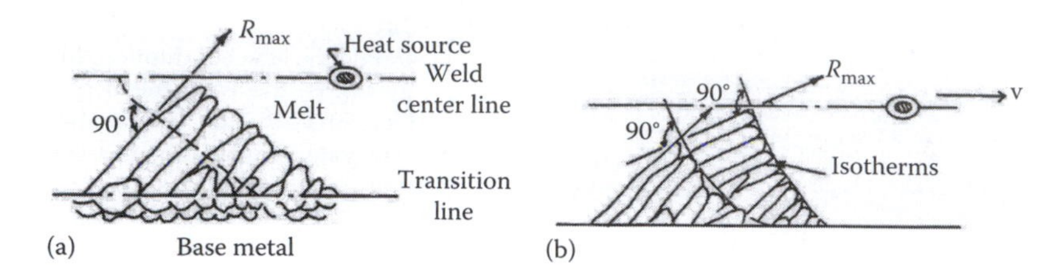

FIGURE 4.48
Illustrating the growth of columnar crystals in the weld, and how growth continues to occur approximately normal to the isotherms.

maximum temperature gradients while still maintaining their preferred <100> growth direction. This often results in sudden changes in growth direction, as illustrated in Figure 4.48a and b.

Few of the grains originating at the base metal survive to reach the weld centerline. The mechanism by which sudden changes in <100> growth direction are brought about is not fully understood. One suggestion is that renucleation occurs by the help of dendritic fragments which have broken away from the growing interface due to turbulence in the weld pool, or simply from melted-off dendrite arms.

Influence of welding speed
An important effect of increasing the welding speed is that the shape of the weld pool changes from an elliptical shape to a narrower, pear shape (see, e.g., Figure 4.49). Since growing crystals try to follow the steepest temperature gradients, the effect of changing the welding speed is to alter the solidification behavior as illustrated in Figure 4.49. As shown in Figure 4.49b, the pear-shaped weld pool maintains fairly constant thermal gradients up to the weld centerline, corresponding to the more angular geometry of the melt in this case. On this basis, growing crystals are not required to change growth direction as at slower speeds (Figure 4.49a). Instead, appropriately oriented crystals stabilize and widen outgrowing crystals of less favorable orientation. The crystal morphology shown in Figure 4.49b is in fact fairly typical of the high production rate welds based

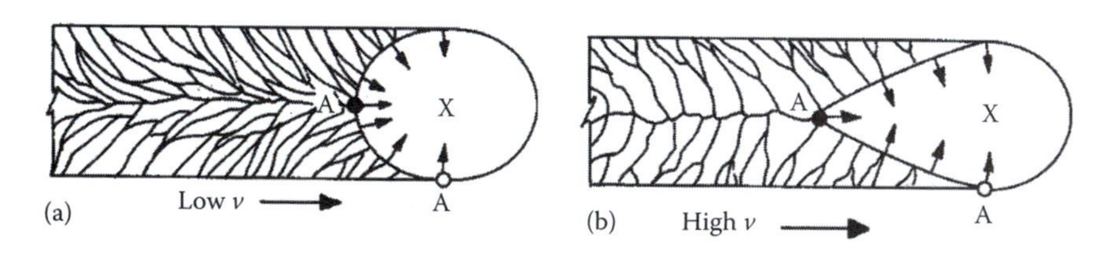

FIGURE 4.49
Illustrating the effect of increasing welding speed on the shape of the melt pool and crystal growth in fusion welds.

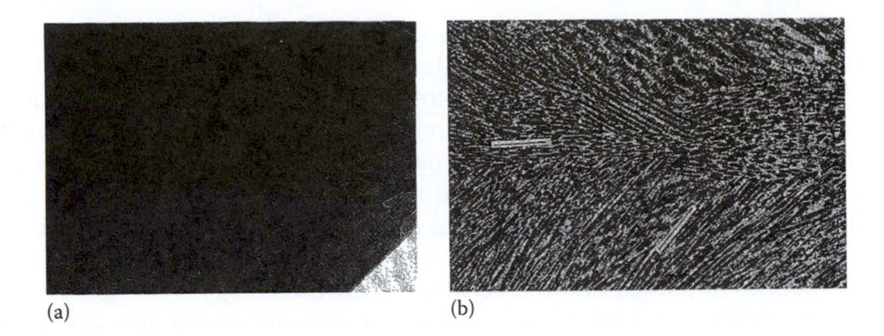

FIGURE 4.50
(a) TIG weld of nickel, illustrating low crystal growth speed (×25). (By Gudrun Keikkala, University of Luleå, Sweden.) (b) Submerged arc weld of steel, illustrating high growth speed (×24). (By H. Åström, University of Luleå, Sweden.)

on modern submerged arc welding. An example of a submerged arc weld is shown in Figure 4.50b, and of a TIG weld of nickel in Figure 4.50a.

While fairly linear dendritic growth is seen to predominate in this figure, it is also observed that dendrites suddenly change direction at the center of the weld by as much as 60°. This feature of high-speed welding will now be clarified.

Geometry of crystal growth
Consider a welding process in which the arc is moving at a speed v. Crystal growth must occur such that it is able to keep pace with the welding speed, and this is illustrated in Figure 4.51. It is seen that for crystal growth rate, R, to keep pace with the welding speed, v, the condition must be met that

$$R = v \cos\theta \tag{4.49}$$

In Figure 4.51, the arrows represent vectors of speed. The vector representing the welding speed, or the speed of movement of the isotherms is constant. On the other hand, the vector representing crystal growth rate must

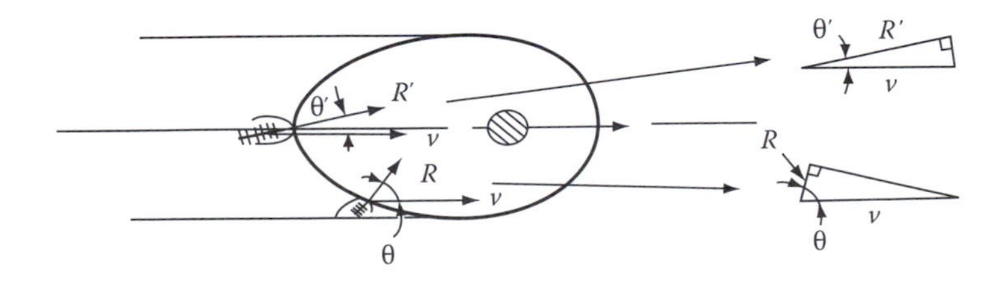

FIGURE 4.51
Illustrating the relationship between crystal growth speed and welding speed in terms of rate vectors.

continuously adjust itself as growth proceeds toward the weld centerline. It thus follows from Equation 4.49 that the solidification rate is greatest when $\theta \simeq 0$, i.e., at the weld centerline, and lowest at the weld edge where θ is a maximum. On this basis the sudden change in crystal direction at the weld centerline illustrated in Figure 4.50 is associated with high growth rates as solidification attempts to keep pace with the moving arc. In addition, the initial low rates of crystal growth are associated with a relatively planar solidification front, and as the growth rate increases, the morphology of the front changes to cellular and then cellular dendritic.

An example of weld solidification rates as measured on stainless steel as a function of different welding speeds is shown in Figure 4.52. In confirmation of Equation 4.49, it was found that completion of weld solidification ($y = 100\%$) corresponds to the highest growth rates. However, the higher welding speeds were associated with a transition from predominantly columnar crystal growth to equiaxed growth at the final stage of solidification.

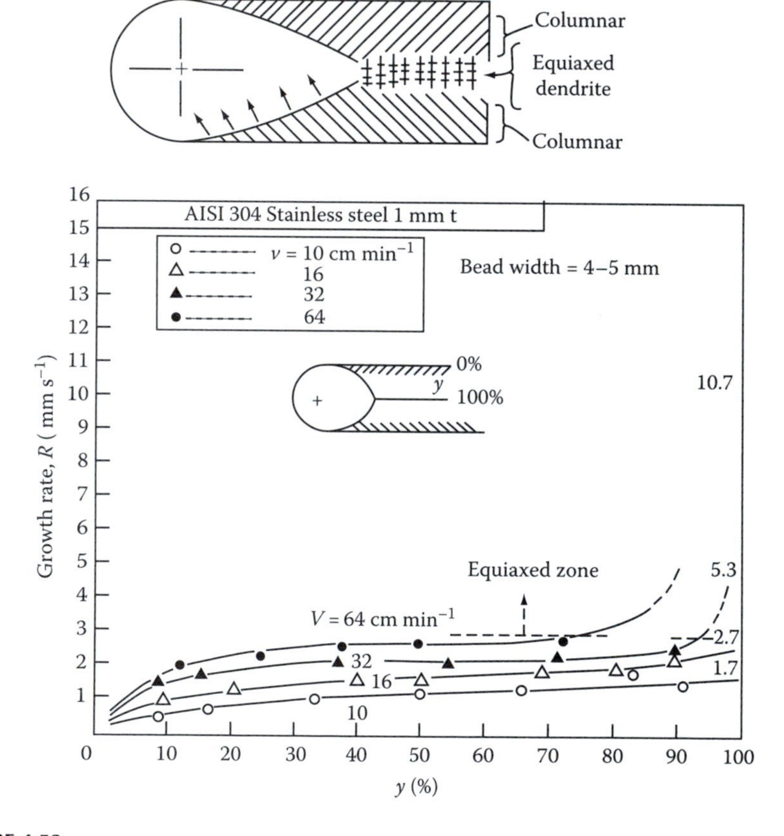

FIGURE 4.52
Measurements of crystal growth rate in stainless steel as a function of percent of weld solidified. (After Senda, T. et al., *Technical Report*, Osaka University, 20, 932, 1970.)

This transition is thought to be due to the high amounts of segregation associated with the final stages of weld solidification. This coupled with the shallow thermal gradient at this stage leads to high degrees of constitutional supercooling and therefore the driving force for random dendritic growth to occur is large. However, it should be noted that in general dendritic and cellular substructures in welds tend to be on a finer scale than in casting, and this is mainly due to the comparatively high solidification rates of weld metal. Since higher welding speeds or thicker base metal give larger rates of solidification, it follows that the finest substructures are associated with these welds (see Equation 4.48).

When the arc is switched off at the completion of a weld run, an elliptical molten pool is left to solidify with a comparatively shallow thermal gradient. This leads to large constitutional supercooling and marked segregation. The final substructure of these weld craters is thus usually equiaxed dendritic.

Summarizing, weld solidification has the following features:

1. Solidification initially occurs epitaxially at the melted-back grains of the base metal.
2. To begin with crystal growth is relatively slow, forming first a planar and then a fine cellular substructure.
3. The intermediate stage of crystal growth is cellular dendritic leading to coarse columnar crystal growth in the <100> direction in the case of cubic crystals.
4. Final solidification at the centerline is associated with rapid crystal growth and marked segregation. Depending on welding conditions, final dendritic structure can be equiaxed.

In many ways, therefore, weld solidification and even continuous casting exhibit essentially different features to those of ingot casting (Exercise 4.22).

4.6 Solidification during Quenching from the Melt

The treatment of solidification presented in this chapter is applicable for cooling rates of less than about 10^3 K s^{-1}. However, solidification can also occur at much higher rates of 10^4 to 10^7 K s^{-1} in such processes as liquid metal atomization, melt spinning, roller-quenching, or plasma spraying, as well as laser or electron beam surface treatment. By quenching melts, it is possible to achieve various metastable solid states not predicted by equilibrium phase diagrams: solid phases with extended solute solubility, new metastable crystalline phases or, if the cooling rate is fast enough, amorphous metallic glasses. Crystalline solidification can occur without microsegregation or with cells or secondary dendrites spaced much more finely than in conventional solidification processes. Whether the solid is crystalline or amorphous, rapid solidification processing offers a way of producing new materials with improved magnetic or mechanical properties.

One consequence of rapid cooling can be that local equilibrium at the solid/liquid interface breaks down. Melts can solidify with no change in composition, i.e., partitionless solidification or solute trapping can occur. The thermodynamic principles involved in partitionless solidification are similar to those for the massive transformation in solids to be treated in Section 5.9.

4.7 Metallic Glasses

Early attempts to produce metallic glasses required rapid quenching of liquid alloys with cooling rates of about 10^6 K s^{-1}. The metastable glassy state obtained represents a noncrystalline amorphous alloy, which takes place below a glass temperature T_g. Upon cooling it is expected to see a decrease in the specific volume. But if cooling is carried out quickly enough so there is not sufficient time for crystallization, the so-called free volume will be quenched in as can be seen in Figure 4.53.

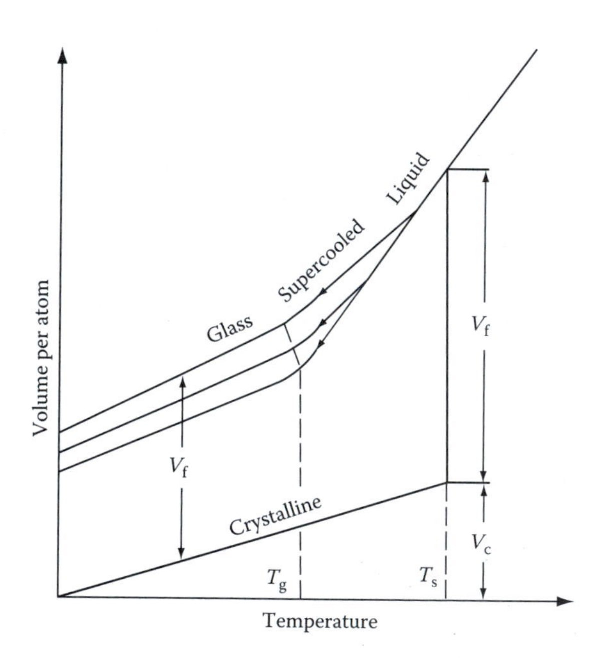

FIGURE 4.53
The decrease of the volume per atom upon cooling.[19] V_c and V_f are the volume per atom in the crystal and the free volume, respectively. Note that the volume per atom in the glassy state is higher than that in the crystal by V_f. (After Haasen, P., *Physical Metallurgy*, 3rd edn., Cambridge University Press, Cambridge, 1997.)

Given the need for such rapid cooling rates to produce metallic glasses, it was only possible to produce thin sections. Recent research on bulk metallic glasses (BMGs), however, yielded an increase in potential alloy compositions which could be cooled at much slower cooling rates where minimum sections of up to 1 cm are possible to obtain. The critical size is defined as the maximum value of the minimum dimension such that a fully glassy sample could be obtained. With these cooling rates, where a simple process such as casting can be used, a near-net-shape product with excellent surface finish can be obtained with reasonable cost as far as processing is concerned. In addition, metallic glasses exhibit very high strength, close to the theoretical strength of their crystalline counterparts, as well as very interesting magnetic and tribological properties. It is however difficult to design a BMG given that no pure metals and very few metallic alloys exhibit the glass-forming ability (GFA). Therefore, it is of importance to understand the physical parameters governing the formation of metallic glasses over the more common crystalline forms.

4.7.1 Thermodynamics and Kinetics

The search for potential BMG systems requires the identification of some physical parameters which can explain the GFA of some alloys. It is however difficult in some cases to explain why minor changes in composition can lead to a very different GFA. Nevertheless, some criteria have been developed in recent years, which is the topic of the current section.

Usually BMGs are multicomponent alloys, at least three elements, with deep- or near-eutectic composition. That is, to say the melting point of BMG-forming liquids is significantly low. The low melting point is an indication that the liquid state is thermodynamically somewhat stable when compared with the crystalline state. The need to use at least three elements stems from the fact that binary alloys tend to form intermetallic compounds which, if formed, will destabilize the glassy state.[20] The so-called confusion principle was apparently coined by H.H. Liebermann where the crystalline state is confused by the introduction of more than two elements in the system. Essentially, the entropy of the system is increased accordingly which helps stabilizing the glass.

It has been reported also that a composition featuring large atomic size mismatch is necessary,[21] reportedly greater than 12%. This has the effect of destabilizing the crystalline solid solution topologically and thus promoting the glassy state. Another important parameter is having a large negative heat of mixing which reduces the atomic mobility in the melt.

However, kinetic effects have to be considered in the design of a BMG as well. The assessment of the nucleation and growth of competing crystalline phases is crucial, in other words, the resistance to crystallization. The time–temperature–transformation (TTT) diagram contains the necessary information regarding not only the formability but also the stability of a glass system.[20] For example, Figure 4.54 shows the experimental TTT

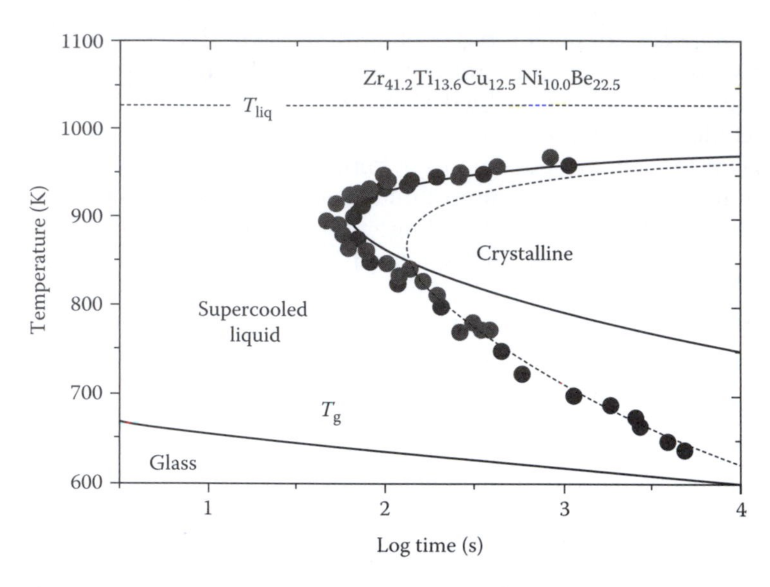

FIGURE 4.54
Measured (dots) TTT diagram of a BMG Vitreloy 1 showing the different states. Calculated curves for the crystalline phase are also shown. Note that for the formation of a glass, applied cooling rate has to be fast enough to avoid intersecting the nose. (After Busch, R., Schroers, J., and Wang, W.H., *MRS Bull.*, 32(8), 620, 2007.)

diagram of the undercooled $Zr_{41.2}Ti_{13.6}Cu_{12.5}Ni_{10.0}Be_{22.5}$ alloy, known as Vitreloy 1 or simply V1. The observed nose shape can be explained as a result of two competing processes, the increase in the driving force for crystallization and the decrease in atomic movement as undercooling increases.[22]

The crystallization process changes from being nucleation-controlled at high temperatures, to growth-controlled at low temperatures. The decrease of atomic movement translates into higher viscosity, a characteristic of BMGs which show highly viscous liquid.

The driving force for crystallization is the Gibbs energy difference between the crystalline and the supercooled liquid state. Figure 4.55 shows the change of the specific heat capacity for several glass-forming systems as a function of temperature.

It is clear from Figure 4.55 that as the BMG-alloys are cooled the specific heat capacity difference, taking that of the crystal as a reference, increases but far less prominently with good glass-forming systems like $Mg_{65}Cu_{25}Y_{10}$, Vitreloy 4 (V4), and Vitreloy 1. Note the more gradual change in C_p as the temperature approaches the glass temperature of these alloys. These alloy systems tend to exhibit little structural changes when undercooled toward T_g.

For a review of newly developed GFA criteria and their limitations, the reader is referred to the chapter written by Lu et al. on the subject.[20]

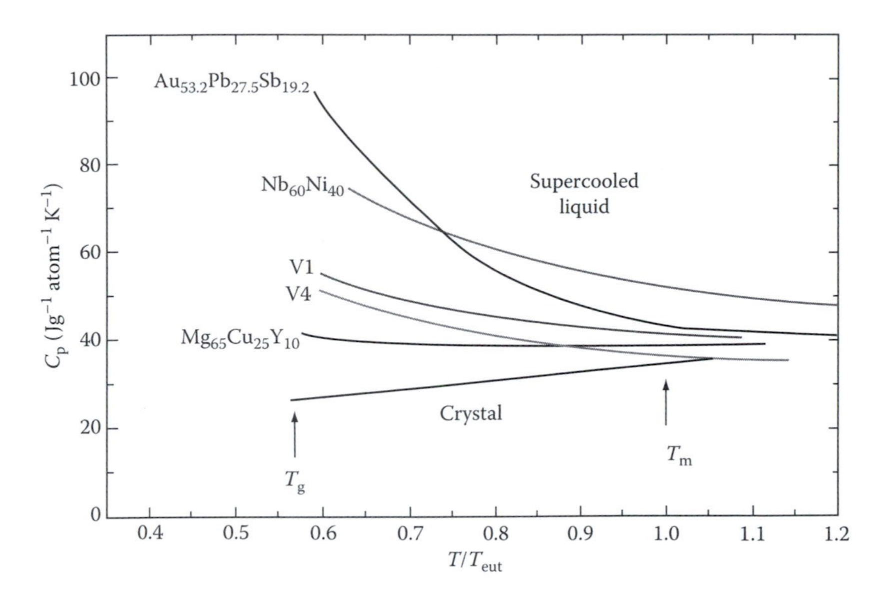

FIGURE 4.55
Specific heat capacities of some glass-forming alloys in the supercooled (undercooled) state. Temperature is presented as normalized to eutectic temperature. T_m and T_g are the melting and glass temperatures respectively. (After Busch, R., Schroers, J., and Wang, W.H., *MRS Bull.*, 32(8), 620, 2007.)

4.8 Case Studies of Some Practical Castings and Welds

4.8.1 Casting of Carbon and Low-Alloy Steels

Typical composition ranges:

$$\left.\begin{array}{l}\left.\begin{array}{l}\text{C:}0.1-1.0\,\text{wt\%}\\ \text{Si:}0.1-0.4\,\text{wt\%}\\ \text{Mn:}0.3-1.5\,\text{wt\%}\end{array}\right\}\text{Carbon steels}\\ \text{Cr:}1.0-1.6\,\text{wt\%}\\ \text{Ni:}1.0-3.5\,\text{wt\%}\\ \text{Mo:}0.1-0.4\,\text{wt\%}\end{array}\right\}\text{Low-alloy steels}$$

Casting processes: castings, ingots, continuous casting.
Relevant phase diagrams: see Figure 4.56.
Solidification transformations:

$$L \rightarrow \delta + L$$
$$\delta + L \rightarrow \delta + \gamma + L \text{ (peritectic } \delta + L \rightarrow \gamma)$$
$$\delta + \gamma + L \rightarrow \gamma$$

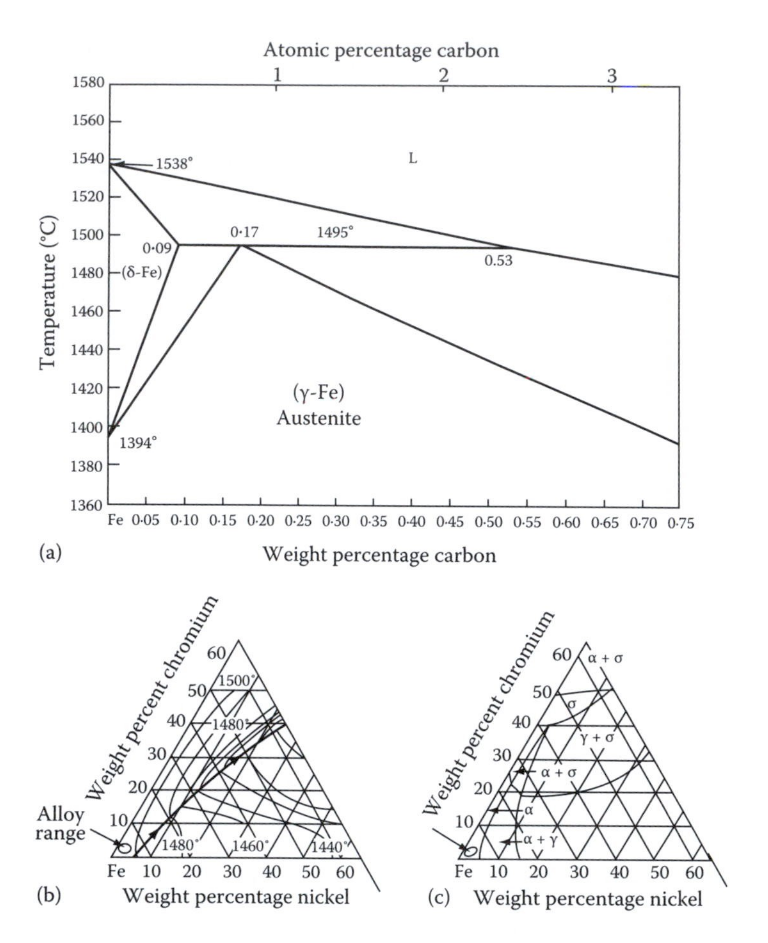

FIGURE 4.56
(a) Part of the iron–carbon phase diagram. (b) Liquidus projection for the Fe–Cr–Ni system, (c) isothermal section (650°C) for the Fe–Cr–Ni system. (From *Metals Handbook*, 8th edn., Vol. 8, American Society for Metals, Materials Park, OH, 1973, (a) p. 276, after D.T. Hawkins and R. Hultgren, (b) p. 424, after G.R. Speich. With permission.)

Subsequent transformations:

$$\gamma \rightarrow \alpha + Fe_3C \text{ (equilibrium structure at ambient temperature)}$$

Microstructures: see Figure 4.57.

Comments

Figure 4.56 shows that alloying with the relatively small amounts of Ni and Cr used in low-alloy steels has little effect on solidification temperature and that the equilibrium structure of the alloy is $\alpha(+Fe_3C)$. Figure 4.57a shows that quenching from the (γ, $\delta + L$) field leaves a structure with

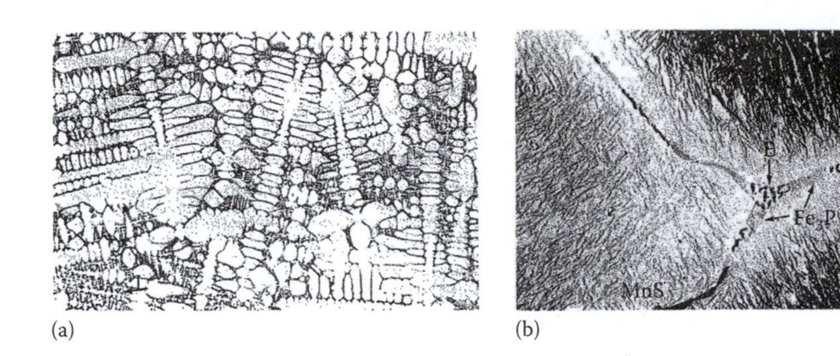

FIGURE 4.57
(a) Alloy quenched from (δ, γ + L) field (×25). (b) Cooled to 20°C. E refers to $\gamma/Fe_3P/Fe_3C$ eutectic (×1000). (From *Guide to the Solidification of Steels*, Jernkontoret, Stockholm, 1977. With permission.)

considerable residual melt between solidified dendrites. As discussed earlier, in practical alloys, the presence of residual melt between dendrite arms is largely due to impurity segregation. The completely solidified structure shown in Figure 4.57b exhibits a residual eutectic between α-Fe dendrites of $\gamma/Fe_3P/Fe_3C$, suggesting that the last liquid to solidify was rich in P and C. The retention of some γ-Fe in the eutectic is possibly due to the high carbon content of the residual iron (the solubility of P in γ-Fe is very low), which, together with the stabilizing effect of Ni, may help to retard the $\gamma \rightarrow \alpha$ transformation. Slower rates of cooling would probably reduce the amount of retained austenite still further. The presence of Mn induces the reaction: $Mn + S \rightarrow MnS$ (see Figure 4.57b). However, this is certainly preferred to FeS, which tends to wet dendrite boundaries more extensively than MnS and is a prime cause of hot cracking.

4.8.2 Casting of High-Speed Steels

Typical composition ranges:

$$\begin{aligned} &C{:}0.5-1.0\,\text{wt\%}\\ &Cr{:}0.5-4.0\,\text{wt\%}\\ &Mo{:}0.5-9.5\,\text{wt\%}\\ &W{:}1.5-6.0\,\text{wt\%}\\ &V{:}0.5-2.0\,\text{wt\%} \end{aligned}$$

Casting processes: ingot.

Special properties: hard, tough, wear-resistant at elevated temperatures.

Relevant phase diagrams: see Figure 4.58.

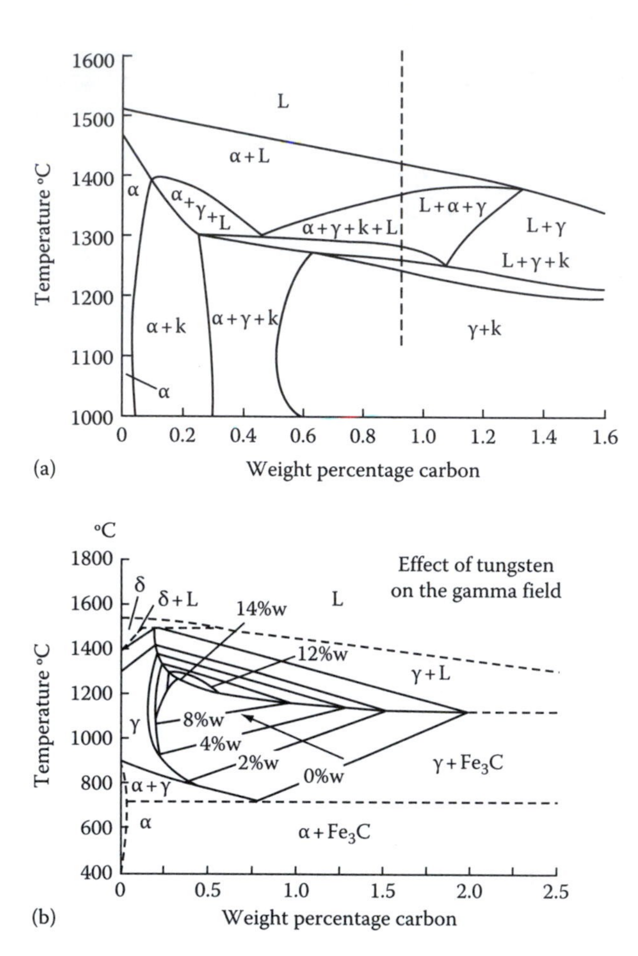

FIGURE 4.58
(a) Phase diagrams for steel with approximately 4 wt% Cr, 5 wt% Mo, 6 wt% W, and 2 wt% V. (After Horn, E. and Brandis, H., *DEW-Techn. Ber.*, 11, 147, 1971.) (b) Effect of W on γ field of steel. (After Cary, R.A. and Henry, R.J., *Metals Handbook*, 8th edn., Vol. 8, American Society for Metals, Materials Park, OH, 1973, p. 416.)

Solidification transformations:

$$L \rightarrow L + \alpha$$
$$L + \alpha \rightarrow L + \alpha + \gamma \text{ (peritectic } L + \delta \rightarrow \gamma)$$
$$L + \alpha + \gamma \rightarrow L + \alpha + \gamma + M_xC$$
$$L + \alpha + \gamma + M_xC \rightarrow \gamma + M_xC \text{ (eventually: } \rightarrow \alpha + M_xC)$$

Microstructures: see Figure 4.59.

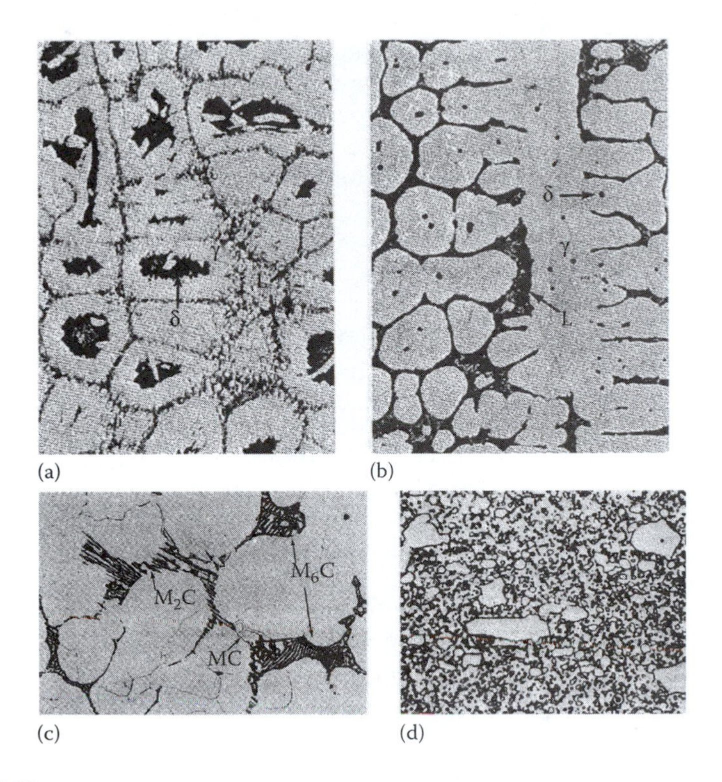

FIGURE 4.59
High-speed tool steel. (a) Quenched from the (L + α (or δ) + γ) field at 1335°C (×150). (b) Quenched from 1245°C (×150). (c) Same alloy after mechanical and thermal heat treatments (×750). (From *A Guide to the Solidification of Steel*, Jernkontoret, Stockholm, 1977.) (d) Final structure after plastic working, austenitizing and double-tempering. (From The ASM Committee on Metallography of Tool Materials, *Metals Handbook*, 8th edn., Vol. 7, American Society for Metals, Materials Park, OH, 1972, p. 121. With permission.)

Comments: Reference to the phase diagrams (Figure 4.58) shows that the presence of W, V, and the other main alloying elements produces a cascade of polyphase fields during cooling of these castings. Solidification occurs initially with the formation of α dendrites, but the γ fields are so extensive that rapid quenching from the $\alpha + L$ field cannot suppress the nucleation and growth of austenite (Figure 4.59a). As expected, the C segregates strongly when α forms but W, Cr, and V are not expected to segregate so markedly in α-Fe. It seems likely that the early formation of γ through the reaction: $\alpha + L \rightarrow \gamma + L$ in these castings is the main cause of the extensive segregation of W, Cr, and V as observed in Figure 4.59b and c. As seen in Figure 4.59b, it is possible that the $\alpha \rightarrow \gamma$ reaction occurs through the rejection of dissolved M back to the melt. Reference to the Fe–Cr, Fe–V, and Fe–W binary phase diagrams shows in all cases very low high-temperature solubility of these elements in γ-Fe. The resulting as-solidified structure (Figure 4.59c) thus consists of α dendrites (following the $\gamma \rightarrow \alpha$ solid-state transformation during cooling) with marked interdendritic segregation. The latter appears in the form of a γ/M_xC eutectic, where M_xC refers to mixed carbides of WC, Cr_2C, VC, etc. The final structure of this type of tool steel (Figure 4.59d) is only reached after further extensive plastic working to break up the eutectic, followed by austenitizing and double-tempering treatments.

Electrode composition range:

$$\begin{aligned}&\text{Cr:}17-19\,\text{wt\%}\\&\text{Ni:}8-10\,\text{wt\%}\\&\text{C:}0.05-0.1\,\text{wt\%}\\&\text{Si:}0.5-1.0\,\text{wt\%}\\&\text{Mn:}0.5-1.5\,\text{wt\%}\\&\left.\begin{array}{l}\text{P:}\\\text{S:}\end{array}\right\}\text{Traces}\end{aligned}$$

Welding process: manual metal arc, gas metal arc.

Relevant phase diagrams: see Figure 4.60.

Phase transformations:

$$\begin{aligned}&L \rightarrow \delta + L\\&\delta + L \rightarrow \delta + \gamma \text{ (approx. peritectic)}\end{aligned}$$

Microstructures: see Figure 4.61.

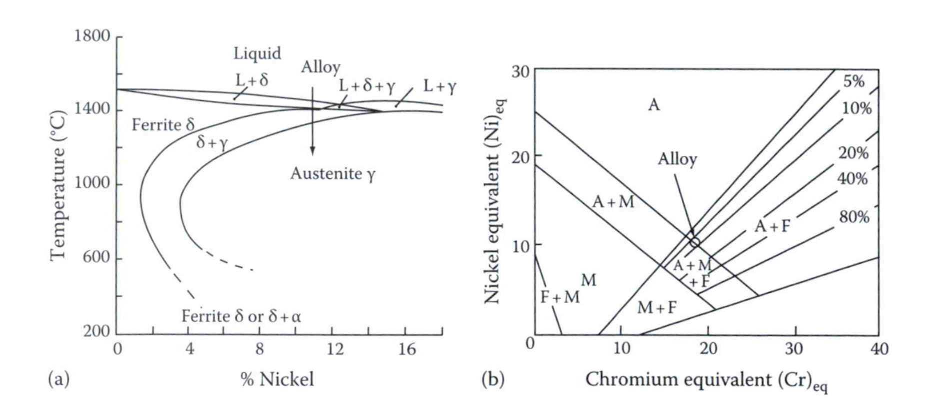

FIGURE 4.60
(a) 18% Cr section of the Fe–Cr–Ni system. (b) Schaeffler diagram indicating the alloy concerned. $(NI)_{cq}$ = %Ni + 30 × %C + 0.5 × %Mn. $(Cr)_{eq}$ = %Cr + %Mo + 1.5 × %Si + 0.5 × %Nb. A, austenite; F, ferrite; M, martensite. (After Castro, R.J. and de Cadenet, J.J., *Welding Metallurgy of Stainless and Heat Resistant Steels*, Cambridge University Press, London, 1974.)

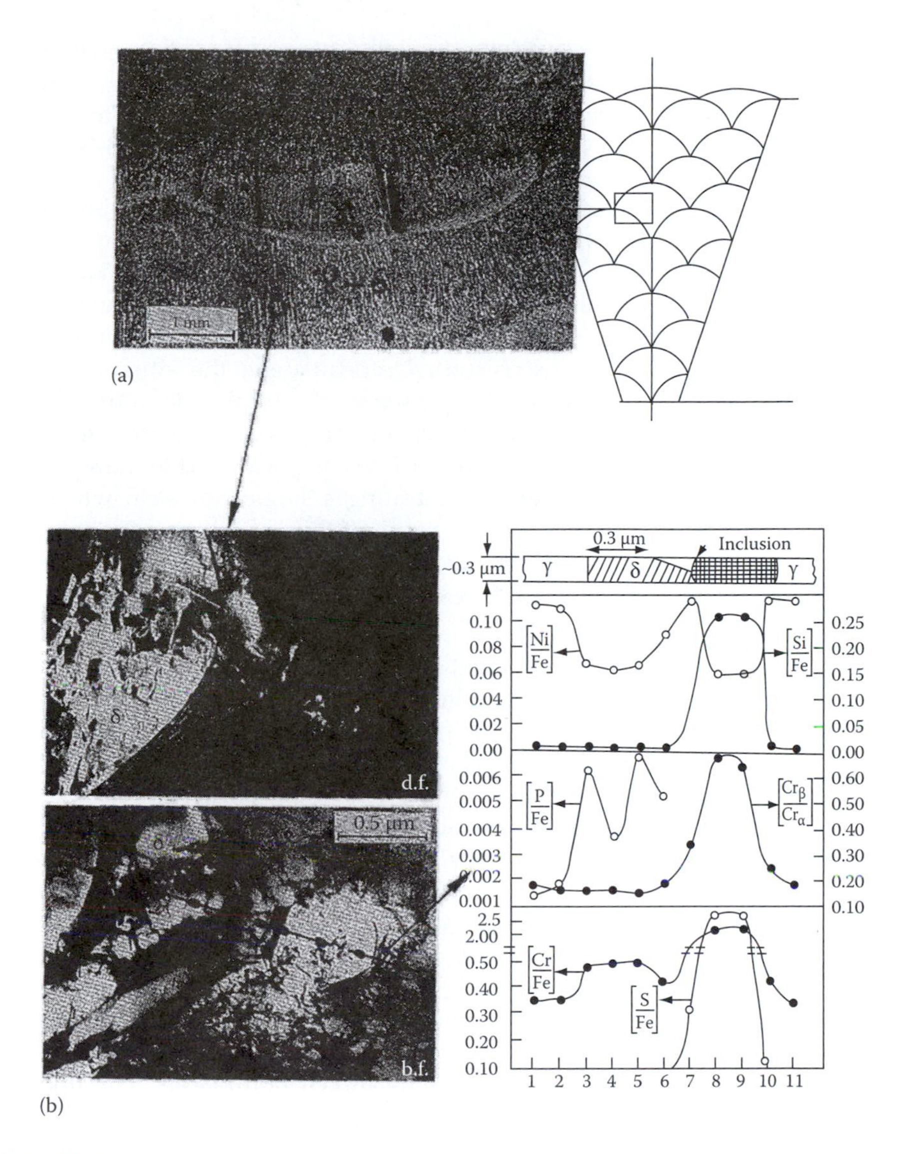

FIGURE 4.61
(a) Illustrating hot cracking in the austenite region of a duplex stainless steel weld deposit. (b) STEM-EDX microanalysis of γ- and δ-Fe and an inclusion. (After Åström, H., et al., *Metal Sci. J.*, 225, 1976.)

Comments:

a. *Phase equilibria*. Figure 4.60a is the 18% Cr vertical section of the Fe–Cr–Ni ternary diagram. The effect of these and other alloying elements on the final microstructure, assuming fairly high quench rates typical of welding, can be predicted with the help of the

Shaeffler diagram (Figure 4.60b). From this diagram it can be seen that if the Ni or Cr contents are reduced much below the nominal analyses given, there is a risk that martensite forms. The most important feature of these alloys with respect to welding, is that some δ-Fe is retained even at ambient temperatures (see Figure 4.60b).

There has been much discussion in the literature as to which phase solidifies first after welding. According to Figure 4.60a it appears that solidification should initiate with the nucleation of δ-Fe. However, if the base metal is fully austenitic at the transition zone, this phase should nucleate first because of the requirement of epitaxial growth (see previous discussion). Unfortunately, the situation is complicated in practice by the presence of carbon and nitrogen, both of which tend to move the peritectic composition toward higher Ni content. An example of the effect of N being admitted to the weld pool is shown in Figure 4.61a illustrating a single run weld which has remained fully austenitic. The result of this is fairly catastrophic, causing hot cracking at the austenite grain boundaries due to increased sulfur and phosphorus segregation in the austenite during solidification.

b. *Microstructure*. It is thought that the first phase to solidify in this alloy is δ-Fe, enriched in Cr and impoverished in Ni, the tendency in either case being to stabilize the ferrite. Further cooling causes γ-Fe to nucleate in the Ni-rich liquid between the δ-Fe dendrites. With the development of this duplex $\gamma + \delta$ structure the peritectic reaction: $L \rightarrow \gamma + \delta$ continues to completion. Cooling of the weld metal to ambient temperature causes the γ-Fe phase to grow at the expense of δ-Fe until only a fine network of δ-Fe remains. The STEM-based x-ray spectrometer microanalysis of the γ, δ, and inclusion phases (Figure 4.61b) indicates that the Cr-rich ferrite has dissolved the phosphorus (one of the inclusion). In this respect, Mn has a double role: both as a deoxidizer and to absorb S through forming MnS. If the weld solidifies directly to γ-Fe, all the Mn remains in solution and thus cannot prevent FeS forming at cell boundaries. The fine duplex $\gamma + \delta$ structure of stainless steel welds thus refines and strengthens the microstructure, and effectively renders S and P harmless. It should be pointed out, however, that the 6–8 vol% retained δ-Fe at ambient temperature should not be exceeded, since higher volume fractions reduce the ductility and toughness of this alloy. In this respect, the Shaeffler diagram (Figure 4.60b) is a useful guide for estimating δ-Fe as a function of equivalent Cr and Ni content. The amount of δ-Fe can also be measured magnetically or metallographically. If the presence of nitrogen is to be accounted for, a modified form of Shaeffler diagram (the DeLong diagram) can be employed (see, e.g., Castro and Cadenet[23]).

Exercises

4.1 Show that differentiation of Equation 4.4 leads to Equations 4.5 and 4.6.

4.2 Use Equations 4.4 and 4.10 to estimate the number of crystal-like clusters in 1 mm^3 of copper at its melting point for spherical clusters containing (a) 10 atoms, (b) 60 atoms. What volume of liquid copper is likely to contain one cluster of 100 atoms? The atomic volume of liquid copper is 1.6×10^{-29} m^3, γ_{SL} is 0.177 J m^{-2}, $k = 1.38 \times 10^{-23}$ J K^{-1}, $T_m = 1356$ K.

4.3 Why does r_{max} in Figure 4.5 vary with ΔT?

4.4 Calculate the homogeneous nucleation rate in liquid copper at undercoolings of 180, 200, and 220 K, using the following data:

$$L = 1.88 \times 10^9 \text{ J m}^{-3}, \quad T_m = 1356 \text{ K}, \quad \gamma_{SL} = 0.177 \text{ J m}^{-2},$$
$$f_0 = 10^{11} \text{ s}^{-1}, \quad C_0 = 6 \times 10^{28} \text{ atoms m}^{-3}, \quad k = 1.38 \times 1$$

4.5 Show that Equation 4.23 applies to homogeneous nucleation and heterogeneous nucleation on a flat mold wall.

4.6 Show that Equation 4.16 follows from Equation 4.15 using the following relationships for a spherical cap:

$$A_{SL} = 2\pi r^2(1 - \cos\theta)$$
$$A_{SM} = \pi r^2 \sin^2\theta$$
$$V_s = \pi r^3(2 + \cos\theta)(1 - \cos\theta)^2/3$$

4.7 Of what importance is the angle of a mold-wall crack in heterogeneous nucleation? Of what importance is the width of the crack at its mouth?

4.8 Under what conditions can solid metal be retained in a mold-wall crevice above T_m?

4.9 If a single crystal is melted by heating to slightly above its melting point and then cooled, it subsequently solidifies with the previous orientation. Likewise a polycrystalline specimen reverts to its original grain size. Can you suggest an explanation for this effect? (See B. Chalmers, *Principles of Solidification*, Wiley, New York, 1964, p. 85.)

4.10 (a) Show that surface melting is to be expected below T_m in gold (1336 K) given $\gamma_{SL} = 0.132$, $\gamma_{LV} = 1.128$, $\gamma_{LV} = 1.400$ J m^{-2}.

(b) Given that the latent heat of fusion of gold is 1.2×10^9 J m^{-3} estimate whether sensible liquid layer thicknesses are feasible at measurably lower temperatures than T_m.

4.11 Use nucleation theory to derive quantitative expressions for the velocity of an atomically smooth interface as a function of undercooling (a) for repeated surface nucleation, and (b) for spiral growth. (See Burton et al., *Philosophical Transactions*, A243: 299, 1950.)

4.12 Draw diagrams to show how the solid/liquid interface temperature varies as a function of position along the bar for Figures 4.20 through 4.22.

4.13 Draw figures corresponding to Figures 4.21 and 4.22 for a dilute binary alloy with $k > 1$.

4.14 Show that Equation 4.35 satisfies Equation 4.34.

4.15 Al–Cu phase diagram is similar to that shown in Figure 4.19 with T_m (Al) = 660°C, $T_E = 548$°C, $X_{max} = 5.65$ wt%, and $X_E = 33$ wt% Cu. The diffusion coefficient for the liquid $D_L = 3 \times 10^{-9}$ m^2 s^{-1}. If an Al–0.5 wt% Cu alloy is solidified with no convection and a planar solid/liquid interface at 5 μm s^{-1}:

(a) What is the interface temperature in the steady state?
(b) What is the thickness of the diffusion layer?
(c) What temperature gradient will be required to maintain a planar interface?
(d) Answer (a), (b), and (c) for an Al–2 wt% Cu alloy solidification under the same conditions.

4.16 (a) Using Equation 4.33 and the data in Exercise 4.15 plot the variation of copper concentration along a unidirectionally solidified bar of an Al–2 wt% Cu alloy assuming no diffusion in the solid and perfect mixing in the liquid.
(b) What fraction of the bar will solidify to a eutectic structure?
(c) How much eutectic would form in an Al–0.5 wt% Cu alloy solidified under the same conditions?

4.17 Explain the experimental observation that in the presence of a convection current cells grow upstream.

4.18 Sketch a possible solidification-front structure for the solidification of an Fe–0.25 wt% C alloy in a shallow temperature gradient. Consider the temperature range 1440°C–1540°C. Assume very rapid diffusion of carbon in δ-Fe.

4.19 Show that the condition $\lambda = 2\lambda^*$ gives (1) the maximum eutectic growth rate for a given undercooling, and (2) a minimum undercooling for a given growth rate (Equation 4.43).

4.20 Calculate the depression of the eutectic temperature for a lamellar eutectic with $\lambda = 0.2$ μm and $\lambda = 1.0$ μm, if $\gamma_{\alpha\beta} = 400$ mJ m^{-2}, $\Delta H/V_m = 800 \times 10^6$ J m^{-3}, $T_E = 1000$ K.

4.21 If it is assumed that the choice of a rod or lamellar eutectic is governed by the minimization of the total α/β interfacial energy it can be shown that for a given λ there is a critical volume fraction of the β-phase (f_c) below which β should be rodlike, and above which it should be lamellar. Assuming the rods are hexagonally arranged and that $\gamma_{\alpha\beta}$ is isotropic, calculate the value of f_c.

4.22 Compare the processes of ingot casting and weld solidification, and show they are in many ways quite different solidification processes. How would you compare continuous casting in this respect?

4.23 What is the influence of welding speed on the solidification structure of welds? How is welding speed likely to affect segregation problems?

References

1. D. Turnbull and R.E. Cech, *J. Appl. Phys.*, 21: 804, 1950.
2. R.W. Cahn, *Nature*, 273: 491, 1978.
3. J. Frenkel, *Kinetic Theory of Liquids*, Dover, New York, 1955.
4. R.M.J. Cotterill et al., *Phil. Mag.*, 31: 245, 1975.
5. Reasons for this crystallographic feature of dendrite growth are discussed for example in B. Chalmers, *Principles of Solidification*, Wiley, New York, 1964, p. 116.
6. V.G. Smith et al., *Can. J. Phys.*, 33: 723, 1955.
7. W.G. Pfann, *Zone Melting*, 2nd edn., John Wiley & Sons, New York, 1966.
8. K.A. Jackson and J.D. Hunt, *Acta Metall.*, 13: 1212, 1965.
9. For a review of eutectic solidification see R. Elliot, Eutectic solidification, Review 219 in *International Metals Reviews*, September 1977, p. 161.
10. See M.C. Flemings, *Solidification Processing*, McGraw-Hill, New York, 1974, p. 101.
11. For a complete treatment of these effects see K.A. Jackson and J.D. Hunt, *Trans. Metall. Soc. AIME*, 236: 1129, 1966.
12. M.H. Burden and J.D. Hunt, The extent of the eutectic range, *J. Cryst. Growth*, 22: 328, 1974.
13. F.D. Lemkey and N.J. Salkind, in H.S. Peiser (Ed.), *Crystal Growth*, Pergamon, Oxford, 1967, p. 171.
14. M.C. Flemings, *Solidification Processing*, McGraw-Hill, New York, 1974, p. 177.
15. A. Ohno, *The Solidification of Metals*, Chijin Shokan, Tokyo, 1976, p. 69.
16. For a detailed treatment see M.C. Flemings, Principles of control of soundness and homogeneity of large ingots, *Scand. J. Metall.*, 5: 1, 1976.
17. D. Rosenthal, *Trans. Am. Soc. Mech. Eng.*, 68: 849, 1946.
18. R.J. Jesseman, in J. Crane (Ed.), *Microalloying'75*, Union Carbide Corporation, Washington DC, 1975, Vol. 3, p. 74.
19. P. Haasen, *Physical Metallurgy*, 3rd edn., Cambridge University Press, Cambridge, 1997.
20. M. Miller and P. Liaw (Eds.), *Bulk Metallic Glasses, An Overview*, 1st edn., Springer, New York, 2008.
21. T. Egami and Y. Waseda, Atomic size effect on the formability of metallic glasses, *J. Non-Cryst. Solids*, 64(1–2): 113–134, 1984.
22. R. Busch, J. Schroers, and W.H. Wang, Thermodynamics and kinetics of bulk metallic glasses, *MRS Bull.*, 32(8): 620–623 (2007).
23. R.J. Castro and J.J. de Cadenet, *Welding Metallurgy of Stainless and Heat Resistant Steels*, Cambridge University Press, London, 1974.
24. T.G. Gray, J. Spence, and T.H. North, *Rational Welding Design*, Newnes-Butterworth, London, 1975.

Further Reading

H. Biloni, Solidification (Chapter 9) and R.W. Cahn, Alloys quenched from the melt (Chapter 28), in R.W. Cahn and P. Haasen (Eds.), *Physical Metallurgy*, North-Holland, Amsterdam, 1983.

G.A. Chadwick, *Metallography of Phase Transformations*, Butterworths, London, 1972.

B. Chalmers, *Principles of Solidification*, John Wiley & Sons, New York, 1964.

S.A. David and J.M. Vitek, Correlation between solidification structures and weld microstructures, *Int. Mater. Rev.*, 34: 213–45, 1989.

G.J. Davies and P. Garland, Solidification structures and properties of fusion welds, *Int. Meter. Rev.*, 20(2): 83–105, 1975.

K.E. Easterling, *Introduction to the Physical Metallurgy of Welding*, 2nd edn., Butterworth-Heinemann, London, 1992.

M.C. Flemings, *Solidification Processing*, McGraw-Hill, New York, 1974.

W. Kurz and D.J. Fisher, *Fundamentals of Solidification*, Trans. Tech. Publications, Clausthol-Zellerfield, Germany, 1984.

H. Ohno, *The Solidification of Metals*, Chijin Shokan, Tokyo, 1976.

5

Diffusional Transformations in Solids

The majority of phase transformations that occur in the solid state take place by thermally activated atomic movements. The transformations that will be dealt with in this chapter are those that are induced by a change of temperature of an alloy that has a fixed bulk composition. Usually we will be concerned with the transformations caused by a temperature change from a single-phase region of a (binary) phase diagram to a region where one or more other phases are stable. The different types of phase transformations that are possible can be roughly divided into the following groups: (a) precipitation reactions, (b) eutectoid transformations, (c) ordering reactions, (d) massive transformations, and (e) polymorphic changes. Figure 5.1 shows several different types of binary phase diagrams that are representative of these transformations.

Precipitation transformations can be expressed in reaction terms as follows:

$$\alpha' \rightarrow \alpha + \beta \tag{5.1}$$

where

α' is a metastable supersaturated solid solution
β is a stable or metastable precipitate
α', but with a composition closer to equilibrium, see Figure 5.1a

Eutectoid transformations involve the replacement of a metastable phase (γ) by a more stable mixture of two other phases ($\alpha + \beta$) and can be expressed as

$$\gamma \rightarrow \alpha + \beta \tag{5.2}$$

This reaction is characteristic of phase diagrams such as that shown in Figure 5.1b.

Both precipitation and eutectoid transformations involve the formation of phases with a different composition to the matrix and therefore long-range diffusion is required. The remaining reaction types can, however, proceed without any composition change or long-range diffusion. Figure 5.1c shows phase diagrams where ordering reactions can occur. In this case the reaction can be simply written

$$\alpha\,(\text{disordered}) \rightarrow \alpha'\,(\text{ordered}) \tag{5.3}$$

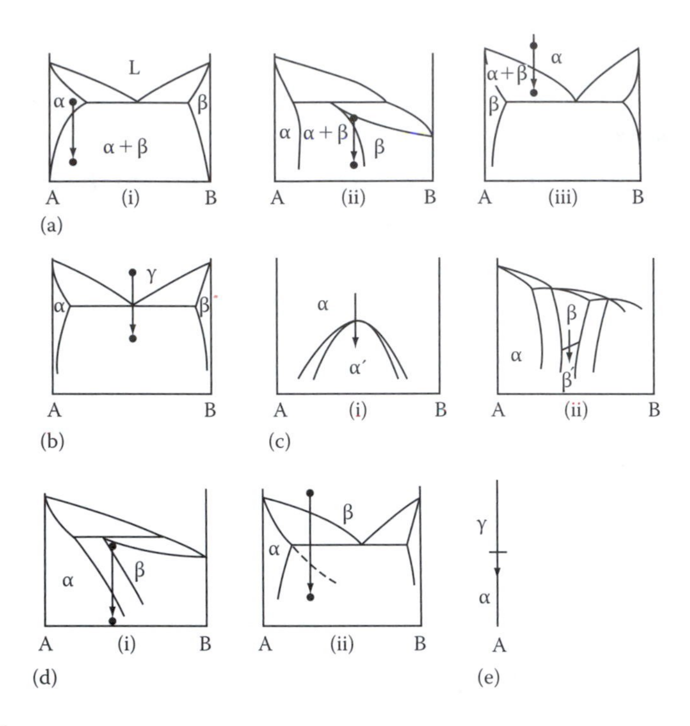

FIGURE 5.1
Examples of different categories of diffusion phase transformations: (a) precipitation, (b) eutectoid, (c) ordering, (d) massive, and (e) polymorphic (single component).

In a massive transformation the original phase decomposes into one or more new phases which have the same composition as the parent phase, but different crystal structures. Figure 5.1d illustrates two simple examples of the type

$$\beta \rightarrow \alpha \tag{5.4}$$

where only one new phase results. Note that the new β-phase can either be stable (Figure 5.1d(i)) or metastable (Figure 5.1d(ii)).

Polymorphic transformations occur in single component systems when different crystal structures are stable over different temperature ranges, Figure 5.1e. The most well-known of these in metallurgy are the transformations between face-centered cubic (fcc)- and body-centered cubic (bcc)-Fe. In practice, however, such transformations are of little practical interest and have not been extensively studied.

Apart from a few exceptions the above transformations all take place by diffusional nucleation and growth. As with solidification, nucleation is usually heterogeneous, but for the sake of simplicity let us begin by considering homogeneous nucleation.

5.1 Homogeneous Nucleation in Solids

To take a specific example consider the precipitation of b-rich β from a supersaturated A-rich α solid solution as shown in Figure 5.1a(i). For the nucleation of β, B-atoms within the α matrix must first diffuse together to form a small volume with the β composition, and then, if necessary, the atoms must rearrange into the β crystal structure. As with the liquid $\rightarrow$ solid transformation an α/β interface must be created during the process and this leads to an activation energy barrier.

The free energy change associated with the nucleation process will have the following three contributions:

1. At temperatures where the β-phase is stable, the creation of a volume V of β will cause a volume free energy reduction of $V\Delta G_v$
2. Assuming for the moment that the α/β interfacial energy is isotropic the creation of an area A of interface will give a free energy increase of $\tilde{A}\gamma$
3. In general the transformed volume will not fit perfectly into the space originally occupied by the matrix and this gives rise to a misfit strain energy ΔG_s per unit volume of β. (It was shown in Chapter 3 that, for both coherent and incoherent inclusions, ΔG_s is proportional to the volume of the inclusion.) Summing all of these gives the total free energy change as

$$\Delta G = -V\Delta G_v + A\gamma + V\Delta G_s \tag{5.5}$$

Apart from the misfit strain energy term, Equation 5.5 is very similar to that derived for the formation of a solid nucleus in a liquid. With solid/liquid interfaces γ can be treated as roughly the same for all interfaces, but for nucleation in solids γ can vary widely from very low values for coherent interfaces to high values for incoherent interfaces. Therefore the $A\gamma$ term in Equation 5.5 should really be replaced by a summation over all surfaces of the nucleus $\Sigma\gamma_i A_i$.

If we ignore the variation of γ with interface orientation and assume the nucleus is spherical with a radius of curvature r Equation 5.5 becomes

$$\Delta G = -\frac{4}{3}\pi r^3(\Delta G_v - \Delta G_s) + 4\pi r^2\gamma \tag{5.6}$$

This is shown as a function of r in Figure 5.2. Note that the effect of the misfit strain energy is to reduce the effective driving force for the transformation to $(\Delta G_v - \Delta G_s)$. Similar curves would in fact be obtained for any nucleus shape as a function of its size. Differentiation of Equation 5.6 yields

$$r^* = \frac{2\gamma}{(\Delta G_v - \Delta G_s)} \tag{5.7}$$

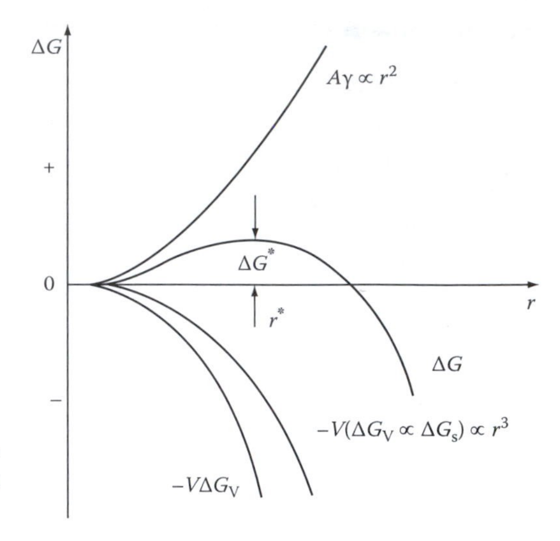

FIGURE 5.2
Variation of ΔG with r for a homogeneous nucleus. There is an activation energy barrier ΔG^*.

$$\Delta G^* = \frac{16\pi\gamma^3}{3(\Delta G_v - \Delta G_s)^2} \tag{5.8}$$

which is very similar to the expressions for solidification, except now the chemical driving force ΔG_v is reduced by a positive strain energy term.

As discussed in Chapter 4 the concentration of critical-sized nuclei C^* will be given by

$$C^* = C_0 \exp(-\Delta G^*/kT) \tag{5.9}$$

where C_0 is the number of atoms per unit volume in the phase. If each nucleus can be made supercritical at a rate of f per second the homogeneous nucleation rate will be given by

$$N_{\text{hom}} = fC^* \tag{5.10}$$

f depends on how frequently a critical nucleus can receive an atom from the α matrix. This will depend on the surface area of the nucleus and the rate at which diffusion can occur. If the activation energy for atomic migration is ΔG_m per atom, f can be written as $\omega \exp(-\Delta G_m/kT)$ where ω is a factor that includes the vibration frequency of the atoms and the area of the critical nucleus. The nucleation rate will therefore be of the form

$$N_{\text{hom}} = \omega C_0 \exp\left(\frac{\Delta G_m}{kT}\right) \exp\left(\frac{\Delta G^*}{kT}\right) \tag{5.11}$$

This is essentially identical to Equation 4.12 except that the temperature dependence of f has been taken into account. In order to evaluate this equation as a function of temperature ω and ΔG_m can be assumed to be constant, but ΔG^* will be strongly temperature dependent. The main factor controlling ΔG^* is the driving force for precipitation ΔG_v, Equation 5.8. Since composition is variable the magnitude of ΔG_v must be obtained from the free energy–composition diagram.

If the alloy X_0 in Figure 5.3, is solution treated at T_1 and then cooled rapidly to T_2 it will become supersaturated with B and will try to precipitate β. When the transformation to $\alpha + \beta$ is complete the free energy of the alloy will have decreased by an amount ΔG_0 per mole as shown in Figure 5.3b. ΔG_0 is therefore the total driving force for the transformation. However, it is not the driving force for nucleation. This is because the first nuclei to appear do not significantly change the α composition from X_0. The free energy released per mole of nuclei formed can be obtained as follows.

If a small amount of material with the nucleus composition (X_B^β) is removed from the α-phase, the total free energy of the system will decrease by ΔG_1:

$$\Delta G_1 = \mu_A^\alpha X_A^\beta + \mu_B^\alpha X_B^\beta \text{ (per mol } \beta \text{ removed)} \tag{5.12}$$

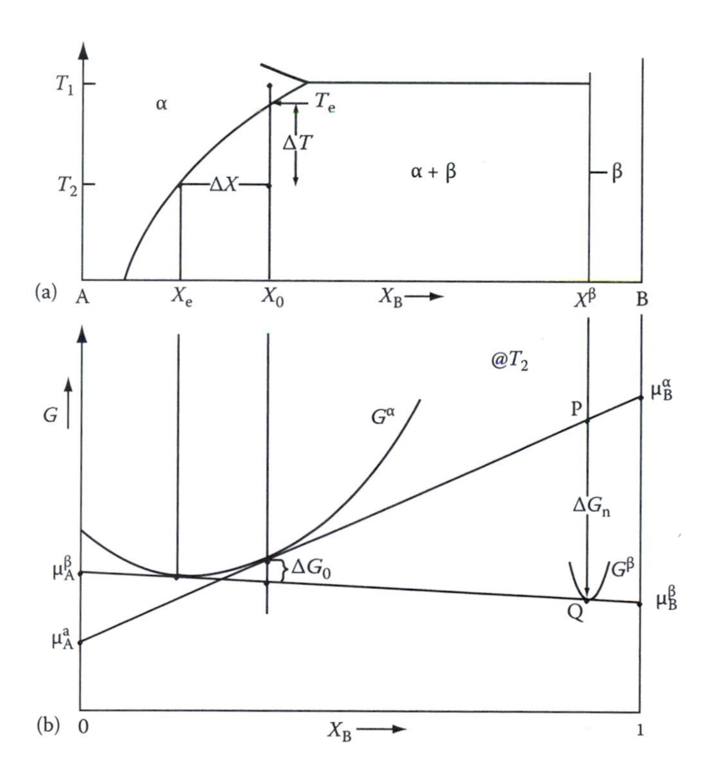

FIGURE 5.3
Free energy changes during precipitation. The driving force for the first precipitates to nucleate is $\Delta G_n = \Delta G_v V_m$. ΔG_0 is the total decrease in free energy when precipitation is complete and equilibrium has been reached.

This follows simply from the definition of chemical potential given by Equation 1.29. ΔG_1 is a quantity represented by point P in Figure 5.3b. If these atoms are now rearranged into the β crystal structure and replaced, the total free energy of the system will increase by an amount

$$\Delta G_2 = \mu_A^\beta X_A^\beta + \mu_B^\beta X_B^\beta \text{ (per mol } \beta \text{ removed)} \tag{5.13}$$

which is given by point Q. Therefore the driving force for nucleation

$$\Delta G_n = \Delta G_2 - \Delta G_1 \text{ per mol of } \beta \tag{5.14}$$

which is just the length PQ in Figure 5.3b. The volume free energy decrease associated with the nucleation event is therefore simply given by

$$\Delta G_v = \frac{\Delta G_n}{V_m} \text{ per unit volume of } \beta \tag{5.15}$$

where V_m is the molar volume of β. For dilute solutions it can be shown that approximately

$$\Delta G_v \propto \Delta X \tag{5.16}$$

where

$$\Delta X = X_0 - X_e \tag{5.17}$$

From Figure 5.3a there it can be seen that the driving force for precipitation increases with increasing undercooling (ΔT) below the equilibrium solvus temperature T_e.

It is now possible to evaluate Equation 5.11 for alloy X_0 as a function of temperature. The variation of ΔG_v with temperature is shown schematically in Figure 5.4b. After taking into account the misfit strain energy term ΔG_s the effective driving force becomes $(\Delta G_v - \Delta G_s)$ and the effective equilibrium temperature is reduced to T_e'. Knowing $(\Delta G_v - \Delta G_s)$ the activation energy ΔG^* can be calculated from Equation 5.8 as shown. Figure 5.4c shows the two exponential terms in Equation 5.11; $\exp(-\Delta G^*/kT)$ is essentially the potential concentration of nuclei and, as with nucleation in liquids, this is essentially zero until a critical undercooling ΔT_c is reached, after which it rises very rapidly. The other term, $\exp(-\Delta G_m/kT)$, is essentially the atomic mobility. Since ΔG_m is constant this decreases rapidly with decreasing temperature. The combination of these terms, i.e., the homogeneous nucleation rate is shown in Figure 5.4d. Note that at undercoolings smaller than ΔT_c, N is negligible because the driving force ΔG_v is too small, whereas at very high undercoolings N is negligible because diffusion is too slow. A maximum nucleation rate is obtained at intermediate undercoolings. For alloys containing less solute the

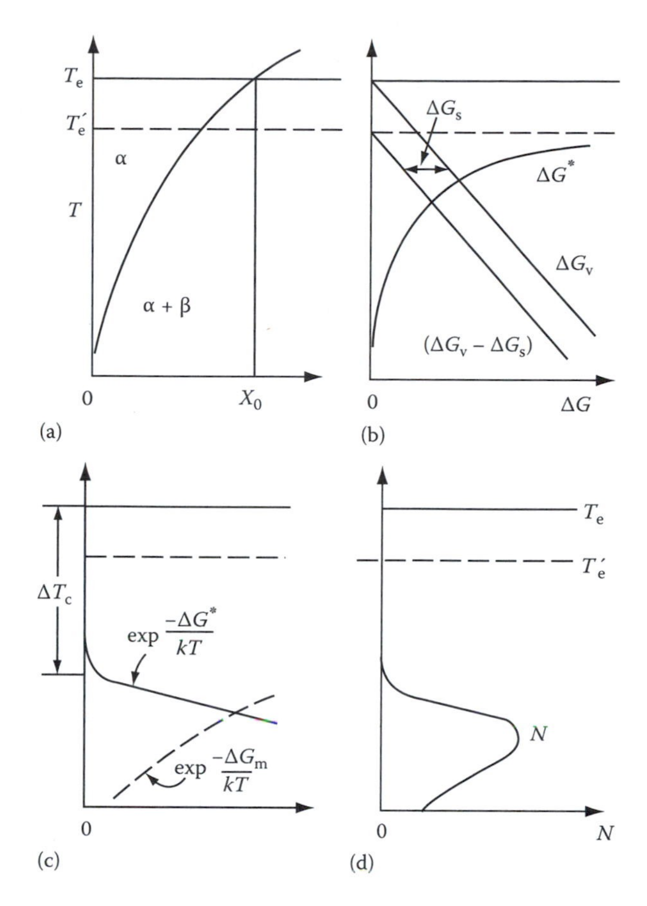

FIGURE 5.4
How the rate of homogeneous nucleation varies with undercooling for alloy X_0. (a) Phase diagram. (b) Effective driving force $(\Delta G_V - \Delta G_S)$ and resultant energy barrier ΔG^*. (c) Two exponential terms that determine N as shown in (d).

critical supercooling will not be reached until lower absolute temperatures where diffusion is slower. The resultant variation of N with T in these alloys will therefore appear as shown in Figure 5.5.

In the above treatment of nucleation it has been assumed that the nucleation rate is constant. In practice, however, the nucleation rate will initially be low, then gradually rise, and finally decrease again as the first nuclei to start growing and thereby reduce the supersaturation of the remaining α.

It has also been assumed that the nuclei are spherical with the equilibrium composition and structure of the β-phase. However, in practice nucleation will be dominated by whatever nucleus has the minimum activation energy barrier ΔG^*. Equation 5.8 shows that by far the most effective way of minimizing ΔG^* is by the formation of nuclei with the smallest total interfacial energy. In fact this criterion is dominating in nucleation processes. Incoherent

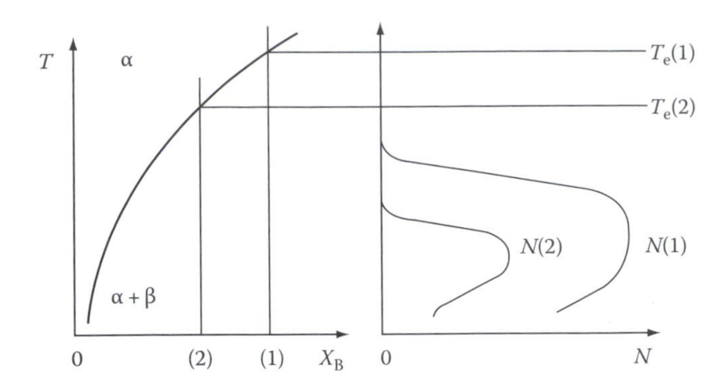

FIGURE 5.5
Effect of alloy composition on the nucleation rate. The nucleation rate in alloy 2 is always less than in alloy 1.

nuclei have such a high value of γ that incoherent homogeneous nucleation is virtually impossible. If, however, the nucleus has an orientation relationship with the matrix, and coherent interfaces are formed, ΔG^* is greatly reduced and homogeneous nucleation becomes feasible. The formation of a coherent nucleus will of course increase ΔG_s which decreases T'_e. But below T'_e the decrease in γ resulting from coherency can more than compensate for the increase in ΔG_s. Also, by choosing a suitable shape it is often possible to minimize ΔG_s as discussed in Section 3.4.3.

In most systems the α and β phases have such different crystal structures that it is impossible to form coherent low-energy interfaces and homogeneous nucleation of the equilibrium β-phase is then impossible. However, it is often possible to form a coherent nucleus of some other metastable phase (β′) which is not present in the equilibrium phase diagram. The most common example of this is the formation of GP zones which will be discussed in more detail later.

There are a few systems in which the equilibrium phase may nucleate homogeneously. For example in the Cu–Co system Cu alloys containing 1%–3% Co can be solution treated and quenched to a temperature where Co precipitates. Both Cu and Co are fcc with only a 2% difference in lattice parameter. Therefore very little coherency strain is associated with the formation of coherent Co particles. The interfacial energy is about 200 mJ m^{-2} and the critical undercooling for measurable homogeneous nucleation is about 40°C. This system has been used to experimentally test the theories of homogeneous nucleation and reasonably close agreement was found.[1]

Another system in which the equilibrium phase is probably formed homogeneously at a few tens of degree undercooling is the precipitation of Ni_3Al in many Ni-rich alloys. Depending on the system the misfit varies up to a maximum of 2%, and γ is probably less than 30 mJ m^{-2}. Most other examples of homogeneous nucleation, however, are limited to metastable phases, usually GP zones (Section 5.5.1).

5.2 Heterogeneous Nucleation

Nucleation in solids, as in liquids, is almost always heterogeneous. Suitable nucleation sites are nonequilibrium defects such as excess vacancies, dislocations, grain boundaries, stacking faults, inclusions, and free surfaces, all of which increase the free energy of the material. If the creation of a nucleus results in the destruction of a defect, some free energy (ΔG_d) will be released thereby reducing (or even removing) the activation energy barrier. The equivalent to Equation 5.5 for heterogeneous nucleation is

$$\Delta G_{het} = -V(\Delta G_v - \Delta G_s) + A\gamma - \Delta G_d \tag{5.18}$$

Nucleation on grain boundaries

Ignoring any misfit strain energy, the optimum embryo shape should be that which minimizes the total interfacial free energy. The optimum shape for an incoherent grain-boundary nucleus will consequently be two abutted spherical caps as shown in Figure 5.6, with θ given by

$$\cos\theta = \gamma_{\alpha\alpha}/2\gamma_{\alpha\beta} \tag{5.19}$$

(assuming $\gamma_{\alpha\beta}$ is isotropic and equal for both grains). The excess free energy associated with the embryo will be given by

$$\Delta G = -VG_v + A_{\alpha\beta}\gamma_{\alpha\beta} - A_{\alpha\alpha}\gamma_{\alpha\alpha} \tag{5.20}$$

where

- V is the volume of the embryo
- $A_{\alpha\beta}$ is the area of α/β interface of energy $\gamma_{\alpha\beta}$ created
- $A_{\alpha\alpha}$ is the area α/α grain boundary of energy $\gamma_{\alpha\alpha}$ destroyed during the process

The last term of the above equation is simply ΔG_d in Equation 5.18.

It can be seen that grain-boundary nucleation is analogous to solidification on a substrate (Section 4.1.3) and the same results will apply. Again the critical radius of the spherical caps will be independent of the grain boundary and given by

$$r^* = 2\gamma_{\alpha\beta}/\Delta G_v \tag{5.21}$$

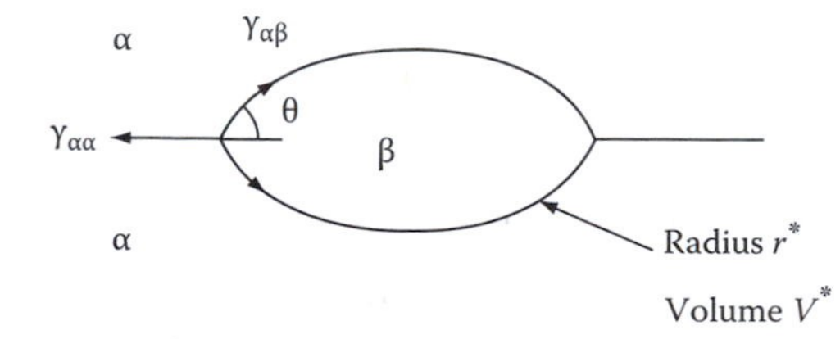

FIGURE 5.6
Critical nucleus size (V^*) for grain-boundary nucleation.

and the activation energy barrier for heterogeneous nucleation will be given by

$$\frac{\Delta G^*_{\text{het}}}{\Delta G^*_{\text{hom}}} = \frac{V^*_{\text{het}}}{V^*_{\text{hom}}} = S(\theta) \tag{5.22}$$

where $S(\theta)$ is a shape factor given by

$$S(\theta) = \frac{1}{2}(2 + \cos\theta)^2(1 - \cos\theta)^2 \tag{5.23}$$

The ability of a grain boundary to reduce ΔG^*_{het}, i.e., its potency as a nucleation site depends on $\cos\theta$, i.e., on the ratio $\gamma_{\alpha\alpha}/2\gamma_{\alpha\beta}$.

V^* and ΔG^* can be reduced even further by nucleation on a grain edge or grain corner, Figures 5.7 and 5.8. Figure 5.9 shows how $\Delta G^*_{\text{het}}/\Delta G^*_{\text{hom}}$ depends on $\cos\theta$ for the various grain-boundary nucleation sites.

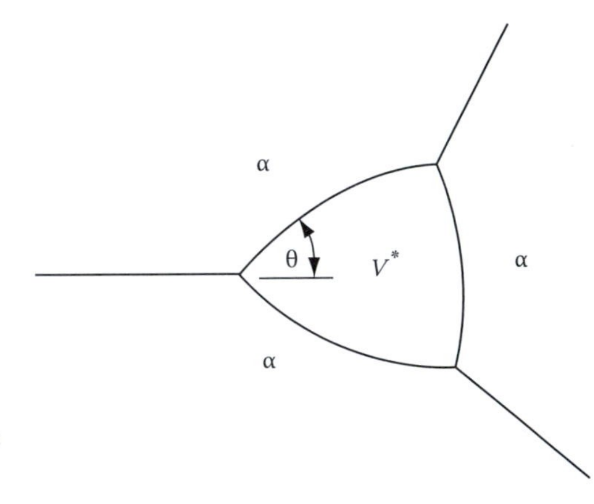

FIGURE 5.7
Critical nucleus shape for nucleation on a grain edge.

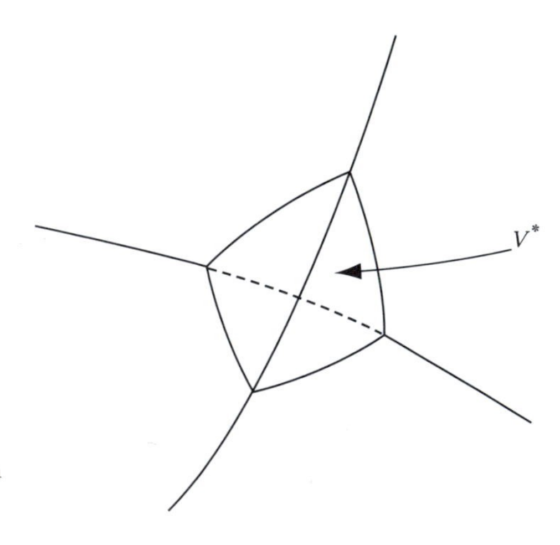

FIGURE 5.8
Critical nucleus shape for nucleation on a grain corner.

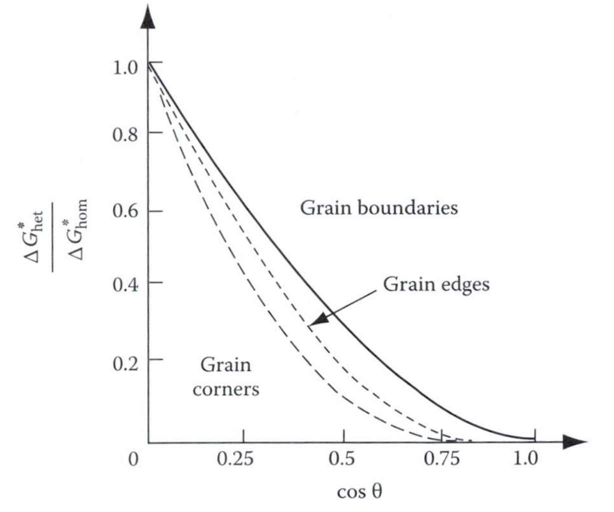

FIGURE 5.9
Effect of θ on the activation energy for grain boundary nucleation relative to homogeneous nucleation. (After Cahn, J.W., *Acta Metall.*, 4, 449, 1956.)

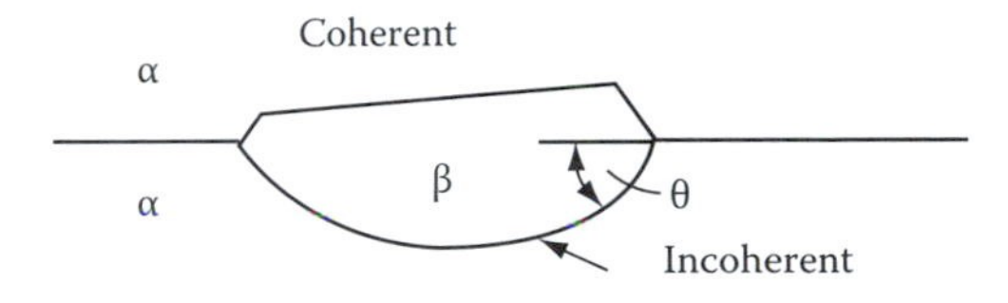

FIGURE 5.10
Critical nucleus size can be reduced even further by forming a low-energy coherent interface with one grain.

High-angle grain boundaries are particularly effective nucleation sites for incoherent precipitates with high $\gamma_{\alpha\beta}$. If the matrix and precipitate are sufficiently compatible to allow the formation of lower energy facets then V^* and ΔG^*_{het} can be further reduced as shown in Figure 5.10. The nucleus will then have an orientation relationship with one of the grains. Such nuclei are to be expected whenever possible, since the most successful nuclei, i.e., those which form most rapidly, will have the smallest nucleation barrier.

Other planar defects such as inclusion/matrix interfaces, stacking faults, and free surfaces can behave in a similar way to grain boundaries in reducing ΔG^*. Note, however, that stacking faults are much less potent sites due to their lower energy in comparison to high-angle boundaries.

Dislocations

The lattice distortion in the vicinity of a dislocation can assist nucleation in several ways. The main effect of dislocations is to reduce the ΔG_s contribution to ΔG^* by reducing the total strain energy of the embryo. A coherent nucleus with a negative misfit, i.e., a smaller volume than the matrix, can reduce its ΔG^* by forming in the region of compressive strain above an edge dislocation, whereas if the misfit is positive it is energetically favorable for it to form below the dislocation.

Nucleation on dislocations may also assisted by solute segregation which can raise the composition of the matrix to nearer that of the precipitate.

The dislocation can also assist in growth of an embryo beyond the critical size by providing a diffusion pipe with a lower ΔG_m.

Dislocations are not very effective for reducing the interfacial energy contribution of ΔG^*. This means that nucleation on dislocations usually requires rather good matching between precipitate and matrix on at least one plane, so that low-energy coherent or semicoherent interfaces can form. Ignoring strain energy effects, the minimum ΔG^* is then achieved when the nucleus shape is the equilibrium shape given by the Wulff construction. When the precipitate and matrix have different crystal structures the critical nucleus should therefore be disklike or needlelike as discussed in Section 3.4.2.

In fcc crystals the $\frac{a}{2}\langle 110 \rangle$ unit dislocations can dissociate to produce a ribbon of stacking fault, for example,

$$\frac{a}{1}[110] \rightarrow \frac{a}{6}[121] + \frac{a}{6}[211]$$

giving a stacking fault on $(1\bar{1}1)$ separated by two Shockley partials. Since the stacking fault is in effect four close-packed layers of hcp crystal (Figure 3.59b) it can act as a very potent nucleation site for an hcp precipitate. This type of nucleation has been observed for the precipitation of the hexagonal transition phase γ' in Al–Ag alloys. Nucleation is achieved simply by the diffusion of silver atoms to the fault. Thus there will automatically be an orientation relationship between the γ' precipitate (fault) and the matrix of the type

$$(0001)_{\gamma'} // (1\bar{1}1)_{\alpha}$$
$$[11\bar{2}0]_{\gamma'} // [110]_{\alpha}$$

which ensures good matching and low-energy interfaces.

It should be noted that even in annealed specimens dislocation densities are often sufficiently high to account for any precipitate dispersion that is resolvable in the light microscope, i.e., $\sim 1\ \mu m^{-2}$. Figure 5.11 shows an example of niobium carbonitride precipitates on dislocations on a ferritic iron matrix. This is a so-called dark-field electron microscope micrograph in which the precipitates are imaged bright and the matrix dark. The precipitates in rows along dislocations.

Excess vacancies

When an age-hardening alloy is quenched from a high temperature, excess vacancies are retained during the quench. These vacancies can assist nucleation by increasing diffusion rates, or by relieving misfit strain energies. They may influence nucleation either individually or collectively by grouping into small clusters.

Since ΔG_d is relatively small for vacancies, nucleation will only take place when a reasonable combination of the following is met: low interfacial energy (i.e., fully coherent nuclei), small volume strain energy, and high

FIGURE 5.11
Rows of niobium carbonitride precipitates on dislocations in ferrite (×108,000). (Dark-field electron micrograph in which the precipitates show up bright.)

driving force. These are essentially the same conditions that must be fulfilled for homogeneous nucleation. Since individual vacancies or small clusters cannot be resolved with conventional transmission electron microscopy (TEM), evidence for the role of vacancies as heterogeneous nucleation sites is indirect (discussed later).

5.2.1 Rate of Heterogeneous Nucleation

If the various nucleation sites are arranged in order increasing ΔG_d, i.e., decreasing ΔG^*, the sequence would be roughly

1. Homogeneous sites
2. Vacancies
3. Dislocations
4. Stacking faults
5. Grain boundaries and interphase boundaries
6. Free surfaces

Nucleation should always occur most rapidly on sites near the bottom of the list. However, the relative importance of these sites in determining the overall rate at which the alloy will transform also depends on the relative

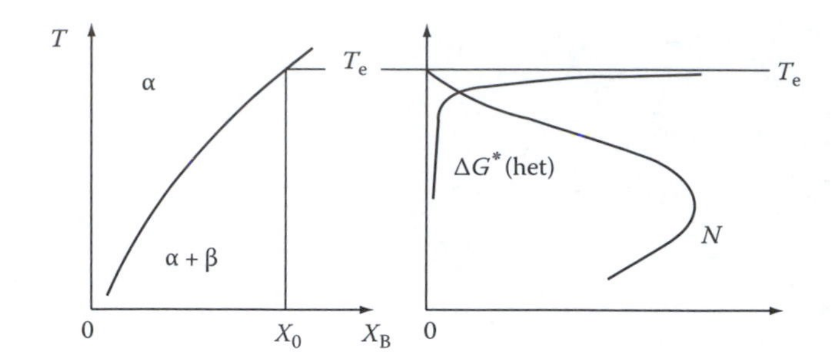

FIGURE 5.12
Rate of heterogeneous nucleation during precipitation of β in alloy X_0 as a function of undercooling.

concentrations of the sites. For homogeneous nucleation every atom is a potential nucleation site, whereas only those atoms on grain boundaries, for example, can take part in boundary-assisted nucleation.

If the concentration of heterogeneous nucleation sites is C_1 per unit volume, the heterogeneous nucleation rate will be given by an equation of the form

$$N_{\text{het}} = \omega C_1 \exp\left(\frac{\Delta G_{\text{m}}}{kT}\right) \cdot \exp\left(\frac{\Delta G^*}{kT}\right) \text{ nuclei m}^{-3}\text{ s}^{-1} \tag{5.24}$$

This is plotted as a function of temperature in Figure 5.12. Note that, as with heterogeneous nucleation in liquids, measurably high nucleation rates can be obtained at very small driving forces. The relative magnitudes of the heterogeneous and homogeneous volume nucleation rates can be obtained by dividing Equation 5.11 by Equation 5.24 giving

$$\frac{N_{\text{het}}}{N_{\text{hom}}} = \frac{C_1}{C_0} \exp\left(\frac{\Delta G^*_{\text{hom}} - \Delta G^*_{\text{het}}}{kT}\right) \tag{5.25}$$

(Differences in ω and ΔG_{m} are not so important and have been ignored.) Since ΔG^* is always smallest for heterogeneous nucleation the exponential factor in the above equation is always a large quantity which favors a high heterogeneous nucleation rate. However, the factor (C_1/C_0) must also be taken into account, i.e., the number of atoms on heterogeneous sites relative to the number within the matrix. For grain-boundary nucleation

$$\frac{C_1}{C_0} = \frac{\delta}{D} \tag{5.26}$$

where

δ is the boundary thickness

D is the grain size

TABLE 5.1

C_1/C_0 for Various Heterogeneous Nucleation Sites

Grain Boundary	Grain Edge	Grain Corner	Dislocations		Excess Vacancies
$D=50\ \mu m$	$D=50\ \mu m$	$D=50\ \mu m$	$10^5\ mm^{-2}$	$10^8\ mm^{-2}$	$X_v=10^{-6}$
10^{-5}	10^{-10}	10^{-15}	10^{-8}	10^{-5}	10^{-6}

For nucleation on grain edges and corners (C_1/C_0) becomes reduced even further to $(\delta/D)^2$ and $(\delta/D)^3$. Therefore for a 50 μm grain size taking δ as 0.5 nm gives $\delta/D\times10^{-5}$. Consequently grain-boundary nucleation will dominate over homogeneous nucleation if the boundary is sufficient potent to make the exponential term in Equation 5.23 greater than 10^5. Values for (C_1/C_0) for other sites are listed in Table 5.1.

In general the type of site which gives the highest volume nucleation rate will depend on the driving force (ΔG_v). At very small driving forces, when activation energy barriers for nucleation are high, the highest nucleation rates will be produced by grain-corner nucleation. As the driving force increases, however, grain edges and then boundaries will dominate the transformation. At very high driving forces it may be possible for the (C_1/C_0) term to dominate and then homogeneous nucleation provides the highest nucleation rates. Similar considerations will apply to the relative importance of other heterogeneous nucleation sites.

The above comments concerned nucleation during isothermal transformations when the specimen is held at a constant temperature. If nucleation occurs during continuous cooling the driving force for nucleation will increase with time. Under these conditions the initial stages of the transformation will be dominated by those nucleation sites which can first produce a measurable volume nucleation rate. Considering only grain boundaries again, if $\gamma_{\alpha\alpha}/\gamma_{\alpha\beta}$ is high, noticeable transformation will begin first at the grain corners, whereas if the grain boundary is less potent ($\gamma_{\alpha\alpha}/\gamma_{\alpha\beta}$ smaller) nucleation may not be possible until such high driving forces are reached that less favorable heterogeneous or even homogeneous nucleation sites dominate. This will not of course exclude precipitation on potent heterogeneous sites, but they will make only a very small contribution to the total nucleation rate.

5.3 Precipitate Growth

As explained above, the successful critical nuclei are those with the smallest nucleation barrier, i.e., the smallest critical volume. In the absence of strain energy effects the precipitate shape satisfying this criterion is that which minimizes the total interfacial free energy. Thus nuclei will usually be bounded by a combination of coherent or semicoherent facets and smoothly curved incoherent interfaces. For the precipitate to grow these interfaces

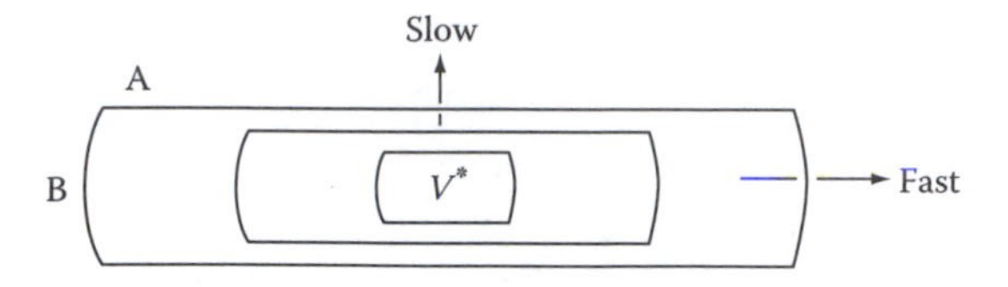

FIGURE 5.13
Effect of interface type on the morphology of a growing precipitate: A, low-mobility semicoherent interfaces; B, high-mobility incoherent interfaces.

must migrate and the shape that develops during growth will be determined by the relative migration rates. As explained in Section 3.5.1, when the two phases have different crystal structures semicoherent interfaces have very low mobility and are forced to migrate by a ledge mechanism. Incoherent interfaces on the other hand are highly mobile. If there are problems in maintaining a constant supply of ledges the incoherent interfaces will be able to advance faster than the semicoherent interface and a nucleus with one plane of good matching should grow into a thin disk or plate as shown in Figure 5.13. This is the origin of the so-called Widmanstätten morphology.[2]

The next few sections will be concerned with developing an approximate quantitative treatment for the ledge mechanism and for the rate of migration of curved incoherent interfaces, but before treating these two cases it is useful to begin with the simpler case of a planar incoherent interface.

5.3.1 Growth behind Planar Incoherent Interfaces

It will be apparent from the above discussion that planar interfaces in crystalline solids will usually not be incoherent. However, one situation where approximately planar incoherent interfaces may be found is after grain-boundary nucleation. If many incoherent nuclei form on a grain boundary they might subsequently grow together to form a slab of β precipitate as shown in Figure 5.14.

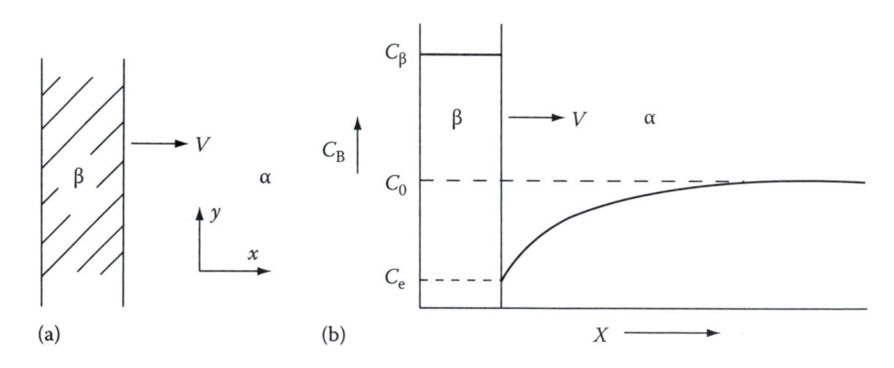

FIGURE 5.14
Diffusion-controlled thickening of a precipitate plate.

Imagine that such a slab of solute-rich precipitate has grown from zero thickness and that the instantaneous growth rate is v. Since the concentration of solute in the precipitate (C_β) is higher than in the bulk (C_0) the matrix adjacent to the precipitate will be depleted to solute as shown. Also since the interface is incoherent diffusion-controlled growth and local equilibrium at the interface can be assumed, i.e., the solute concentration in the matrix adjacent to the β will be the equilibrium value C_e. The growth rate (v) will depend on the concentration gradient at the interface dC/dx.

For unit area of interface to advance a distance dx a volume of material $1 \cdot dx$ must be converted from α containing C_e to β containing C_β moles of B per unit volume, i.e., $(C_\beta - C_e)dx$ moles of B must be supplied by diffusion through the α. The flux of B through unit area in time dt is given by $D(dC/dx)dt$, where D is the interdiffusion coefficient (or interstitial diffusion coefficient). Equating these two quantities gives

$$v = \frac{dx}{dt} = \frac{D}{C_\beta - C_e} \cdot \frac{dC}{dx} \tag{5.27}$$

As the precipitate grows solute must be depleted from an ever-increasing volume of matrix so that dC/dx in Equation 5.27 decreases with time. To make this quantitative, consider a simplified approach originally due to Zener.[3] If the concentration profile is simplified to that shown in Figure 5.15 dC/dx is given $\Delta C_0/L$ where $\Delta C_0 = C_0 - C_e$. The width of the diffusion zone L can be obtained by noting that the conservation of solute requires the two shaded areas in Figure 5.15 to be equal, i.e.,

$$(C_\beta - C_0)x = L\Delta C_0/2$$

where x is the thickness of the slab. The growth rate therefore becomes

$$v = \frac{D(\Delta C_0)^2}{2(C_\beta - C_e)(C_\beta - C_0)x} \tag{5.28}$$

If it is assumed that the molar volume (V_m) is a constant, the concentrations in the above equation can be replaced by mole fractions ($X = CV_m$). Furthermore,

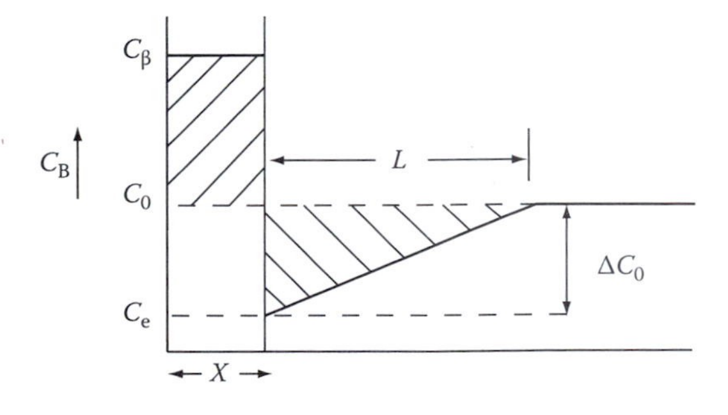

FIGURE 5.15
Simplification of the concentration profile.

for the sake of simplicity it can often be assumed that $(C_\beta - C_0)$ $(C_\beta - C_e)$. Integration of Equation 5.28 then gives

$$x = \frac{\Delta X_0}{\sqrt{(X_\beta - X_e)}}\sqrt{(Dt)} \tag{5.29}$$

and

$$v = \frac{\Delta X_0}{2(X_\beta - X_e)}\sqrt{\frac{D}{t}} \tag{5.30}$$

where $\Delta X_0 = X_0 - X_e$ (Figure 5.16) is the supersaturation prior to precipitation.

The following points are important to note regarding these equations:

1. $x \propto \sqrt{(D/t)}$, i.e., precipitate thickening obeys a parabolic growth law
2. $v \propto \Delta X_0$, i.e., for a given time the growth rate is proportional to the supersaturation
3. $v \propto \sqrt{(D/t)}$

The effect of alloy composition and temperature on growth rate is illustrated in Figure 5.16. Growth rates are low at small undercoolings due to small supersaturation ΔX_0 but are also low at large undercoolings due to slow diffusion. A maximum growth rate will occur at some intermediate undercooling.

When the diffusion fields to separate precipitates begin to overlap Equation 5.30 will no longer apply, but growth will decelerate more rapidly and finally cease when the matrix concentration is X_e everywhere, Figure 5.17.

Although these equations are only approximate and were derived for a planar interface, the conclusions are not significantly altered by more thorough treatments or by allowing curved interfaces. Thus it can be shown that any linear dimension of a spheroidal precipitate increases as $\sqrt{(Dt)}$ provided all interfaces migrate under volume diffusion control.

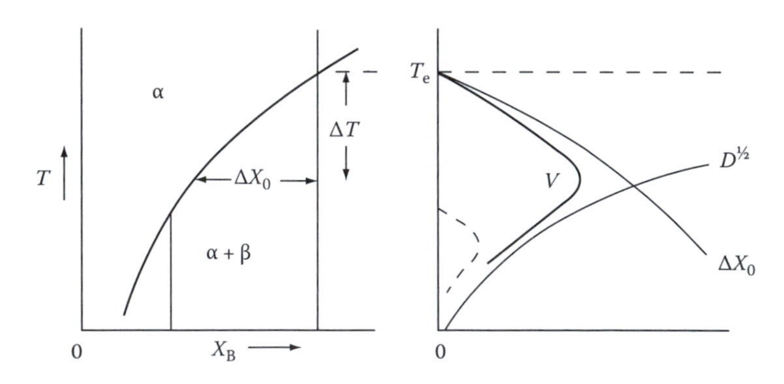

FIGURE 5.16
Effect of temperature and position on growth rate, v.

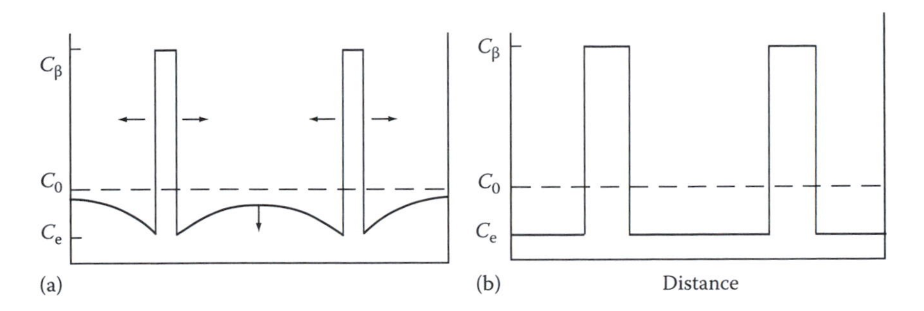

FIGURE 5.17
(a) Interference of growing precipitates due to overlapping diffusion fields at later stage of growth. (b) Precipitate has stopped growing.

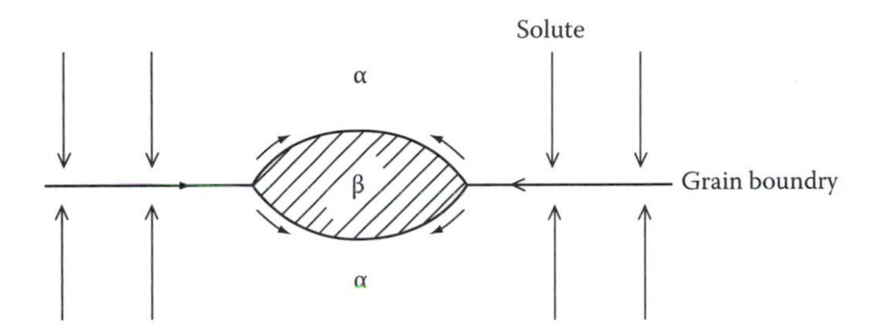

FIGURE 5.18
Grain-boundary diffusion can lead to rapid lengthening and thickening of grain-boundary precipitates.

Usually grain-boundary precipitates do not form a continuous layer along the boundary but remain as isolated particles. The growth of such precipitates can occur at rates far greater than allowed by volume diffusion. The reason for this is that the grain boundary can act as a collector plate for solute as shown in Figure 5.18.[4] Growth of such a so-called grain-boundary allotriomorph involves three steps: (1) volume diffusion of solute to the grain boundary, (2) diffusion of solute along the grain boundary with some attachment at the precipitate rim, and (3) diffusion along the α/β interfaces allowing accelerated thickening. This mechanism is of greatest significance when substitutional diffusion is involved. In the case of interstitial solutions diffusion short circuits are comparatively unimportant due to the high volume diffusion rates.

5.3.2 Diffusion-Controlled Lengthening of Plates or Needles

Imagine now that the β precipitate is a plate of constant thickness having a cylindrically curved incoherent edge of radius r as shown in Figure 5.19a. Again the concentration profile across the curved interface will appear as shown in Figure 5.19b, but now, due to the Gibbs–Thomson effect, the equilibrium concentration in the matrix adjacent to the edge will be increased to C_r. The concentration gradient available to drive diffusion to the advancing

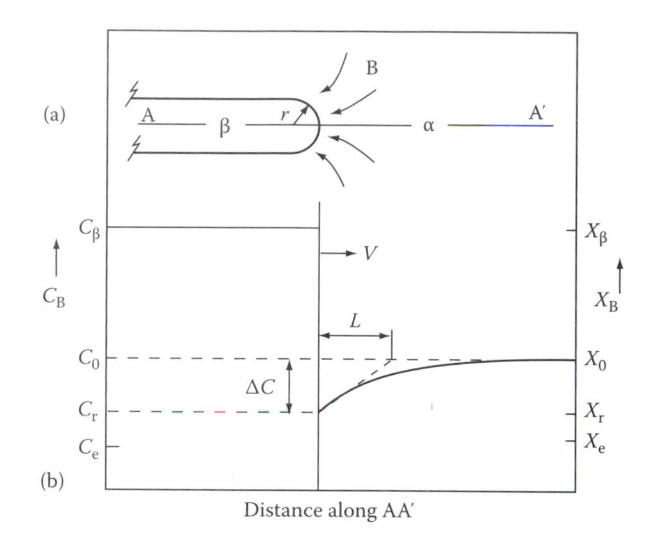

FIGURE 5.19
(a) Edge of a platelike precipitate. (b) Concentration profile along AA′ in (a).

edge is therefore reduced to $\Delta C/L$ where $\Delta C = C_0 - C_r$ and L is a characteristic diffusion distance. The diffusion problem in this case is more complex as diffusion occurs radially. However, solution of the relevant equations shows that L is given by kr where k is a numerical constant (~1). By analogy with Equation 5.27, therefore, the lengthening rate will be given by

$$v = \frac{D}{C_\beta - C_r} \cdot \frac{\Delta C}{kr} \tag{5.31}$$

The composition difference available to drive diffusion will depend on the tip radius as shown in Figure 5.20. With certain simplifying assumptions it can be shown that

$$\Delta X = \Delta X_0 \left(1 - \frac{r^*}{r}\right) \tag{5.32}$$

where
$\Delta X = X_0 - X_r$
$\Delta X_0 = X_0 - X_e$
r^* is the critical nucleus, radius, i.e., the value of r required to reduce ΔX to zero

Again, assuming constant molar volume, Equations 5.31 and 5.32 can be combined to give

$$v = \frac{D \Delta X_0}{k(X_\beta - X_r)} \cdot \frac{1}{r} \left(1 - \frac{r^*}{r}\right) \tag{5.33}$$

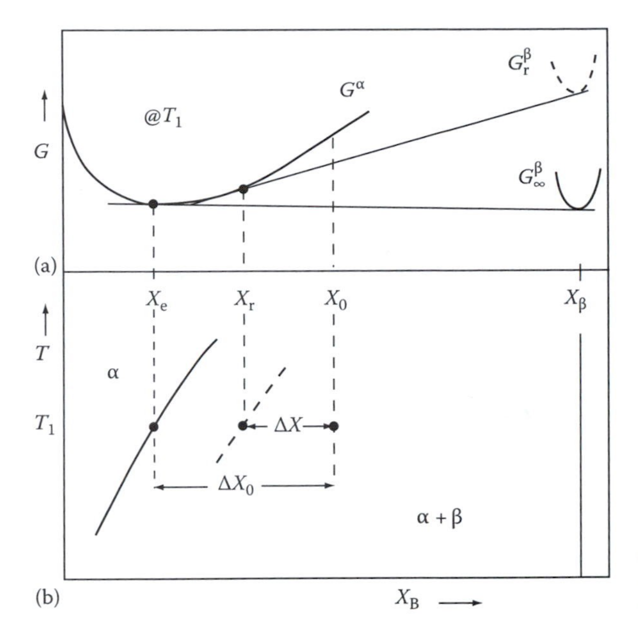

FIGURE 5.20
Gibbs–Thomson effect. (a) Free energy curves at T_1. (b) Corresponding phase diagram.

This equation will apply as long as there is no decrease in supersaturation far from the interface due to other precipitates. The difference between Equation 5.33 and Equation 5.30 is that for a given plate thickness the lengthening rate should be constant, i.e., $x \propto t$ (linear growth).

Although the above equations (Equations 5.31 through 5.33) were developed for the lengthening of a plate, the same equations can be derived for the lengthening of a needle under diffusion-controlled growth. The only difference is that the edge of a needle has a spherical tip so that the Gibbs–Thomson increase in free energy is $2\gamma V_m/r$ instead of $\gamma V_m/r$. The value of r^* in Equation 5.33 will, therefore, be different for a plate and a needle.

The above treatment only applies to plates or needles that lengthen by a volume diffusion-controlled continuous growth process. This is a reasonable assumption for the curved ends of needles, but in the case of platelike precipitates the edges are often faceted and are observed to migrate by a ledge mechanism. Atoms can then only attach at the ledges and new equations must be derived as discussed below.

Another source of deviation between theory and practice is if solute can be transported to the advancing precipitate edges by short-circuit diffusion in the broad faces of the precipitate plate.

5.3.3 Thickening of Platelike Precipitates

The treatment given in Section 5.3.1 for a planar incoherent interface is only valid for interfaces with high accommodation factors. In general this will not be the case for the broad faces of platelike precipitates which are semicoherent and are restricted to migrate by the lateral movement of ledges.

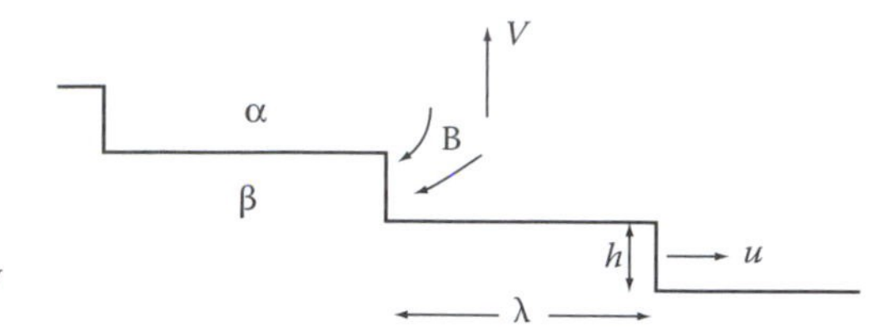

FIGURE 5.21
Thickening of platelike precipitates by ledge mechanism.

For simplicity, imagine a platelike precipitate that is thickening by the lateral movement of linear ledges of constant spacing λ and height h, Figure 5.21. It can readily be seen that the half-thickness of the plate should increase at a rate v given by

$$v = \frac{uh}{\lambda} \tag{5.34}$$

where u is the rate of lateral migration.

The problem of ledge migration is very similar to that of plate lengthening. The necessary composition changes required for precipitate growth must be achieved by long-range diffusion to and from the ledges as shown in Figure 5.21. If the edges of the ledges are incoherent the matrix composition in contact with the ledges will be X_e and growth will be diffusion controlled. A similar treatment to that given in Section 5.3.2 then gives the rate of lateral migration as[5]

$$u = \frac{D\Delta X_0}{k(X_\beta - X_e)h} \tag{5.35}$$

This is essentially the same as Equation 5.33 for the lengthening of a plate with $h \simeq r$ and $X_r = X_e$, i.e., no Gibbs–Thomson effect. Combining Equations 5.34 and 5.35 shows that the thickening rate is independent of h and given by

$$v = \frac{D\Delta X_0}{k(X_\beta - X_e)\lambda} \tag{5.36}$$

Thus provided the diffusion fields of different precipitates do not overlap, the rate at which plates thicken will be inversely proportional to the interledge spacing λ. The validity of Equation 5.36 is dependent on there being a constant supply of ledges. As with faceted solid/liquid interfaces, new ledges can be generated by various mechanisms such as repeated surface nucleation, spiral growth, nucleation at the precipitate edges, or from intersections with other precipitates. With the exception of spiral growth, nucleation at the precipitate edges, or from intersections with other precipitates. With the exception of spiral growth, however, none of these mechanisms can maintain a supply of ledges with constant λ.

By using hot-stage TEM it is possible to measure the thickening rates of individual precipitate plates. Figure 5.22 shows results obtained from a

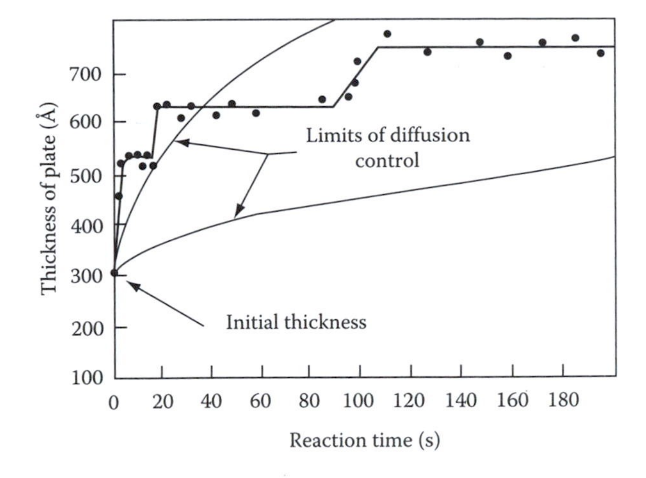

FIGURE 5.22
Thickening of a γ plate in an Al–15 wt% Ag alloy at 400°C. (From Laird, C. and Aaronson, H.I., *Acta Metall.*, 17, 505, 1969. With permission.)

γ plate in the Al–Ag system.[6] It can be seen that there are appreciable intervals of time when there is no perceptible increase in plate thickness followed by periods when the thickness increases rapidly as an interfacial ledge passes. The two smooth lines in the figure are upper and lower limits for the rate of thickening for a planar incoherent interface in the same system, assuming diffusion control. The ledge mechanism is clearly a very different process. The fact that there is no perceptible increase in thickness except when ledges pass is strong evidence in favor of the immobility of semicoherent interfaces. It can also be seen that the thickening rate is not constant implying that ledge nucleation is rate controlling.

Measurements on precipitates in other systems indicate that even within the same system the thickness/time relationship can vary greatly from plate to plate, presumably depending on differences in the ease of nucleation of new ledges.

5.4 Overall Transformation Kinetics: TTT Diagrams

The progress of an isothermal phase transformation can be conveniently represented by plotting the fraction transformation (f) as a function of time and temperature, i.e., a TTT diagram as shown in Figure 5.23a for example. For transformations of the type $\alpha \rightarrow \beta$, f is just the volume fraction of β at any time. For precipitation reactions $\alpha' \rightarrow \alpha + \beta$, f can be defined as the volume of β at time t divided by the final volume of β. In both cases f varies from 0 to 1 from the beginning to the end of the transformation, Figure 5.23b.

Among the factors that determine $f(t, T)$ are the nucleation rate, the growth rate, the density and distribution of nucleation sites, the overlap of diffusion fields from adjacent transformed volumes, and the impingement of adjacent

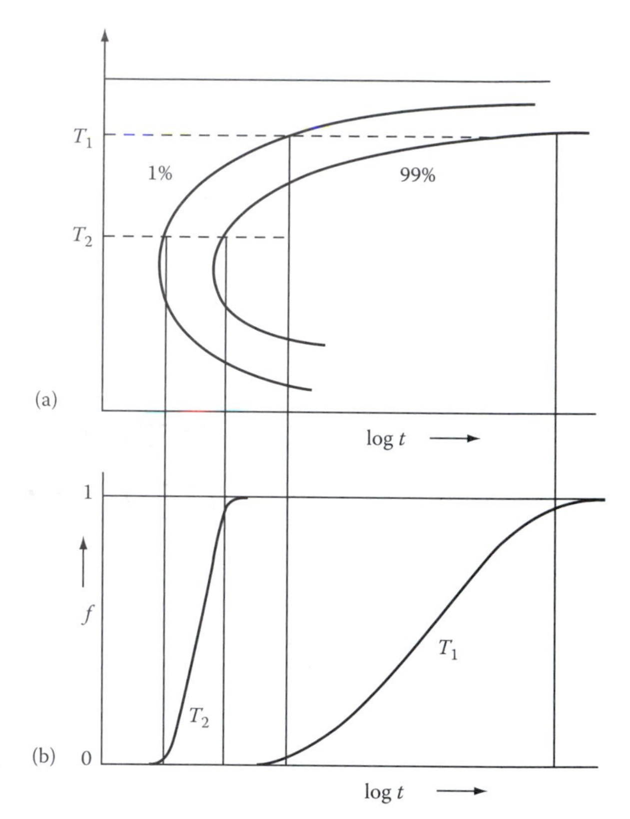

FIGURE 5.23
Percentage transformation versus time for different transformation temperatures.

transformed volumes. Some of the problems involved are illustrated in Figure 5.24. After quenching to the transformation temperature the metastable α-phase will contain many nucleation sites (usually heterogeneous). One possible sequence of events, Figure 5.24a, is that nuclei form throughout the transformation so that a wide range of particle sizes exists at any time. Another possibility is that all nuclei form right at the beginning of transformation, Figure 5.24b. If all potential nucleation sites are consumed in the process this is known as site saturation. In Figure 5.24a, f will on the nucleation rate and the growth rate. In Figure 5.24b, f will only depend on the number of nucleation sites and the growth rate. For transformations of the type $\alpha \rightarrow \beta$ or $\alpha \rightarrow \beta + \gamma$ (known collectively as cellular transformations) all of the parent phase is consumed by the transformation product, Figure 5.24c. In these cases the transformation does not terminate by the gradual reduction in the growth rate, but by the impingement of adjacent cells growing with a constant velocity. Pearlite, cellular precipitation, massive transformations, and recrystallization belong to this category.

As a simple example of the derivation of $f(t, T)$ consider a cellular transformation ($\alpha \rightarrow \beta$) in which β cells are continuously nucleated throughout the transformation at a constant rate N.[7] If the cells grow as spheres at a constant rate v, the volume of a cell nucleated at time zero will be given by

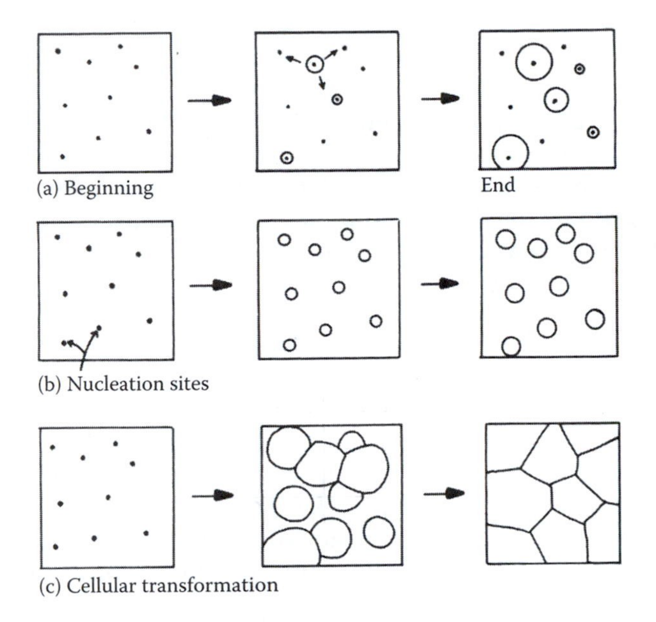

FIGURE 5.24
(a) Nucleation at a constant rate during the whole transformation. (b) Site saturation—all nucleation occurs at the beginning of transformation. (c) Cellular transformation.

$$V = \frac{4}{3}\pi r^3 = \frac{4}{3}\pi(vt)^3$$

A cell which does not nucleate until time τ will have a volume

$$(\bar{r})^3 - r_0^3 = kt$$

The number of nuclei that formed in a time increment of $\mathrm{d}\tau$ will be $N\,\mathrm{d}\tau$ per unit volume of untransformed α. Thus if the particles do not impinge on one another, for a unit total volume

$$f = \sum V' = \frac{4}{3}\pi N v^3 \int_0^t (t-\tau)^3 \mathrm{d}\tau$$

i.e.,

$$f = \frac{\pi}{3} N v^3 t^4 \tag{5.37}$$

This equation will only be valid for $f \ll 1$. As time passes the β cells will eventually impinge on one another and the rate of transformation will decrease again. The equation valid for randomly distributed nuclei for both long and short times is[8]

$$f = 1 - \exp\left(-\frac{\pi}{3}Nv^3t^4\right) \tag{5.38}$$

Note that this is the same as Equation 5.37 for short times, since $1 - \exp(-z) \cdot z$ when $z \ll 1$. It is also reasonable for long times since as $t \to \infty, f \to 1$.

Equation 5.58 is known as a Johnson–Mehl–Avrami equation. In general, depending on the assumptions made regarding the nucleation and growth processes, a variety of similar equations can be obtained with the form

$$f = 1 - \exp(-kt^n) \tag{5.39}$$

where n is a numerical exponent whose value can vary from $\sim$1 to 4. Provided there is no change in the nucleation mechanism n is independent of temperature k, on the other hand, depends on the nucleation and growth rates and is therefore very sensitive to temperature. For example, in the case above, $k = \pi Nv^3/3$ and both N and v are very temperature sensitive.

Since exp $(-0.7) = 0.5$ the time for 50% transformation ($t_{0.5}$) is given by $kt_{0.5}^n = 0.7$, i.e.,

$$t_{0.5} = \frac{0.7}{k^{1/n}} \tag{5.40}$$

For the case discussed above

$$t_{0.5} = \frac{0.9}{N^{1/4}v^{3/4}} \tag{5.41}$$

Consequently it can be seen that rapid transformations are associated with large values of k, i.e., rapid nucleation and growth rates, as expected.

Civilian transformations that occur on cooling are typified by C-shaped TTT curves as shown in Figure 5.23a. This can be explained on the basis of the variation of nucleation and growth rates with increasing undercooling. At temperatures close to T_e the driving force for transformation is very small so that both nucleation and subsequent growth rates are slow and a long time is required for transformation. When ΔT is very large, on the other hand, slow diffusion rates limit the rate of transformation. A maximum rate is, therefore, obtained at intermediate temperatures.

5.5 Precipitation in Age-Hardening Alloys

The theory of nucleation and growth that has been described above is able to provide general guidelines for understanding civilian transformations. Let us now turn to a consideration of some examples of the great variety of civilian transformations that can occur in solids, and begin with alloys that can be age-hardened. These alloys are characterized by phase diagrams such as that shown in Figure 5.1a(i). Two extensively researched and illustrative examples are aluminum–copper and aluminum–silver alloys.

5.5.1 Precipitation in Aluminum–Copper Alloys

GP zones

Figure 5.25 shows the Al-rich end of the Al–Cu phase diagram. If an alloy with the composition Al–4 wt% Cu (1.7 atomic%) is heated to a temperature of about 540°C all copper will be in solid solution as a stable fcc α-phase, and by quenching the specimen rapidly into water there is no time for any transformation to occur so that the solid solution is retained largely unchanged to room temperature. However, the solid solution is now supersaturated with Cu and there is a driving force for precipitation of the equilibrium θ-phase, $CuAl_2$.

If the alloy is now aged by holding for a period of time at room temperature or some other temperature below about 180°C it is found that the first precipitate to nucleate is not θ but coherent Cu-rich GP zones. (Copper-rich zones in Al–Cu alloys were detected independently in 1938 by Guinier and Preston from streaks in x-ray diffraction patterns.) The reason for this can be understood on the basis of the relative activation energy barriers for nucleation as discussed earlier. GP zones are fully coherent with the matrix and therefore have a very low interfacial energy, whereas the θ-phase has a complex tetragonal crystal structure which can only form with high-energy

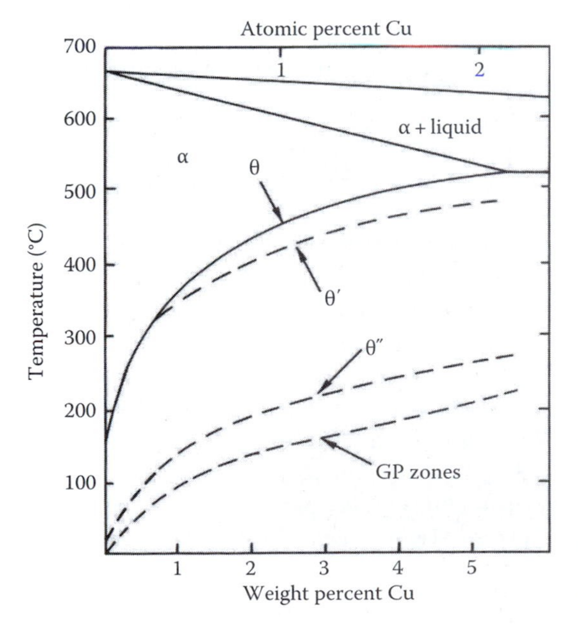

FIGURE 5.25
Al–Cu phase diagram showing the metastable GP zone. θ″ and θ′ solvuses. (Reproduced from Lorimer, G., in *Precipitation Processes in Solids*, The Metallurgical Society of AMIE, Warrendele, PA, 1978, 87.)

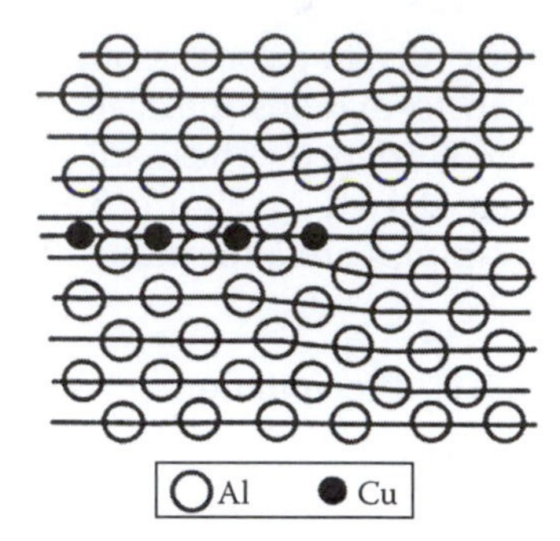

FIGURE 5.26
Section through a GP zone parallel to the (200) plane. (Based on the work of Gerold, V., *Zeitschrift für Metallkunde*, 45, 599, 1954.)

incoherent interfaces. In addition, the zones minimize their strain energy by choosing a disk-shape perpendicular to the elastically soft ⟨100⟩ directions in the fcc matrix, Figure 5.26. Therefore, despite the fact that the driving force for precipitation of GP zones ($\Delta G_v - \Delta G_s$) is less than for the equilibrium phase, the barrier to nucleation (ΔG^*) is still less, and the zones nucleate most rapidly. The microstructure of an Al–Cu alloy aged to produce GP zones is shown in Figure 5.30a. These zones are about two atomic layers thick and 10 nm in diameter with a spacing of ~10 nm. The zones themselves are not resolved. The contrast in the image is due to the coherency misfit strain perpendicular to the zones. This distorts the lattice causing local variations in the intensity of electron diffraction, which in turn shows as variations in the image intensity. Microstructurally, the zones appear to be homogeneously nucleated, however excess vacancies are thought to play an important role in their formation. This point will be returned to later.

GP zones are formed as the first precipitate during low-temperature aging of many technologically important alloys, notably those based on aluminum (Tables 5.2 and 5.3). In dilute Al–Zn and Al–Ag alloys Zn-rich and Ag-rich GP zones are found. In these cases there is very little misfit strain and ΔG^* is

TABLE 5.2
Some Precipitation-Hardening Sequences

Base Metal	Alloy	Precipitation Sequence
Aluminum	Al–Ag	GPZ (spheres) → γ' (plates) → γ (Ag_2Al)
	Al–Cu	GPZ (disks) → θ'' (disks) → θ' (plates) → θ ($CuAl_2$)
	Al–Cu–Mg	GPZ (rods) → S′ (laths) → S ($CuMgAl_2$) (laths)
	Al–Zn–Mg	GPZ (spheres) → η' (plates) → η ($MgZn_2$) (plates or rods)
	Al–Mg–Si	GPZ (rods) → β' (rods) → β (Mg_2Si) (plates)
Copper	Cu–Be	GPZ (disks) → γ' → γ (CuBe)
	Cu–Co	GPZ (spheres) → β (Co) (plates)
Iron	Fe–C	ε-carbide (disks) → Fe_3C (plates)
	Fe–N	α'' (disks) → Fe_4N
Nickel	Ni–Cr–Ti–Al	γ' (cubes or spheres)

Source: Mainly from Martin, J.W., in *Precipitation Hardening*, Pergamon Press, Oxford, 1968.

TABLE 5.3

Mechanical Properties of Some Commercial Precipitation-Hardening Alloys

Base Metal	Alloy	Composition (wt%)	Precipitate	YS[a] (MPa)	UTS[a] (MPa)	Elongation[a]
Aluminum	2024	Cu (4.5) Mg (1.5) Mn (0.6)	S′ (Al_2CuMg)	390	500	13
	6061	Mg (1.0) Si (0.6) Cu (0.25) Cr (0.2)	β′ (Mg_2Si)	280	315	12
	7075	Zn (5.6) Mg (2.5) Cu (1.6) Mn (0.2) Cr (0.3)	η′ ($MgZn_2$)	500	570	11
Copper	Cu–Be	Be (1.9) Co (0.5)	Zones	770	1160	5
Nickel	Nimonic 105	Co (20) Cr (15) Mo (5)	γ′ (Ni_3TiAl)	750[b]	1100[b]	25[b]
		Al (4.5) Ti (1.0) C (0.15)				
Iron	Maraging	Ni (18) Co (9) Mo (5)	σ (FeMo)	1000	1900	4
	Steel	Ti (0.7) Al (0.1)	Ni_3Ti			

[a] At peak hardness tested at room temperature.
[b] Tested at 600°C.

minimized by the formation of spherical zones with a minimum interfacial energy, Figure 3.34.

Transition phases

The formation of GP zones is usually followed by the precipitation of the so-called transition phases. In the case of Al–Cu alloys the equilibrium θ-phase is preceded by θ″ and θ′. The total precipitation process can be written

$$\alpha_0 \rightarrow \alpha_1 + \text{GP zones} \rightarrow \alpha_2 + \theta'' \rightarrow \alpha_3 + \theta' \rightarrow \alpha_4 + \theta$$

where

α_0 is the original supersaturated solid solution
α_1 is the composition of the matrix in equilibrium with GP zones
α_2 is the composition in equilibrium with θ″, etc.

Figure 5.27 shows a schematic free energy diagram for the above phases. Since GP zones and the matrix have the same crystal structure they lie on the same free energy curve (ignoring strain energy effects—see Section 5.5.5). The transition phases θ″ and θ′ are less stable than the equilibrium θ-phase and consequently have higher free energies as shown. The compositions of the matrix in equilibrium with each phase—α_1, α_2, α_3, and α_4—are given by the common tangent construction. These compositions correspond to points on the solvus lines for GP θ″, θ′, and θ shown in Figure 5.25. The free energy of the alloy undergoing the above precipitation sequence decreases as

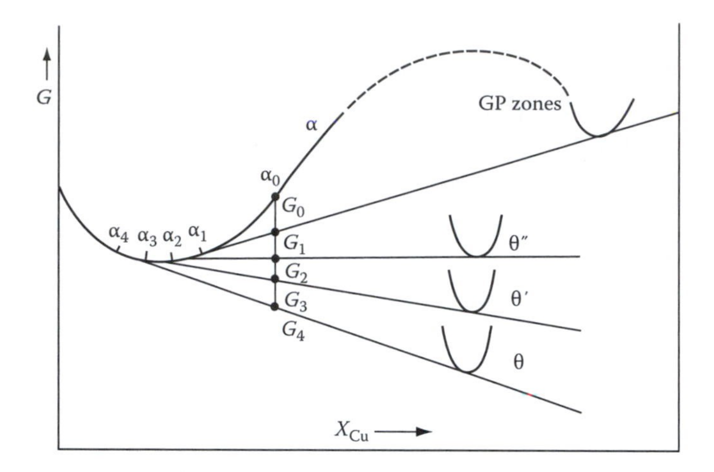

FIGURE 5.27
Schematic molar free energy diagram for the Al–Cu system.

$$G_0 \rightarrow G_1 \rightarrow G_2 \rightarrow G_3 \rightarrow G_4$$

as shown in Figure 5.27. Transformation stops when the minimum free energy equilibrium state G_4 is reached, i.e., $\alpha_4 + \theta$.

Transition phases form because, like GP zones, they have a lower activation energy barrier for nucleation than the equilibrium phase, Figure 5.28a.

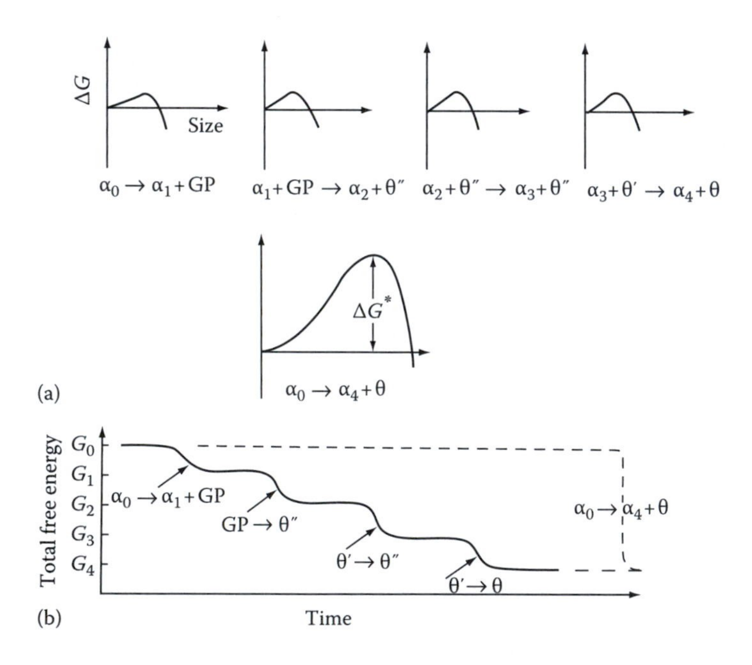

FIGURE 5.28
(a) The activation energy barrier to the formation of each transition phase is very small in comparison to the barrier against the direct precipitation of the equilibrium phase. (b) Schematic diagram showing the total free energy of the alloy versus time.

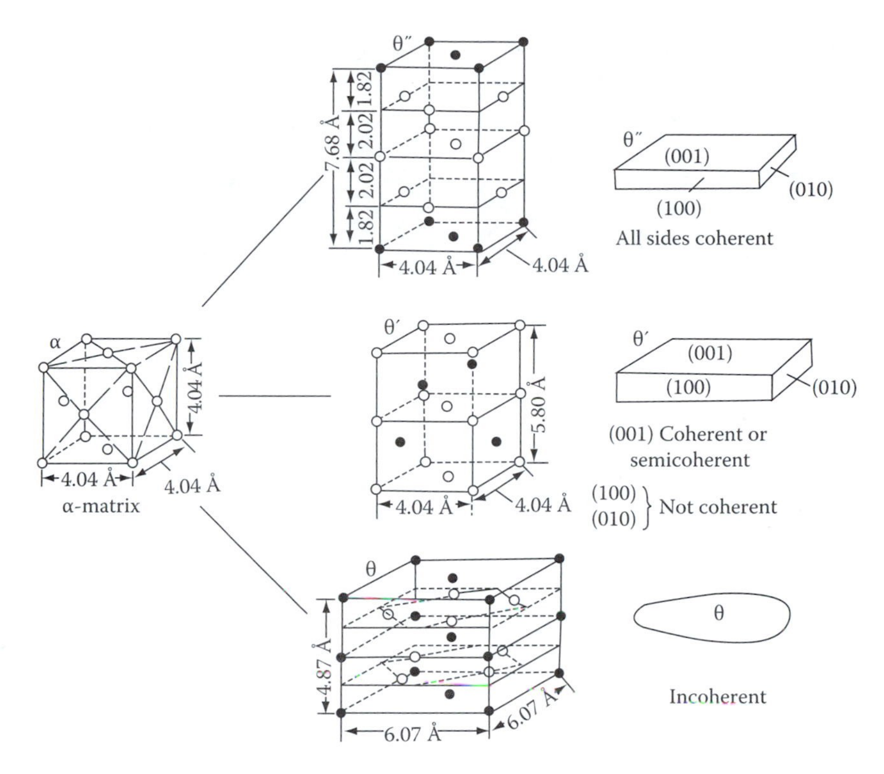

FIGURE 5.29
Structure and morphology of θ″, θ′, and θ in Al–Cu (○ Al, ● Cu).

The free energy of the alloy therefore decreases more rapidly via the transition phases than by direct transformation to the equilibrium phase, Figure 5.28b.

The lower activation energy barriers are achieved because the crystal structures of the transition phases are intermediate between those of the matrix and the equilibrium phase. In this way the transition phases can achieve a high degree of coherence and thus a low interfacial energy contribution to ΔG^*. The equilibrium phase on the other hand usually has a complex crystal structure that is incompatible with the matrix and results in high-energy interfaces and high ΔG^*.

The crystal structures of θ″, θ′, and θ are shown in Figure 5.29 along with that of the fcc matrix for comparison. θ″ has a tetragonal unit cell which is essentially a distorted fcc structure in which the copper and aluminum atoms are ordered on (001) planes as shown. Note that the atomic structure of the (001) planes is identical to that in the matrix, and the (010) and (100) planes are very similar, apart from a small distortion in the [001] direction. θ″ forms as fully coherent platelike precipitates with a $\{001\}_\alpha$ habit plane and the following orientation relationship to the matrix:

$$(001)_{\theta''} \parallel (001)_\alpha$$
$$[100]_{\theta''} \parallel [100]_\alpha$$

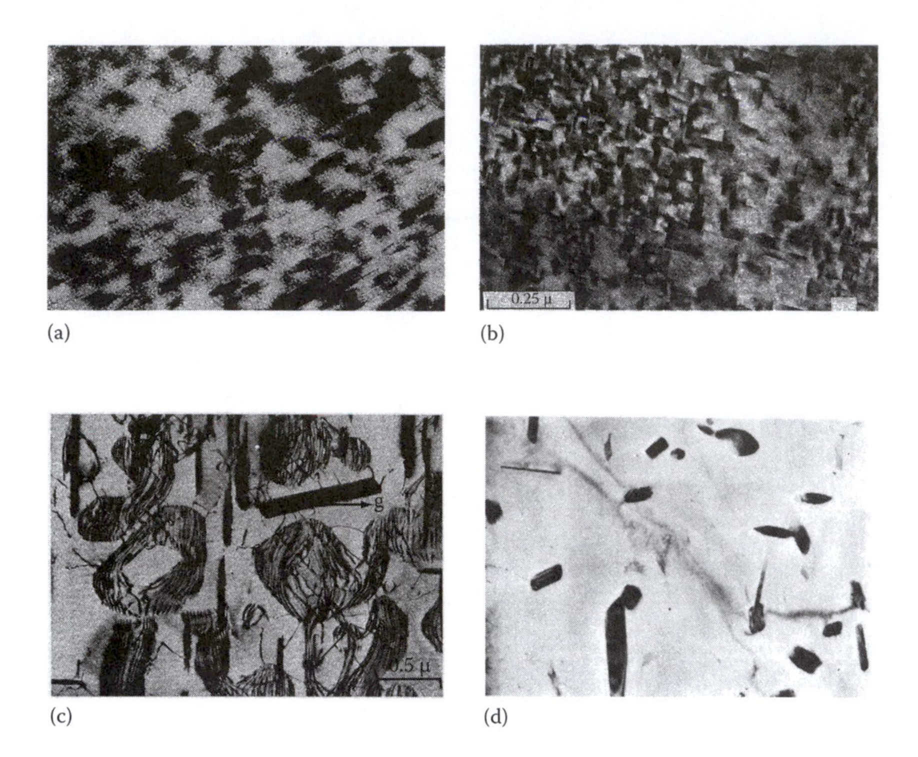

FIGURE 5.30
Microstructures at different stages during aging of Al–Cu alloys. (a) GP zones × 720,000. (After Nicholson, R.B. and Nutting, J., *Philos. Mag.*, 3, 531, 1958.) (b) θ'' × 63,000. (After Nicholson, R.B., Thomas, G., and Nutting, J., *J. Inst. Met.*, 87, 431, 1958–1959.) (c) θ' × 18,000. (After Weatherly, G.C. and Nicholson, R.B., *Philos. Mag.*, 17, 813, 1968.) (d) θ × 8,000. (After Chadwick, G.A., in *Metallography of Phase Transformations*, Butterworths, London, 1972, from C. Laird.)

A high magnification TEM micrograph of an alloy aged to produce θ'' precipitates is shown in Figure 5.30b. Like the GP zones in Figure 5.30a, the θ'' precipitates are visible by virtue of the coherency-strain fields caused by the misfit perpendicular to the plates. θ'' precipitates are larger than GP zones being up to ~10 nm thick and 100 nm in diameter.

θ' is also tetragonal with an approximate composition $CuAl_2$ and again has (001) planes that are identical with $\{001\}_\alpha$. The (100) and (010) planes, however, have a different crystal structure to the matrix and a large misfit in the [001] direction. θ' therefore forms as plates on $\{001\}_\alpha$ with the same orientation relationship as θ''. The broad faces of the plates are initially fully coherent but lose coherency as the plates grow, while the edges of the plates are either incoherent or have a complex semicoherent structure. A TEM micrograph of θ' plates ~1 μm diameter is shown in Figure 5.30c. Note the presence of misfit dislocations in the broad faces of the precipitates.

Note also that since the edges of the plates are not coherent there are no long-range coherency-strain fields.

The equilibrium θ-phase has the approximate composition $CuAl_2$ and a complex body-centered tetragonal structure as shown in Figure 5.29. There are no planes of good matching with the matrix and only incoherent, or at best complex semicoherent interfaces are possible. The microstructure at this final stage of aging is shown in Figure 5.30d. Note the large size and coarse distribution of the precipitates.

The transformation from GP zones to θ″ occurs by the in situ transformation of the zones, which can be considered as very potent nucleation sites for θ″. After longer ageing times the θ′-phase nucleates on matrix dislocations with two orientations of θ′ plates on any one $\frac{a}{2}\langle 110 \rangle$ dislocation. This is because the strain field of such a dislocation is able to reduce the misfit in two $\langle 100 \rangle$ matrix directions. Figure 5.31a shows θ′ plates that have nucleated on dislocations. Note that as the θ′ grows the surrounding, less-stable θ″ can be seen to dissolve. After still longer aging times the equilibrium θ-phase nucleates either on grain boundaries, Figure 5.31b, or at θ′/matrix interfaces, Figure 5.31c. The choice of these nucleation sites is governed by the need to reduce the large interfacial energy contribution to ΔG^* for this phase.

The full sequence of GP zones and transition precipitates is only possible when the alloy is aged at a temperature below the GP zones solvus.

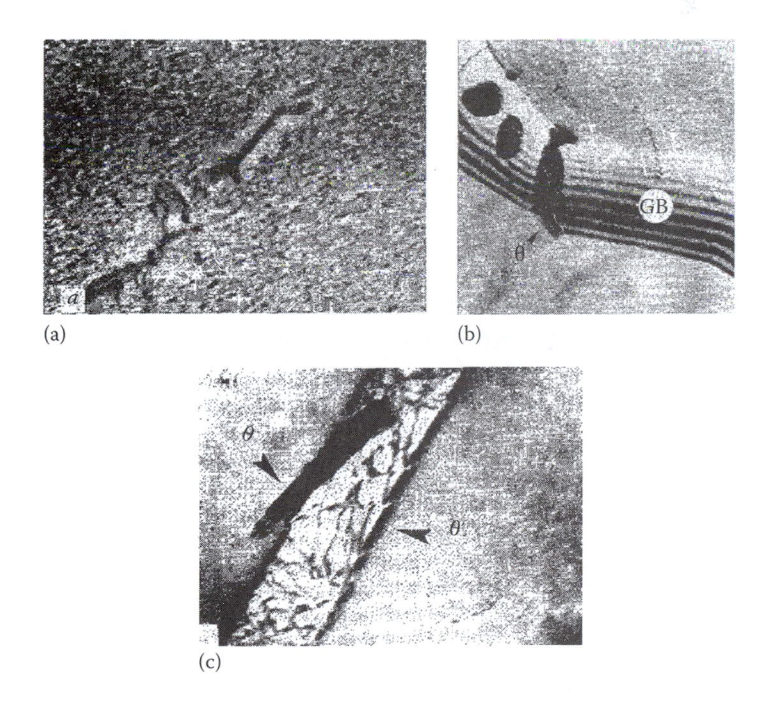

FIGURE 5.31
Electron micrographs showing nucleation sites in Al–Cu alloys. (a) θ″ → θ′. θ′ nucleates at dislocation (×70,000). (b) θ nucleation on grain boundary (GB) (×56,000). (c) θ′ → θ. θ nucleates at θ′/matrix interface (×70,000). (After Haasen, P., in *Physical Metallurgy*, Cambridge University Press, Cambridge, 1978.)

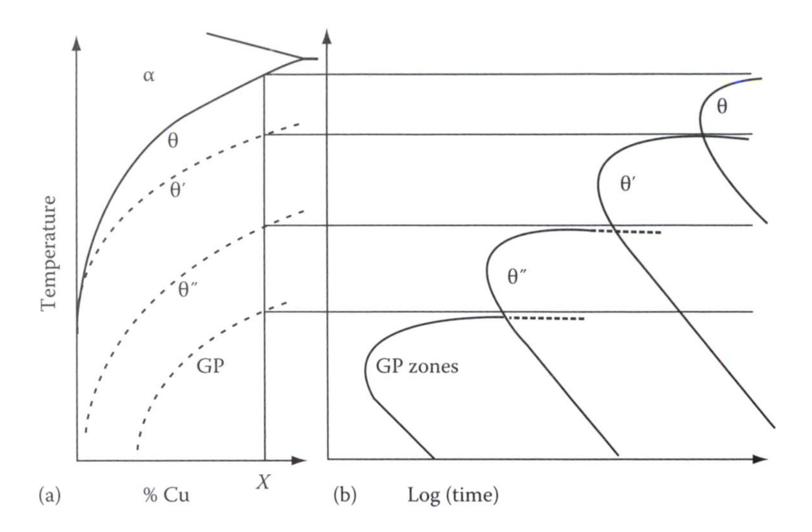

FIGURE 5.32
(a) Metastable solvus lines in Al–Cu (schematic). (b) Time for start of precipitation at different temperatures for alloy X in (a).

For example, if aging is carried out at a temperature above the θ″ solvus but below the θ′ solvus, Figure 5.25, the first precipitate will be θ′, heterogeneously nucleated on dislocations. If ageing is carried out above the θ′ solvus, the only precipitate that is possible is θ which nucleates and grows at grain boundaries. Also, if an alloy containing GP zones is heated to above the GP zones solvus the zones will dissolve. This is known as reversion.

The effect of aging temperature on the sequence of precipitates is illustrated by a schematic TTT diagram in Figure 5.32. The fastest transformation rates are associated with the highest nucleation rates and therefore the finest precipitate distributions. There is consequently an increasing coarseness of microstructure through the sequence of precipitates as can be seen in Figure 5.30.

The mechanism whereby a more stable precipitate grows at the expense of a less stable precipitate is illustrated in Figure 5.33 for the case θ″/θ′. Figure 5.27 shows that the Cu concentration in the matrix close to the θ″ precipitates (α_2) will be higher than that close to θ′ (α_3). Therefore Cu will tend to diffuse through the matrix way form θ″, which thereby dissolves, and toward θ′, which grows.

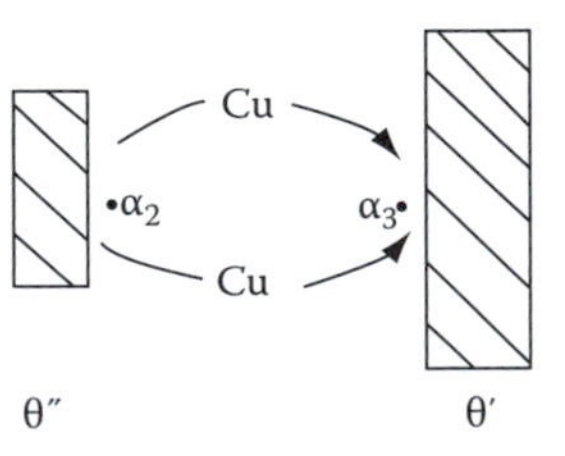

FIGURE 5.33
Matrix in equilibrium with θ″ (α_2) contains more Cu than matrix in equilibrium with θ′ (α_3). Cu diffuses as shown causing θ″ to shrink and θ′ to grow.

5.5.2 Precipitation in Aluminum–Silver Alloys

Figure 5.34 shows the Al–Ag phase diagram. If alloys containing up to about 23 atomic Ag are solution treated, quenched, and given a low-temperature aging treatment the precipitation sequence is

$$\alpha_0 \rightarrow \alpha_1 + \text{GP zones} \rightarrow \alpha_2 + \gamma' \rightarrow \alpha_3 + \gamma$$

As discussed earlier, the GP zones in this system are spherical. γ' is a close-packed hexagonal transition phase with an orientation relationship to the matrix of

$$(0001)'_{\gamma} // (111)_{\alpha}$$
$$[11\bar{2}0]_{\gamma} // [1\bar{1}0]_{\alpha}$$

γ' is heterogeneously nucleated on helical dislocations by the enrichment of stacking faults with silver as discussed in Section 5.2. The equilibrium γ-phase has the composition Ag_2Al, is hexagonal and has the same orientation relationship with the matrix as γ'. It forms as platelike precipitates with

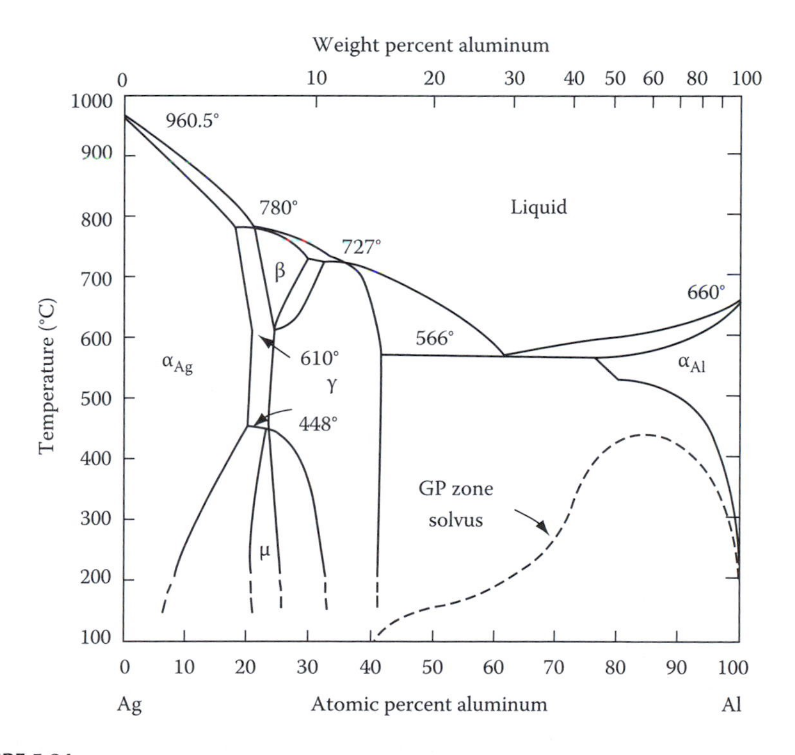

FIGURE 5.34
Al–Ag phase diagram showing metastable two-phase field corresponding to GP zones. (After Baur, R. and Gerold, V., *Zeitschrift für Metallkunde*, 52, 671, 1961.)

(111) habit planes. γ can be formed from γ' by the latter acquiring misfit dislocations. It can also be separately nucleated at grain boundaries and grow by a cellular mechanism (Section 5.7).

5.5.3 Quenched-In Vacancies

It was shown in Chapter 1 that the equilibrium concentration of vacancies increases exponentially with temperature. Thus the equilibrium vacancy concentration will be relatively high at the solution treatment temperature and much lower at the aging temperature. However, when the alloy is rapidly quenched from the high temperature there will be no time for the new equilibrium concentration to be established and the high vacancy concentration becomes quenched-in. Given time, those vacancies in excess of the equilibrium concentration will anneal out. There will be a tendency for vacancies to be attracted together into vacancy clusters, and some clusters collapse into dislocation loops which can grow by absorbing more vacancies. The dislocations that are already present can also absorb vacancies by climbing. In this way straight screw dislocations can become converted into longer helical edge dislocations. There are many ways, therefore, in which excess vacancies are able to provide heterogeneous nucleation sites.

Another effect of quenched-in vacancies is to greatly increase the rate at which atoms can diffuse at the aging temperatures. This in turn speeds up the process of nucleation and growth. Indeed the only way of explaining the rapid formation of GP zones at the relatively low aging temperatures used is by the presence of excess vacancies.

If GP zones are separated by a mean spacing λ, the mean diffusion distance for the solute atoms is $\lambda/2$. Therefore, if the zones are observed to form in a time t, the effective diffusion coefficient is roughly given by x^2/t, i.e.,

$$D \cdot \frac{\lambda^2}{4t}$$

If high-temperature diffusion data are extrapolated down to the aging temperature, the values obtained are orders of magnitude smaller than the above value. The difference can, however, be explained by a quenched-in vacancy concentration that is orders of magnitude greater than the equilibrium value. In Al–Cu alloys, for example, GP zones can form by aging at room temperature, which would not be feasible without assistance from excess vacancies.

There is other evidence for the role of quenched-in vacancies in enhancing diffusion rates. If the alloy is quenched from different solution treatment temperature and aged at the same temperature, the initial rate of zone formation is highest in the specimens quenched from the highest temperatures. Also, if the quench is interrupted at an intermediate temperature, so that a new equilibrium concentration can be established, the rate of transformation is reduced. Reducing the rate of cooling from the solution treatment temperature produces a similar effect by allowing more time for vacancies to be lost during the quench. This is important when large parts are to be heat treated as the cooling rate varies greatly from the surface to the center when the specimen is water-quenched for example.

Apart from dislocations, the main sinks for excess vacancies are the grain boundaries and other interfaces within the specimen. Since vacancies have such a high diffusivity it is difficult to avoid losing vacancies in the vicinity of grain boundaries and interfaces. This has important effects on the distribution of precipitates that form in the vicinity of grain boundaries on subsequent aging. Figure 5.35a shows the vacancy concentration profiles that should be produced by vacancy diffusion to grain boundaries during quenching. Close to the boundary the vacancy concentration will be the equilibrium value for the aging temperature, while away from the boundary it will be that for the solution treatment temperature. On aging these alloys it is found that a precipitate-free zone (PFZ) is formed as shown in Figure 5.35b. The solute concentration within the zone is largely unchanged, but no nucleation has occurred. The reason for this is that a critical vacancy supersaturation must be exceeded for nucleation to occur. The width of the PFZ is determined by the vacancy concentration as shown in Figure 5.35c. At low temperatures, where the driving force for precipitation is high, the critical vacancy supersaturation is lower and narrower PFZs are formed. High quench rates will also produce narrow PFZs by reducing the width of the

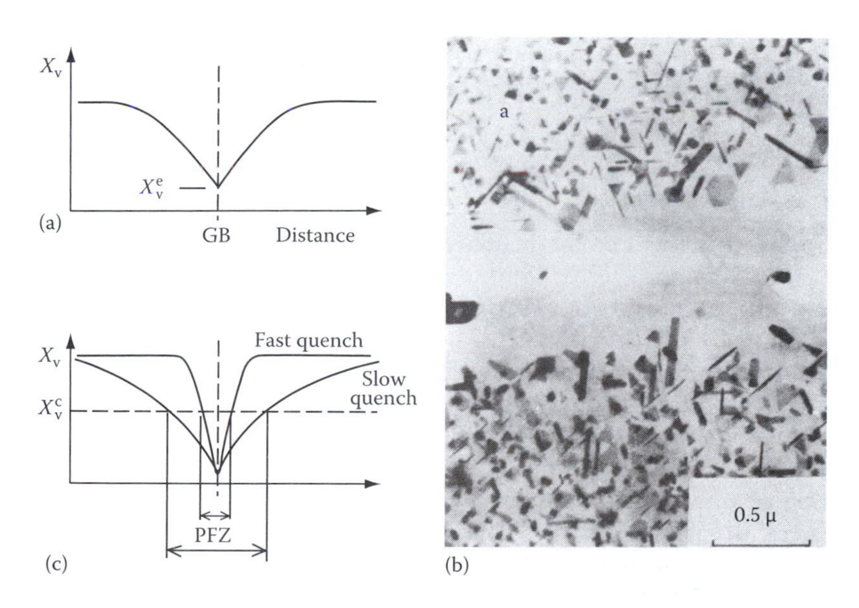

FIGURE 5.35
PFZ due to vacancy diffusion to a grain boundary during quenching. (a) Vacancy concentration profile. (b) PFZ in an Al–Ge alloy (×20,000). (After Lorimer, G., in *Precipitation in Solids*, The Metallurgical Society of AIME, Warrendale, PA, 1978.) (c) Dependence of PFZ width on critical vacancy concentration X_v^c and rate of quenching.

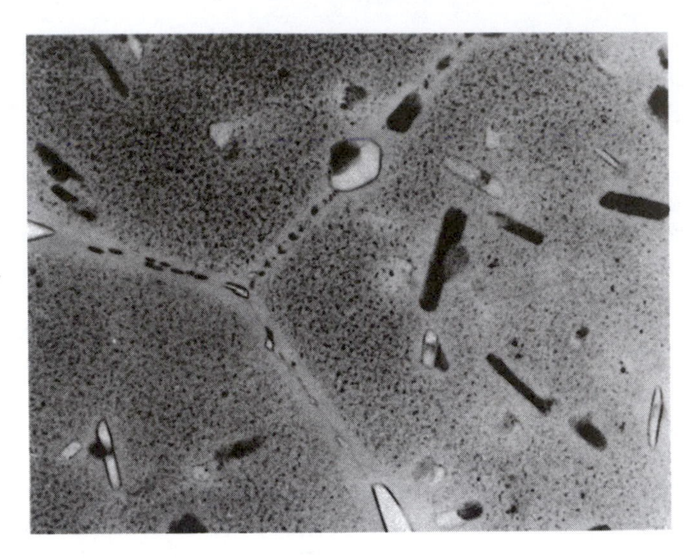

FIGURE 5.36
PFZs around grain boundaries in a high-strength commercial Al–Zn–Mg– Cu alloy. Precipitates on grain boundaries have extracted solute from surrounding matrix (×59,200).

vacancy concentration profile. Similar PFZs can also form at inclusions and dislocations.

Finally, it should be mentioned that another cause of PFZs can be the nucleation and growth of grain-boundary precipitates during cooling from the solution treatment temperature. This causes solute to be drained from the surrounding matrix and a PFZ results. An example of this type of PFZ is shown in Figure 5.36.

5.5.4 Age Hardening

The reason for the interest in alloy systems that show transition phase precipitation is that great improvement in the mechanical properties of these alloys can be achieved by suitable solution treatment and aging operations.

This is illustrated for various Al–Cu alloys in Figure 5.37. The alloys were solution treated in the single-phase α region of the phase diagram, quenched to room temperature and aged at either 130°C (Figure 5.37a) or 190°C (Figure 5.37b). The curves show how the hardness of the specimens varies as a function of time and the range of time over which GP zones, θ'', and θ' appear in the microstructure. Immediately after quenching the main resistance to dislocation movement is solid solution. The specimen is relatively easily deformed at this stage and the hardness is low. As GP zones from the hardness increases due to the extra stress required to force dislocations through the coherent zones.

The hardness continues to increase with the formation of the coherent θ'' precipitates because now the dislocations must also be forced through the

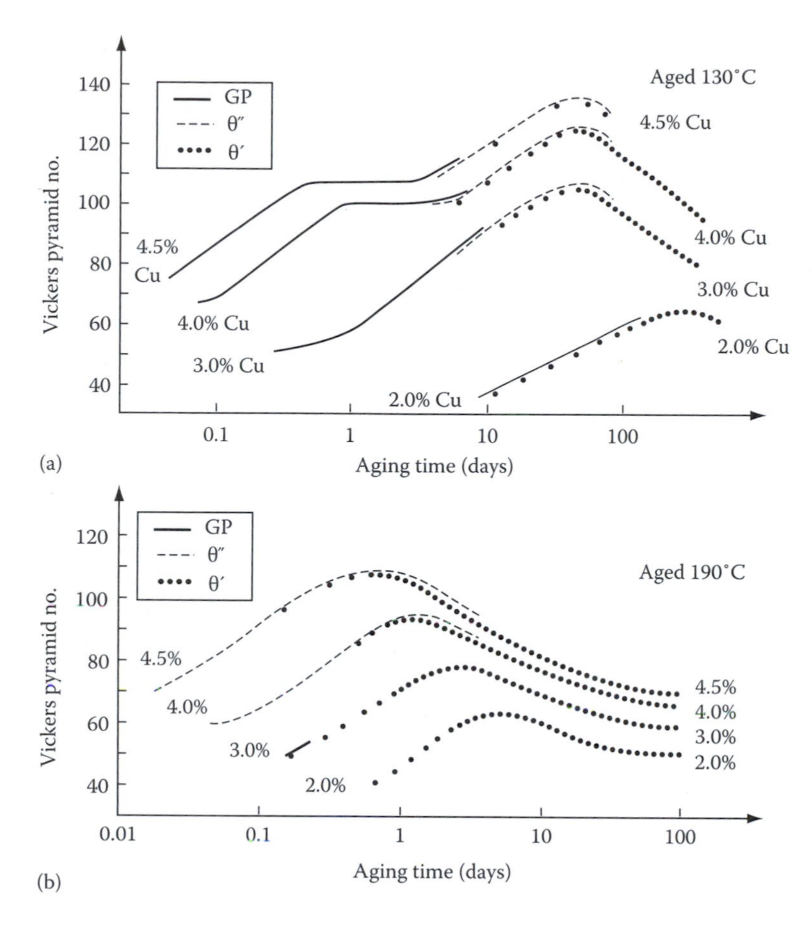

FIGURE 5.37
Hardness versus time for various Al–Cu alloys at (a) 130°C and (b) 190°C. (After Silcock, J.M., Heal, T.J., and Hardy, H.K., *J. Inst. Met.*, 82, 239, 1953–1954.)

highly strained matrix that results from the misfit perpendicular to the θ″ plates (Figure 5.30b). Eventually, with the formation of θ′ the spacing between the precipitates becomes so large that the dislocations are able to bow between the precipitates and the hardness begins to decrease. Maximum increases the distance between the precipitates making dislocation bowing referred to as overaged.

If Al–4.5 wt% Cu is aged at 190°C, GP zones are unstable and the first precipitate to form is θ″. The volume fraction of θ″ increases with time causing the hardness to increase as shown in Figure 5.37b. However, at 190°C the θ″ nucleates under the influence of a smaller driving force than at 130°C and the resultant precipitate dispersion is therefore coarser. Also the maximum volume fraction of θ″ is reduced. Both of these factors contribute to a lower peak hardness on aging at the higher temperature (compare

Figure 5.37a and b). However, diffusion rats are faster at higher temperatures and peak hardness is therefore achieved after shorter aging times.

It can be seen that at 130°C peak hardness in the Al–4.5 wt% Cu alloy is not reached for several tens of days. The temperatures that can be used in the heat treatment of commercial alloys are limited by economic considerations to those which produce the desired properties within a reasonable period of time, usually up to ~24 h. In some high-strength alloys use is, therefore, made of a double aging treatment whereby aging is carried out in two steps: first at a relatively low temperature below the GP zone solvus, and then at a higher temperature. In this way a fine dispersion of GP zones obtained during the first stage can act as heterogeneous nucleation sites for precipitation at the higher temperature. This type of treatment can lead to a finer precipitate distribution than would be obtained from a single aging treatment at the higher temperature.

Another treatment used commercially is to give the alloy a controlled deformation either before a single-stage or between the two stages of a double-aging treatment. The strength of the alloy after this treatment can be increased by a higher precipitate density, resulting from a higher nucleation rate, and by the retained dislocation networks which also act as a barrier to further deformation. However, deformation prior to aging does not always result in an improvement in properties. In some cases deformation can lead to a coarser precipitate distribution.

Precipitation hardening is common to many alloy systems. Some of the more important systems are listed in Table 5.2. Some commercial alloys are listed in Table 5.3, along with their mechanical properties. In many of these systems it is possible to come very close to the maximum theoretical strength of the matrix, i.e., about $\sim\mu/30$. However, engineering alloys are not heat treated for maximum strength alone. Consideration must also be given to toughness, stress corrosion resistance, fatigue, etc., when deciding on the best heat treatment in practice.

5.5.5 Spinodal Decomposition

It was mentioned at the beginning of this chapter that there are certain transformations where there is no barrier to nucleation. One of these is the spinodal mode of transformation. Consider a phase diagram with a miscibility gap as shown in Figure 5.38a. If an alloy with composition X_0 is solution treated at a high temperature T_1 and then quenched to a lower temperature T_2 the composition will initially be the same everywhere and its free energy will be G_0 on the G curve in Figure 5.38b. However, the alloy will be immediately unstable because small fluctuations in composition that produce A-rich and B-rich regions will cause the total free energy to decrease, Therefore "up-hill" diffusion takes place as shown in Figure 5.39 until the equilibrium compositions X_1 and X_2 are reached.

The above process can occur for any alloy composition where the free energy curve has a negative curvature i.e.,

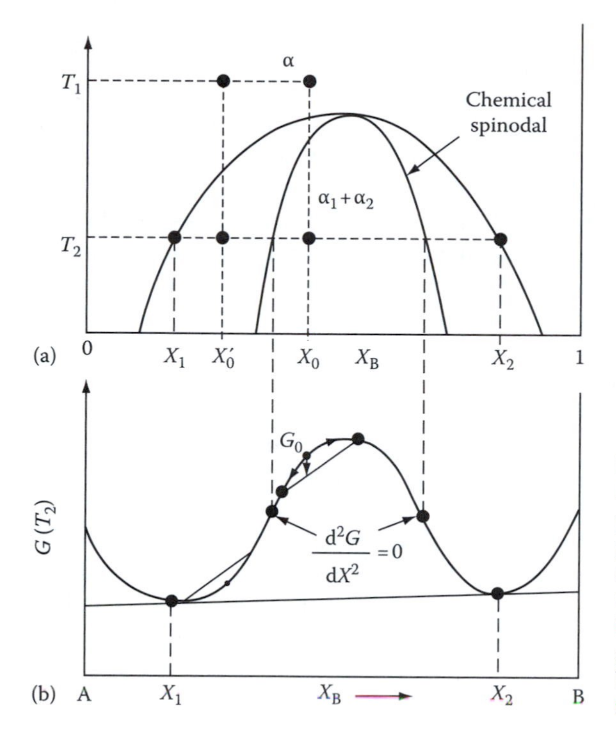

FIGURE 5.38
Alloys between the spinodal points are unstable and can decompose into two coherent phase α_1 and α_2 without overcoming an activation energy barrier. Alloys between the coherent miscibility gaps and the spinodal are metastable and can decompose only after nucleation of the other phase.

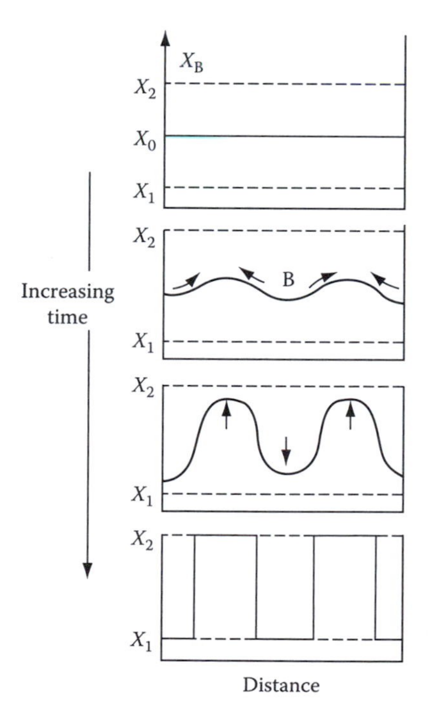

FIGURE 5.39
Schematic composition profiles at increasing times in an alloy quenched into the spinodal region (X_0 in Figure 5.38).

$$\frac{d^2G}{dX^2} < 0 \tag{5.42}$$

Therefore the alloy must lie between the two points of inflection on the free energy curve. The locus of the points on the phase diagram, Figure 5.32a, is known as the chemical spinodal.

If the alloy lies outside the spinodal, small variations in composition lead to an increase in free energy and the alloy is therefore metastable. The free energy of the system can only be decreased in this case if nuclei are formed with a composition very different from the matrix. Therefore, outside the spinodal the transformation must proceed by a process of nucleation and growth. Normal down-hill diffusion occurs in this case as shown in Figure 5.40.

The rate of spinodal transformation is controlled by the interdiffusion coefficient, D. Within the spinodal $D < 0$ and the composition fluctuations shown in Figure 5.39 will therefore increase exponentially with time, with a characteristic time constant $\tau = -\lambda^2/4\pi^2 D$, where λ is the wavelength of the composition modulations (assumed one-dimensional). The rate of transformation can therefore become very high by making λ as small as possible. However, as will be shown below, there is a minimum value of λ below which spinodal decomposition cannot occur.

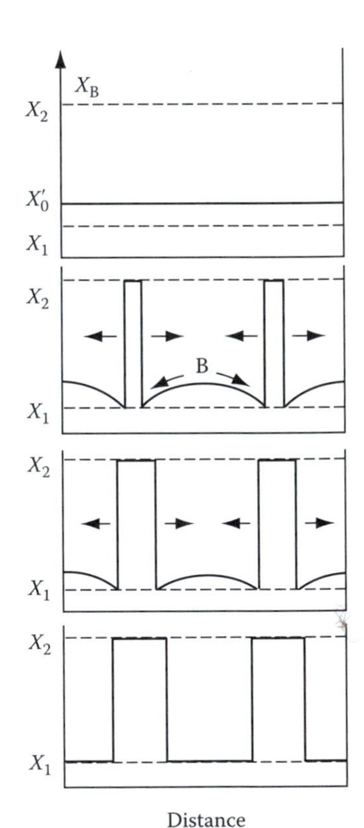

FIGURE 5.40
Schematic composition profiles at increasing times in an alloy outside the spinodal points (X'_0 in Figure 5.38).

In order to be able to calculate the wavelength of the composition fluctuations that develop in practice it is necessary to consider two important factors that have been omitted from the above discussion: (1) interfacial energy effects and (2) coherency strain energy effects.

If a homogeneous alloy of composition X_0 decomposes into two parts one with composition $X_0 + \Delta X$ and the other with composition $X_0 - \Delta X$, it can be shown that[9] the total chemical free energy will change by an amount ΔG_c given by

$$\Delta G_c = \frac{1}{2}\frac{d^2 G}{2dX^2}(\Delta X)^2 \tag{5.43}$$

If, however, the two regions are finely dispersed and coherent with each other there will be an additional energy change due to interfacial energy effects. Although, during the early stages of spinodal decomposition, the interface between A-rich and B-rich regions is not sharp but very diffuse, there is still an effective interfacial energy contribution. The magnitude of this energy depends on the composition gradient across the interface, and for this reason it is known as a "gradient energy." In solid solutions which tend to cluster the energy of like atom pairs is less than that of unlike pairs. Thus the origin of the gradient energy is the increased number of unlike nearest neighbors in a solution containing composition gradients compared to a homogeneous solution. For a sinusoidal composition modulation of wavelength λ and amplitude ΔX the maximum composition gradient is proportional to $(\Delta X/\lambda)$ and the gradient energy term ΔG_γ is given by

$$\Delta G_\gamma = K\left(\frac{\Delta X}{\lambda}\right)^2 \tag{5.44}$$

where K is a proportionality constant dependent on the difference in the bond energies of like and unlike atom pairs.

If the size of the atoms making up the solid solution are different, the generation of composition differences will introduce a coherency strain energy term, ΔG_s. If the misfit between the A-rich and B-rich regions is δ, $\Delta G_s \propto E\delta^2$ where E is Young's modulus. For a total composition difference ΔX, δ will be given by $(da/dX)\ \Delta X/a$, where a is the lattice parameter. An exact treatment of the elastic strain energy shows that

$$\Delta G_s = \eta^2(\Delta X)^2 E' V_m \tag{5.45}$$

where

$$\eta = \frac{1}{a}\left(\frac{da}{dX}\right) \tag{5.46}$$

i.e., η is the fractional change in lattice parameter per unit composition change $E' = E/(1 - v)$, where v is Poisson's ratio, and V_m is the molar volume. Note that ΔG_s is independent of λ.

If all of the above contributions to the total free energy change accompanying the formation of a composition fluctuation are summed we have

$$\Delta G = \left(\frac{d^2G}{dX^2} + \frac{2K}{\lambda^2} + 2\eta^2 E' V_m\right)\frac{(\Delta X)^2}{2} \tag{5.47}$$

It can be seen therefore that the condition for a homogeneous solid solution to be unstable and decompose spinodally is that

$$-\frac{d^2G}{dX^2} > \frac{2K}{\lambda^2} + 2\eta^2 E' V_m \tag{5.48}$$

Thus the limits of temperature and composition within which spinodal decomposition is possible are given by the conditions $\lambda = \infty$ and

$$\frac{d^2G}{dX^2} = -2\eta^2 E' V_m \tag{5.49}$$

The line in the phase diagram defined by this condition is know as the coherent spinodal and it lies entirely within the chemical spinodal ($d^2G/dX^2 = 0$) as shown in Figure 5.41. It can be seen from Equation 5.48

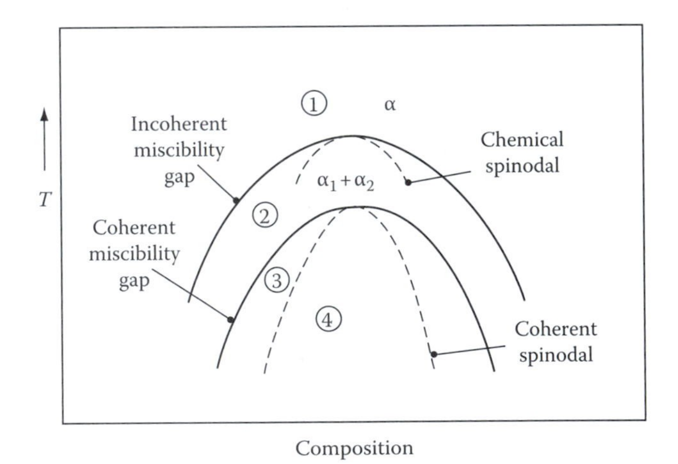

FIGURE 5.41
Schematic phase diagram for a clustering system. Region 1: homogeneous α stable. Region 2: homogeneous α metastable, only incoherent phases can nucleate. Region 3: homogeneous α metastable, coherent phase can nucleate. Region 4: homogeneous α unstable, no nucleation barrier, spinodal decomposition occurs.

that the wavelength of the composition modulations that can develop inside the coherent spinodal must satisfy the condition

$$\lambda^2 > \frac{-2K}{(d^2G/dX^2) + 2\eta^2 E' V_m} \tag{5.50}$$

Thus the minimum possible wavelength decreases with increasing undercooling below the coherent spinodal.

Figure 5.41 also shows the coherent miscibility gap. This is the line defining the equilibrium compositions of the coherent phases that result from spinodal decomposition (X_1 and X_2 in Figure 5.39). The miscibility gap that normally appears on an equilibrium phase diagram is the incoherent (or equilibrium) miscibility gap. This corresponds to the equilibrium compositions of incoherent phases, i.e., in the absence of strain fields. The chemical spinodal is also shown in Figure 5.41 for comparison, but it is of no practical importance.

Spinodal decomposition is not only limited to systems containing a stable miscibility gap. All systems in which GP zones form, for example, contain a metastable coherent miscibility gap, i.e., the GP zone solvus (see the Al–Ag system in Figure 5.34 for example). Thus it is possible that at high supersaturations GP zones are able to form by the spinodal mechanism. If aging is carried out below the coherent solvus but outside the spinodal, GP zones can only form by a process of nucleation and growth, Figure 5.40. Between the incoherent and the coherent miscibility gaps, Figure 5.41, $\Delta G_v - \Delta G_s < 0$ and only incoherent strain-free nuclei can form.

The difference in temperature between the coherent and the incoherent miscibility gaps, or the chemical and coherent spinodals in Figure 5.41, is dependent on the magnitude of $|\eta|$. When there is a large atomic size different $|\eta|$ is large and a large undercooling is required to overcome the strain energy effects. As discussed earlier large value of $|\eta|$ in cubic metals can be mitigated if the misfit strains are accommodated in the elastically soft $\langle 100 \rangle$ directions. This is achieved by the composition modulations building up parallel to $\{100\}$.

Figure 5.42 shows a spinodal structure in a specimen of Al–22.5 Zn–0.1 Mg (atomic%) solution treated at 400°C and aged 20 h at 100°C. The wavelength in the structure is 25 nm. But this is greater than the initial microstructure due to coarsening which occurs on holding long times at high temperature.

5.5.6 Particle Coarsening[10]

The microstructure of a two-phase alloy is always unstable if the total interfacial free energy is not a minimum. Therefore a high density of small precipitates will tend to coarsen into a lower density of larger particles with a smaller total interfacial area. However, such coarsening often produces an undesirable degradation of properties such as a loss of strength or the

FIGURE 5.42
Coarsened spinodal microstructure in Al–22.5 at% Zn–0.1 at% Mg solution treated 2 h at 400°C and aged 20 h at 100°C. Thin-foil electron micrograph (×314,000). (After Rundman, K.B., in *Metals Handbook*, 8th edn., Vol. 8, American Society for Metals, Metals Park, OH, 1973, p. 184.)

disappearance of grain-boundary pinning effects (Section 3.3.5). As with grain growth, the rate of coarsening increases with temperature and is of particular concern in the design of materials for high temperature applications.

In any precipitation-hardened specimen there will be a range of particle sizes due to differences in the time of nucleation and rate of growth. Consider two adjacent spherical precipitates[11] with different diameters as shown in Figure 5.43. Due to the Gibbs–Thomson effect, the solute concentration in the matrix adjacent to a particle will increase as the radius of curvature decreases (Figure 5.43b). Therefore there will be concentration gradients in the matrix which will cause solute to diffuse in the direction of the largest particles away from the smallest, so that the small particles shrink and disappear while large particles grow. The overall result is that that total number of particles decreases and the mean radius ($\bar{r}$) increases with time. By assuming volume diffusion is the rate-controlling factor it has been shown[12] that the following relationship should be obeyed:

$$(\bar{r})^3 - r_0^3 = kt \tag{5.51}$$

where

$$k \propto D\gamma X_e$$

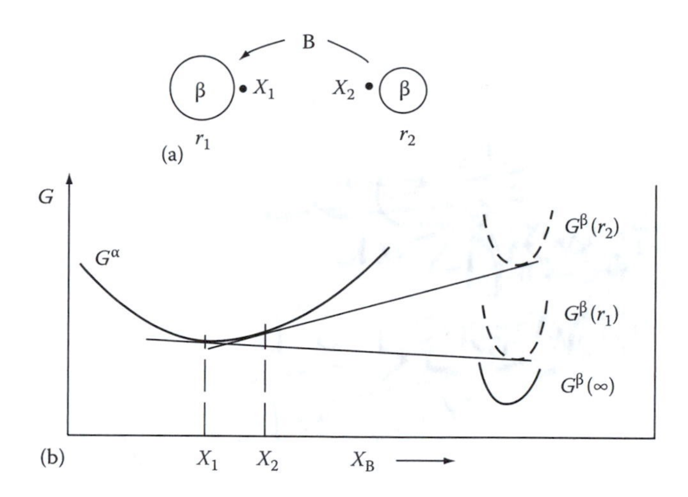

FIGURE 5.43
Origin of particle coarsening. β with a small radius of curvature (r_2) has a higher molar free energy than β with a large radius of curvature (r_1). The concentration of solute is therefore highest outside the smallest particles.

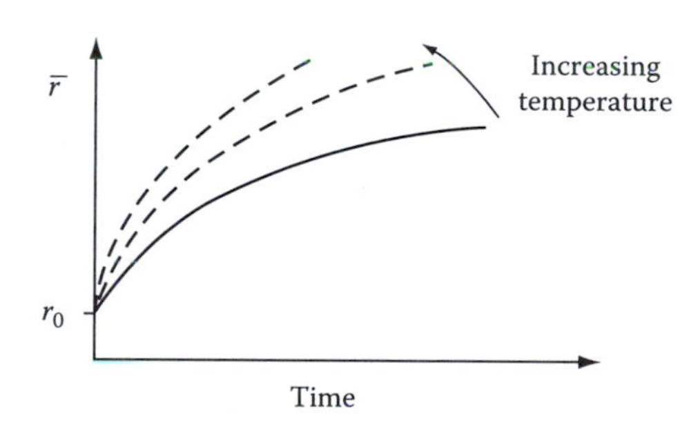

FIGURE 5.44
Schematic diagram illustrating how the mean particle radius $\bar{r}$ increases with time at different temperatures.

where
- r_0 is the mean radius at time $t = 0$
- D is the diffusion coefficient
- γ is the interfacial energy
- X_e is the equilibrium solubility of very large particles

Since D and X_c increase exponentially with temperature, the rate of coarsening will increase rapidly with increasing temperature, Figure 5.44. Note that the rate of coarsening

$$\frac{d\bar{r}}{dt} \propto \frac{k}{\bar{r}^2} \tag{5.52}$$

so that distributions of small precipitates coarsen most rapidly.

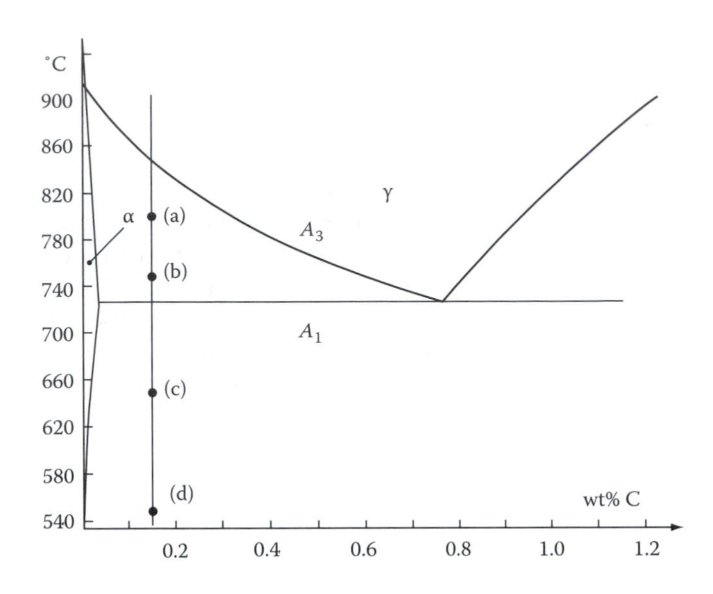

FIGURE 5.45
Holding temperatures for steel in Figure 5.46.

In practice the rate at which particles coarsen may not follow a linear $r^3 - t$ relationship. Deviations from this relationship can be caused by diffusion short-circuits such as dislocations, or grain boundaries. Also the coarsening rate may be interface-controlled. Nevertheless, apart from the case of interface control, the rate of coarsening should depend on the product $D\gamma X_e$ (k in Equation 5.51). Therefore high-temperature alloys whose strength depends on a fine precipitate dispersion must have a low value for at least one of γ, X_e, or D. Let us consider examples of each of these.

Low γ

The heat-resistant Nimonic alloys based on Ni–Cr with additions of Al and Ti obtain their high strength from a fine dispersion of the ordered fcc phase $Ni_3(TiAl)$ (γ') which precipitates in the fcc Ni-rich matrix. The Ni/γ' interfaces are fully coherent and the interfacial energy is exceptionally low (~10–30 mJ m^{-2}) which enables the alloys to maintain a fine structure at high temperature. The misfit between the precipitates and matrix varies between zero and about 0.2% depending on composition. It is interesting that the total creep-rupture life of these alloys can be increased by a factor of 50× by careful control of composition to give zero misfit as compared to 0.2% misfit. The reason for this may be that during creep deformation the particles with the slightly higher misfits lose coherency with the result that γ is increased thereby increasing the rate of coarsening.

Low X_e

High strength at high temperature can also be obtained with fine oxide dispersions in a metal matrix. For example W and Ni can be strengthened for high temperature use by fine dispersions of thoria ThO_2. In general, oxides are very insoluble in metals and the stability of these microstructure at high temperatures can be attributed to a low value of X_e in the product $D\gamma X_e$.

Low *D*

Cementite dispersions in tempered steels coarsen very quickly due to the high diffusivity of interstitial carbon. However, if the steel contains a substitutional alloying element that segregates to the carbide, the rate of coarsening becomes limited by the much slower rate at which substitutional diffusion can occur. If the carbide-forming element is present in high concentrations more stable carbides are formed which have the additional advantage of a lower solubility (X_e). Therefore low-alloy steels used for medium-temperature creep resistance often have additions of strong carbide-forming elements.

5.6 Precipitation of Ferrite from Austenite

In this section we will be concerned with phase transformations in which the first phase to appear is that given by the equilibrium phase diagram. The discussion will be illustrated by reference to the diffusional transformation of Fe–C–austenite into ferrite. However, many of the principles are quite general and have analogues in other systems where the equilibrium phases are not preceded by the precipitation of transition phases. Under these conditions the most important nucleation sites are grain boundaries and the surfaces of inclusions.

Consider an Fe–0.15 wt% C alloy which, after austenitizing, is allowed to partially transform to ferrite at various temperatures below A_3 (Figure 5.45) and then quenched into water. The resultant microstructures are shown in Figure 5.46. The white areas are ferrite (α). The gray areas are martensite that formed from the untransformed austenite (γ) during quenching. At small undercooling below A_3, Figure 5.46a, the ferrite nucleates on austenite grain boundaries and grows in a "blockey" manner to form what are known as grain-boundary allotriomorphs. Note that both smoothly curved, presumably incoherent, α/γ interfaces as well as faceted, semicoherent interfaces are present. At larger undercoolings there is an increasing tendency for the ferrite to grow from the grain boundaries as plates, so-called Widmanstätten sideplates, which become finer with increasing undercooling, Figure 5.46b through d.

Experimental measurements on Widmanstätten ferrite in other ferrous alloys show that the habit planes are irrational, scattered 4° to 20° from $\{111\}_\gamma$, and that orientation relationships close to the Nishiyama–Wasserman or Kurdjumov–Sachs type are usually found. High-resolution TEM has also shown that the habit planes have a complex semicoherent structure,

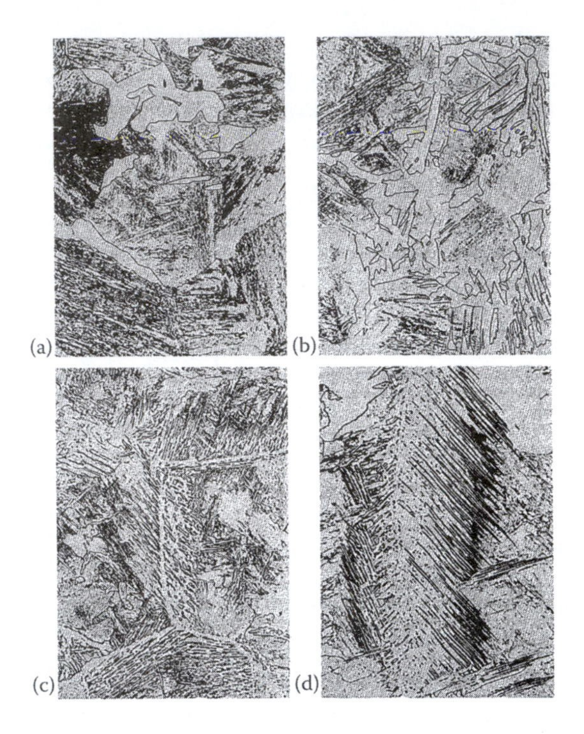

FIGURE 5.46
Microstructures of an Fe–0.15% C alloy. The specimens were austenitized, held at an intermediate temperature to give some ferrite, and then quenched to room temperature to give some ferrite, and then quenched to room temperature. The ferrite is white. The gray, fine constituent is a mixture of ferrite and carbide formed on quenching. All photographs are ×100 except (d). (a) 800°C for 150 s—primarily ferrite allotriomorphs with a few plates. (b) 750°C for 40 s—many more plates, mostly growing from grain boundaries. (c) 650°C for 9 s—relatively fine. Note common direction of plates along each boundary. (d) 550°C for 2 s (×300). (After Shewmon, P.G., in *Transformations in Metals*, McGraw-Hill, New York, 1969, after H.I. Aaronson.)

containing structural ledges and misfit dislocations, similar to that described in Section 3.4.1.[13]

As explained previously, the need to minimize ΔG^* leads to the creation of semicoherent interfaces and orientation relationships, even in the case of grain-boundary nucleation. A critical nucleus could therefore appear as shown in Figure 3.45b with faceted (planar) coherent (or semicoherent) interfaces and smoothly curved incoherent interfaces. For certain misorientations across the grain boundary it may even be possible for low-energy facets to form with both grains. Due to their low mobility faceted interfaces will tend to persist during growth while incoherent interfaces will be able to grow continuously and thereby retain a smooth curvature. Thus it is possible to explain the presence of smoothly curved and faceted interfaces in Figure 5.46a.

The reason for the transition from grain boundary allotriomorphs to Widmanstätten side-plates with increasing undercooling is not fully understood. It has been suggested by Aaronson and coworkers[2] that the relative

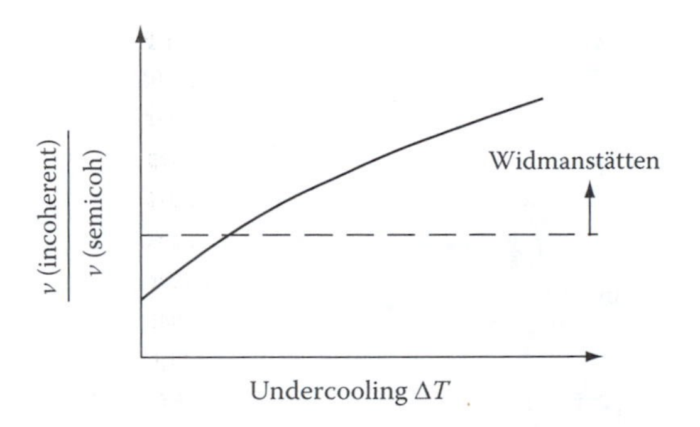

FIGURE 5.47
Possible variation of the relative velocity of incoherent and semicoherent interfaces at different undercoolings. Above a certain ratio Widmanstätten morphologies should develop, as shown in Figure 5.13. (After Aaronson, H.I., in *Decomposition of Austenite by Diffusional Processes*, Interscience, New York, 1962.)

rates at which semicoherent and incoherent interfaces can migrate vary with undercooling as shown in Figure 5.47. At small undercoolings it is proposed that both semicoherent and incoherent interfaces can migrate at similar rats, while at large undercoolings only incoherent can migrate at similar rates, and make full use of the increased driving force. Consideration of Figure 5.13 thus shows that approximately equiaxed morphologies should develop at low undercoolings while platelike morphologies, with ever-increasing aspect ratios, should develop at high undercoolings. Another factor which may contribute to the increased fineness of the Widmanstätten morphologies with decreasing temperature is that the minimum plate-tip radius r^* is inversely proportional to the undercooling.

It can be seen in Figure 5.46 that ferrite can also precipitate within the austenite grains (intragranular ferrite). Suitable heterogeneous nucleation sites are thought to be inclusions and dislocations. These precipitates are generally equiaxed at low undercoolings and more platelike at higher undercoolings.

In general the nucleation rate within grains will be less than on grain boundaries. Therefore, whether or not intragranular precipitates are observed depends on the grain size of the specimen. In fine-grained austenite for example, the ferrite that forms on grain boundaries will rapidly raise the carbon concentration within the middle of the grains, thereby reducing the undercooling and making nucleation even more difficult. In a large-grained specimen, however, it takes a longer time for the carbon rejected from the ferrite to reach the centers of the grains and meanwhile there will be time for nucleation to occur on the less favorable intragranular sites.

A TTT diagram for the precipitation of ferrite in a hypoeutectoid steel will have a typical C shape as shown in Figure 5.48. The $\gamma \rightarrow \alpha$ transformation should be approximately described by Equation 5.39 and the time for a given percentage transformation will decrease as the constant k increase, for

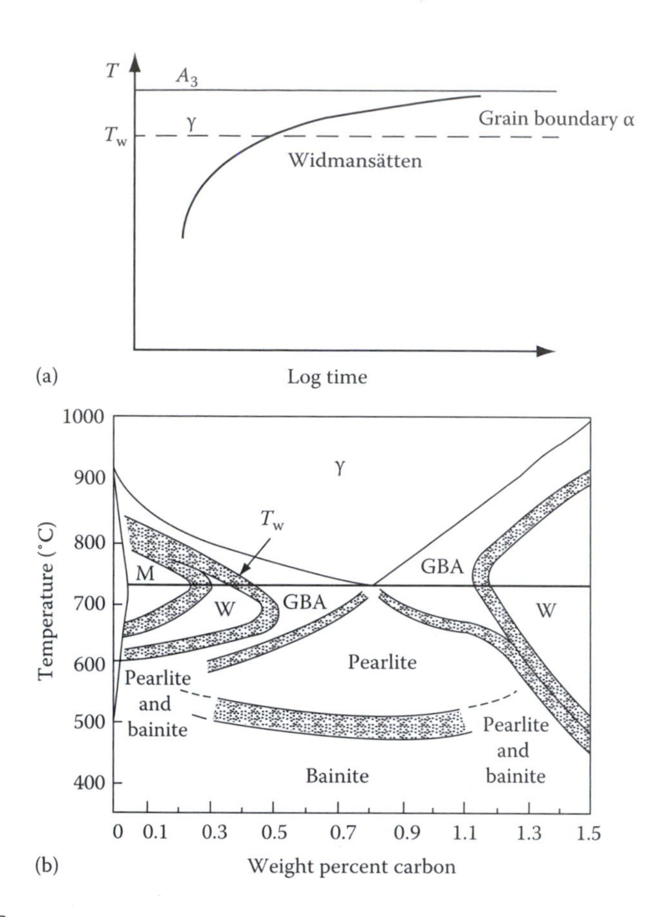

FIGURE 5.48
(a) Typical TTT curve for $\gamma \rightarrow \alpha$ transformation. (b) Temperature-composition regions in which the various morphologies are dominant at late reaction times in specimens with ASTM grain size Nos. 0–1. GBA = grain boundary allotriomorphs, W = Widmanstätten sideplates and/or intragranular plates, M = massive ferrite. See Section 5.9. (After Aaronson, H.I., in *Decomposition of Austenite by Diffusional Processes*, Interscience, New York, 1962.)

example, Equation 5.40. As usual, k increases with small increases in T due to increased nucleation and growth rates—k is also raised by an increase in the total number of nucleation sites. Thus decreasing the austenite grain size has the effect of shifting the C curve to shorter transformation times.

It is possible to mark a temperature T_w below which the ferrite forms as predominantly Widmanstätten plates and above which it is mainly in the form of grain boundary allotriomorphs. For alloys of different carbon content A_3 and T_w vary as shown on the phase diagram in Figure 5.47b.

During practical heat treatments, such as normalizing or annealing, transformation occurs continuously during cooling. Under these circumstances the final microstructure will depend on the cooling rate. If the specimen is cooled very slowly there will be time for nucleation to occur at small undercoolings on grain corners, edges, and boundaries. As these nuclei grow the carbon rejected into the austenite will have time to diffuse over large distances and the austenite grain should maintain a uniform composition given by the equilibrium phase diagram. Finally the austenite reaches the eutectoid composition and transforms to pearlite. Furnace cooling corresponds fairly close to these conditions and an example is shown in Figure 5.49c. The final proportions of ferrite and pearlite should be as determined by the equilibrium phase diagram.

The microstructure that results from more rapid cooling will depend on the grain size and the cooling rate. If the rate of cooling is moderately high the specimen will not remain long enough at high temperatures for nucleation to occur. Thus nuclei will not be formed until higher supersaturations are reached. The nucleation rate will then be rapid and large areas of grain boundary will become covered with nuclei. If the temperature is below T_W the ferrite will grow into the austenite as Widmanstätten side-plates with a spacing that becomes finer with decreasing temperature.

The nuclei that form at the highest temperatures will be on grain corners which will be followed by edges at lower temperatures and finally grain boundaries at still lower temperatures. In a small-grained specimen where there are a large number of grain corner and edge sites a large number of nuclei can be formed above the T_W temperature and grow as grain-boundary

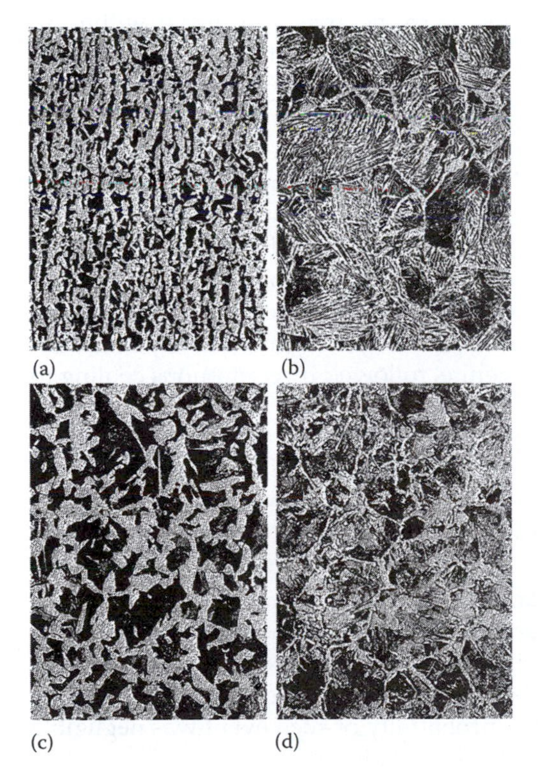

FIGURE 5.49
Microstructures obtained from different heat treatments in plain carbon steels (×60). 0.23 wt% C 1.2% Mn air-cooled, showing influence of prior austenite grain size: (a) austenitized at 900°C, (b) austenitized at 1150°C. 0.4% C showing effect of cooling rate for same grain size, (c) furnace-cooled (annealed), and (d) air-cooled (normalized). (After P.G. Shewmon, *Transformations in Metals*, McGraw-Hill, New York. 1969: (a) and (b) after R. Yoe, U.S. Steel Company, (c) and (d) after K. Zurlippe, J. and L. Steel Company.)

allotriomorphs. In a large-grained specimen, on the other hand, relatively few nuclei will form at high temperatures and the austenite far from these particles will remain supersaturated until lower temperatures, below T_W, when ferrite will be able to nucleate on grain boundary sites and grow as Widmanstätten side-plates. The effect of cooling rate and grain size is illustrated in Figure 5.49. Note also that the total volume fraction of ferrite decreases as the transformation temperature decreases. This point will be returned to later.

If the austenite contains more than about 0.8 wt% C, the first phase to form will be cementite. This also nucleates and grows with an orientation relationship to the austenite, producing similar morphologies to ferrite—grain boundary allotriomorphs at high temperatures and Widmanstätten side-plates at lower temperatures as shown in Figure 5.48b.

5.6.1 Case Study: Ferrite Nucleation and Growth

Given the scientific and technological importance of the austenite to ferrite transformation in steel, extensive experimental work has been carried out to assess this transformation. Furthermore, transformation models have been proposed, which rely on the classical nucleation theory (CNT),[14] as well as the growth law suggested by Zener,[15] as described in Sections 5.1 through 5.3. Therefore, the Zener classical growth model can be used to model the behavior of individual ferrite grains assuming that they grow within an infinite austenite matrix. In other words, the concentration of the solute atoms in the parent phase far from the austenite/ferrite interface is assumed to be the starting solute concentration. This assumption has the implication that when the transformation progresses and there is overlapping of diffusion fields (soft impingement) and/or (hard impingement) due to growing nearby grains, the theory cannot be applied.

A recent report[14] investigating ferrite nucleation and growth from austenite during continuous cooling of a 0.21 wt% carbon construction steel showed much smaller activation energy for nucleation compared with values predicted by thermodynamic models. The steel was annealed at 900°C for 10 min to allow the formation of austenite, followed by continuous cooling to 600°C in 1 h.

The findings were possible through the use of high-energy x-ray microscopes at synchrotron sources which enabled the study of individual grains during phase transformation in the bulk. Ferrite nuclei density was determined by counting the number of valid ferrite spots on the two-dimensional detector, see Figure 5.50.

It is important to mention that, as described by Aaronson et al. more recently,[16] a single diffraction spot on a two-dimensional detector might represent a number of ferrite grains if they are close enough with almost identical crystallographic orientation. That is probable theoretically, however, Offerman et al. argued that the probability of such event was negligible in their experiment.[17]

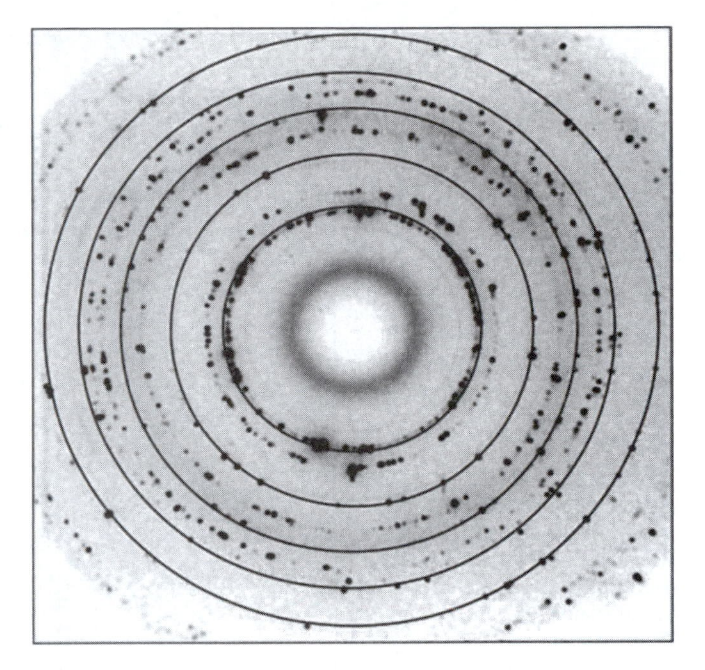

FIGURE 5.50
X-ray diffraction pattern exhibiting austenite and ferrite reflections. Pattern acquired at 763°C. Solid circles exhibit the ferrite grains scattering angles. (From Offerman, S.E., van Dijk, N.H., Sietsma, J., Grigull, S., Lauridsen, E.M., Margulies, L., Poulsen, H.F., Rekveldt, M.Th., and van der Zwaag, S., *Science*, 298, 1003, 2002. Reprinted with permission from AAAS.)

Figure 5.51b shows the predicted nucleation rate calculated using the CNT versus measured data. Although the shape of the calculated curve agrees with measurements, the actual nucleation rate apparently peaked at a higher temperature. The authors reported that the activation energy for nucleation was at least two orders of magnitude smaller compared with prediction. However, it was not possible to establish the location at which ferrite grains nucleated. As mentioned in Section 5.2, the activation energy barrier could considerably be reduced, or even removed, if the formation of a nucleus results in the destruction of a defect. Offerman et al., however, questioned the applicability of the current continuum thermodynamics which predict ΔG_V given the small size of the critical nucleus which is only about 10–100 atoms.[18]

Offerman et al. also reported new types of ferrite grain growth which were not in line with Zener's theory. These results are shown in Figure 5.52 where predictions made using Zener's model are presented as straight lines.

Figure 5.52 shows the evolution of individual ferrite grains radius, as well as in pearlite colonies, as the steel is continuously cooled. By measuring the intensity of diffraction spots, the ferrite grain volume could be determined. Combined with the assumption that the grains are spherical, therefore, the grain radius was calculated. It is important to note that the spatial resolution of the experimental technique is limited to about 2 μm only making the

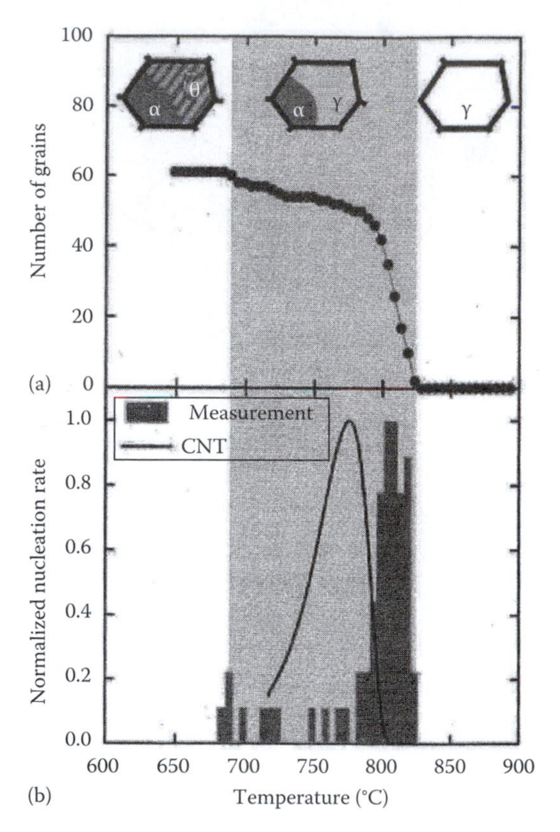

FIGURE 5.51
(a) Evolution of ferrite (α) from austenite (γ) during continuous cooling. (b) Measured versus calculated (curve) nucleation rate using the CNT. Data were normalized to the maximum nucleation rate. (θ) stands for cementite. (From Offerman, S.E., van Dijk, N.H., Sietsma, J., Grigull, S., Lauridsen, E.M., Margulies, L., Poulsen, H.F., Rekveldt, M.Th., and van der Zwaag, S., *Science*, 298, 1003, 2002. Reprinted with permission from AAAS.)

detection of smaller grains not possible. Each set of symbols in Figure 5.52 represents a single ferrite grain.

As can be seen in Figure 5.52a, there is agreement with Zener's theory in early stages of grain growth even though there was a deviation in the later stages of growth which could be related to different impingement conditions for the measured grains. Figure 5.52b shows a type of growth where some ferrite grains continued to grow in a pearlite colony formed at a later stage of transformation with the same crystallographic orientation. This finding was also observed by Thompson and Howell.[19] The third type of growth is shown in Figure 5.52c, where ferrite grain soft impingement apparently took place resulting in deceleration of the growth process. Figure 5.52d shows what the authors termed "complex behavior" in that some grains actually shrink during cooling. The authors explained this behavior in terms of grains hard impingement after which the grain growth mechanism described in Section 3.3.4 can actually lead to the shrinking of some grains.

Ferrite/ferrite grain-boundary migration has recently been modeled by Li et al.[20] where the mesoscale deterministic cellular automaton (CA) method as well as a probabilistic mesoscale Monte Carlo (MC) model were utilized. Both simulations were carried out independent from each other to allow

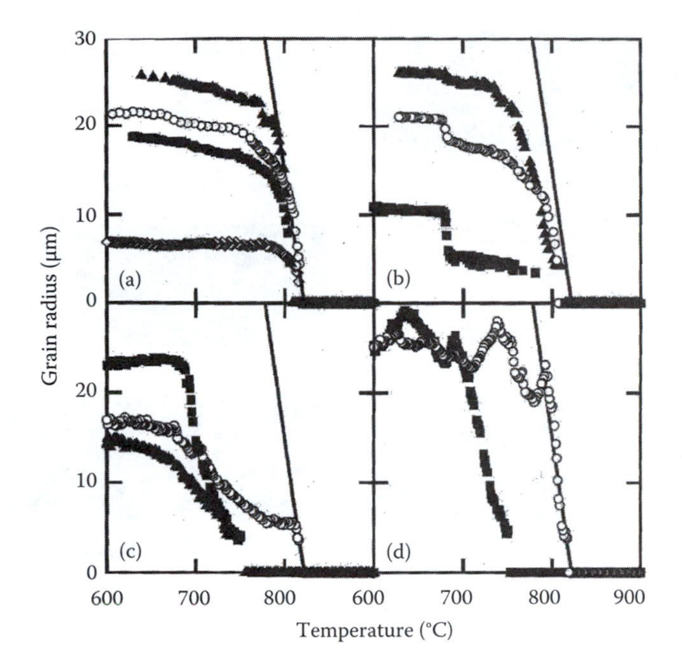

FIGURE 5.52
Four distinctive types of ferrite grain growth. (a) Initial grain growth is in agreement with Zener law with latter deviations explained in the light of various grain impingement conditions. (b) Ferrite grains continued to grow in a pearlite colony with the same crystallographic orientation. (c) Decelerated ferrite grain growth caused by ferrite grain soft-impingement. (d) Ferrite grains growth showing "complex behavior" where some grains actually shrink in the course of growth. See text for details. (From Offerman, S.E., van Dijk, N.H., Sietsma, J., Grigull, S., Lauridsen, E.M., Margulies, L., Poulsen, H.F., Rekveldt, M.Th., and van der Zwaag, S., *Science*, 298, 1003, 2002. Reprinted with permission from AAAS.)

the comparison of the results. The models were also two-dimensional representing a cross section of a three-dimensional microstructure. The mesoscale represents the length scale of the features in the microstructure.[21] Thereby, it was possible to predict the evolution of the microstructure during transformation in terms of average grain sizes, phase fractions, phase morphology, grains distribution, the evolution of solute concentration fields, etc. The models were applied on a steel with the composition Fe-0.132C (wt%) isothermally transformed at 1037 K. The starting austenite microstructure was created by a normal growth process by which the average austenite grain diameter was 80 μm.

As shown in Figure 5.53 (CA results) and 5.27 (MC results), Li et al. observed six different growth modes during monitoring the individual growth behavior of ferrite grains. Note that carbon-poor regions with concentration lower than 0.025 wt% represents the ferrite phase whereas carbon-rich regions with carbon content no less than that of the initial concentration of 0.132 wt% were considered as the retained austenite phase. Both Figures 5.53a and 5.54a represent the "parabolic" ferrite grain growth mode, which

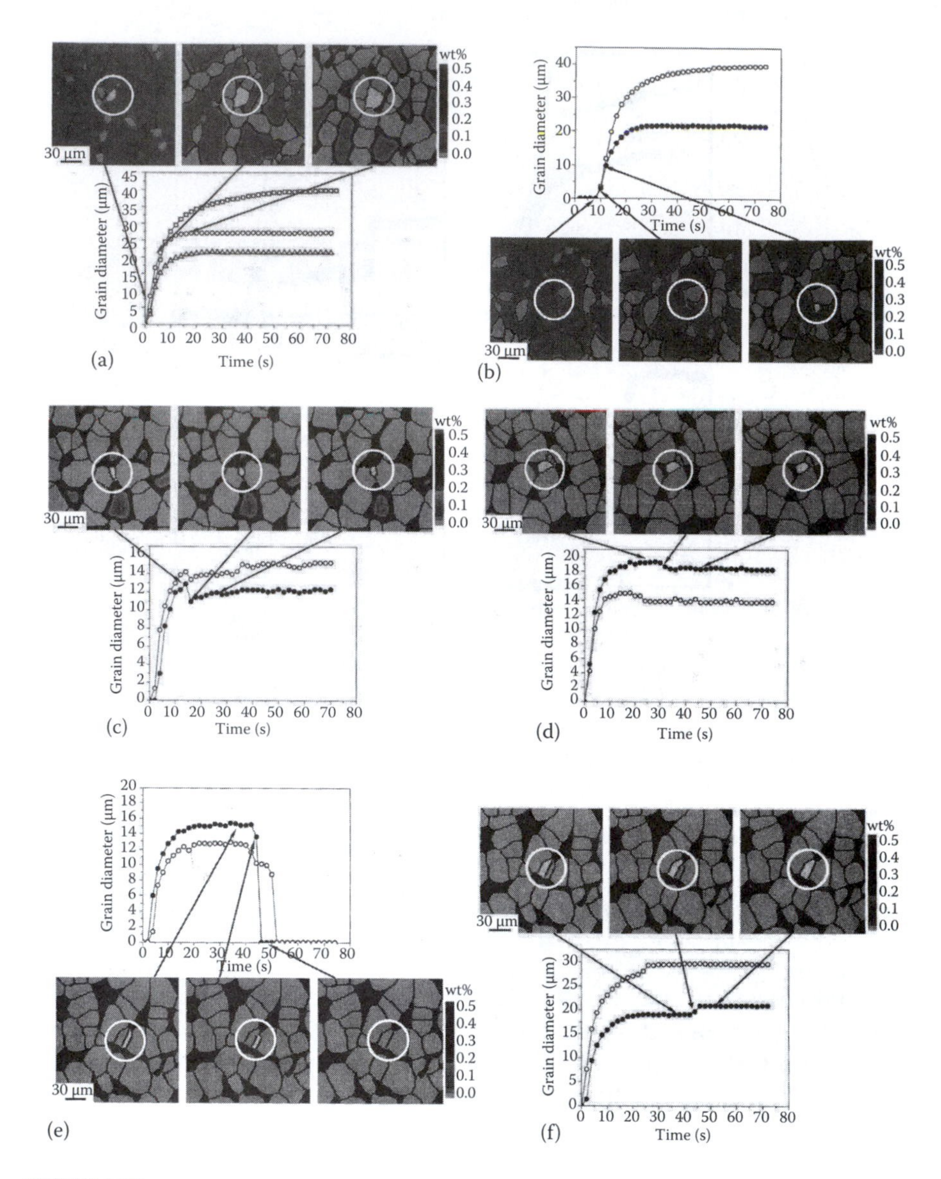

FIGURE 5.53
Six types of ferrite grain growth, or coarsening, simulated using the cellular automaton method: (a) parabolic grain growth, (b) retarded nucleation and growth, (c) "temporary shrinkage" growth mode, (d) partial shrinkage, (e) complete shrinkage, and (f) accelerated growth. The scale to the right of the simulated microstructures indicates the solute atom (carbon) concentration. (Reprinted from Li D.Z., Xiao N.M., Lan Y.J., Zheng C.W., and Li Y.Y., *Acta Mater*. 55, 6234, 2007. With permission from Elsevier.)

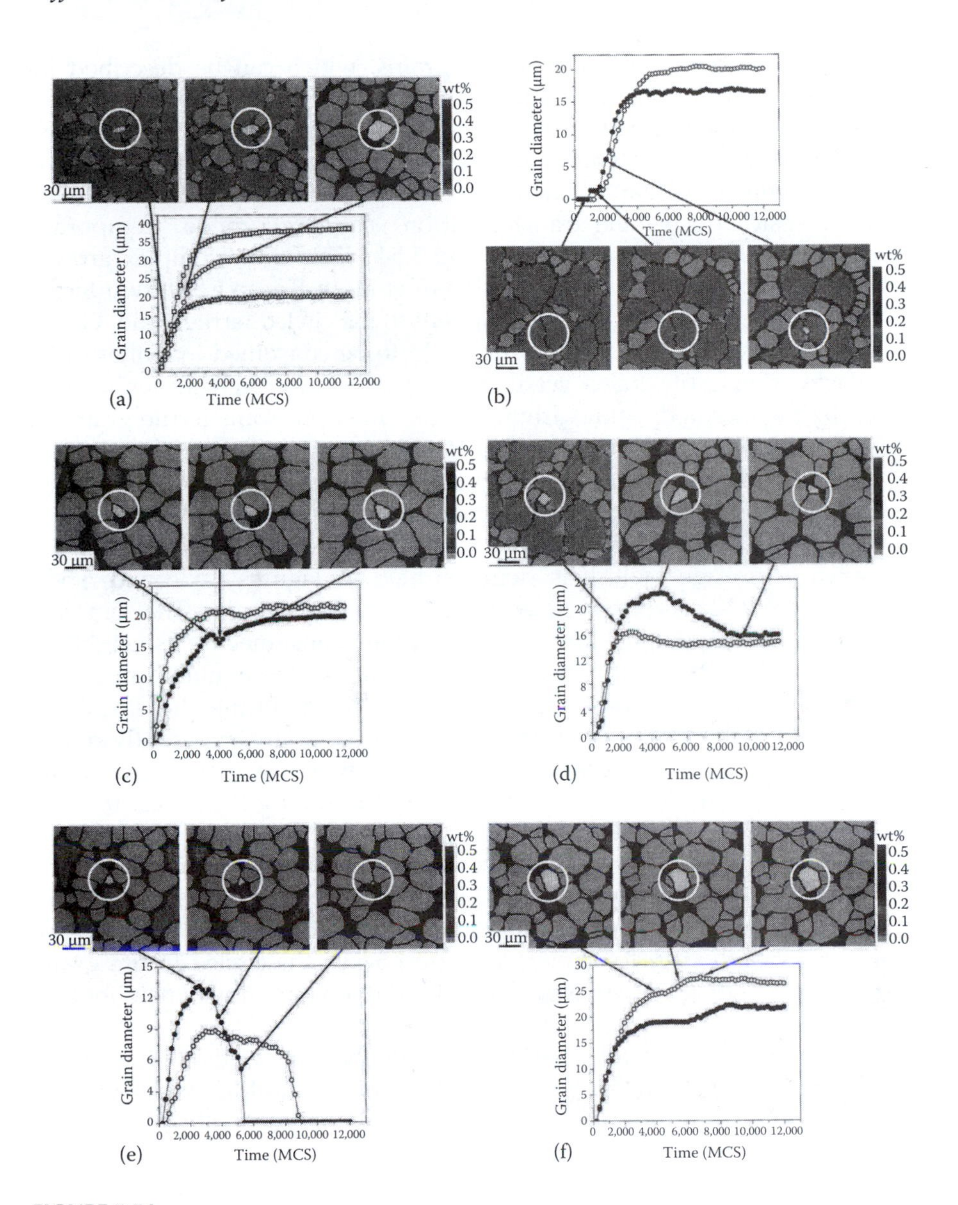

FIGURE 5.54
Ferrite grain growth simulated using the Monte Carlo model: (a) parabolic grain growth, (b) retarded nucleation and growth, (c) "temporary shrinkage" growth mode, (d) partial shrinkage, (e) complete shrinkage, and (f) accelerated growth. The scale to the right of the simulated microstructures indicates the solute atom (carbon) concentration. (Reprinted from Li, D.Z., Xiao, N.M., Lan, Y.J., Zheng, C.W., and Li, Y.Y., *Acta Mater.*, 55, 6234, 2007. With permission from Elsevier.)

was the case for about 55% of ferrite grains, which can be described by Zener model. Both simulations showed that upon reaching thermodynamic equilibrium only about 0.81 ferrite phase fraction was formed. Figures 5.53b and 5.24b show the "retarded nucleation and growth" mode where the carbon enrichment of the retained austenite phase ahead of the austenite/ferrite interface delays the transformation. The third mode "temporary shrinkage" is shown in Figures 5.53c and 5.54c. This is the complex growth behavior reported in the work of Offerman et al. In Figure 5.53c, two ferrite grains were to the right and the left adjacent to the circled ferrite grain. Grain-boundary pressures caused the ferrite grain to be consumed by either of its two neighboring ferrite grains, yet, it grows into the lower right austenite grain driven by the thermodynamic driving force. However, some ferrite grains in the simulations (reported to be around 4% of the population) exhibited "partial shrinkage mode" where the grains ceased to grow following shrinkage, presumably due to the local impingement conditions. This growth mode is shown in Figures 5.53d and 5.54d. Another mode of shrinkage was the "complete shrinkage" which is demonstrated in Figures 5.53e and 5.54e. This growth mode was observed in about 40% of ferrite grains. The complete shrinkage indicates that the ferrite grain is totally consumed by its neighbors following its initial growth. This was interpreted as the equilibrium conditions at the triple junction could not be sustained during the shrinkage process. The authors suggested that most of these grains apparently nucleated at middle and later stages of transformation where overlapping of solute atoms fields was more prominent which led to limiting their size. In turn, compared with their ferrite neighboring grains, their interface curvature is small as a consequence of their small size which facilitates their consumption by the coarser neighboring ferrite grains. Additionally, a small number of ferrite grains exhibited a growth mode "accelerated growth", shown in Figures 5.53f and 5.54f. As shown in Figure 5.53e, the circled ferrite grain, the ferrite coarsening process was caused by the consumption of neighboring ferrite grains.

The CA method was adopted along with the MC method due to the probabilistic nature of the rules in the MC model. The MC model yielded quite higher values for nucleation and coarsening rates of ferrite grains when compared with those predicted by the CA model. Nevertheless, similar results were observed in both models which exhibited the same six ferrite growth modes.

5.7 Cellular Precipitation

Grain-boundary precipitation does not always result in grain-boundary allotriomorphs or Widmanstätten side-plates or needles. In same cases it can result in a different mode of transformation, know as cellular precipitation.

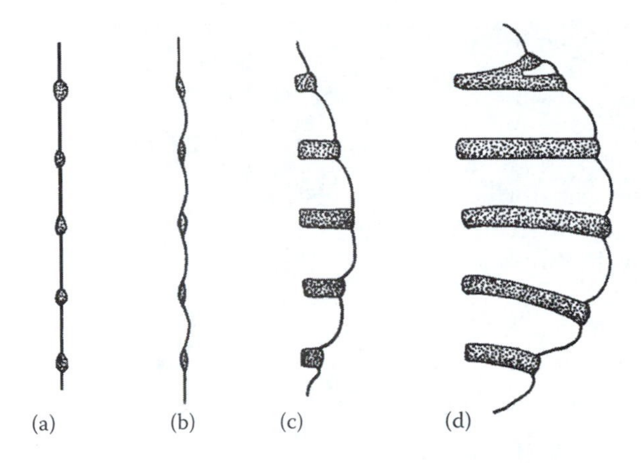

FIGURE 5.55
Schematic diagram showing a possible sequence of steps during the development of cellular precipitation.

The essential feature of this type of transformation is that the boundary moves with the growing tips of the precipitates as shown in Figure 5.55. Morphologically the transformation is very similar to the eutectoid reaction.

However, in this case the reaction can be written

$$\alpha' \rightarrow \alpha + \beta$$

where

α' is the supersaturated matrix
α is the same phase but with a lower thermodynamic excess of solute
β is the equilibrium precipitate

The mechanism whereby grain-boundary nucleation develops into cellular precipitation differs from one alloy to another and is not always fully understood. The reason why cells develop in some alloys and not in others is also unclear.

Figure 5.56 shows an example of cellular precipitation in an Mg–9 atomic % Al alloy. The β-phase in this case is the equilibrium precipitate $Mg_{17}Al_{12}$ indicated in the phase diagram, Figure 5.57. It can be seen in Figure 5.56 that the $Mg_{17}Al_{12}$ forms as lamellae embedded in a Mg-rich matrix. The grain boundary between grains I and II was originally straight along AA but has been displaced, and the cell matrix and grain I are the same grain.

Figure 5.58 shows another specimen which has been given a two-stage heat treatment. After solution treating at 410°C the specimen was quenched to a temperature of 220°C for 20 min followed by 90 s at 277°C and finally water quenched. It is apparent that the mean interlamellar spacing is higher at higher aging temperatures. As with eutectic solidification this is because less free energy is available for the formation of α/β interfaces when the total driving force for transformation is reduced.

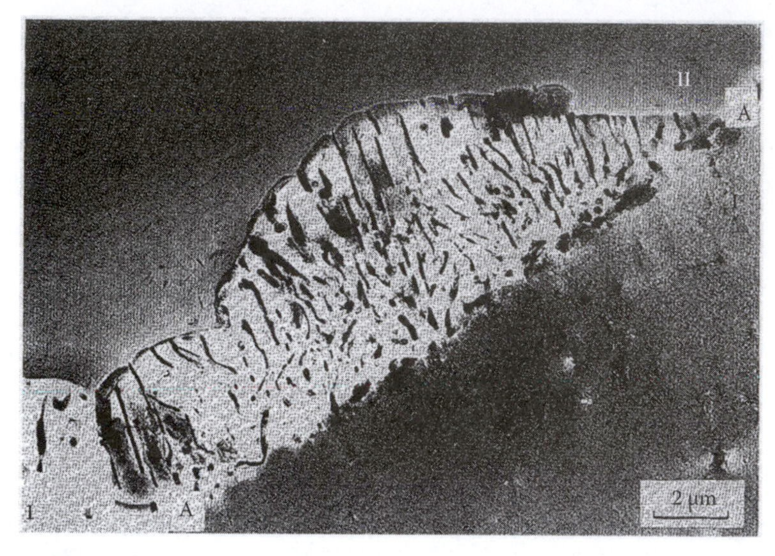

FIGURE 5.56
Cellular precipitation of $Mg_{17}Al_{12}$ in an Mg–9 at % Al alloy solution treated and aged 1 h at 220°C followed by 2 min at 310°C. Some general $Mg_{17}Al_{12}$ precipitation has also occurred on dislocations within the grains.

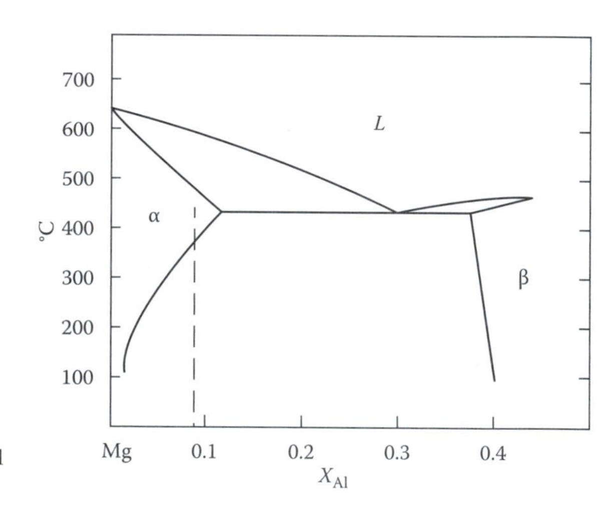

FIGURE 5.57
The relevant part of the Mg–Al phase diagram.

The growth of cellular precipitates requires the portioning of solute to the tips of the precipitates in contact with the advancing grain boundary. This can occur in one of two ways: either by diffusion through the lattice ahead of the advancing cell front, or by diffusion in the moving boundary. Partitioning by lattice diffusion would require solute concentration gradients ahead of the cell front while, if the grain boundary is the most effective diffusion route, the matrix composition should remain unchanged right up to the cell front. In the case of the Mg–Al alloy it has been possible to do microanalysis with

FIGURE 5.58
A cell formed during aging at two temperatures; 30 min at 220°C followed by 30 min at 277°C and water quenched. Note the change in interlamellar spacing caused by the change in undercooling.

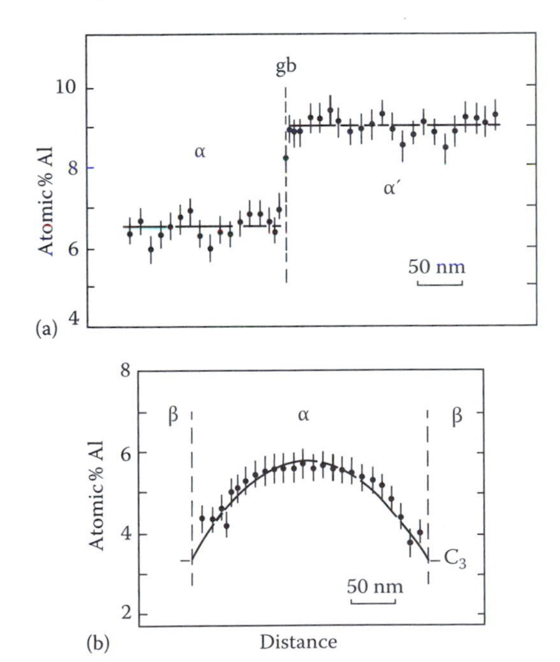

FIGURE 5.59
(a) Variation of aluminum concentration across an advancing grain boundary midway between two precipitate lamellae. (b) Similar profile along a line such as S in Figure 5.58.

sufficiently high spatial resolution to resolve these possibilities directly. (The technique used was electron energy loss spectroscopy using plasmon losses.[22]) The results of such measurements, Figure 5.59a, clearly indicate that the matrix composition remains unchanged to within 10 nm of the advancing cell front so that partitioning must be taking place within the boundary itself. This is to be expected since precipitation is occurring at relatively low temperatures where solute transport tends to become more effective via grain boundaries than through the lattice.

Figure 5.59b shows the aluminum concentration in the α-matrix along a line between the b ($Mg_{17}Al_{12}$) lamellae. This is essentially a replica of a similar concentration profile that must exist within the advancing grain boundary. Therefore apart from the matrix in contact with the β precipitate, the cell matrix is still supersaturated with respect to equilibrium.

Cellular precipitation is also known as discontinuous precipitation because the composition of the matrix changes discontinuously as the cell front passes. Precipitation that is not cellular is referred to as general or continuous because it occurs generally throughout the matrix on dislocations or grain boundaries, etc., and the matrix composition at a given point decreases continuously with time. Often general precipitation leads to a finely distributed intermediate precipitate that is associated with good mechanical properties. The cellular reaction is then unwanted because the intermediate precipitates will dissolve as they are overgrown and replaced by the coarse equilibrium precipitates within the cells.

5.8 Eutectoid Transformations

5.8.1 Pearlite Reaction in Fe–C Alloys

When austenite containing about 0.8 wt% C is cooled below the A_1 temperature it becomes simultaneously supersaturated with respect to ferrite and cementite and a eutectoid transformation results, i.e.,

$$\gamma \rightarrow \alpha + Fe_3C$$

The manner in which this reaction occurs is very similar to a eutectic transformation where the original phase is a liquid instead of a solid. In the case of Fe–C alloys the resultant microstructure comprises lamellae, or sheets, of cementite embedded in ferrite as shown in Figure 5.60. This is known as pearlite. Both cementite and ferrite form directly in contact with the austenite as shown.

Pearlite nodules nucleate on grain boundaries and grow with a roughly constant radial velocity into the surrounding austenite grains. At small under cooling below A_1 the number of pearlite nodules that nucleate is relatively small, and the nodules can grow as hemispheres or spheres without interfering with each other. At larger undercoolings the nucleation rate is much higher and site saturation occurs, that is all boundaries become quickly covered with nodules which grow together forming layers of pearlite outlining the prior austenite grain boundaries, Figure 5.61.

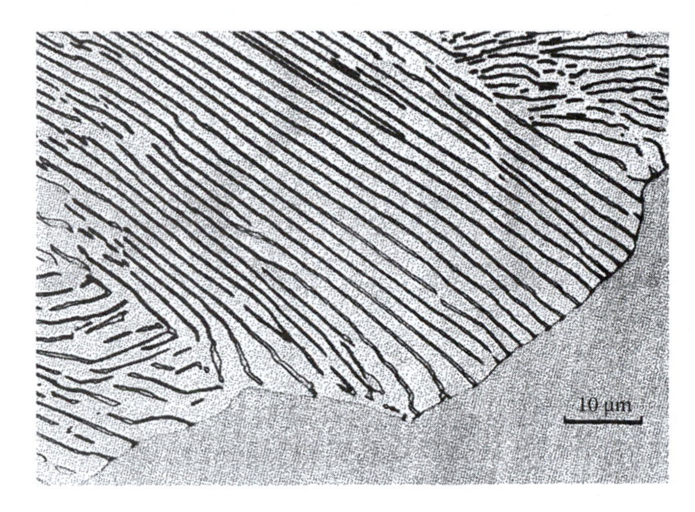

FIGURE 5.60
Pearlite colony advancing into an austenite grain. (After Darken, L.S. and Fisher, R.M., in *Decomposition of Austenite by Diffusional Processes*, Interscience, New York, 1962.)

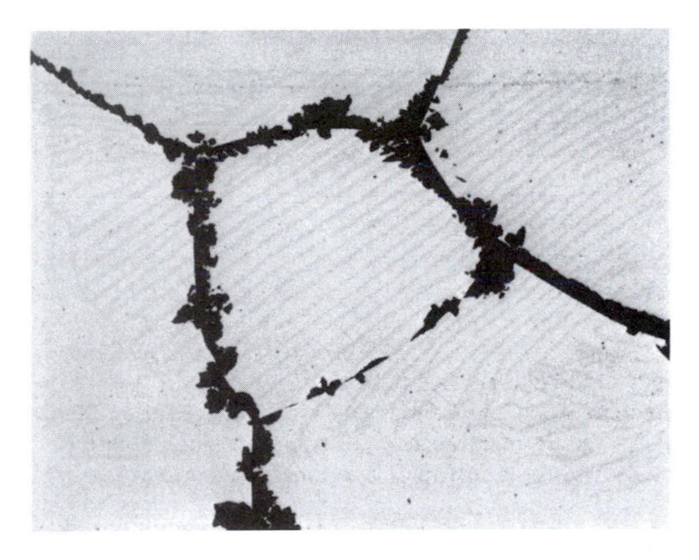

FIGURE 5.61
Partially transformed eutectoid steel. Pearlite has nucleated on grain boundaries and inclusions (×100). (After Cahn, J.W. and Hagel, W.C., in *Decomposition of Austenite by Diffusional Processes*, Interscience, New York, 1962.)

Nucleation of pearlite

The first stage in the formation of pearlite is the nucleation of either cementite or ferrite on an austenite grain boundary. Which phase nucleates first will depend on the grain-boundary structure and composition. Suppose that it is cementite. The cementite will try to minimize the activation

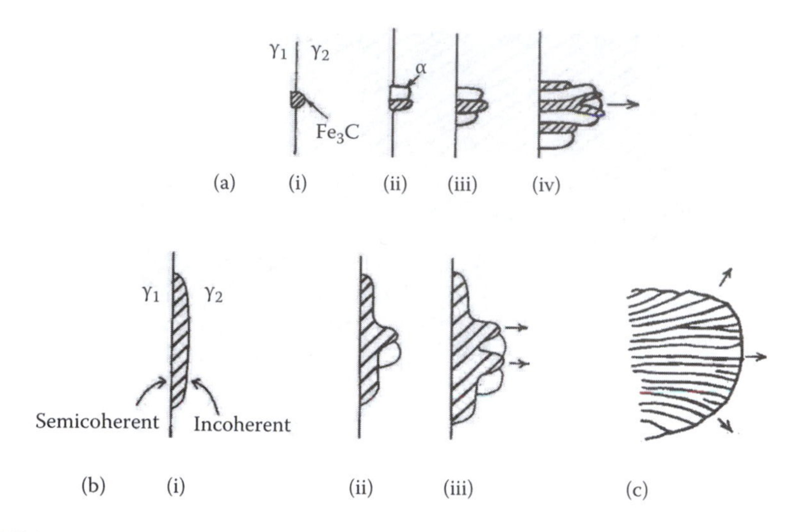

FIGURE 5.62
Nucleation and growth of pearlite. (a) On a "clean" grain boundary. (i) Cementite nucleates on grain boundary with coherent interface and orientation relationship with γ_1 and incoherent interface with γ_2. (ii) α nucleates adjacent to cementite also with a coherent interface and orientation relationship with γ_1. (This also produces an orientation relationship between the cementite and the ferrite.) (iii) The nucleation process repeats sideways, while incoherent interfaces grow into γ_2. (iv) New plates can also form by a branching mechanism. (b) When a proeutectoid phase (cementite or ferrite) already exists on that boundary, pearlite will nucleate and grow on the incoherent side. A different orientation relationship between the cementite and the ferrite results in this case. (c) Pearlite colony at a latest stage of growth.

energy barrier to nucleation by forming with an orientation relationship to one of the austenite grains, γ_1 in Figure 5.62a. (The crystal structure of cementite is orthorhombic and the orientation relationship is close to $(100)_c//(1\bar{1}1)_\gamma$, $(010)_c//(110)_\gamma$, $(001)_c//(\bar{1}12)_\gamma$.) Therefore the nucleus will have a semicoherent, low-mobility interface with γ_1 and an incoherent mobile interface with γ_2. The austenite surrounding this nucleus will become depleted of carbon which will increase the driving force for the precipitation of ferrite, and a ferrite nucleus forms adjacent to the cementite nucleus also with an orientation relationship to γ_1 (the Kurdjumov–Sachs relationship). This process can be repeated causing the colony to spread sideways along the grain boundary. After nucleation of both phases the colony can grow edgewise by the movement of the incoherent interfaces, that is, pearlite grows into the austenite grain with which it does not have an orientation relationship. The carbon rejected from the growing ferrite diffuses through the austenite to in front of the cementite, as with eutectic solidification.

If the alloy composition does not perfectly correspond to the eutectoid composition the grain boundaries may already be covered with a proeutectoid ferrite or cementite phase. If, for example, the grain boundary already contains a layer of cementite, the first ferrite nucleus will form with an orientation relationship to this cementite on the mobile incoherent side of the allotriomorphs as shown in Figure 5.62b. Again due to the higher mobility of the incoherent interfaces the pearlite will grow into the austenite with which there is no orientation relationship.

Whatever the pearlite nucleation mechanism, new cementite lamellae are able to form by the branching of a single lamella into two new lamellae as shown in Figure 5.62a(iv) or c. The resultant pearlite colony is effectively two interpenetrating single crystals.

It can be seen that the nucleation of pearlite requires the establishment of cooperative growth of the two phases. It takes time for this cooperation to be established and the rate of colony nucleation therefore increases with time. In some cases cooperation is not established and the ferrite and cementite grow in a nonlamellar manner producing so-called degenerate pearlite.[23]

Pearlite growth

The growth of pearlite in binary Fe–C alloys is analogous to the growth of a lamellar eutectic with austenite replacing the liquid. Carbon can diffuse interstitially through the austenite to the tips of the advancing cementite lamellae so that the equations developed in Section 4.3.2 should apply equally well to pearlite. Consequently the minimum possible interlamellar spacing (S^*) should vary inversely with undercooling below the eutectoid temperature (A_1), and assuming the observed spacing (S_0) is proportional to S^* gives

$$S_0 \propto S^* \propto (\Delta T)^{-1} \tag{5.53}$$

Similarly the grow rate of pearlite colonies should be constant and given by a relationship of the type

$$v = kD_c^{\gamma}(\Delta T)^2 \tag{5.54}$$

where k is a thermodynamic term which is roughly constant.

Observed spacings are found to obey Equation 5.53, varying from ~1 μm at high temperature to ~0.1 μm at the lowest temperatures of growth.[24] However, it is found that S_0 is usually greater than $2S^*$, i.e., the observed spacing is not determined by the maximum growth rate criterion. Instead it may be determined by the need to create new cementite lamellae as the perimeter of the pearlite nodules increases. This can occur either by the nucleation of new cementite lamellae, or by the branching of existing lamellae, Figure 5.62c.

In the case of binary Fe–C alloys, observed growth rates are found to agree rather well with the assumption that the growth velocity is controlled by the diffusion of carbon in the austenite. Figure 5.63 shows measured and calculated growth rates as a function of temperature. The calculated line is based on an equation similar to Equation 5.54 and shows that the measured growth rates are reasonably consistent with volume–diffusion control. However, it is also possible that some carbon diffusion takes place through the γ/α and

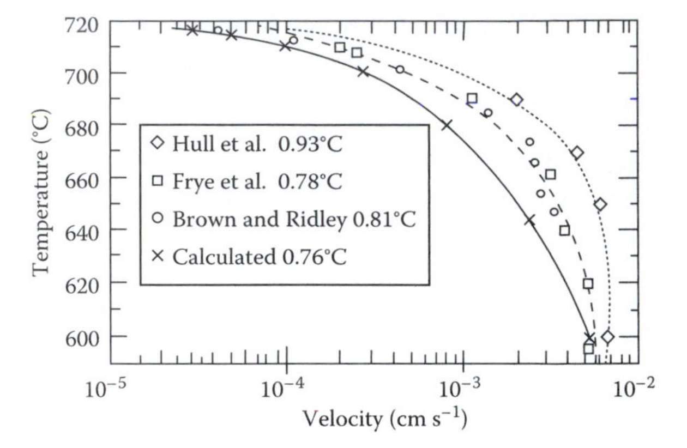

FIGURE 5.63
Pearlite growth rate versus temperature for plain carbon steels. (After Puls, M.P. and Kirkaldy, J.S., *Metall. Trans.*, 3, 2777, 1972.)

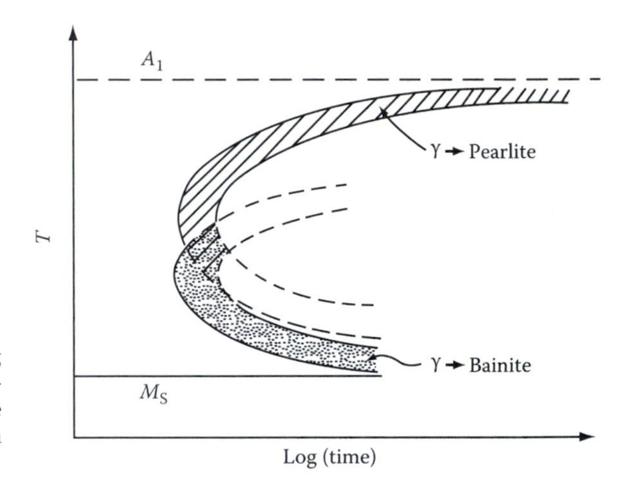

FIGURE 5.64
Schematic diagram showing relative positions of the transformation curves for pearlite and bainite in plain carbon eutectoid steels.

γ/cementite interfaces, which could account for the fact that the predicted growth rates shown in Figure 5.63 are consistently too low.

A schematic TTT diagram for the pearlite reaction in eutectoid Fe–C alloys is shown in Figure 5.64. Note the "C" shape typical of diffusional transformations that occur on cooling. The maximum rate of transformation occurs at about 550°C. At lower temperatures another type of transformation product, namely Bainite, can grow faster than pearlite. This transformation is dealt with in Section 5.8.2.

Eutectoid transformations are found in many alloys besides Ee–C. In alloys where all elements are in substitutional solid solution, lattice diffusion is found to be too slow to account for observed growth rates. In these cases diffusion occurs instead through the colony/matrix interface. Consideration of the diffusion problem in this case leads to a relationship of the type

$$v = kD_B(\Delta T)^3 \tag{5.55}$$

where

k is a thermodynamic constant

D_B is the boundary diffusion coefficient

Pearlite in off-eutectoid Fe–C alloys

When austenite containing more or less carbon than the eutectoid composition is isothermally transformed below the A_1 temperature the formation of pearlite is usually preceded by the precipitation of proeutectoid ferrite or cementite. However, if the undercooling is large enough and the departure form the eutectoid composition is not too great it is possible for austenite of noneutectoid composition to transform directly to pearlite. The region in which this is possible corresponds approximately to the condition that the austenite is simultaneously saturated with respect to both cementite and ferrite, i.e., the hatched region in Figure 5.65. (See also Figure 5.53.) Thus a 0.6% c alloy, for example can be transformed to ~100% pearlite provided the temperature is low enough to bring the austenite into the hatched region of Figure 5.65 (but not so low that bainite forms). At intermediate undercoolings some proeutectoid ferrite will form but less than predicted by the equilibrium phase diagram.

Similar considerations apply to transformations during continuous cooling—larger gain sizes and faster cooling rates favor low volume fractions of ferrite. Compare Figure 5.54c and d.

5.8.2 Bainite Transformation

When austenite is cooled to large supersaturations below the nose of the pearlite transformation curve a new eutectoid product called bainite is produced. Like pearlite, bainite is a mixture of ferrite and carbide, but it is microstructurally quite distinct from pearlite and can be characterized by its own C curve on a TTT diagram. In plain carbon steels this curve overlaps with the pearlite curve (Figure 5.64) so that at temperatures around 500°C both pearlite and bainite form competitively. In some alloy steels, however, the two curves are separated as shown in Figure 5.73.

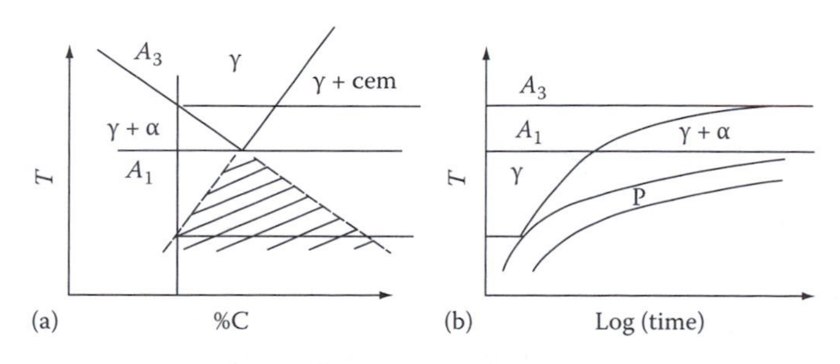

FIGURE 5.65
Effect of transformation temperature on the volume fraction of proeutectoid ferrite.

The microstructure of bainite depends mainly on the temperature at which it forms.[25]

Upper bainite

At high temperature (350°C–550°C) bainite consists of needles or laths of ferrite with cementite precipitates between the laths as shown in Figure 5.66. This is known as upper bainite. Figure 5.66a shows the ferrite laths growing into partially transformed austenite. The light contrast is due to the cementite. Figure 5.66b illustrates schematically how this microstructure is thought to develop. The ferrite laths grow into the austenite in a similar way to Widmanstätten side-plates. The ferrite nucleates on a grain boundary with

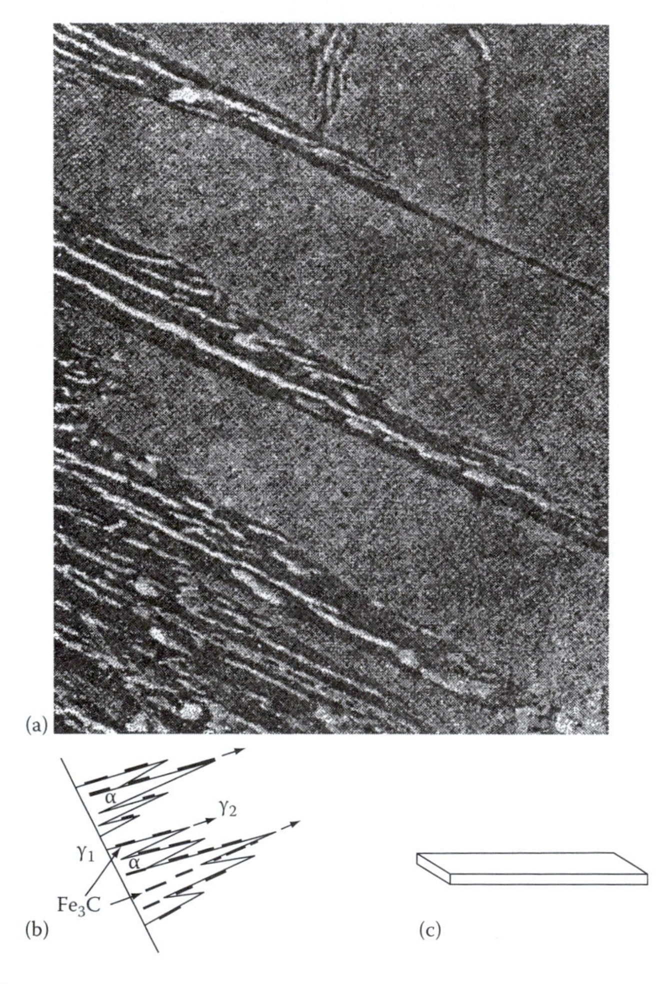

FIGURE 5.66
(a) Upper bainite in medium-carbon steel (replica × 13,000) (by permission of the Metals Society). (b) Schematic of growth mechanism. Widmanstätten ferrite laths growth into γ_2. (α and γ_2 have Kurdjumov–Sachs orientation relationship.) Cementite plates nucleate in carbon-enriched austenite. (c) Illustrating the shape of a "lath."

a Kurdjumov–Sachs orientation relationship with one of the austenite grains, γ_2, say. Since the undercooling is very large the nucleus grows most rapidly into the γ_2 grain forming ferrite laths with how energy semicoherent interfaces. This takes place at several sites along the boundary so that a group of finely spaced laths develops. As the laths thicken the carbon content of the austenite increases and finally reaches such a level that cementite nucleates and grows.

At the higher temperatures of formation upper bainite closely resembles finely spaced Widmanstätten side-plates, Figure 5.51d. As the temperature decreases the bainitic laths become narrower so that individual laths may only be resolved by electron microscopy.

At the highest temperatures where pearlite and bainite grow competitively in the same specimen it can be difficult to distinguish the pearlite colonies from the upper bainite. Both appear as alternate layers of cementite in ferrite. The discontinuous nature of the bainitic carbides does not reveal the difference since pearlitic cementite can also appear as broken lamellae. However, the two microstructures have formed in quite different ways. The greatest difference between the two constituents lies in their crystallography. In the case of pearlite the cementite and ferrite have no specific orientation relationship to the austenite grain in which they are growing, whereas the cementite and ferrite in bainite do have an orientation relationship with the grain in which they are growing. This point is illustrated in Figure 5.67. The micrograph is from a hypoeutectoid steel (0.6% C) which has been partially transformed at 710°C and then quenched to room temperature, whereupon the untransformed austenite was converted into martensite. The quench however, was not fast enough to prevent further transformation at the γ/α interface. The dark constituent is very fine pearlite which was nucleated on the incoherent α/γ interface, across which there is no

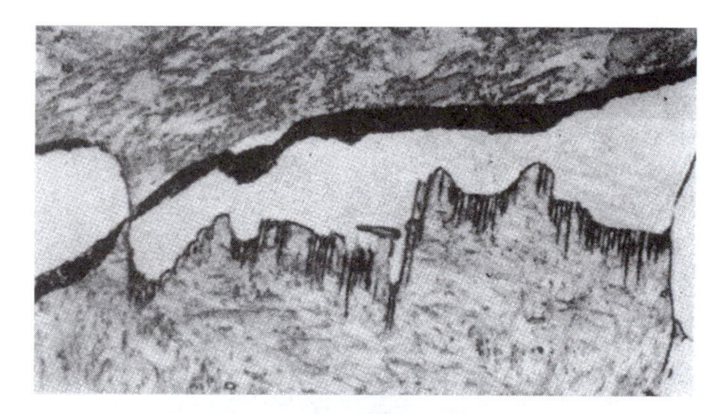

FIGURE 5.67
Hypoeutectoid steel (0.6% C) partially transformed for 30 min at 710°C, inefficiently quenched. Bainitic growth into lower grain of austenite and pearlitic growth into upper grain during quench (×1800). (After Hillert, M., in *Decomposition of Austenite by Diffusional Processes*, Interscience, New York, 1962.)

orientation relationship. The ferrite and lower austenite grain, however, have an orientation relationship which has led to bainite formation.

Lower bainite

At sufficiently low temperatures the microstructure of bainite changes form laths into plates and the carbide dispersion becomes much finer, rather like in tempered martensite. The temperature at which the transition to lower bainite occurs depends on the carbon content in a complex manner. For carbon levels below about 0.5 wt% the transition temperature increases with increasing carbon, from 0.5 to 0.7 wt% C it decreases and above approximately 0.7 wt% C it is constant at about 350°C. At the temperatures where lower bainite form the diffusion of carbon is slow, especially in the austenite and carbides precipitate in the ferrite with an orientation relationship. The carbides are either cementite or metastable transition carbides such as ε-carbide and they are aligned at approximately the same angle to the plane of the ferrite plate (Figure 5.68). The habit plane of the ferrite plates in lower bainite is the same as that of the martensite that forms at lower temperatures in the same alloy. As with upper bainite, some carbides can also be found between the ferrite plates.

The different modes of formation of upper and lower bainite result in different transformation kinetics and separate C curves in the TTT diagram. An example, the case of a low-alloy steel, is shown in Figure 5.76.

Transformation shears

If a polished specimen of austenite is transformed to bainite (upper or lower) it is found that the growth of bainite laths or plates produces a surface relief effect like that of martensite plates. For example, Figure 5.69 shows the surface tilts that result from the growth of lower bainite plates. This has been interpreted as suggesting that the bainite plates form by a shear mechanism

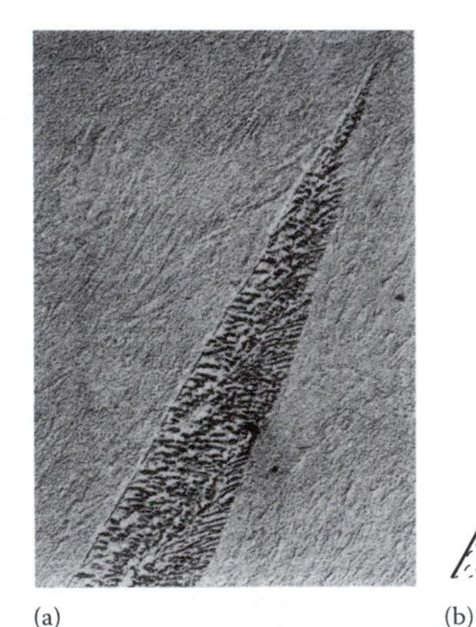

(a)

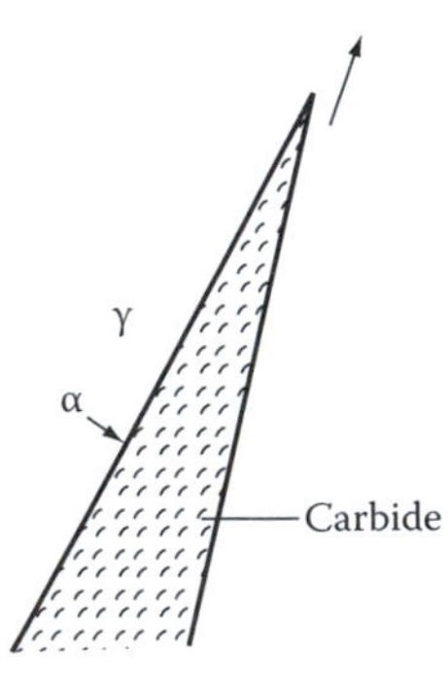

(b)

FIGURE 5.68
(a) Lower bainite in 0.69 wt% C low-alloy steel (replica × 1100). (After Heheman, R.F., in *Metals Handbook,* 8th edn., Vol. 8, American Society for Metals, Metals Park, OH, 1973, 196.) (b) A possible growth mechanism. α/γ interface advances as fast as carbides precipitate at interface thereby removing the excess carbon in front of the α.

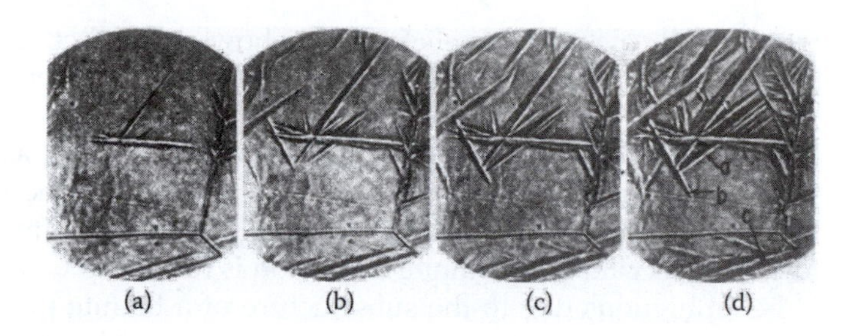

FIGURE 5.69
Photos taken with a hot-state microscope showing the nucleation and growth of bainite plates at 350°C. The contrast is due to surface relief. (a) 14.75 min; (b) 16.2 min; (c) 17.25 min; and (d) 19.2 min. (After Speich, G.R., in *Decomposition of Austenite by Diffusional Processes*, Interscience, New York, 1962.)

in the same way as the growth of martensite plates (Chapter 6). In other words it is supposed that the iron atoms are transferred across the ferrite/austenite interface in an ordered military manner.

As demonstrated in Figure 5.70, plastic relaxation of the shape change caused by bainite transformation may take place in the adjacent austenite

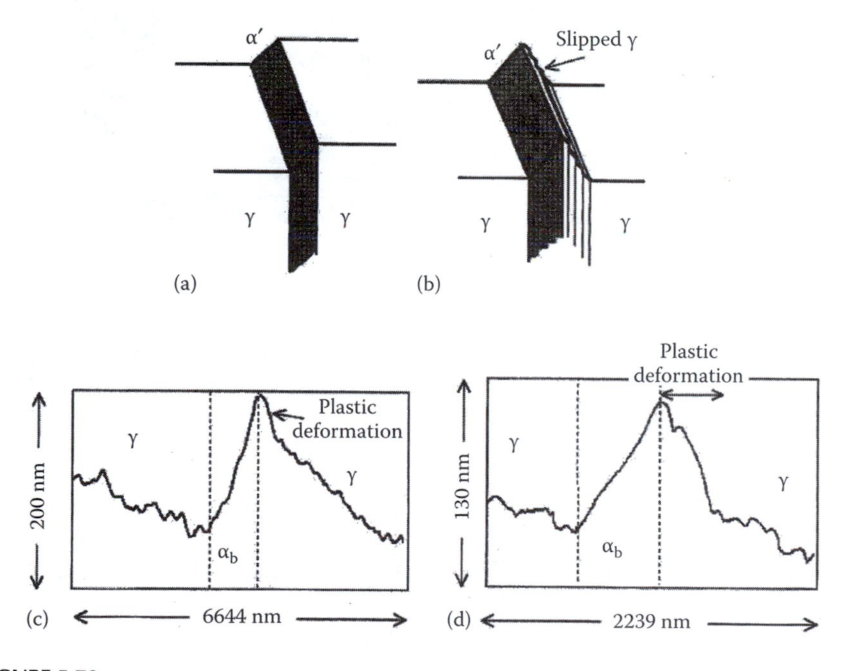

FIGURE 5.70
(a) Surface relief caused by a "perfect" invariant plane strain. (b) Plastic relaxation occurs in the adjacent matrix. (After Bhadeshia, H.K.D.H., in *Bainite in Steels*, 2nd edn., IOM Communications Ltd., London, 2001.) (c) and (d) are AFM scans across a surface relief caused by a bainite subunit. (Originally appeared in Swallow, E. and Bhadeshia, H.K.D.H., *Mater. Sci. Tech.-Lond.*, 12, 121, 1996.)

matrix, without which the surface relief would show a "perfect" invariant plane strain[26] (see Section 6.2 for more information about the *invariant plane strain*).

Fang et al. used scanning tunneling microscopy (STM) as well as atomic force microscopy (AFM) to study the surface relief caused by the smallest bainite unit they could observe.[27–29] Their reasoning being that, often, observations of surface relief caused by bainite formation is related to whole plates which could be ambiguous due to the substructure of a bainite plate. For a single sub-subunit (the smallest bainitic ferrite unit observed by the authors, Figure 5.71), a surface relief of a "tent shape" was reported whereby the possibility of a surface relief caused by an invariant plane strain was ruled out. A schematic which shows the two types of surface relief is presented in Figure 5.72. The ultrafine bainitic microstructure was observed in both lower and upper bainite.

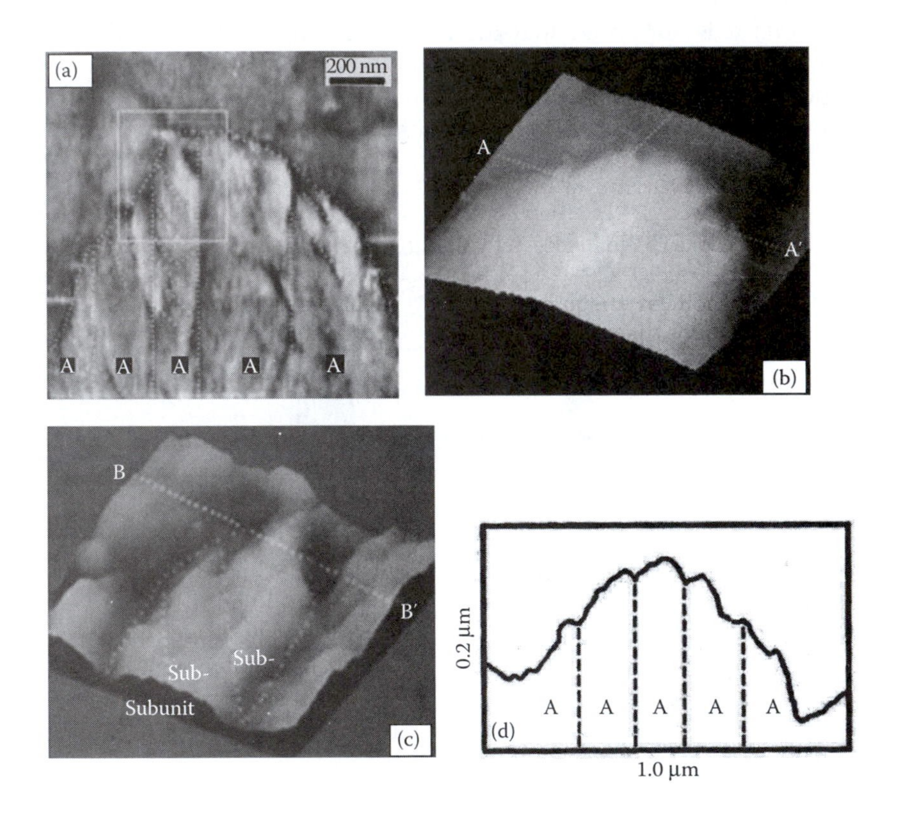

FIGURE 5.71
(a) STM image of the surface relief of the tip of a lower bainite plate in an austenitic matrix. (b) AFM three-dimensional image of the plate tip in (a). (c) Rectangle shown in (a) magnified where sub-subunits are delineated. (d) AFM scan across AA′ in (b); AFM scan across BB′ in (c). (Reprinted from Fang, H.-S., Yang, J.-B., Yang, Z.-G., and Bai, B.-Z., *Scripta Mater.* 47, 157, 2002. With permission from Elsevier; Originally appeared in Yang, Z.-G., Fang, H.-S., Wang, J.J., and Zheng, Y.K., *J. Mater. Sci. Lett.*, 15, 721, 1996.)

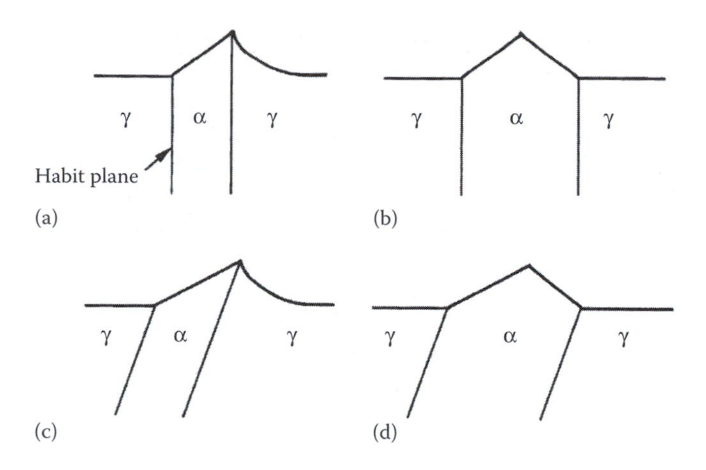

FIGURE 5.72
(a) Invariant plane strain. (b) Tent-shape surface reliefs where the habit plane is normal to the free surface. (c) and (d) same as (a) and (b) but with the habit plane not normal to the free surface. (Reprinted from Yang, Z.-G., Zhang, C., Bai, B.-Z., and Fang, H.-S., *Mater. Lett.*, 48, 292, 2001. With permission from Elsevier.)

Furthermore, Fang et al. reported that the surface relief caused by bainitic transformation was not accompanied by any significant plastic deformation of the adjacent austenite matrix.[30] Ohmori et al. did indeed observe tent shape surface reliefs, however, they interpreted them as pairs of plates exhibiting an opposite shape strain.[31] Aaronson cited some references where monocrystalline ferrite laths were reported to form in different systems (Ti–X alloys and Cu–Cr alloy) exhibiting "tents," consistent with a proposed ledgewise diffusional growth mechanism.[32] For more information about this growth mechanism the reader is referred to Refs. [29,32].

However, the growth rate of the bainite plates is controlled by the rate at which carbon can diffuse away from the interface, or by the rate at which carbides can precipitate behind the interface, whereas martensite plates are able to advance without any carbon diffusion, and the plates can grow as fast as the glissile interfaces can advance.

From the above discussion, it is apparent that there is much uncertainty regarding the mechanism by which bainitic ferrite grows, and the nature of the austenite–ferrite interface during bainite transformation. In fact the formation of Widmanstätten side-plates also leads to surface tilts of the type produced by a shear transformation. Also the phenomenological theory of martensite is able to account for the observed orientation relationships and habit planes found in Widmanstätten plates as well as bainite and martensite.

Bhadeshia reported that the nature of the bainite nucleus is similar to that of Widmanstätten ferrite.[26] However, unlike Widmanstätten ferrite, carbon diffuses into the residual austenite from the carbon-saturated bainitic ferrite not during but *after* bainite plate growth. Therefore, the thermodynamic conditions accompanying growth must be the determining factor for a nucleus to evolve either to Widmanstätten ferrite or bainite. If diffusionless growth cannot be sustained, the formation of Widmanstätten ferrite takes place instead of bainite. A stored energy in Widmanstätten ferrite of about 50 J mol^{-1} was assumed,[33] compared with that for bainite at around 400 J mol^{-1}. The chemical free energy change caused by undercooling has to be sufficient as to exceed the stored energy for a transformation to take place. This is to say, when undercooling is large enough to account for the proposed stored energy in bainite, the Widmanstätten ferrite does not form. The much higher stored energy in bainite was rationalized on the basis of the absence of favorable strain interactions "within" bainite sheaves.[33]

Hillert argued that the assumed high stored energy in bainite requires high growth velocity which has not been observed experimentally, consequently, carbon supersaturation in bainitic ferrite bears "no conclusive evidence."[34] However, Oblak and Hehemann proposed that bainite growth is rapid though it occurs in short steps resulting in an overall low growth rate.[35]

Hillert adopted the basic bainite transformation model suggested by Hultgren,[36] in which, bainite was assumed to form initially as Widmanstätten ferrite plates "followed by the formation of a mixture of ferrite and cementite in the interjacent spaces."[34] He argued that in eutectoid Fe–C alloys, at low temperatures, bainite is promoted over pearlite not due to a martensite-like transformation mode but rather the high asymmetry in the Fe–C phase diagram. It was suggested that "The edgewise growth mechanism [of bainite] is the same as for Widmanstätten ferrite" with carbon diffusion controlling growth velocity. As the transformation temperature is lowered, the formation of carbides is enhanced which "may speed up the edgewise growth." The carbides precipitation occurs in the vicinity of advancing bainite plate-tips which results in a "shorter diffusion distances for carbon away from the advancing tip."

Aaronson et al. argued in favor of the short-range diffusion of carbon to be the predominant aspect of the bainite transformation mechanism.[37] For a review of some of the historical theories of bainite transformation the reader may consult Ref. [38].

It is interesting to point out to another bainite formation mechanism hypothesis where bainite transformation was assumed to be martensitic but with carbon diffusion across the austenite/ferrite interface believed to change the free energy available for transformation thereby facilitating the formation of bainite above the M_s temperature.[39]

Speer et al. suggested that bainite may grow by a martensitic diffusionless interface migration mechanism whereas the overall growth kinetics are controlled by carbon diffusion.[40,41] In other words, the growth mechanism is both " 'fully' displacive and 'fully' diffusional." This growth mechanism was also proposed by Muddle and Nie[42] and Saha et al.[43]

It can be seen, therefore, that some phase transformations are not exclusively military or civilian, but show characteristics common to both types of transformation.

5.8.3 Effect of Alloying Elements on Hardenability

The primary aim of adding alloying elements to steels is to increase the hardenability, that is, to delay the time required for the decomposition into ferrite and pearlite. This allows slower cooling rates to produce fully martensitic structures. Figure 5.73 shows some examples of TTT diagrams for various low-alloy steels containing Mn, Cr, Mo, and Ni in various combinations and concentrations. Note the appearance of two separate C curves for pearlite and bainite, and the increasing time for transformation as the alloy content increases.

Basically there are two ways in which alloying elements can reduce the rate of austenite decomposition. They can reduce either the growth rate or the nucleation rate of ferrite, pearlite, or bainite.

The main factor limiting hardenability is the rate of formation of pearlite at the nose of the C curve in the TTT diagram. To discuss the effects of alloy elements on pearlite growth it is necessary to distinguish between austenite stabilizers (e.g., Mn, Ni, Cu) and ferrite stabilizers (e.g., Cr, Mo, Si). Austenite stabilizers depress the A_1 temperature, while ferrite stabilizers have the opposite effect. All of these elements are substitutionally dissolved in the austenite and ferrite.

At equilibrium an alloy element X will have different concentrations in cementite and ferrite, i.e., it will partition between the two phases. Carbide-forming elements such as Cr, Mo, Mn will concentrate in the carbide while elements like Si will concentrate in the ferrite. When pearlite forms close to the A_1 temperature the driving force for growth will only be positive if the equilibrium partitioning occurs. Since X will be homogeneously distributed within the austenite, the pearlite will only be able to grow as fast as substitutional diffusion of X allows partitioning to occur. The most likely diffusion route for substitutional elements is through the γ/α and γ/cementite interfaces. However, it will be much slower than the interstitial diffusion of carbon and will therefore reduce the pearlite growth rate.

When X is a ferrite stabilizer there are thermodynamic considerations that suggest that X will partition even at large undercoolings close to the nose of the C curve. Thus Si, for example, will increase the hardenability by diffusing along the austenite/pearlite interface into the ferrite. The partitioning of alloying elements in a Fe–0.6 wt% C–0.85% Cr–0.66% Mn–0.26% Si steel transformed at 597°C for 2 min is shown in Figure 5.74.

When X is an austenite stabilizer such as Ni, it is possible, at sufficiently high undercoolings, for pearlite to grow without partitioning. The ferrite and cementite simply inherit the Ni content of the austenite and there is no need for substitutional diffusion. Pearlite can then grow as fast as diffusion of carbon alloys. However, the growth rate will still be lower than in binary Fe–C alloys since the nonequilibrium concentration of X in the ferrite and cementite will raiser their free energies, thereby lowering the eutectoid temperature, Figure 5.75, and reducing the total driving force. For the same

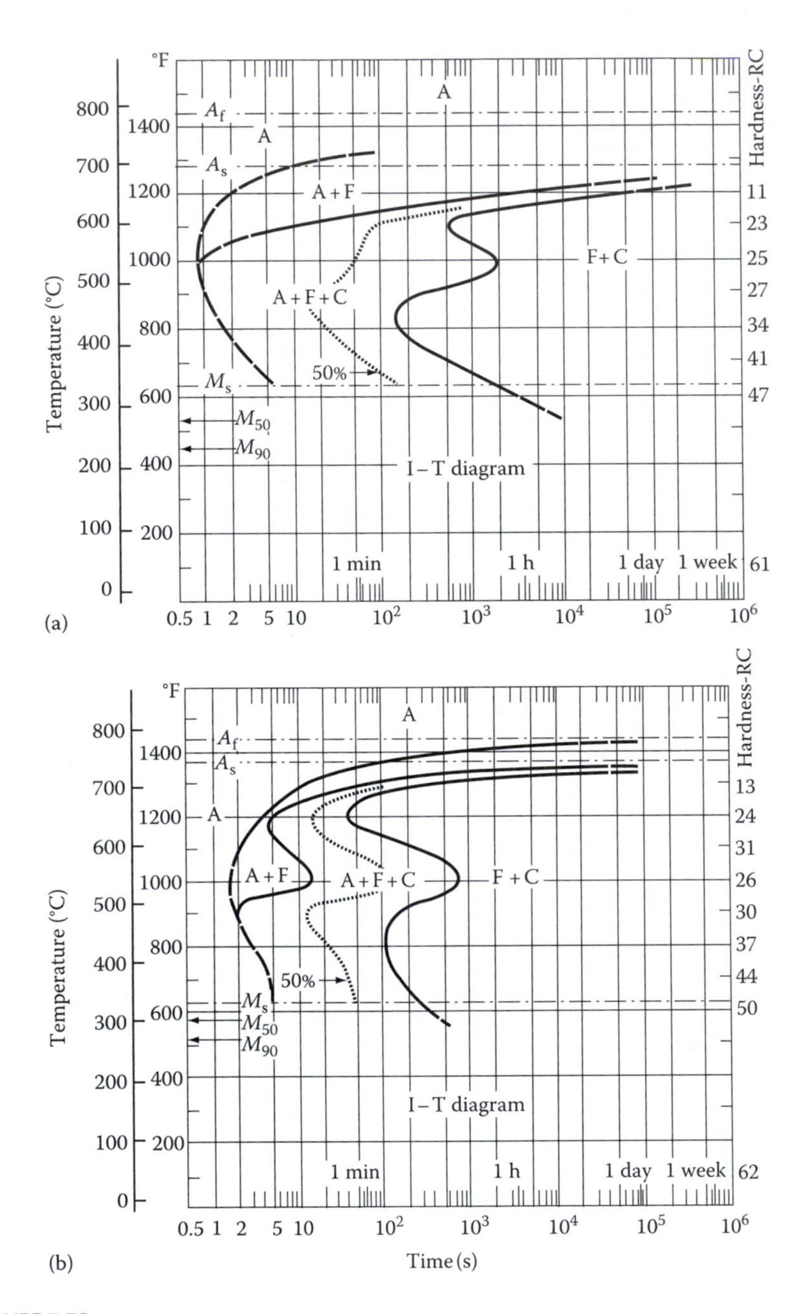

FIGURE 5.73
TTT diagrams for four commercial low-alloy steels all of which contain roughly 0.4% C and 1% Mn. In addition (b) contains 0.9% Cr,

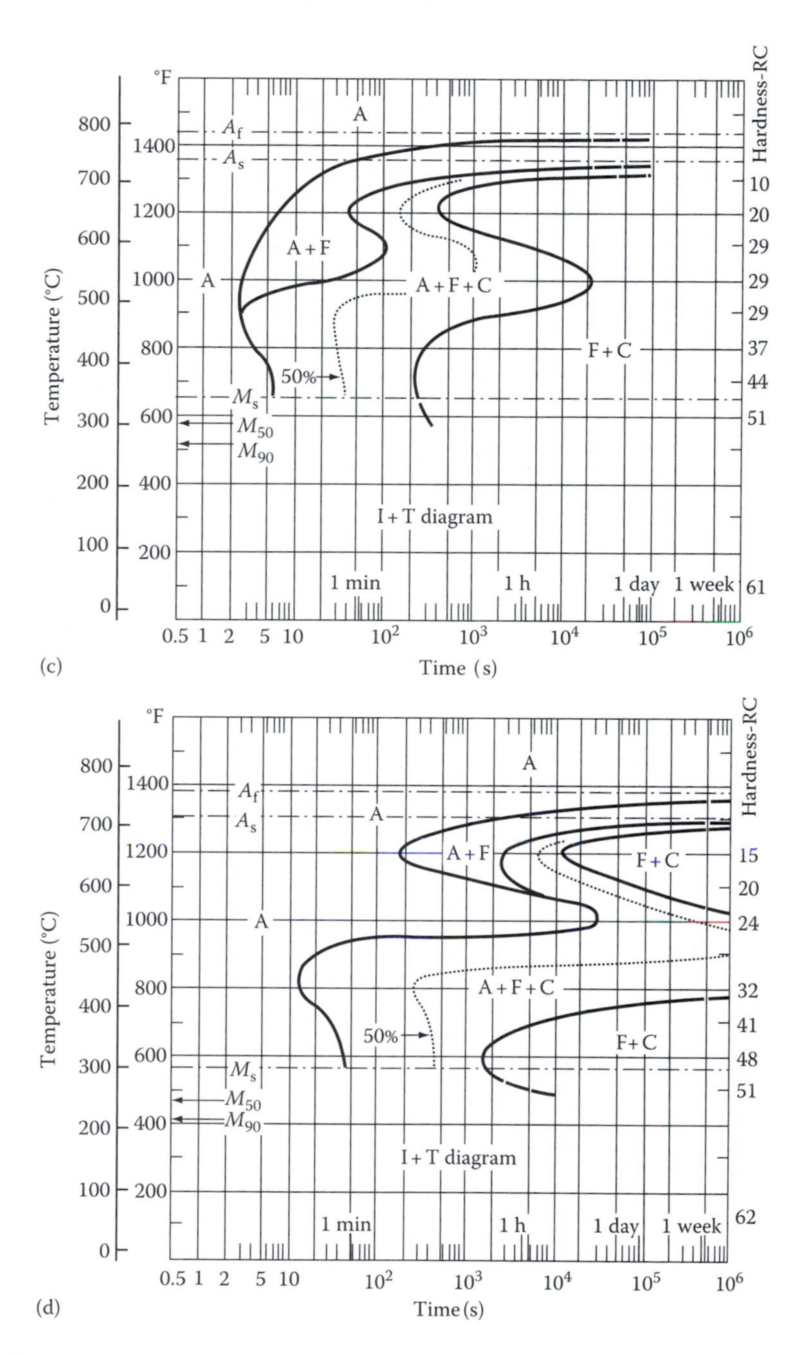

FIGURE 5.73 (continued)
(c) contains 1.0% Cr and 0.2% Mo, and (d) contains 0.8% Cr, 0.3% Mo, and 1.8% Ni. Note the tendency to form two distinct knees, one for pearlite formation and one for bainite formation. (From *Atlas of Isothermal Transformation and Cooling Transformation Diagrams*, American Society for Metals, Metals Park, OH, 1977. With permission.)

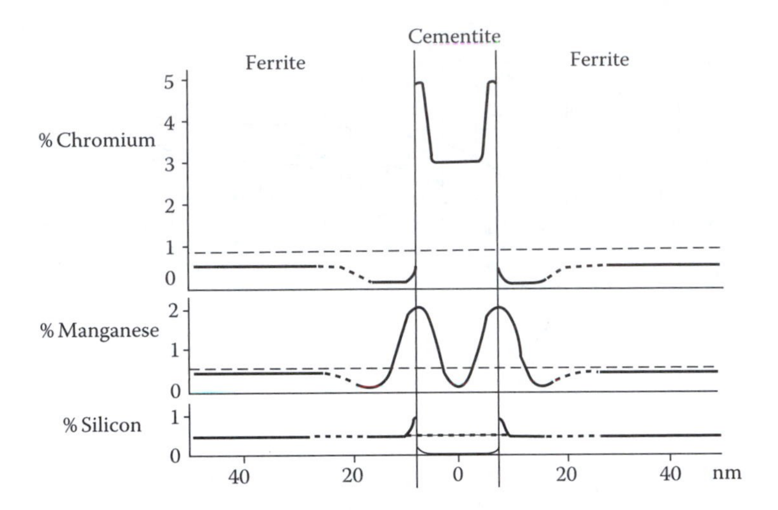

FIGURE 5.74
Schematic diagram showing the measured variations of alloying elements in pearlite. These measurements were made using a time-of-flight atom probe. (From Williams, P.R., Miller, M.K., Beavan, P.A., and Smith, G.D.W., in *Phase Transformations*, Vol. 2, Institute of Metallurgists, London, 1979, 98. With permission.)

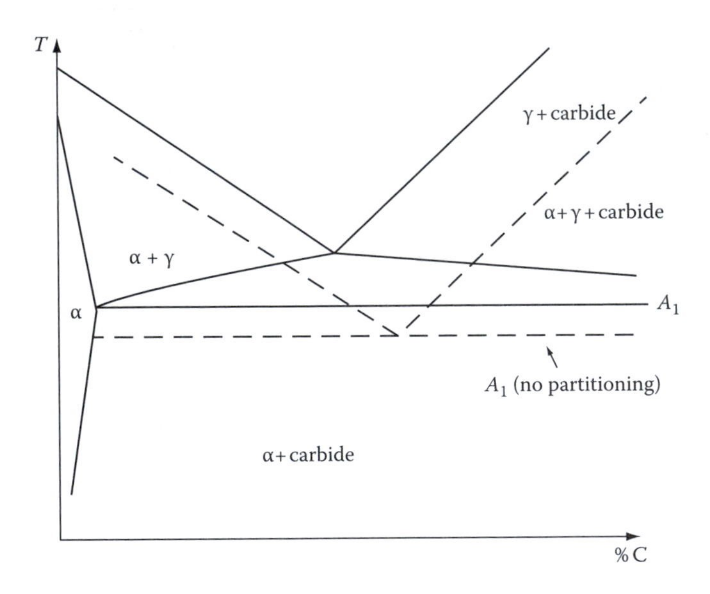

FIGURE 5.75
Schematic phase diagram for Fe–C–X alloy where X is a substitutional element. Between the solid and dashed lines, precipitation can occur in the austenite only if X is partitioned between the phases.

reasons zero-partitioning is only possible at temperatures below the metastable eutectoid as shown in Figure 5.75.

When X is a strong carbide-forming element such as Mo or Cr it has been suggested[44] that it can reduce the rate of growth of pearlite, as well as proeutectoid ferrite, by a solute-drag effect on the moving γ/α interface. These elements also partition to cementite as shown in Figure 5.74.

Hardenability is not solely due to growth-rate effects. It is also possible that the alloying elements affect the rate of nucleation of cementite or ferrite. For example, it has been suggested[45] that the "bay" at ~500°C in the TTT diagrams of steels containing Cr, Mo, and B (Figure 5.73) may be due to the poisoning of ferrite nucleation sites by the precipitation of X-carbide clusters in grain boundaries.

The diagrams shown in Figure 5.73 are not entirely accurate especially with regard to the bainite transformation at temperatures in the vicinity of the M_s temperatures. It has been found that below the M_s temperature the bainite transformation rate is greatly increased by the martensite-transformation strains. The TTT diagram for the bainite transformation in Figure 5.73d has recently been redetermined using a new experimental technique based on magnetic permeability measurements[46] and the results are shown in Figure 5.76. The acceleration of the transformation close to M_s and the existence of separate C curves for upper and lower bainite are apparent.

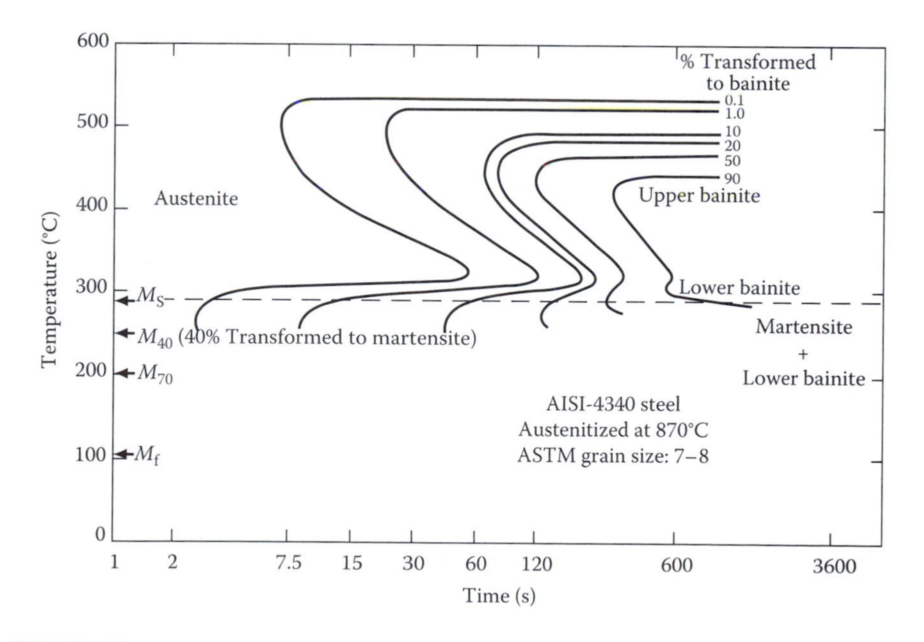

FIGURE 5.76
TTT diagrams for steel in Figure 5.73d determined by a magnetic permeability technique. (After Zackay, V.F. and Parker, E.R., in *Alloy and Microstructural Design*, Academic Press, New York, 1976.)

5.8.4 Continuous Cooling Diagrams

Isothermal transformation (TTT) diagrams are obtained by rapidly quenching to a given temperature and then measuring the volume fraction of the various constituents that form as a function of time at that temperature. Practical heat treatments, however, are usually concerned with transformations that occur during continuous cooling, and under these conditions TTT diagrams cannot be used to give the time and temperatures of the various transformations. A continuous cooling transformation (CCT) diagram must be used instead.

To a first approximation the CCT diagram is the TTT diagram shifted to lower temperatures and longer times. This can be understood as follows. In a specimen held at a constant temperature the transformation starts when the product (Nt) reaches a certain value, α, say. In a continuously cooled sample, time near the start of cooling is not very effective since N is low at low supercoolings. Therefore when the cooling curve reaches the TTT start curve the total value of Nt will be less than α and further time (and therefore cooling) will be required before the start of the CCT diagram is reached. Similarly the end of the reaction will be displaced to lower temperatures and longer times. The relationship between a CCT and an TTT diagram for a eutectoid steel is shown in Figure 5.77. Note that whereas the TTT diagram is interpreted by reading from left to right at a constant temperature the CCT diagram is read along the cooling curves from the top left to bottom right. The cooling curves in Figure 5.77 refer to various distances from the quenched end of a Jominy end-quench specimen. Transformation occurs along the hatched parts of the lines. Figure 5.77 is in simplified and cooling along B would lead to the production of some bainite. But otherwise it can be seen that point B will transform partly to fine pearlite at high temperatures around 500°C–450°C. Between 450°C and 200°C the remaining austenite will be unable to transform and below 200°C the remaining austenite will be unable to transform and below 200°C transformation to martensite occurs.

The above relationship between TTT and CCT diagrams is only approximate. There are several features of CCT diagrams that have no counterpart in TTT diagrams especially in alloy steels. These include the following: (1) a depression of the M_s temperature at slow cooling rates, (2) the tempering of martensite that takes place on cooling from M_s to about 200°C, and (3) a greater variety of microstructures.

Figure 5.78 shows more complete CCT diagrams for a medium-carbon steel with different Mn contents. These diagrams were obtained with a high-speed dilatometer using programmed linear cooling rates for all except the highest quench rates. For each cooling curve the cooling rate and volume fractions of ferrite and pearlite are indicated. Note how the volume fraction of pearlite increases as the cooling rate is increased from 2.5°F/min to 2300°F/min in the low-Mn steel. In practical heat treatments the cooling curves will not be linear but will depend on the transfer of heat from the specimen to the quenching medium and the rate of release of latent heat

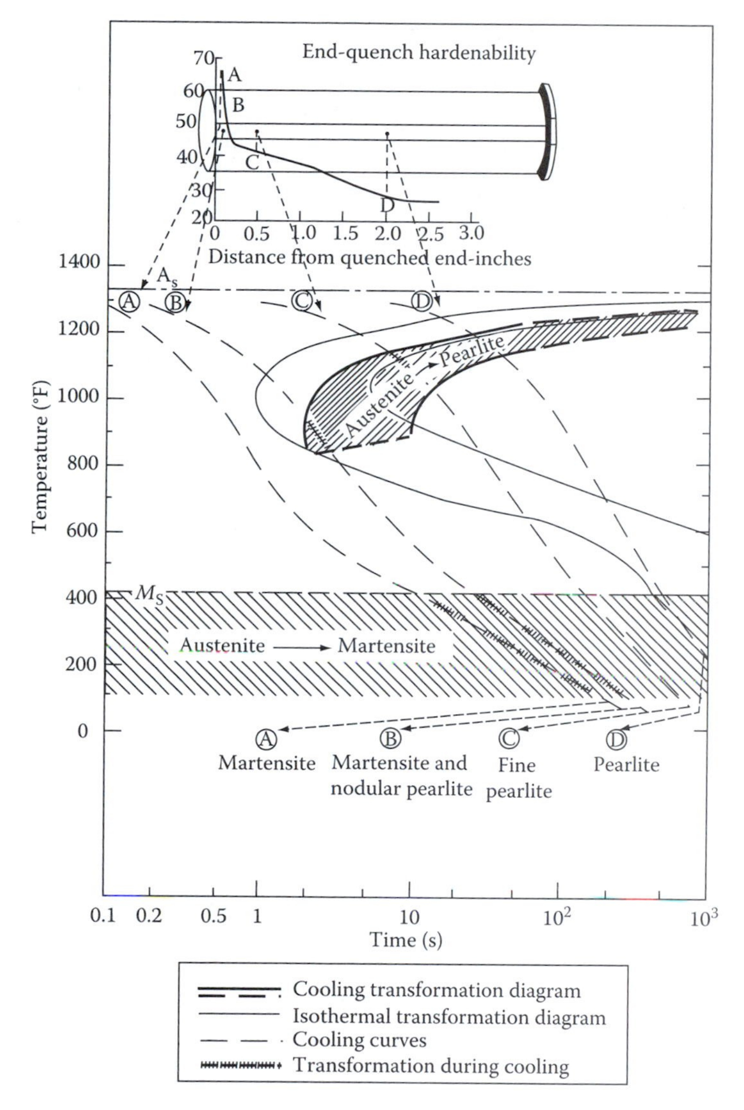

FIGURE 5.77
Correlation of CCT and isothermal transformation with end-quench hardenability test data for eutectoid carbon steel. (From *Atlas of Isothermal Transformation and Cooling Transformation Diagrams*, American Society for Metals, Metals Park, OH, 1977, 376. With permission.)

during transformation. In general, the evolution of latent reduces the rate of cooling during the transformation range and can even lead to a rise temperature, i.e., recalescence. Recalescence is often associated with the pearlite transformation when the growth rate is very high, for example, in unalloyed steels, but the effect can also be seen quite clearly for a cooling rate of 4100°F/min in Figure 5.78a.

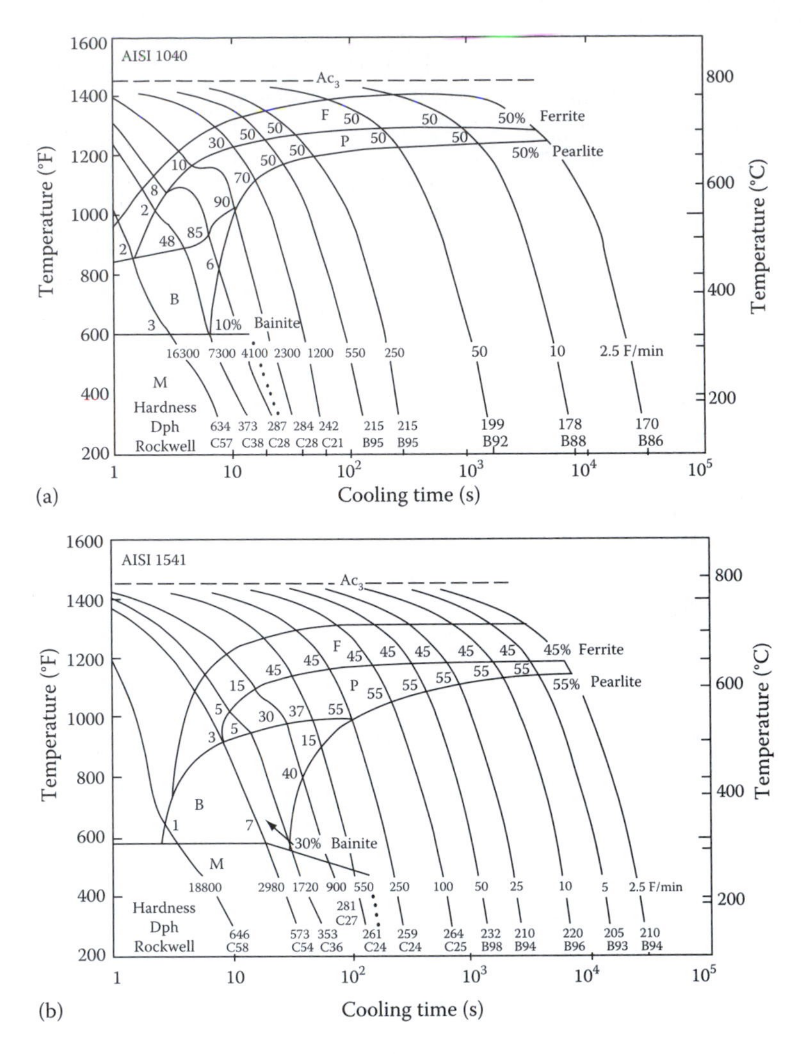

FIGURE 5.78
CCT diagrams showing the influence of Mn on a 0.4 wt% C steel. (a) 0.39 C, 0.72 Mn, 0.23 Si, 0.018 S, 0.010 P. $Ac_1 = 728°C$, $Ac_3 = 786°C$. Grain size, ASTM No. 7–8. (b) 1.6 Mn: 0.39 C, 1.56 Mn, 0.21 Si, 0.024 S, 0.010 P. $Ac_1 = 716°C$. $Ac_3 = 788°C$. Grain size, ASTM No. 8. F, ferrite; P, pearlite; B, bainite; M, martensite. (From *Atlas of Isothermal Transformation and Cooling Transformation Diagrams*, American Society of Metals, Metals Park, OH, 1977, 414. With permission.)

5.8.5 Fibrous and Interphase Precipitation in Alloy Steels

When a few percent of a strong-carbide-forming element (e.g., Mo, W, Cr, Ti, and V) is alloyed with steel, cementite is entirely replaced by a more stable carbide. When such steels are isothermally transformed at temperatures where the substitutional alloying element has appreciable mobility (~600°C–750°C) two new alloy-carbide morphologies can form.

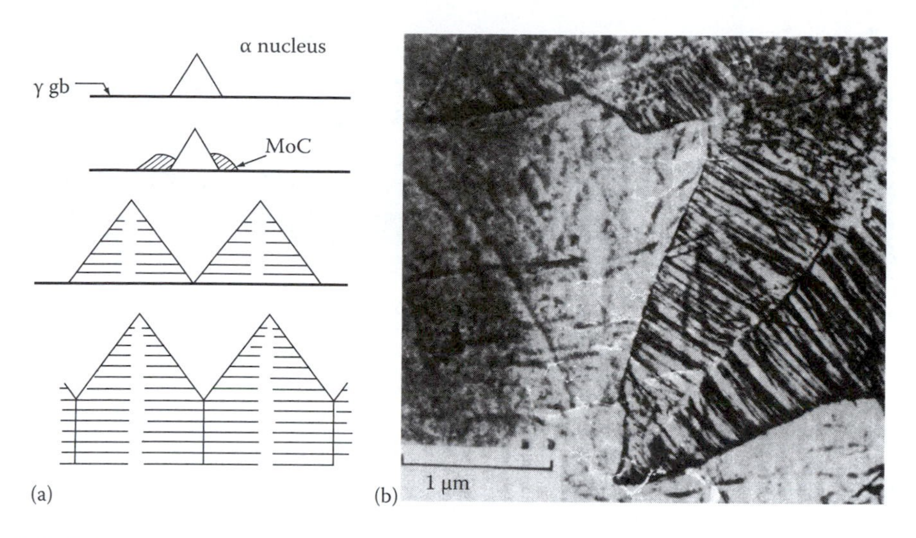

FIGURE 5.79
Fe–4% Mo–0.2% C transformed 2 h at 650°C. (a) Schematic of possible nucleation and growth mechanism. (b) Thin-foil electron micrograph (×32,000). (After Honeycombe, R.W.K., *Metall. Trans.*, 7A, 91, 1976, after D.V. Edmonds.)

Sometimes a fibrous morphology can be formed as illustrated in Figure 5.79. This is a mixture of Mo_2C fibers in ferrite. The interfiber spacings are about an order of magnitude less than found in pearlite with fiber diameters ~10–50 nm.

In other cases planar arrays of alloy carbides in ferrite are produced, Figure 5.80. The spacing of the sheets of precipitates decreases with

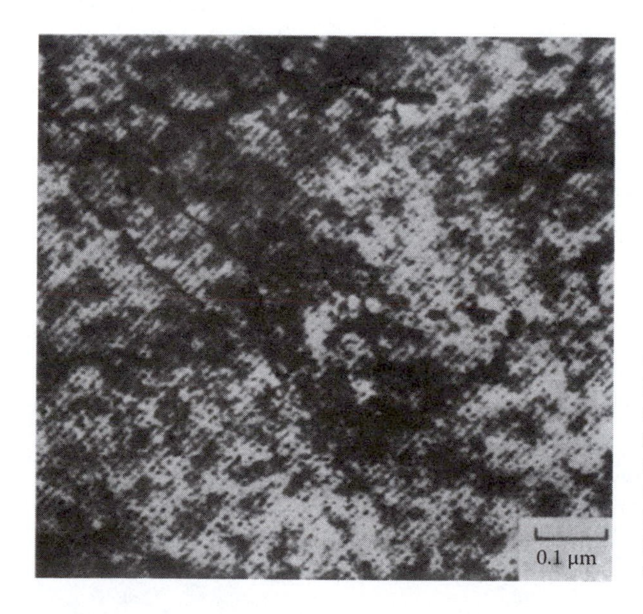

FIGURE 5.80
Fe–0.75% V–0.15% C after 5 min at 725°C. Thin-foil electron micrograph showing sheets of vanadium carbide precipitates (interphase precipitation). (After Honeycombe, R.W.K., *Metall. Trans.*, 7A, 91, 1976, after A.D. Batte.)

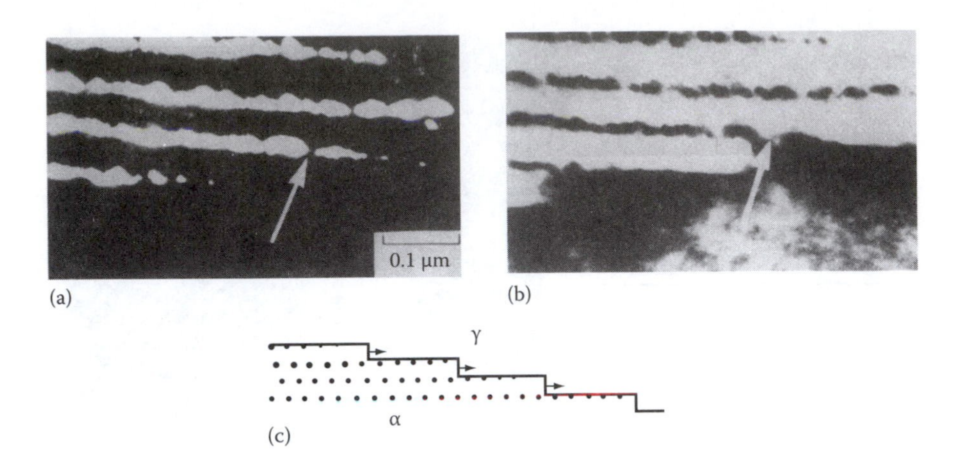

FIGURE 5.81
Fe–12% Cr%–0.2% C transformed 30 min at 650°C. Interphase precipitation of $Cr_{23}C_6$ at α/γ interface. (a) Dark-field micrograph showing bright precipitates appear dark. (c) Schematic of nucleation and growth mechanism for interphase precipitation. (After Honeycombe, R.W.K., *Metall. Trans.* 7A, 91, 1976, after K. Campbell.)

decreasing temperature of transformation, being of the order of 10–50 nm. The sheets of precipitate are parallel to successive positions of the γ/α interface, hence this type of precipitation is known as interphase precipitation. The mechanism by which the microstructure develops is shown in the thin-foil electron micrographs in Figure 5.81a and b and schematically in Figure 5.81c. The α/γ interface can be seen to advance by the ledge mechanism, whereby mobile incoherent ledges migrate across immobile semicoherent facets. Note that these growth ledges are ~100 atom layers high in contrast to the structural ledges discussed in Section 3.4.1 which are only a few atom layers high at most. Normally the incoherent risers would be energetically favorable sites for precipitation, but in this case the alloy carbides nucleate on the low-energy facets. This is because the ledges are moving too fast for nucleation to occur. As can be seen in Figure 5.81a and as shown schematically in Figure 5.81c, the precipitate size increases with distance behind the step, indicating that nucleation occurs on the semicoherent facets just ahead of the steps.

5.8.6 Rule of Scheil

The *additive reaction rule* of Scheil can be used to convert between isothermal phase transformation data obtained from TTT and CCT diagrams. The rule can only be justified if the reaction of interest is isokinetic, in other words, the transformation rate depends on the transformed phase fraction and temperature only.

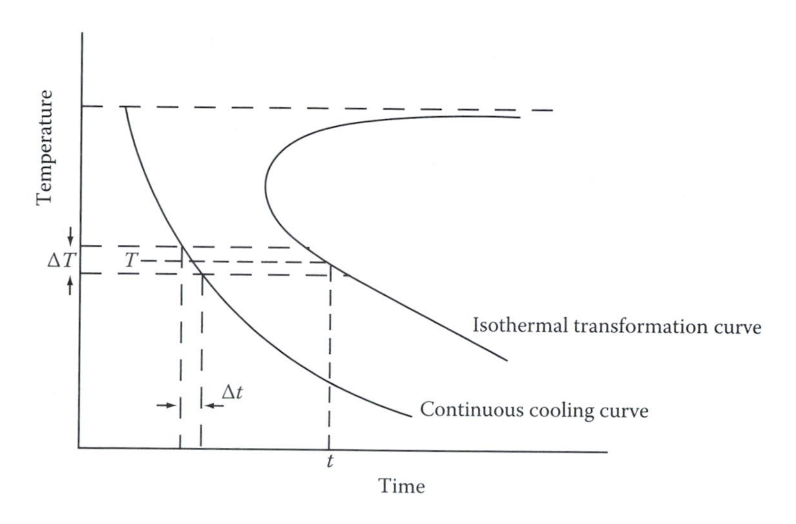

FIGURE 5.82
A schematic representing the additivity rule of Scheil. (Based on Cottrell, A., *An Introduction to Metallurgy*, 2nd edn., The Institute of Materials, London, 1975, 374–375.)

Figure 5.82 shows the application of the rule to an isothermal transformation curve from which the corresponding continuous cooling curve will be calculated as follows. First, the temperature range is divided into steps ΔT with the corresponding time steps Δt representing the time spent at each step. The fraction of the time spent at a step is calculated by dividing the time step by the time of the isothermal transformation t which corresponds to the average temperature in the step. The "cumulative" phase fraction of the transformed phase at a temperature T is then determined by adding the fractions of time such that[47]

$$f(T) = \sum_{T_e}^{T} \frac{\Delta t}{t} \tag{5.56}$$

Therefore, the transformation begins when $f \approx 1$.

It is important to note that not all phase transformations can be expressed using the additivity rule. For example, in the work of Hsu et al. on pearlite transformed during continuous cooling, the pearlite phase fraction from nucleation was additive but not that from growth.[48–50]

5.9 Massive Transformations

Consider the Cu–Zn alloys in Figure 5.83 containing approximately 38 atomic% Zn. The most stable state for such alloys is β above ~800°C,

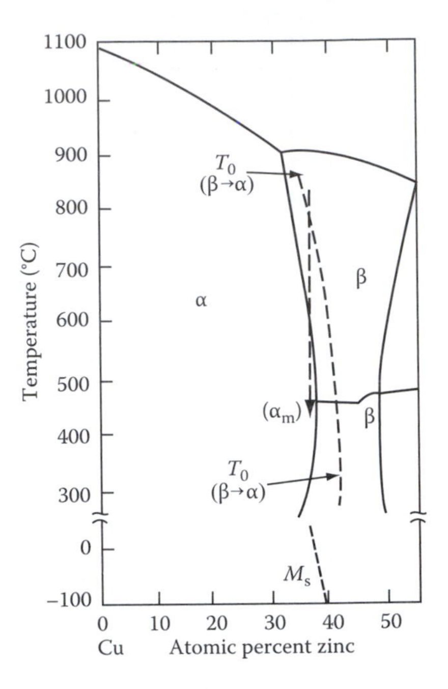

FIGURE 5.83
A part of the Cu–Zn phase diagram showing the α/β equilibrium. The temperature at which $G^\alpha = G^\beta$ is marked as T_0. M_s marks the beginning of the martensite transformation in rapidly quenched specimens. (After Massalski, T.B., in *Phase Transformations*, American Society for Metals, Metals Park, OH, 1970.)

α below ~500°C, and a mixture of αβ with compositions given by the equilibrium phase diagram in between. The type of transformation that occurs on cooling the β-phase depends on the cooling rate. At slow to moderate cooling rates α precipitates in a similar way to the precipitation of ferrite from austenite in Fe–C alloys: slow cooling favors transformation at small undercooling and the formation of equiaxed α; higher cooling rates result in transformation at lower temperatures and Widmanstätten α needles precipitate. According to the phase diagram, the α that precipitates will be richer in Cu than the parent β-phase, and therefore the growth of the α-phase requires the long-range diffusion of Zn away from the advancing α/β interfaces. This process is relatively slow, especially since the Cu and Zn form substitutional solid solutions, and consequently the C curve for the α precipitation on a TTT or CCT diagram will be located at relatively long times. A possible CCT diagram is shown schematically in Figure 5.84.

If the alloy is cooled fast enough, by quenching in brine for example, there is no time for the precipitation of α, and the β-phase can be retained to temperatures below 500°C where it is possible for β to transform into α with the same composition. The result of such a transformation is a new massive transformation product, Figure 5.85.

Massive α grains nucleate at grain boundaries and grow rapidly into the surrounding β. Note also that because of the rapid growth the α/β boundaries have a characteristic irregular appearance. Since both the α and β phases have the same composition, massive α(α_m) can grow as fast as the Cu and Zn atoms can cross the α/β interface, without the need for long-range diffusion.

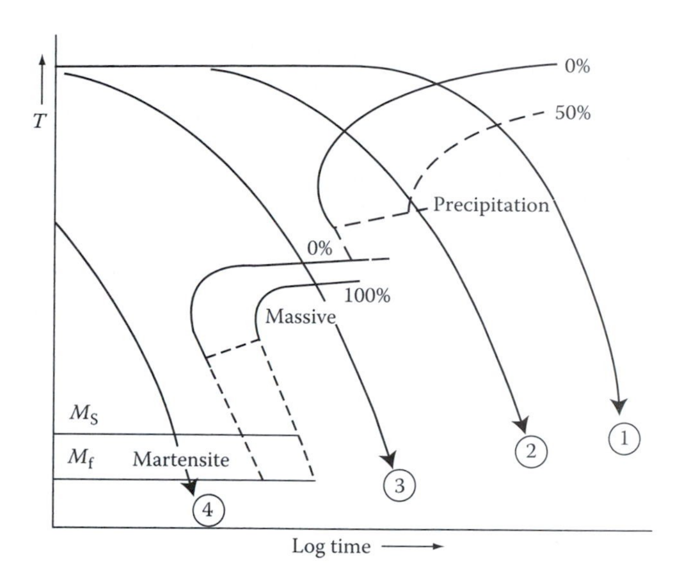

FIGURE 5.84
A possible CCT diagram for systems showing a massive transformation. Slow cooling (1) produces equiaxed α. Widmanstätten morphologies result from faster cooling (2). Moderately rapid quenching (3) produces the massive transformation, while the highest quench rate (4) leads to a martensitic transformation. Compare with Figure 5.88.

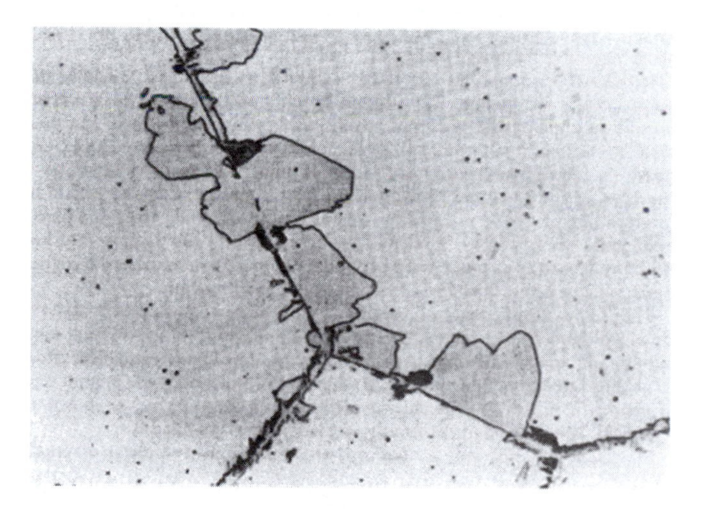

FIGURE 5.85
Massive α formed at the grain boundaries of β in Cu–38.7 wt% Zn quenched from 850°C in brine at 0°C. Some high temperature precipitation has also occurred on the boundaries. (From Hull, D. and Garwood, K., *The Mechanism of Phase Transformations in Metals*, Institute of Metals, London, 1956. With permission.)

Since growth only involves thermally activated jumping across the α/β interface, the massive transformation can be defined as a diffusionless civilian transformation and it is characterized by its own C curve on TTT or CCT diagrams as shown in Figure 5.84. The migration of the α/β interfaces is very similar to the migration of grain boundaries during recrystallization of single-phase material. However, in the case of the massive transformation the driving force is orders of magnitude greater than for recrystallization, which explains why the transformation is so rapid.

Massive transformations should not be confused with martensite. Although the martensitic transformation also produces a change of crystal structure without a change in composition, the transformation mechanism is quite different. Martensite growth is a diffusionless military transformation, i.e., β is sheared into α by the cooperative movement of atoms across a glissile interface, whereas the growth of massive α involves thermally activated interface migration. Systems showing massive transformations will generally also transform martensitically if sufficiently high quench rates are used to suppress the nucleation of the massive product, Figure 5.84. However, Figure 5.83 shows that for the Cu–Zn alloys the M_s temperature is below 0°C and some β-phase is therefore retained after quenching to room temperature, as can be seen in Figure 5.85.

It was stated above that β can transform massively into α provided the β-phase could be cooled into the stable α-phase field without precipitation at a higher temperature. Thermodynamically, however, it is possible for the transformation to occur at higher temperatures. The condition that must be satisfied for a massive transformation is that the free energy of the new phase must be lower than the parent phase, both phases having the same composition. In the case of Cu–38 atomic% Zn therefore, it can be seen from Figure 5.86 that there is a temperature ~700°C below which G^α becomes less than G^β. This temperature is marked as T_0 in Figure 5.83 and the locus of T_0 is

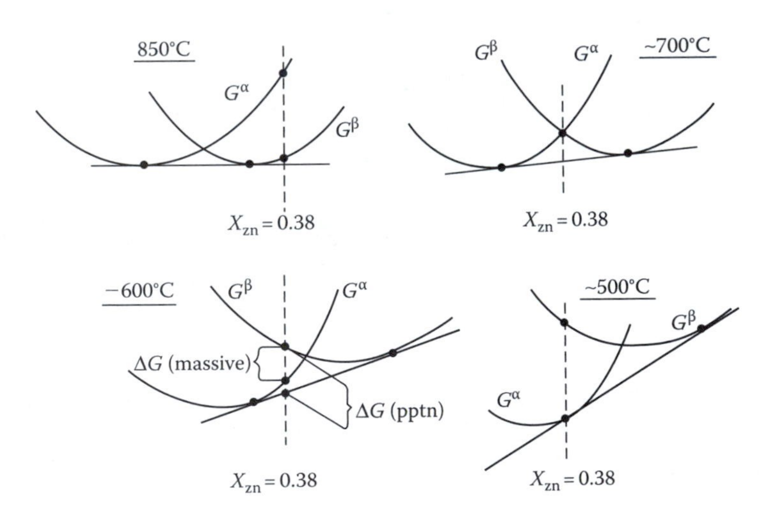

FIGURE 5.86
Schematic representation of the free energy–composition curves for α and β in the Cu–Zn system at various temperatures.

also shown for other alloy compositions. Therefore it may be possible for a massive transformation to occur within the two-phase region of the phase diagram anywhere below the T_0 temperature. In practice, however, there is evidence that massive transformations usually occur only within the single-phase region of the phase diagram.

Massive transformations are found in many alloy systems. Usually the interfaces are incoherent and migrate by continuous growth in a similar manner to a high-angle grain boundary, but in some cases growth can take place by the lateral movement of ledges faceted interfaces. The Cu–Al phase diagram is similar to that shown in Figure 5.83. Figure 5.87 shows a specimen of Cu–20 atomic% Al that has been quenched from the β field to produce almost 100% massive α. Again characteristically irregular phase boundaries are apparent. In both Figures 5.85 and 5.87 the cooling rate has been insufficient to prevent some precipitation on grain boundaries at higher temperatures before the start of the massive transformation.

The $\gamma \rightarrow \alpha$ transformation in iron and its alloys can also occur massively provided the γ is quenched sufficiently rapidly to avoid transformation near equilibrium, but slow enough to avoid the formation of martensite. The effect of cooling rate on the temperature at which transformation starts in pure iron is shown in Figure 5.88. The microstructure of massive ferrite is shown in Figure 5.89. Note the characteristically irregular grain boundaries.

Massive transformations are not restricted to systems with phase diagrams like that shown in Figure 5.83. Metastable phases can also form massively as

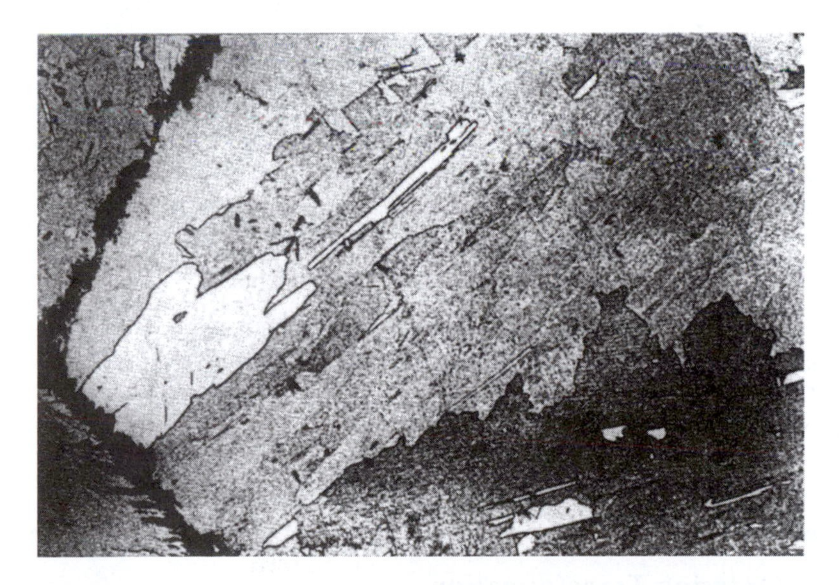

FIGURE 5.87
Massive α in Cu–20 atomic % Al after quenching from the β field at 1027°C into iced brine. Note the irregular α/α boundaries. Some other transformation (possibly bainitic) has occurred on the grain boundaries. (After Chadwick, G.A., in *Metallography of Phase Transformations*, Butterworths, London, 1972.)

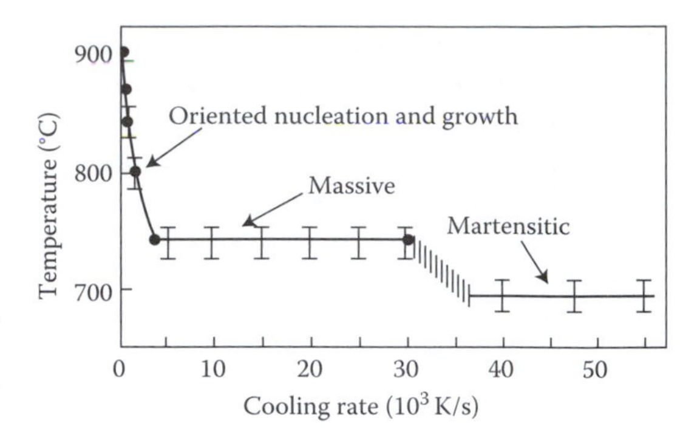

FIGURE 5.88
Effect of cooling rate on the transformation temperature of pure iron. (After Bibby, M.J. and Part, J.G., *J. Iron Steel Inst.* 202, 100, 1964.)

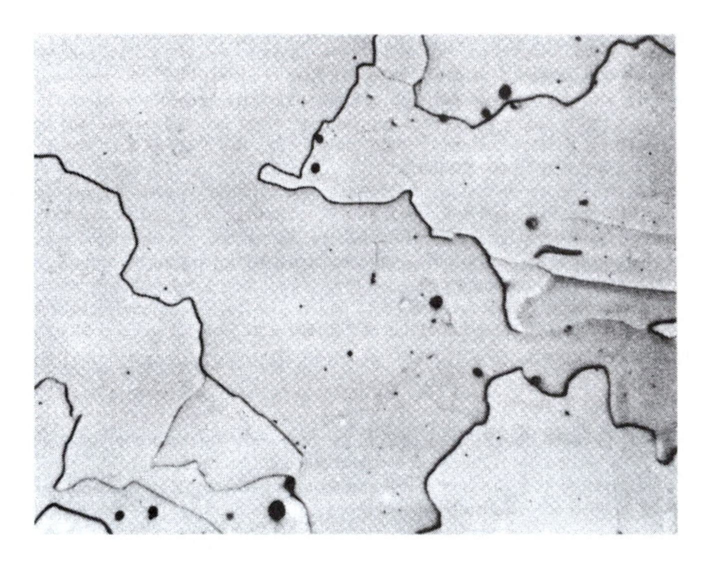

FIGURE 5.89
Massive α in an Fe–0.002 wt% C quenched into iced brine from 1000°C. Note the irregular α/α boundaries. (After Massalski, T.B., in *Metals Handbook*, 8th edn., Vol. 8, American Society for Metals, Metals Park, OH, 1973, 186.)

shown in Figure 5.1d(ii) for example. It is not even necessary for the transformation product to be a single phase: two phases, at least one of which must be metastable, can form simultaneously provided they have the same composition as the parent phase.

5.10 Ordering Transformations

The structure of ordered phases has already been briefly discussed in Section 1.3.7. To recap, solid solutions which have a negative enthalpy of mixing ($V < 0$) prefer unlike nearest neighbors and therefore show a tendency to form ordered phases at low temperatures. The five main types of ordered solutions are shown in Figure 1.22. An example of a phase diagram containing low-temperature ordering reactions is the Au–Cu diagram in Figure 1.21. Another example is the ordering of bcc β-brass below ~460°C to the so-called $L2_0$ (or B2) superlattice, Figure 5.83. The bcc (or the so-called A2) lattice can be considered as two interpenetrating simple cubic lattices: one containing the corners of the bcc until cell and the other containing the body-centering sites. If these two sublattices are denoted as A and B the formation of a perfectly ordered β′ superlattice involves segregation of all Cu atoms to the A sublattice, say, and Zn to the B sublattice. This is not feasible in practice, however, as the β′ does not have the ideal CuZn composition. There are two ways of forming ordered structures in nonstoichiometric phases: either some atom sites can be left vacant or some atoms can be located on wrong sties. In the case of β(CuZn) the excess Cu atoms are located on some of the Zn sites.

Let us begin the discussion of ordering transformations by considering what happens when a completely ordered single crystal such as CuZn or Cu_3Au is heated from low temperatures to above the disordering temperature. To do this it is useful to quantify the degree of order in the crystal by defining a long-range order parameter L such that $L = 1$ for a fully ordered alloy where all atoms occupy their "correct" sites and $L = 0$ for a completely random distribution. A suitable definition of L is given by

$$L = \frac{r_A - X_A}{1 - X_A} \quad \text{or} \quad \frac{r_B - X_B}{1 - X_B}$$

where

X_A is the mole fraction of A in the alloy
r_A is the probability that an A sublattice site is occupied by the "right" kind of atom

At absolute zero the crystal will minimize its free energy by choosing the most highly ordered arrangement ($L = 1$) which corresponds to the lowest internal energy. The configurational entropy of such an arrangement, however, is zero and at higher temperatures the minimum free energy state will contain some disorder, i.e., some atoms will interchange positions by diffusion so that they are located on "wrong" sites. Entropy effects become increasingly more important with rising temperature so that L continuously decreases until above some critical temperature (T_c) $L = 0$. By choosing a suitable model, such as the quasichemical model discussed in Section 1.3.4, it is possible to calculate how L varies with temperature for different superlattices. The results of such a calculation for the CuZn and Cu_3Au superlattices are shown in Figure 5.90. It can be seen that the way in which L decreases to zero is different for the different superlattices. In the equiatomic CuZn case L decreases continuously with temperature up to T_c, whereas in Cu_3Au L decreases only slightly up to T_c and then abruptly drops to zero above T_c. This difference in behavior is a consequence of the different atomic configurations in the two superlattices.

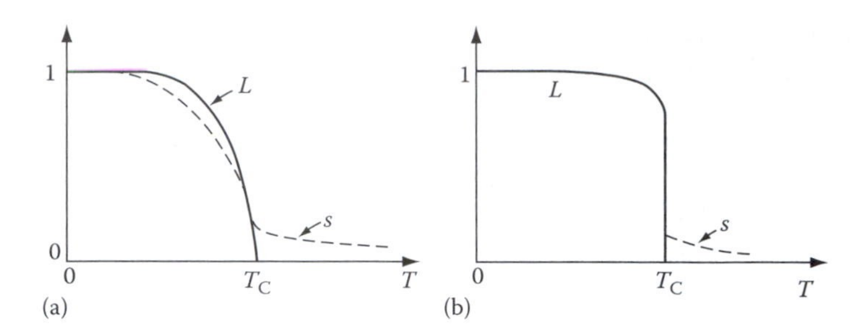

FIGURE 5.90
Variation of long-range order (L) and short-range order (s) for (a) CuZn-type and (b) Cu_3Au-type transformations (schematic).

Above T_c it is impossible to distinguish separate sublattices extending over long distances and $L = 0$. However, since $V < 0$ there is still a tendency for atoms to attract unlike atoms as nearest neighbors, i.e., there is a tendency for atoms to order over short distances. The degree of short-range order (s) is defined in Section 1.3.7. The variation of s with temperature is shown as the dashed lines in Figure 5.90.

The majority of phase transformations that have been discussed in this book have been the so-called first-order transformations. This means that at the equilibrium transformation temperature the first derivatives of the Gibbs free energy $\partial G/\partial T$ and $\partial G/\partial P$ are discontinuous. The melting of a solid is such a transformation, Figure 5.91a. Since $\partial G/\partial T = -S$ and $\partial G/\partial P = V$,

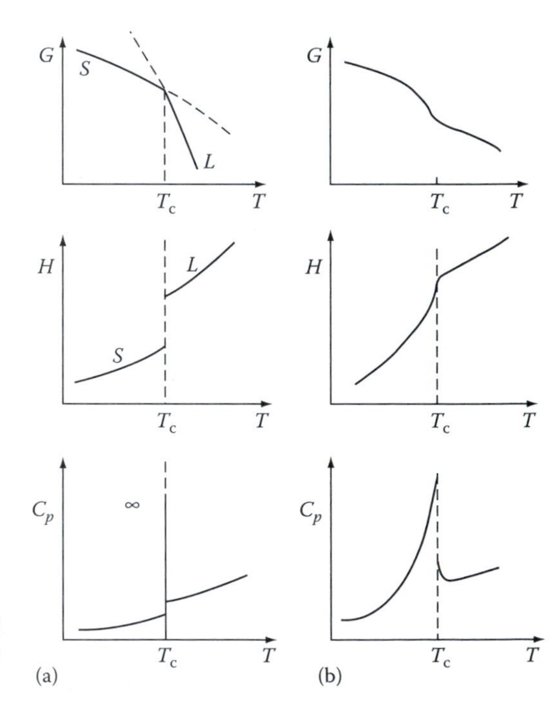

FIGURE 5.91
Thermodynamic characteristics of (a) first-order and (b) second-order phase transformations.

first-order transformations are characterized by discontinuous changes in S and V, first-order transformations are characterized by discontinuous changes in S and V. There is also a discontinuous change in enthalpy H corresponding to the evolution of a latent heat of transformation. The specific heat of the system is effectively infinite at the transformation temperature because the addition of a small quantity of heat converts more solid into liquid without raising the temperature.

Figure 5.91b illustrates the characteristics of a second-order transformation. For such a transformation the second derivatives of Gibbs free energy $\partial^2 G/\partial T^2$ and $\partial^2 G/\partial P^2$ are discontinuous. The first derivatives, however, are continuous which means that H is also continuous. Consequently since

$$\left(\frac{\partial^2 G}{\partial T^2}\right)_P = -\left(\frac{\partial S}{\partial T}\right)_P = \frac{1}{T}\left(\frac{\partial H}{\partial T}\right)_P = \frac{C_P}{T}$$

there is no latent heat, only a high specific heat, associated with the transformation.

Returning to a consideration of order–disorder transformations it can be seen from Figure 5.90 that the loss of long-range order in the $\beta' \rightarrow \beta$ (CuZn) transformation corresponds to a gradual disordering of the structure over a range of temperatures. There is no sudden change in order at T_c and consequently the internal energy and enthalpy (H) will be continuous across T_c. The $\beta' \rightarrow \beta$ transformation is therefore a second-order transformation. In the case of Cu_3Au, on the other hand, a substantial change in order takes place discontinuously at T_c. Since the disordered state will have a higher internal energy (and enthalpy) than the ordered state, on account of the greater number of high-energy like–like atom bonds, there will be a discontinuous change in H at T_c, i.e., the transformation is first order.

So far we have been concerned with the disordering transformation that takes place on heating a fully ordered single crystal. The mechanism by which order is lost is most likely the interchange of atoms by diffusional processes occurring homogeneously throughout the crystal. The same changes will of course take place in every grain of a polycrystal. Let us now turn to the reverse transformation that occurs on cooling a single crystal, i.e., disorder $\rightarrow$ order.

There are two possible mechanisms for creating an ordered superlattice from a disordered solution. (1) There can be a continuous increase in short-range order by local rearrangements occurring homogeneously throughout the crystal which finally leads to long-range order. (2) There may be an energy barrier to the formation of ordered domains, in which case the transformation must take place by a process of nucleation and growth. These two alternative mechanisms are equivalent to spinodal decomposition and precipitation as mechanisms for the formation of coherent zones in alloys with positive heats of mixing ($V > 0$). The first mechanism may only be able to operate in second-order transformations or at very high supercoolings below T_c. The second mechanism is generally believed to be more common.

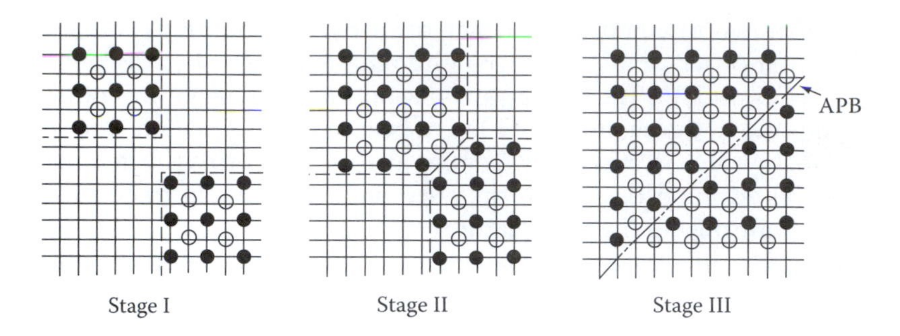

FIGURE 5.92
Formation of an APB when out-of-phase ordered domains grow together. The diagram could represent a {100} plane in Cu_3Au in which case the black and white atoms could represent Cu and Au. (After Haasen, P., *Physical Metallurgy*, Cambridge University Press, Cambridge, 1978.)

The nucleation and growth process is illustrated in Figure 5.92. The disordered lattice is represented by the cross-grid of lines. Within this lattice two sublattices are marked by heavy and faint lines. Atoms are located at each intersection but only atoms within the ordered regions, or domains, are marked; the unmarked sites are disordered. The diagram is only schematic, but could represent a [100] plane of the Cu_3Au superlattice. Since the two types of atoms can order on either the A or B sublattice, the independently nucleated domains will often be "out of phase" as shown. When these domains subsequently grow together a boundary will form (known as an antiphase domain boundary or APB) across which the atoms will have the wrong kind of neighbors. APBs are therefore high-energy regions of the lattice and are associated with an APB energy.

Even at rather low undercoolings below T_c the activation energy barrier to the nucleation of ordered domains ΔG^* should be rather small because both nucleus and matrix have essentially the same crystal structure and are therefore coherent with a low interfacial energy. Also, provided the alloy has a stoichiometric composition, both nucleus and matrix have the same composition so that there should not be large strain energies to be overcome. Consequently, it is to be expected that nucleation will be homogeneous, independent of lattice defects such as dislocations and grain boundaries. Figure 5.93 shows evidence for the existence of a nucleation and growth mechanism during ordering in CoPt. This is a field ion micrograph showing that the two types of atoms are ordered in a regular manner in the upper part but disordered in the lower part of the micrograph.

At low ΔT the nucleation rate will be low and a large mean domain size results, whereas higher values of ΔT should increase the nucleation rate and diminish the initial domain size. The degree of long-range order in a given domain will vary with temperature according to Figure 5.90 and with decreasing temperature the degree of order is increased by homogeneous diffusive rearrangements among the atoms within the domain. Within the crystal as a

FIGURE 5.93
Field ion micrograph of the boundary between an ordered domain (above) and disordered matrix (below) in CoPt. (After Müller, E.W. and Tsong, T.T., *Field Ion Microscopy, Principles and Applications*, Elsevier, New York, 1969.)

whole, the degree of long-range order will initially be very small because there are likely to be equal numbers of domains ordered on both A and B lattices. The only way for long-range order to be established throughout the entire crystal is by the coarsening of the APB structure. The rate at which this occurs depends on the type of superlattice.

In the CuZn-type superlattice ($L2_0$) there are only two sublattices on which the Cu atoms, say, can order and therefore only two distinct types of ordered domain are possible. A consequence of this is that it is impossible for a metastable APB structure to form. It is therefore easy for the APB structure to coarsen in this type of ordered alloy. Figure 5.94 shows an electron micrograph of APBs in AlFe ($L2_0$ superlattice) along with a schematic diagram to illustrate the two different types of domain. The Cu_3Au ($L1_2$) superlattice is different to the above in that there are four different ways in which ordered domains can be formed from the disordered fcc lattice: the au atoms can be located either at the corners of the unit cell, Figure 1.20c, or at the one of the three distinct face-centered sites. The Cu_3Au APBs are therefore more complex than the CuZn type, and a consequence of this is that it is possible for the APBs to develop a metastable, so-called foam structure, Figure 5.95. Another interesting feature of this microstructure is that the APBs tend to align parallel to {100} planes in order to minimize the number of high-energy Au–Au bonds.

The rate at which ordering occurs varies greatly from one alloy to another. For example the ordering of β (CuZn) is so rapid that it is almost impossible to quench-in the disordered bcc structure. This is because the transformation is second order and can occur by a rapid continuous ordering process.

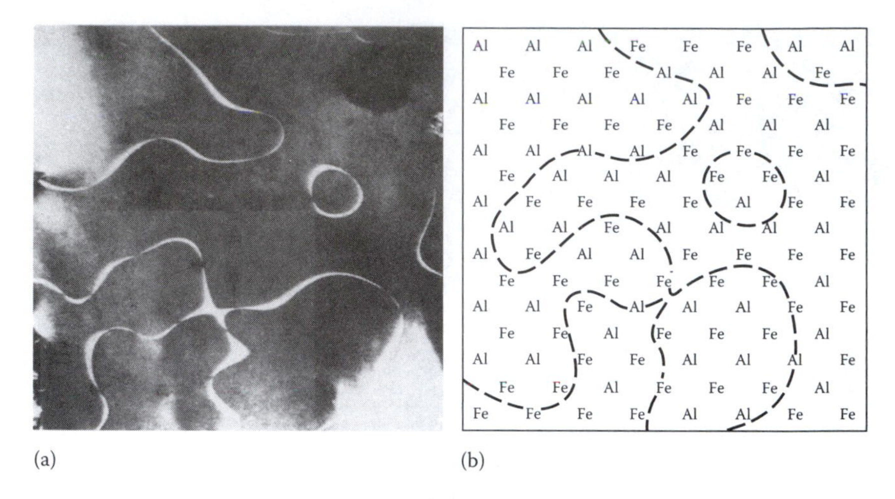

FIGURE 5.94
(a) Thin-foil electron micrograph showing APBs in an ordered AlFe alloy (×17,000). (b) Schematic representation of the atomic configurations comprising the APB structure in (a). (After Marcinkowski, M.J., in *Metals Handbook*, 8th edn., Vol. 8, American Society for Metals, Metals Park, OH, 1973, 205.)

FIGURE 5.95
APBs in ordered Cu_3Au. Thin-Au. Thin-foil electron micrograph ×53,000. Note that due to the method of imaging about one-third of the APBs are invisible. (After Marcinkowski, M.J., in *Metals Handbook*, 8th edn., Vol. 8, American Society for Metals, Metals Park, OH, 1973, 205.)

Ordering of Cu_3Au on the other hand is relatively slow requiring several hours for completion, despite the fact that the atomic mobilities ought to be similar to those in the CuZn transformation. This transformation, however, is second order and proceeds by nucleation and growth. Also the development of long-range order is impeded by the formation of metastable APB networks.

The above comments have been concerned primarily with alloys of stoichiometric composition. However, it has already been pointed out that ordering is often associated with nonstoichiometric alloys.

In the case of first-order transformations there is always a two-phase region at nonstoichiometric compositions, Figure 1.21, so that the transformation can be expressed as: disordered phase → ordered precipitates + disordered matrix. There is then a change in composition on ordering and long-range diffusion must be involved. Second-order transformations on the other hand do not involve a two-phase region even at nonstoichiometric compositions, Figure 5.83.

5.11 Case Studies

5.11.1 Titanium Forging Alloys

Composition: Ti–6 wt% Al–4 wt% V.
Phase diagrams: Binary Ti–Al and Ti–V diagrams in Figure 5.96.

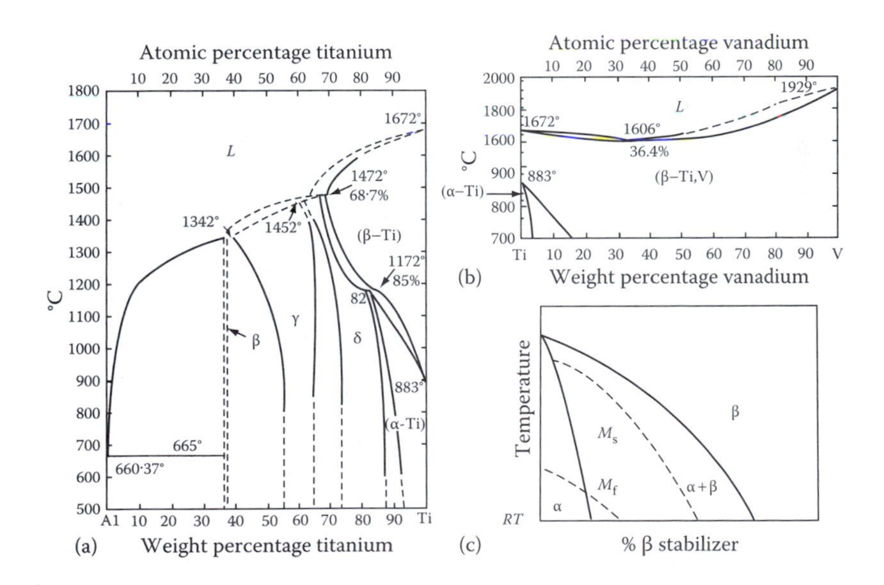

FIGURE 5.96
Phase diagrams: (a) Ti–Al, (b) Ti–V, and (c) a schematic diagram showing the variation of M_s and M_f with percentages β stabilizer. (After Margelin, H., in *Metals Handbook*, 8th edn., Vol. 8, American Society for Metals, Metals Park, OH, 1973, 264.)

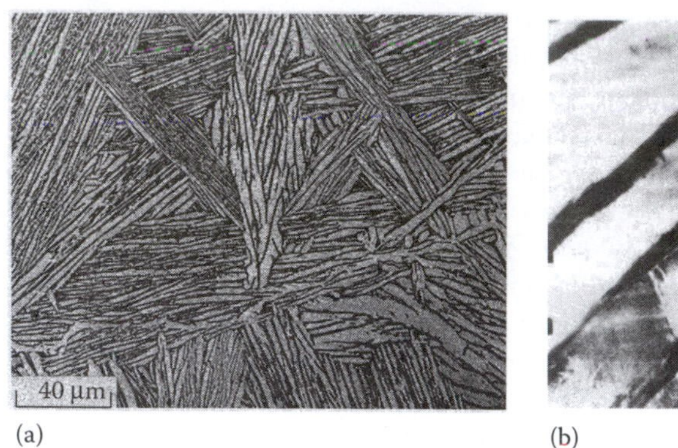

(a)

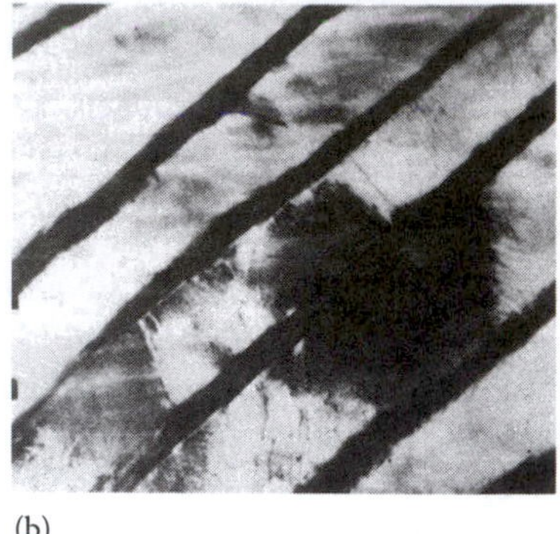
(b)

FIGURE 5.97
(a) Widmanstätten α (light) and β (dark) in a Ti–6 Al–4 alloy air-cooled from 1037°C. (From Copley, S.M. and Williams, J.C., in *Alloy and Microstructural Design*, Academic Press, New York, 1976. With permission.) (b) Alternate layers of α (light) and β (dark) in a Widmanstätten microstructure. Ti–6 Al–4 V forged at 1038°C above the β transus, air-cooled, annealed 2 h at 704°C, air-cooled. Thin-foil electron micrograph (×15,000). (From The ASM Committee on Metallography of Titanium and Titanium Alloys, *Metals Handbook*, 8th edn., Vol. 7, American Society for Metals, Metals Park, OH, 329, 1972. With permission.)

Important phases: α-hcp, β-bcc
Microstructures: See Figures 5.97 through 5.100.

Applications: As a result of the high cost of titanium, uses are restricted to applications where high performance is required and high strength to weight ratio is important, for example, gas turbine aero engines and airframe structures.

(a)

(b)

FIGURE 5.98
(a) α′ Martensite in Ti–6 Al–4 V held above the β transus at 1066°C and water quenched. Prior β grain boundaries are visible (×370). (From *Metals Handbook*, 8th edn., Vol. 7, American Society for Metals, 1972, 328.) (b) β-precipitates that have formed during the tempering of α′ martensite in Ti–6 Al–4 V. Specimen quenched from 1100°C and aged 24 h at 600°C. Thin-foiled electron micrograph. (After Copley, S.M. and Williams, J.C., in *Alloy and Microstructural Design*, Academic Press, New York, 1976.)

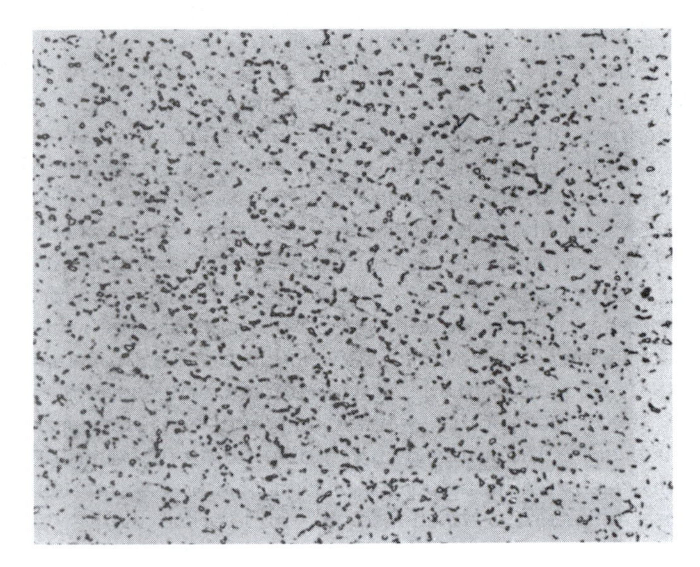

FIGURE 5.99
Microstructure of hot-worked and annealed Ti–6 Al–4 V (×540). (From Morton, P.H., in *Rosenhain Centenary Conference*, The Royal Society, London, 1976. With permission.)

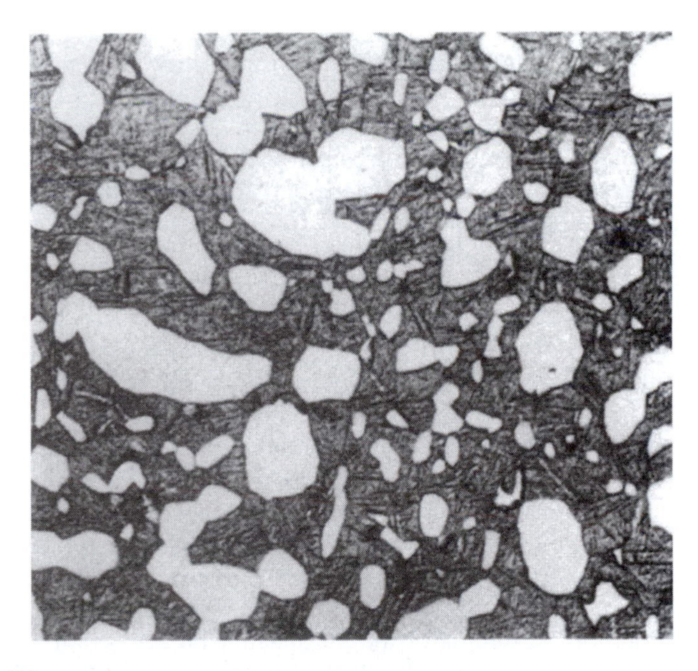

FIGURE 5.100
Microstructure of hardened Ti–6Al–4V. Solution treated at 954°C (high in the $\alpha + \beta$ range), water quenched, aged 4 h at 5.38°C. Equiaxed "primary" α grains (light) in an aged martensitic matrix. (From The ASM Committee on Metallography of Titanium and Titanium Alloys, *Metals Handbook*, 8th edn., Vol. 7, American Society for Metals, Metals Park, OH, 1972, 329. With permission.)

Comments: At low temperatures pure titanium exists as the hcp α-phase, but above 883°C up to the melting point (~1672°C) the bcc β-phase is stable. Figure 5.96 shows that Al is an α stabilizer, i.e., it raises the α/β transition temperature, whereas *V* is a β stabilizer which lowers the transition temperature. A wide range of titanium alloys are available. These can be classified as either α, α β, or β alloys. The Ti–6Al–4V alloy to be discussed here belongs to the α + β group of alloys. For simplicity, the phase diagram relevant to these alloys can be envisaged as that shown in Figure 5.96c. Two principal types of transformation are of interest. The first of these is the precipitation of α from β on cooling from above the β transus into the α + β field. This is in principle the same as the formation of ferrite during the cooling of austenite in Fe–C alloys. However, in this case the Widmanstätten morphology predominates at all practical cooling rates, Figure 5.97a. Figure 5.97b is a thin-foil electron micrograph of a similar structure and shows more clearly the two phases present after air cooling. The β-phase remains as a thin layer between the Widmanstätten α plates. Furnace cooling produces similar though coarser microstructures. The α plates and the β matrix are oriented such that

$$(0001)_\alpha // (110)_\beta$$

$$[11\bar{2}0]_\alpha // [1\bar{1}1]_\beta$$

The second important transformation is the formation of martensite that takes place when β is rapidly cooled by water quenching. The transformation can be written $\beta \rightarrow \alpha'$ where α' is a supersaturated hcp α-phase. M_f for β containing 6 wt% Al–4 wt% *V* is above room temperature so that quenching from above the β transus produces a fully martensitic structure, Figure 5.98a. The martensite can be aged by heating to temperatures where appreciable diffusion can occur, in which case the supersaturated α' can decompose by the precipitation of β on the martensite plate boundaries and dislocations, Figure 5.98b.

Alloys for engineering applications are not usually used in the above conditions but are hot worked in the α + β region of the phase diagram in order to break up the structure and distribute the α-phase in a finely divided form. This is usually followed by annealing at 700°C which produces a structure of mainly α with finely distributed retained β, Figure 5.99. The advantage of this structure is that it is more ductile than when the α is present in a Widmanstätten form. When additional strength is required the alloys are hardened by heating to high temperatures in the α + β range (~940°C) so that a large volume fraction of β is produced, followed by a water quench to convert the β into α' martensite, and then heating to obtain precipitation hardening of the martensite (Figure 5.100). Mechanical properties that can be obtained after these treatments are given in Table 5.4. If the alloy is held lower in the α + β field before quenching the β-phase that forms can be so rich in vanadium that the M_s temperature is depressed to below room temperature and quenching results in retained β, see Figure 5.96c.

TABLE 5.4

Room-Temperature Mechanical Properties of Ti–6 wt% Al–4 wt% V Alloys

Condition	YS (MPa)	UTS (MPa)	Elongation (%)
Annealed	930	990	15
Hardened	950	1030	14

5.11.2 Weldability of Low-Carbon and Microalloyed Rolled Steels

Composition: $C \leq 0.22$ wt%, Si 0.3%, Mn = 1.0%–1.5%, $P \leq 0.04\%$, $S \leq 0.04\%$. C_{eq} (see text) * 0.4%.

Possible microalloying elements: Al, Nb, Ti, V, with possible additions of Zr and/or N. The total amount of microalloying elements does not usually exceed 0.15%.

Phase diagrams: Fe–C binary

Modified CCT diagrams (see below)

Welding nomographs (see text)

Microstructure: Depends on type of steel, e.g., whether quench and tempered, microalloyed—fine grained, plain rolled C–Mn, see, for example, Figure 5.54a.

Applications: Constitutional steels for building frames, bridges, pressure vessels, ships, oil platforms, etc.

Comments: Steels used for heavy, high-strength constructions are nowadays rather sophisticated, relying for their high strength and toughness on having a fine and uniform grain size. When fusion welding plates together, the steel is subjected to an extremely severe thermal cycle, and at the fusion line the temperature attains the melting point of the alloy. Because the steel plate provides an effective thermal sink (Section 4.5) the cooling rate is very high for most types of welding process as illustrated in Figure 5.101. This thermal cycle causes changes in properties of the base material in the heat-affected zone due to the combination of phase changes and thermal/mechanical stresses. Typical microstructural changes experienced by a C–Mn steel are illustrated in Figure 5.102, showing that recrystallization, grain growth, and even aging are occurring in the heat-affected zone. Of these changes, grain growth is potentially the most troublesome in decreasing the strength and toughness of these steels particularly since in most cases high-energy submerged arc welding is used with its associated relatively long dwell-time at peak temperatures. In order to avert the problem of grain growth at high temperatures, new steels have recently been introduced containing a fine dispersion of TiN precipitates. These precipitates remain fairly stable at temperatures as high as 1500°C and, at their optimum size of about 10 nm, act as a barrier to grain growth during welding.

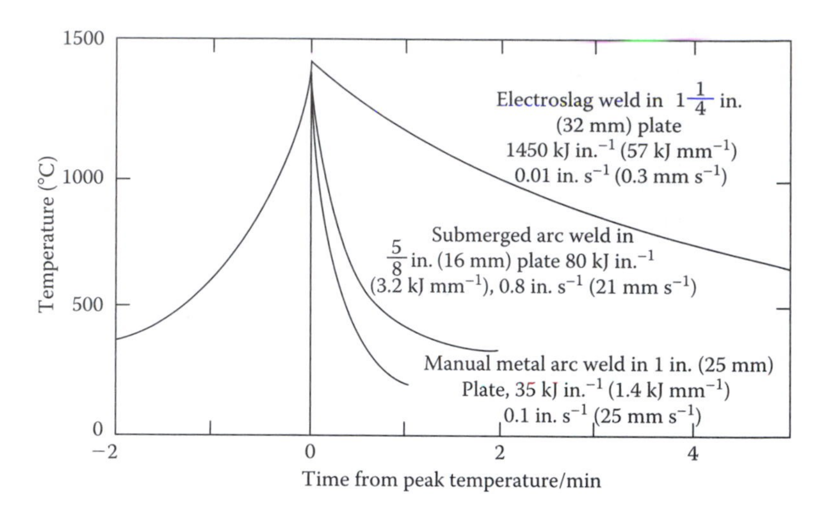

FIGURE 5.101
Comparison of thermal cycles experienced in the heat-affected zone when welding steel by different processes. (After Baker, R.G., in *Rosenhain Centenary Conference*, The Royal Society, London, 1976.)

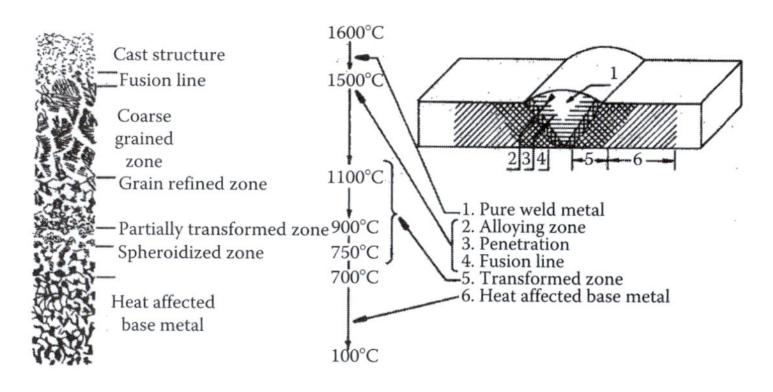

FIGURE 5.102
Weld and HAZ structures in fusion welding. (After Räsänen, E. and Tenkula, J., *Scand. J. Metall.*, 1, 75, 1972.)

Another important problem in welding high-strength steels concerns the formation of martensite. The reason for this is that it is very difficult in welding to avoid the presence of hydrogen. This is because hydrogen-containing compounds are invariably present in fluxes, the electrode material or even in the environment if welding is done outside. In this way atomic hydrogen is absorbed into the molten metal of the fusion weld where it then diffuses rapidly into the heat-affected base metal. If during subsequent cooling, martensite forms, hydrogen (whose solubility in martensite is lower than in ferrite) is forced out of the martensite where it concentrates at the martensite–ferrite phase boundary, or at inclusion boundaries. Thus in combination with weld residual stresses the hydrogen weakens the iron lattice and may initiate cracks. This phenomenon is known as cold cracking. It is found vital in welding to exert a close control over the amount of residual hydrogen in welds and to avoid martensite, particularly in cases where residual stresses may be high. Since it is usually difficult to totally avoid the presence of hydrogen, special CCT diagrams are employed in conjunction with estimated cooling rates in the heat-affected material as shown in Figure 5.103. The essential feature of this types of CCT diagram is that the phase boundaries need to be plotted under conditions of actual welding, or weld simulation, in which both thermal and residual stresses are present.[51] In the case of weld simulation, a special equipment is employed in which it is possible to program in the appropriate thermal and stress cycles. As illustrated in Figure 5.103, the various cooling curves 1–8 represent different heat inputs corresponding to different welding

Marking	1	2	3	4	5	6	7	8
Heat input J cm^{-1}	6,500	10,000	20,000	30,000	32,500	50,000	40,000	45,000
t_r(800°C to 500°C) s	2.3	5	10.2	16.5	23.5	32	44.5	57
Working temperature (°C)	Amb	Amb	Amb	Amb	Amb	Amb	200	200

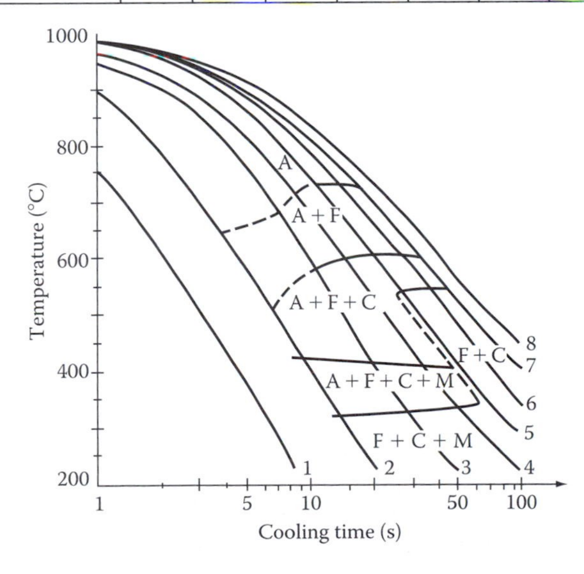

FIGURE 5.103
CCT diagrams for a 0.19% C–1.52 Mn–0.55 Si steel, with superimposed cooling rates corresponding to the weld hat inputs given in the table. (From IIW's Doc. 115/11W-382-71, 1971. With permission.)

processes or parameters. The parameter t_r in the table refers to the time in seconds for cooling through the temperature range: 800°C to 500°C, this being almost a constant within the heat-affected zone, and is thus considered a useful parameter in welding in helping to predict microstructure as a function of welding input energy. The working temperature in the table refers to whether or not preheating was employed. Thus in Figure 5.103, martensite is predicted to occur for all welding energies below 37,500 J cm^{-1} (curve 5), this corresponding in practice to a weld deposit on a 20 mm thick plate of the composition given. In practice of course it is more useful if microstructural or cold cracking predictions could be made as a function of differing chemical composition, plate thickness, peak temperature, preheating, and welding variables. This obviously requires much more complex diagrams than that of Figure 5.103, and will therefore be correspondingly less accurate, although such diagrams, or welding nomographs as they are called, have been developed for certain applications. The composition variations are estimated using a so-called carbon equivalent,[52] in which the effect of the various elements present is empirically expressed as a composition corresponding to a certain carbon content. This is then used to estimate possible martensite formation for the welding conditions given.

5.11.3 Very Low-Carbon Bainitic Steel with High Strength and Toughness

Manufacturing machine and automotive parts from nonheat-treated steels with carbon content in the range 0.2–0.5 wt% is advantageous in terms of cost and processing time given that the quench-tempering process is eliminated.[53] Yet these steels which are ferritic–pearlitic do not have sufficient toughness and as such their use is limited. Nonheat-treated bainitic steels, on the other hand, possess higher strength and toughness than their ferritic–pearlitic counterparts but given their low yield-ratio (yield strength to ultimate tensile strength) their yield point is relatively low.

This case study aims at introducing a new type of bainitic steel with extremely low carbon content, yet it exhibits high toughness and strength.[53] The high toughness can be ascribed to the very low carbon content of the steel. This steel also shows high weldability given its very low carbon content. Precipitation hardening was the strengthening mechanism induced by the so-called thermomechanical precipitation control process (TPCP).

To ensure the formation of a fully bainitic microstructure it is crucial to select a steel composition which does not show much dependence on cooling rates applied following hot deformation. This enables the production of thick sections with a homogeneous microstructure and mechanical properties.

The design of this steel requires having sufficient hardenability which is very important to avoid undesired high temperature austenite transformation products like pearlite. This could be achieved through the addition of an element like boron. Also the carbon content was optimized with a concentration below 0.02 wt%, at which, carbon partitioning from ferrite into the remaining austenite does not take place.

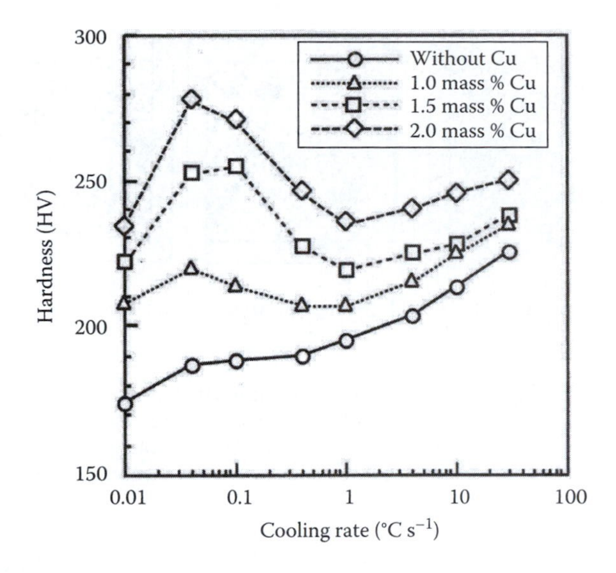

FIGURE 5.104
The change in steel hardness as cooling rate decreased. Note that for Cu-bearing very low carbon bainitic steels, unlike the variant without Cu, the hardness tend to increase at lower cooling rates sufficient enough for Cu-precipitates to form. The increase in hardness is beneficial as the tensile strength will increase as a result. (After Hase, K., Hoshino, T., and Amano, K., *Kawasaki Steel Tech. Rep.*, (47), 35–41, 2002.)

The strengthening mechanism in this steel was achieved via the control of copper precipitates during a slow cooling step following hot rolling. In comparison, the use of copper precipitates as a strengthening mechanism has been successfully used in maraging steels, for example, by applying an aging heat-treatment step following quenching.

The steel composition was suggested to be Fe-0.009C-0.26Si-1.99Mn-0.015P-0.015S-0.034Al in addition to other elements such as niobium, titanium, and boron.

Figure 5.104 shows the change in hardness of the very low carbon bainitic steel as a function of cooling rate. As expected, the variant not alloyed with copper showed a decrease in hardness with slower cooling rates unlike the case of copper-containing steels.

Figure 5.105 presents three-dimension atom probe which shows clustering of copper atoms (ε-Cu precipitate) in the structure as cooling rate decreased.

5.11.4 Very Fine Bainite

During the last two decades, ultrafine- (<500 nm) and nanograined (<100 nm) polycrystalline materials with exceptional mechanical properties have been discovered.[54,55] However, due to production difficulties and ductility shortfalls, their technological viability is limited.

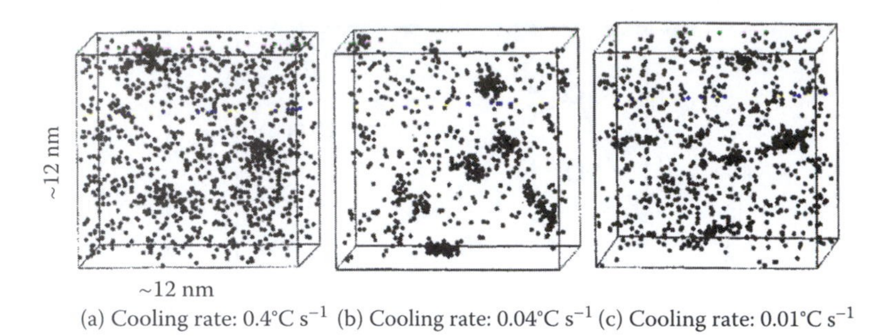

FIGURE 5.105
Copper distribution maps obtained using three-dimensional atom probe. (a) some clusters of Cu atoms can be seen (GP zones, see Section 3.4.2), however, cooling rate was not sufficient for Cu-precipitates leaving more Cu atoms in solid solution. As the cooling rate is lowered clustering becomes more prominent (b) and (c). (After Hase, K., Hoshino, T., and Amano, K., *Kawasaki Steel Tech. Rep.*, (47), 35–41, 2002.)

While utilization of conventional powder metallurgy techniques suffers from limitations such as the expected increase in grain boundary chemical reactivity of very fine powders and consequently limited thermal stability, additionally, contamination and possible residual porosity could be problematic.[56] On the other hand, severe plastic deformation techniques have yet to mature with respect to the shape and geometry of the components produced.

Caballero and Bhadeshia[57,58] developed a high-carbon high-silicon carbide-free bainite with very thin austenite films on the scale of 50 nm. Figure 5.106 shows a TEM micrograph which demonstrates the fineness of the microstructure achieved in this steel.

The steel microstructure yielded an excellent combination of strength and ductility. The strength comes from the fine scale of the structure whereas the

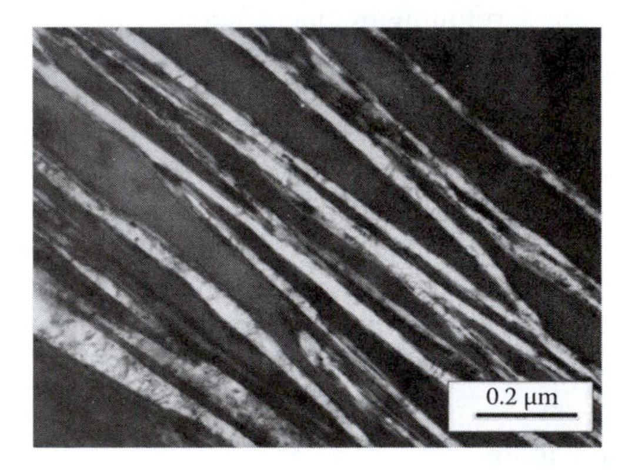

FIGURE 5.106
TEM micrograph exhibiting the developed nanostructured bainitic steel. The dark phase represents the austenite whereas the whitish plates are those of bainite. (Reprinted from Caballero, F.G. and Bhadeshia, H.K.D.H., *Curr. Opin. Solid State Mater. Sci.*, 8, 255, 2004. With permission from Elsevier.)

increased ductility is associated with the presence of a chemically stable retained austenite content which transforms in a progressive manner upon deformation. Austenite stability was achieved through alloying with carbide-suppressing elements such as silicon and aluminum whereby carbon partitioned off bainite plates instead of precipitating as carbides enriches the adjacent austenite matrix. On the other hand, carbon content in retained austenite, which is crucial for its chemical stability, is generally found to decrease with an increase in the transformation temperature.[59,60]

The steel production involves a conventional heat treatment process. This is believed to be a significant advantage over other production techniques usually adopted for the synthesis of nanostructured materials.

The fineness of the microstructure is associated with the low transformation temperatures as shown in Figure 5.107.

Figure 5.108 shows that, for a given volume fraction of retained austenite, there is an increase in the strength with a decrease in transformation temperature.

This class of bainitic steels has shown interesting properties, in particular, its strength–ductility balance. Uniaxial tensile testing showed an almost entirely uniform elongation with the onset of necking being delayed thanks to its superior work-hardening properties. In addition to the low cost of alloying elements needed for alloying and its conventional production route, the steel also exhibited high hardness of about 600 HV with K_{ic} toughness in excess of 30–40 MPa $m^{1/2}$. For a review on this steel class and its design philosophy the reader may consult Ref. [57].

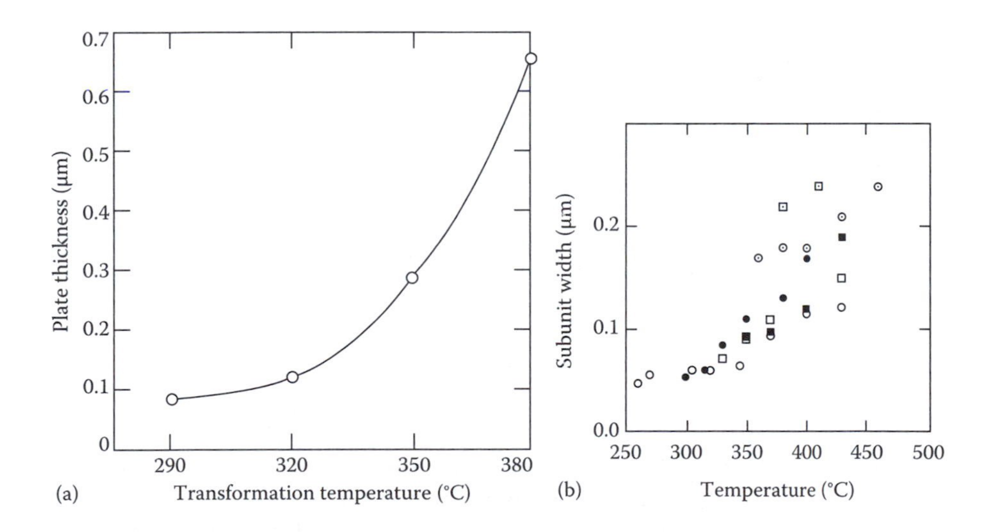

FIGURE 5.107
Bainite plate width versus transformation temperature. (a) Steels containing 0.65–0.99 C and 2–2.78 Si (wt%), with and without Cr. (After Sandvik, B.P.J. and Nevalainen, H.P., *Met. Technol.*, 213, 1981.) (b) Several steel compositions. (After Chang, L.C. and Bhadeshia, H.K.D.H., *Mater. Sci. Tech.-Lond.*, 11, 874, 1995.)

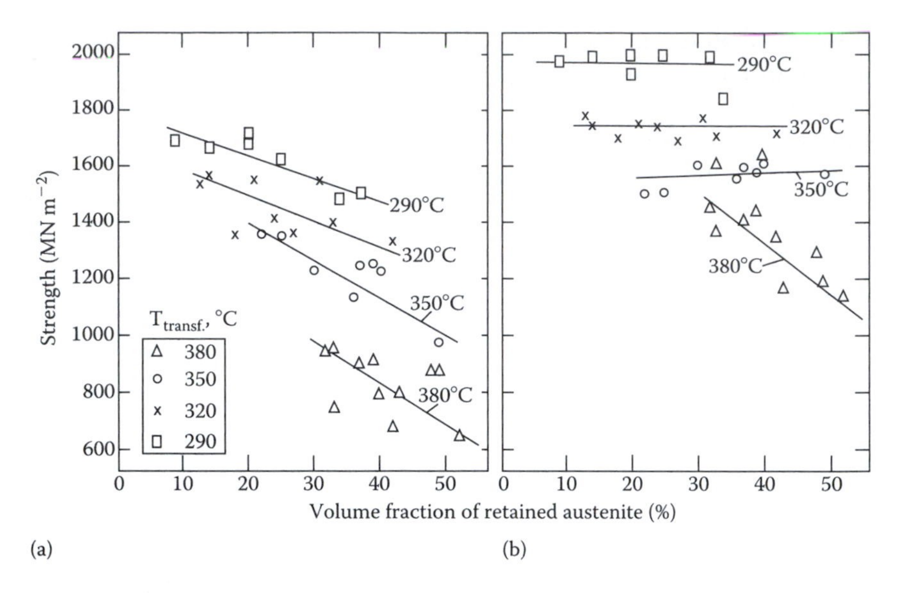

FIGURE 5.108
Effect of volume fraction of retained austenite and bainite transformation temperature on (a) yield strength and (b) ultimate tensile strength. Note that the ultimate tensile strength does not seem to depend solely on the retained austenite content. This was explained in terms of the work-hardening properties of the structure components. (After Sandvik, B.P.J. and Nevalainen, H.P., *Met. Technol.*, 213, 1981.)

Exercises

5.1 An approximate expression for the total driving force for precipitation in a regular solution (ΔG_0 in Figure 5.3) is

$$\Delta G_0 = RT\left[X_0 \ln\frac{X_0}{X_e} + (1 - X_0)\ln\frac{(1 - X_0)}{(1 - X_e)}\right] - \Omega(X_0 - X_e)^2$$

where X_0 and X_e are the mole fractions of solute defined in Figure 5.3.

a. Use this equation to estimate the total free energy released when $\alpha' \rightarrow \alpha + \beta$ at 600 K if $X_0 = 0.1$, $X_e = 0.02$, and $\Omega = 0$ (ideal solution) ($R = 8.31$ J mol^{-1} K^{-1}).
b. Estimate the volume fraction of precipitate at equilibrium if β is pure solute ($X_B^\beta = 1$). (Assume the molar volume is constant.)
c. If the alloy is heat treated to produce a precipitate dispersion with a spacing of 50 nm estimate the total α/β interfacial area m^{-3} of alloy. (Assume a simple cubic array.)
d. If $\gamma_{\alpha\beta} = 200$ mJ m^{-2}, what is the total interfacial energy m^{-3} of alloy? mol^{-1} of alloy? ($V_m = 10^{-5}$ m^3).

e. What fraction of the total driving force would remain as interfacial energy in the above case?
f. Repeat c–e for a dispersion of 1 μm spacing.

5.2 Use the methods of Chapter 1 to derive the expression for ΔG_0 in Exercise 5.1.

5.3 In dilute or ideal solutions the driving force for precipitate nucleation (assuming $X_B^\beta = 1$) is given approximately by

$$\Delta G_n = RT \ln \frac{X_0}{X_e} \text{ per mole of precipitate}$$

where X_0 and X_e are the mole fractions of solute defined in Figure 5.3.
a. Evaluate ΔG_n for the precipitate in Exercise 5.1.
b. Assuming homogeneous nucleation, what will be the critical nucleus radius?
c. How does the mean precipitate size in Exercise 5.1c compare with the size of the critical nucleus?

5.4 Derive the expression for ΔG_n in Exercise 5.3. (Use Equation 1.68.)

5.5 a. Calculate θ in Figure 5.6 if $\gamma_{\alpha\beta} = 500$ and $\gamma_{\alpha\alpha} = 600$ mJ m^{-2}.
b. Evaluate the magnitude to the shape factor $S(\theta)$ for this nucleus.

5.6 Imagine the Fe–0.15 wt% C alloy in Figure 5.50 is austenitized above A_3, and then quenched to 800°C where ferrite nucleates and covers the austenite grain boundaries.
a. Draw a composition profile normal to the α/γ interface after partial transformation assuming diffusion-controlled growth.
b. Derive an approximate expression for the thickness of the ferrite slabs as a function of time.
c. Given that D_C^γ (800°C) $= 3 \times 10^{-12}$ m^2 s^{-1} plot the thickness as a function of time.
d. If the austenite grain size is 300 μm extend the above curve to long times. (State any simplifying assumptions you make.)

5.7 Derive Equation 5.34.

5.8 a. By considering short transformation times derive expressions for λ and n in Equation 5.39 for the pearlite transformation when nucleation is restricted to grain corners and all nuclei form at time zero (site saturation). Assume spherical pearlite nodules and a cubic grain structure with a cube side d.
b. Repeat the above for grain-boundary nucleation again assuming site saturation. In this case pearlite grows as grain.

5.9 Draw schematic diagrams to show how growth rate and nucleation rate should vary with temperature for civilian transformations that are induced by an increase in temperature.

5.10 A and B form a regular solution with a positive heat of mixing so that the A–B phase diagram contains a miscibility gap.

a. Starting from Equation 1.39 derive an equation for d^2G/dX_B^2, assuming $G_A = G_B = 0$.
b. Use the above equation to calculate the temperature at the top of the miscibility gap T_c in terms of Ω.
c. Plot the miscibility gap for this system. Hint: the limits of solubility for this simple case are given by $dG/dX_B = 0$.
d. On the same diagram plot the locus of $d^2G/dX_B^2 = 0$, i.e., the chemical spinodal.

5.11 By expressing G as a Taylor series, i.e.,

$$G(X_0 + \Delta X) = G(X_0) + \frac{dG}{dX}(\Delta X) + \frac{d^2G}{dX^2}\frac{(\Delta X)^2}{2} + \cdots$$

show that Equation 5.43 is valid for small values of ΔX.

5.12 How should alloy composition affect the initial wavelength of a spinodally decomposed microstructure at a given temperature?

5.13 a. Account for the location of massive transformations in Table 3.5.
b. Why do massive transformations generally occur at lower temperatures but higher rates than precipitation transformations?

References

1. I.S. Servi and D. Turnbull, *Acta Metall.*, 14: 161, 1966.
2. H.I. Aaroson, C. Laird, and K.K. Kinsman, *Phase Transformations*. American Society for Metals, Metals Park, OH, 1970, Chapter 8.
3. C. Zenzer, *J. Appl. Phys.*, 20: 950, 1949.
4. H.B. Aaron and H.I. Aaronson, *Acta Metall.*, 16: 789, 1968.
5. G.J. Jones and R. Trivedi, *J. Appl. Phys.*, 42: 4299, 1971.
6. C. Laird and H.I. Aaronson, *Acta Metall.*, 17: 505, 1969.
7. P.G. Shewmon, *Transformations in Metals*, McGraw-Hill, New York, 1969, p. 92.
8. For a rigorous derivation of this see J. Cahn and W. Hagel, in V. Zackay and H.I. Aaronson (Eds.), *Decomposition of Austenite by Diffusional Processes*. Interscience, New York, 1962, p. 134.
9. This treatment is based on a series of papers by J. Cahn. See for example *Acta Metall.*, 9: 795, 1961 and *Trans. Metall. Soc. AIME*, 242: 168, 1968.
10. J.W. Martin and R.D. Doherty, *Stability of Microstructure in Metallic Systems*. Cambridge University Press, Cambridge, 1976, Chapter 4.
11. For a treatment of plate-shaped precipitates see M. Ferrante and R.D. Doherty, *Acta Metall.*, 27: 1603, 1979.
12. C. Wagner, *Z. Elektrochem.*, 65: 581, 1961, and I.M. Lifshitz and V.V. Slyozov, *J. Phys. Chem. Solids*, 19: 35, 1961.
13. J.M. Rigsbee and H.I. Aaronson, *Acta Metall.*, 27: 351–363 and 365–376, 1979.
14. J.W. Christian, *The Theory of Transformations in Metals and Alloys*, Pergamon, Oxford, 1981.

15. C. Zener, *J. Appl. Phys.*, 20: 950, 1949.
16. H.I. Aaronson, W.F. Lange III, and G.R. Purdy, *Scripta Mater.*, 51: 931, 2004.
17. S.E. Offerman, N.H. van Dijk, J. Sietsma, S. van der Zwaag, E.M. Lauridsen, L. Margulies, S. Grigull, and H.F. Poulsen, *Scripta Mater.*, 51: 937, 2004.
18. S.E. Offerman, N.H. van Dijk, J. Sietsma, S. Grigull, E.M. Lauridsen, L. Margulies, H.F. Poulsen, M.Th. Rekveldt, and S. van der Zwaag, *Science*, 298: 1003, 2002.
19. S.W. Thompson and P.R. Howell, *Scripta Metall.*, 22: 1775, 1988.
20. D.Z. Li, N.M. Xiao, Y.J. Lan, C.W. Zheng, and Y.Y. Li, *Acta Mater.*, 55: 6234, 2007.
21. M. Militzer, *ISIJ Int.*, 47: 1, 2007.
22. D.A. Porter and J.W. Edington, *Proc. R. Soc., Lond.*, A358: 335–350, 1977.
23. See M. Hillert, in V.F. Zackay and H.I. Aaronson (Eds.), *The Decomposition of Austenite by Diffusional Processes*. Interscience, New York, 1962, p. 197.
24. For a review of the pearlite transformation see M.P. Puls and J.S. Kirkaldy, *Metall. Trans.*, 3: 2777, 1972.
25. For detailed description of bainite microstructures see R.F. Hehemann, *Phase Transformations*. American Society for Metals, Metals Park, OH, 1970, p. 397.
26. H.K.D.H. Bhadeshia, *Bainite in Steels*, 2nd edn., IOM Communications Ltd, London, 2001.
27. H.-S. Fang, Z.-G. Yang, and Y.K. Zheng, *J. Phys. IV*, 5(12): C8521, 1995.
28. Z.-G. Yang, H.-S. Fang, J.J. Wang, and Y.K. Zheng, *J. Mater. Sci. Lett.*, 15: 721, 1996.
29. H.-S. Fang, J.-B. Yang, Z.-G. Yang, and B.-Z. Bai, *Scripta Mater.*, 47: 157, 2002.
30. Z.-G. Yang, C. Zhang, B.-Z. Bai, and H.-S. Fang, *Mater. Lett.*, 48: 292, 2001.
31. Y. Ohmori, H. Ohtsubo, Y.-C. Jung, S. Okaguchi, and H. Ohtani, *Met. Mater. Trans.*, 25A: 1981, 1994.
32. H.I. Aaronson, J.M. Rigsbee, B.C. Muddle, and J.F. Nie, *Scripta Mater.*, 47: 207, 2002.
33. H.K.D.H. Bhadeshia, A rationalisation of shear transformations in steels, *Acta Metall.*, 29: 1117, 1981.
34. M. Hillert, Paradigm shift for bainite, *Scripta Mater.*, 47: 175, 2002.
35. J.M. Oblak and R.F. Hehemann, *Transformations and Hardenability in Steels*. Climax Molybdenum Corporation, Ann Arbor, MI, 1967, pp. 15–30.
36. A. Hultgren, *Trans. ASM*, 39: 915, 1947.
37. H.I. Aaronson, J.M. Rigsbee, B.C. Muddle, and J.F. Nie, Aspects of the surface relief definition of bainite, *Scripta Mater.*, 47: 207, 2002.
38. M. Hillert, The nature of bainite, *ISIJ Int.*, 35: 1134, 1995.
39. G.B. Olson, H.K.D.H. Bhadeshia, and M. Cohen, Coupled diffusional/displacive transformations, *Acta Metall.*, 37: 381, 1989.
40. J. Speer, D.K. Matlock, B.C. De Cooman, and J.G. Schroth, Carbon partitioning into austenite after martensite transformation, *Acta Mater.*, 51: 2611, 2003.
41. J.G. Speer, D.V. Edmonds, F.C. Rizzo, and D.K. Matlock, Partitioning of carbon from supersaturated plates of ferrite, with application to steel processing and fundamentals of the bainite transformation, *Curr. Opin. Solid State Mater. Sci.*, 8: 219, 2004.
42. B.C. Muddle and J.F. Nie, Formation of bainite as a diffusional-displacive phase transformation, *Scripta Mater.*, 47: 187, 2002.
43. A. Saha, G. Ghosh, and G.B. Olson, An assessment of interfacial dissipation effects at reconstructive ferrite–austenite interfaces, *Acta Mater.*, 53: 141, 2005.
44. See for example D.E. Coates, *Metall. Trans.*, 4: 2313, 1973.
45. R.C. Sharma and G.R. Purdy, *Metall. Trans.*, 4: 2303, 1973.
46. B.N.P. Babu, M.S. Bhat, E.R. Parker, and V.F. Zackay, *Metall. Trans.*, 7A: 17, 1976.

47. A. Cottrell, *An Introduction to Metallurgy*, 2nd edn., The Institute of Materials, London, 1975, pp. 374–375.
48. J.S. Ye, T.Y. Hsu, and X. Zuyao, *ISIJ Int.*, 44: 777, 2004.
49. J.S. Ye, H.B. Chang, T.Y. Hsu, and X. Zuyao, *Metall. Mater. Trans. A*, 34: 1259, 2003.
50. T.Y. Hsu, *Curr. Opin. Solid State Mater. Sci.*, 9: 256, 2005.
51. See, for example, H.C. Cotton, Material requirements for offshore structures, in *Rosenhain Centenary Conference*. The Metals Society and Royal Society, 1975, p. 19.
52. From IIW's Doc.IIS/IIW-382-71, Guide to the weldability of C–Mn steels and C–Mn microalloyed steels, 1971.
53. K. Hase, T. Hoshino, and K. Amano, New extremely low carbon bainitic high-strength steel bar having excellent machinability and toughness produced by TPCP technology, *Kawasaki Steel Tech. Rep.*, 47: 35, 2002.
54. R. Valiev, Nanomaterial advantage, *Nature*, 419: 887, 2002.
55. Y. Wang, M. Chen, F. Zhou, and E. Ma, High tensile ductility in a nanostructured metal, *Nature*, 419: 912, 2002.
56. R.Z. Valiev, R.K. Islamgaliev, and I.V. Alexandrov, Bulk nanostructured materials from severe plastic deformation, *Prog. Mater Sci.*, 45: 103, 2000.
57. F.G. Caballero and H.K.D.H. Bhadeshia, Very strong bainite, *Curr. Opin. Solid State Mater. Sci.*, 8: 251, 2004.
58. F.G. Caballero, H.K.D.H. Bhadeshia, K.J.A. Mawella, and D.G. Jones, Very strong low temperature bainite, *Mater. Sci. Tech.-Lond.*, 18: 279, 2002.
59. F.B. Pickering, *Transformation and Hardenability in Steels*. Climax Molybdenum Corporation, Ann Arbor, MI, 1967.
60. V.M. Pivovarov, I.A. Tananko, and A.A. Levechenko, *Phys. Met. Metallogr.*, 33: 116, 1972.
61. B.P.J. Sandvik and H.P. Nevalainen, Structure–property relationships in commercial low-alloy bainitic–austenitic steel with high strength, ductility, and toughness, *Met. Technol.*, 213–220, June 1981.

Further Reading

H.I. Aaronson (Ed.), *Lectures on the Theory of Phase Transformations*, Metallurgical Society of AIME, Warrendale, PA, 1975.

H.K.D.H. Bhadeshia and J.W. Christian, Bainite in steels, *Metall. Trans., A*, 21A: 767, 1990.

J. Burke, *The Kinetics of Phase Transformations in Metals*, Pergamon Press Oxford, 1965.

G.A. Chadwick, *Metallography of Phase Transformations*, Butterworths, London, 1972, Chapters 6 and 7.

J.W. Christian, *The Theory of Transformations in Metals and Alloys*, 2nd edn., Pergamon Press, New York, 1975, Chapters 10–12.

R.D. Doherty, Diffusive phase transformations in the solid state, in R.W. Cahn and P. Haasen (Eds.), *Physical Metallurgy*, North-Holland Physics Pub. Co., New York, 1983.

K.E. Easterling, *Introduction to the Physical Metallurgy of Welding*, 2nd edn., Butterworth-Heinemann, London, 1992, Chapter 14.

M.E. Fine, *Introduction to Phase Transformations in Condensed Systems*, Macmillan, New York, 1964.
R.W.K. Honeycombe, *Steels—Microstructure and Properties*, Edward Arnold, London, 1980.
E. Hornbogen, Physical metallurgy of steels, in R.W. Cahn and P. Haasen (Eds.), *Physical Metallurgy*, North-Holland, 1983, Chapter 16.
A. Kelly and R.B. Nicholson, Precipitation hardening, *Progress in Materials Science*, Vol. 10. Macmillan, New York, 1963, pp. 149–392.
A. Kelly and R.B. Nicholson (Eds.), *Strengthening Methods in Crystals*, Applied Science, London, 1971.
J.W. Martin, *Precipitation Hardening*, Pergamon Press, Oxford, 1968.
I. Polmear, *Light Alloys*, 2nd edn., Edward Arnold, 1989.
K.C. Russell and H.I. Aaroson (Eds.), *Precipitation Processes Solids*, The Metallurgical Society of AIME, Warrendale, PA, 1978.
I. Tamura, C. Ouchi, T. Tanaka, and H. Sekine, *Thermomechanical Processing of High Strength Low Alloy Steels*, Butterworths, London, 1988.
V.F. Zackay and H.I. Aaronson (Eds.), *Decomposition of Austenite by Diffusional Process*, Interscience, New York, 1962.

6

Diffusionless Transformations

One of the most important technological processes is the hardening of steel by quenching. If steel is quenched rapidly enough from the austenitic field, there is insufficient time for eutectoidal diffusion-controlled decomposition processes to occur, and the steel transforms to martensite—or in some cases martensite with a few percent of retained austenite. This transformation is important and best known in connection with certain types of stainless steels, quenched and tempered steels, and ball bearing steels. Important recent developments involving the martensitic transformation in steels include maraging steels (precipitation-hardened martensite), TRIP steels (transformation-induced plastic deformation), ausforming steels (plastically deformed austenite prior to quenching), and dual phase steels (a mixture of ferrite + martensite obtained by quenching from the $\lambda + \alpha$ field).

Because of the technological importance of hardened steel we shall mainly be concerned with this transformation, although martensite is a term used in physical metallurgy to describe any diffusionless transformation product, i.e., any transformation in which from start to completion of the transformation individual atomic movements are less than one interatomic spacing. The regimented manner in which atoms change position in this transformation has led to it being termed ''military,'' in contrast to diffusion-controlled transformations which are termed ''civilian.'' In principle, all metals and alloys can be made to undergo diffusionless transformations provided the cooling rate or heating rate is rapid enough to prevent transformation by an alternative mechanism involving the diffusional movement of atoms. Martensitic transformations can thus occur in many types of metallic and nonmetallic crystals, minerals, and compounds. In the case of martensite steel, the cooling rate is such that the majority of carbon atoms in solution in the fcc γ-Fe remain in solution in the α-Fe phase. Steel martensite is thus simply a supersaturated solid solution of carbon in α-Fe. The way in which this transformation occurs, however, is a complex process and even today the transformation mechanism, at least in steels, is not properly understood. The main purpose of this chapter is to consider some of the characteristics of martensitic transformations including a brief study of their crystallography, and to examine possible theories of how the phase nucleates and grows. We shall then consider the process of tempering steel martensites and finally give some examples of engineering materials based on martensitic transformations.

6.1 Characteristics of Diffusionless Transformations

There have been a number of excellent reviews of martensitic transformations, and the most complete treatments to date have been given by Christian[1] and Nishiyama.[2] The formation of martensite appears from micrographs to be a random process and the way it is observed to develop is illustrated schematically in Figure 6.1a and b. As seen from Figure 6.1a, the

FIGURE 6.1
(a, b) Growth of martensite with increasing cooling below M_s. (c–e) Different martensite morphologies in iron alloys: (c) low C (lath), (d) medium C (plate), and (e) Fe–Ni (plate).

martensitic phase (designated α') is often in the shape of a lens and spans initially an entire grain diameter. The density of plates does not appear to be a function of the grain size of the austenite. For example, it is observed to form randomly throughout a sample with a plate density which appears to be independent of grain size. Where the plates intersect the surface of a polished specimen and bring about an elastic deformation, or tilting of the surface as shown in Figure 6.2. Observations have shown that, at least macroscopically, the transformed regions appear coherent with the surrounding austenite. This means that intersection of the lenses with the surface of the specimen does not result in any discontinuity. Thus, lines of a polished surface are displaced, as illustrated in Figure 6.2a, but remain continuous after the transformation. It has been shown that a fully grown plate spanning a whole grain may form within $\sim 10^{-7}$ s which means that the α'/γ interface reaches almost the speed of sound in the solid. Martensite is thus able to grow independently of thermal activation, although some Fe–Ni alloys do exhibit isothermal growth characteristics. This great speed of formation makes martensite nucleation and growth a difficult process to study experimentally.

It is seen in Figure 6.1a and b that the volume fraction of martensite increases by the systematic transformation of the austenite remaining between the plates that have already formed. The first plates form at the M_s (martensite start) temperature. This temperature is associated with a certain driving force for the diffusionless transformation of λ into α' as shown in Figure 6.3a and b. In low-carbon steels, $M_s \approx 500°C$ (Figure 6.3c), but increasing C contents progressively decrease the M_s temperature as shown. The M_f temperature (martensite finish) corresponds to that temperature below which further

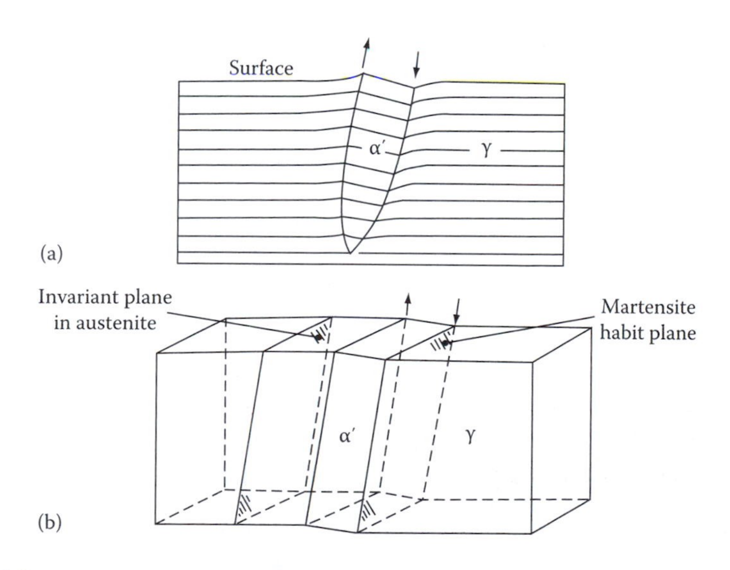

FIGURE 6.2
Illustrating how a martensite plate remains macroscopically coherent with the surrounding austenite and even the surface it intersects.

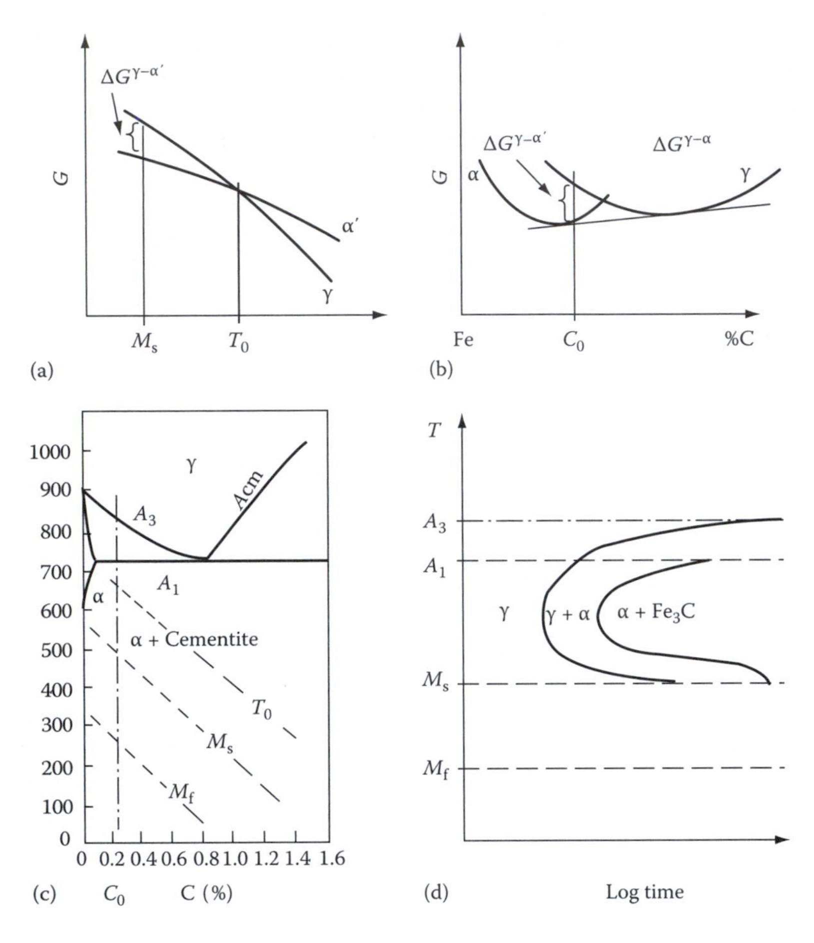

FIGURE 6.3
Various ways of showing the martensite transformation. (a) Free energy–temperature diagram for austensite and martensite of fixed carbon concentration (C_0 in (b)). (b) Free energy–composition diagram for the austensite and martensite phases at the M_s temperature. (c) Iron–carbon phase diagram with T_0 as defined in (a), M_s and M_f superimposed. (d) M_s and M_f in relation to the TTT diagram for alloy C_0 in (c).

cooling does not increase the amount of martensite. In practice, the M_f may not correspond to 100% martensite, and some retained austenite can be left even below M_f. The retention of austenite in such cases may be due to the high elastic stresses between the last martensite plates to form, which tend to suppress further growth or thickening of existing plates. As much as 10%–15% retained austenite is a common feature of especially the higher C content alloys such as those used for ball bearing steels. Figure 6.3d shows a TTT diagram used for estimating the speed of quench necessary to obtain a given microstructure. This diagram can be plotted and used in technological applications for any one particular alloy, and that illustrated for example applies to only one carbon content, as shown.

TABLE 6.1

Comparisons of Calorimetric Measurements of Enthalpy and Undercooling in Some Martensitic Alloys

Alloy	$\Delta H^{\gamma\to\alpha'}$ (J mol^{-1})	$T_0 - M_s$(K)	$-\Delta G^{\gamma\to\alpha'}$ (J mol^{-1})
Ti–Ni	1550	20	92
Cu–Al	170–270	20–60	19.3 ± 7.6
Au–Cd	290	10	11.8
Fe–Ni 28%	1930	140	840
Fe–C			1260
Fe–Pt 24% Ordered	340	10	17
Fe–Pt Disordered	2390	~150	~1260

Source: From Guénin, G., PhD thesis, Polytechnical Institute of Lyon, 1979.

By analogy with Equation 1.17, the driving force for the nucleation of martensite at the M_s temperature should be given by

$$\Delta G^{\gamma\to\alpha'} = \Delta H^{\gamma\to\alpha'} \frac{(T_0 - M_s)}{T_0} \tag{6.1}$$

where T_0 and M_s are defined in Figure 6.3a. Some calorimetric measurements of ΔH are given in Table 6.1 for a number of alloys exhibiting martensitic transformations, together with the corresponding amounts of undercooling and free energy changes. Note especially in this table the large differences in $\Delta G^{\gamma\to\alpha'}$ between ordered and disordered alloys, the ordered alloys exhibiting a relatively small undercooling. We shall now examine the atomic structures of steel austenite and martensite in more detail.

6.1.1 Solid Solution of Carbon in Iron

In an fcc (or hcp) lattice structure, there are two possible positions for accommodating interstitial atoms as shown in Figure 6.4. These are the tetrahedral site which is surrounded by four atoms and the octahedral site

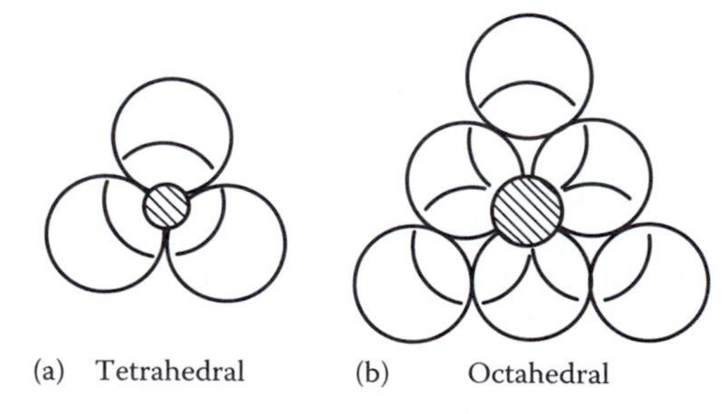

FIGURE 6.4
Illustrating possible sites for interstitial atoms in the fcc or hcp lattices.

which has six nearest neighbors. The sizes of the largest atoms that can be accommodated in these holes without distorting the surrounding matrix atoms can be calculated by assuming that the atoms are close-packed hard spheres. Such a calculation gives

$$\begin{aligned}\text{Tetrahedral interstice } d_4 &= 0.225D\\ \text{Octahedral interstice } d_6 &= 0.414D\end{aligned} \tag{6.2a}$$

where
D is the diameter of the parent atoms
d_4 and d_6 are the maximum interstitial diameters of the two types of site

In the case of γ-iron, at ambient temperature $D = 2.52$ Å, so that interstitial atoms of diameter 0.568 or 1.044 Å can be contained in tetrahedral and octahedral interstices without distorting the lattice. However, the diameter of a carbon atom is 1.54 Å. This means that considerable distortion of the austenite lattice must occur to contain carbon atoms in solution and that the octahedral interstices should be the most favorable.

The possible positions of interstitials in the bcc lattice are shown in Figure 6.5a. It is seen that there are three possible octahedral positions $\left(\frac{1}{2}[100], \frac{1}{2}[010], \frac{1}{2}[001]\right)$, and six possible tetrahedral spaces for each unit cell. In this case, the maximum sizes of interstitials that can be accommodated without distorting the lattice are as follows:

$$\begin{aligned}d_4 &= 0.291D\\ d_6 &= 0.155D\end{aligned} \tag{6.2b}$$

The interesting feature of the bcc lattice is that although there is more free space than the close-packed lattices, the larger number of possible interstitial positions means that the space available per interstitial is less than for the fcc structure (compare Equations 6.2a and b). In spite of the fact that $d_6 < d_4$, measurements of carbon and nitrogen in solution in iron show that these interstitials in fact prefer to occupy the octahedral positions in the bcc lattice. This causes considerable distortion to the bcc lattice as illustrated in Figure 6.5b. It is conjectured that the bcc lattice is weaker in the $\langle 100 \rangle$ directions due to the lower number of near and next nearest neighbors compared to the tetrahedral interstitial position (see, e.g., Cottrell[3]). The estimated atomic diameters of pure carbon and nitrogen are 1.54 and 1.44 Å, respectively, although these values are very approximate. It should also be remembered that in a given steel relatively few $\langle \frac{1}{2}00 \rangle$ sites are occupied. Nevertheless, the martensitic Fe–C lattice is distorted to a bct structure as shown in Figure 6.5c. These measurements, made by x-ray diffraction at −100°C to avoid carbon diffusion, show that the c/a ratio of the bct lattice is

$$c/a = 1.005 + 0.045(\text{wt\% C}) \tag{6.3}$$

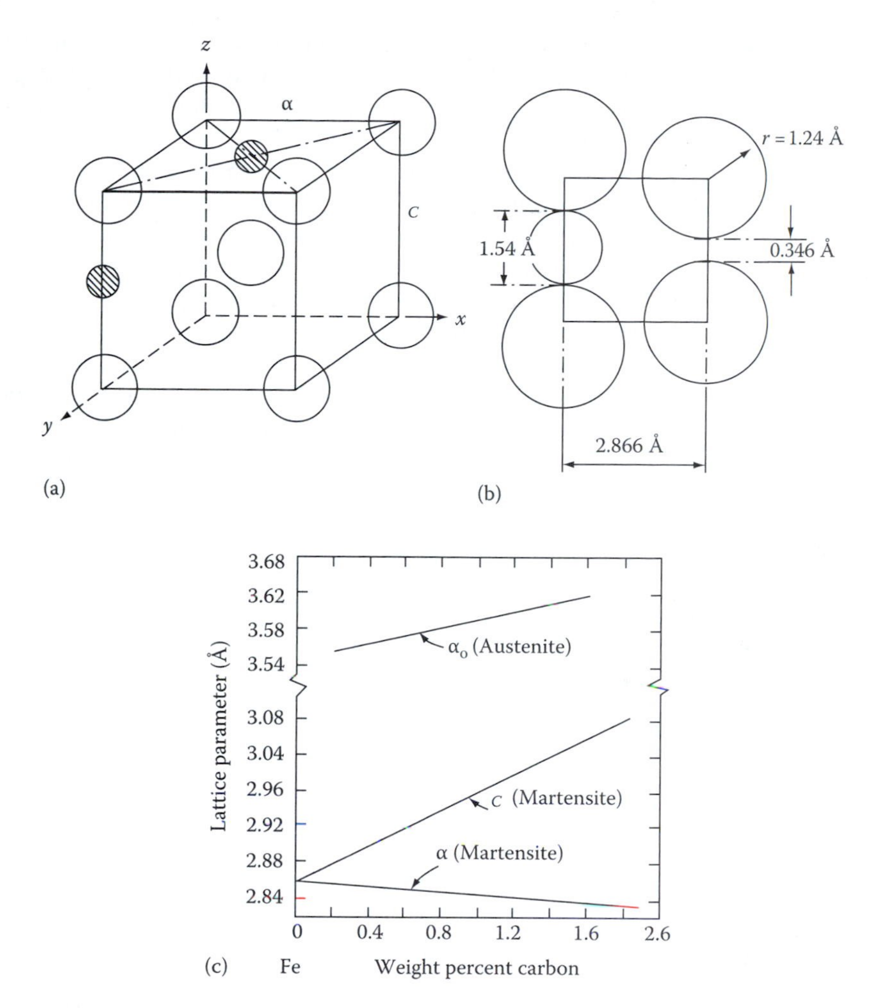

FIGURE 6.5
Illustrating (a) possible sites for interstitial atoms in bcc lattice, and (b) the large distortion necessary to accommodate a carbon atom (1.54 Å diameter) compared with the space available (0.346 Å). (c) Variation of a and c as a function of carbon content. (After Roberts, C.S., *Trans. AIME*, 191, 203, 1953.)

As seen by these results, the distortion of the lattice in one direction (z) causes a contraction in the two directions normal to $z(x, y)$. In fact, these measurements suggest a certain long-range order in the distribution of the carbon interstitials.

6.2 Martensite Crystallography[4]

A feature of the microstructures shown in Figure 6.1 is the obvious crystallographic dependence of martensite plate information. Within a given grain, all the plates grow in a limited number of orientations. In the case of iron alloys, for example, the orientation variants and even plate morphology chosen turn out to be dependent upon alloy content, particularly carbon or nickel, as illustrated in Table 6.2.

The irrational nature of the growth planes of high-carbon or high-nickel martensites has been the subject of much discussion in the literature for the following reason: if martensite is able to grow at speeds approaching the speed of sound, then some sort of highly mobile dislocation interface is required. The problem is then to explain the high mobility of an interface moving on austenite planes not always associated with dislocation glide. Yet another is that the growth or habit plane of martensite is observed to be macroscopically undistorted, i.e., the habit plane is a plane which is common to both the austenite and martensite in which all directions and angular separations in the plane are unchanged during the transformation. That this is so can be reasoned in conjunction with Figure 6.2. The absence of

TABLE 6.2

Orientation Relationships and Habit Planes of Some Martensites

Alloy	Transformation	Orientation Relation	Habit Plane	Defect Structure
Fe–(0–0.4 wt% C)	fcc → bct	$(111)_\gamma \,\|\|\, (011)_{\alpha'}$ $[10\bar{1}]_\gamma \,\|\|\, [\bar{1}1\bar{1}]_{\alpha'}$	$\{111\}_\gamma$	Needles or laths, high dislocation density
Fe–(0.5–1.4 wt% C)	fcc → bct	$(111)_\gamma \,\|\|\, (011)_{\alpha'}$ $[10\bar{1}]_\gamma \,\|\|\, [\bar{1}1\bar{1}]_{\alpha'}$	$\{225\}_\gamma$	Mixed twins and dislocations
Fe–(1.5–1.8 wt% C)	fcc → bct	$(111)_\gamma \,\|\|\, (011)_{\alpha'}$ $[10\bar{1}]_\gamma \,\|\|\, [\bar{1}1\bar{1}]_{\alpha'}$	$\{259\}_\gamma$	Mainly twinned
Fe–(27–34 wt% Ni)	fcc → bct	$(111)_\gamma \,\|\|\, (011)_{\alpha'}$ $[10\bar{1}]_\gamma \,\|\|\, [111]_{\alpha'}$	$\{259\}_\gamma$	Mainly twinned
Fe–(11%–29% Ni)–(0.4%–1.2% C)	fcc → bct	$(111)_\gamma \,\|\|\, (011)_{\alpha'}$ $[10\bar{1}]_\gamma \,\|\|\, [111]_{\alpha'}$	$\{259\}_\gamma$	Mainly twinned
Fe–(7%–10% Al)–2% C	fcc → bct	$(111)_\gamma \,\|\|\, (011)_{\alpha'}$ $[10\bar{1}]_\gamma \,\|\|\, [111]_{\alpha'}$	$\{3\ 10\ 15\}_\gamma$	Twinned
Fe–(2.8%–8% Cr)–(1.1%–1.5% C)	fcc → bct	Probably as for Fe–C alloys	$\{225\}_\gamma$	Mixed twins and dislocations
Fe–4.5 wt% Cu	fcc → bcc	As for Fe–C alloys	$\{112\}_\gamma$	Laths, dislocations
Co	fcc → hcp	$(111)_\gamma \,\|\|\, (0001)_{\alpha'}$ $[11\bar{2}]_\gamma \,\|\|\, [1\bar{1}00]_{\alpha'}$ $[1\bar{1}0] \,\|\|\, [11\bar{2}0]_{\alpha'}$	$\{111\}_\gamma$	Stacking faults
Ti	bcc → hcp	$(110)_\gamma \,\|\|\, (0001)_{\alpha'}$	$\{89, 12\}_\gamma$, $\{133\}_\gamma$	Mixed twins and dislocations
Zr	bcc → hcp	$(110)_\gamma \,\|\|\, (0001)_{\alpha'}$	$\{569\}_\gamma$, $\{145\}_\gamma$	—
Li	bcc → hcp	$(110)_\gamma \,\|\|\, (0001)_{\alpha'}$	$\{441\}$	—

Source: Data mainly from Nishiyama, Z., *Martensitic Transformation*, Academic Press, New York, 1978, pp. 115–123.

plastic deformation in the form of a discontinuity surface shows that the shape strain does not cause any significant rotation of the habit plane. If the habit plane had been rotated, plastic deformation would be necessary to maintain coherence between the martensite and parent austenite and this would have resulted in additional displacements of the surface, or of the lines traversing the plate. In order that the habit plane is left undistorted, the martensitic transformation appears to occur by a homogeneous shear parallel to the habit plane (see Figure 6.2). Since the $\gamma \rightarrow \alpha'$ transformation is also associated with ~4% expansion, this implies in turn that the dilatation in question must take place normal to the habit plane, i.e., normal to the lens. However, some homogeneous dilatation of the habit plane may be necessary.

The question now arises: can bct martensite lattice structure be generated by simple shear parallel to the habit plane, together with a small dilatation normal to the plane? In order to answer this question adequately we must consider the crystallography of the $\gamma \rightarrow \alpha'$ transformation in more detail.

It has been stated that the habit plane of a martensite plate remains undistorted following the transformation. An analogous situation is found in twinning as illustrated in Figure 6.6a and b. It is convenient to consider the $(111)_\gamma$ $\langle 112 \rangle_\gamma$ twinning reaction illustrated in Figure 6.6a in terms of the homogeneous shear of a sphere, Figure 6.6b. In the shearing plane *K*, the lattice is undistorted, i.e., it is invariant. Let us assume first that the equivalent macroscopic shape change in the formation of a martensite plate is a twinning shear occurring parallel to the habit (or twinning) plane, plus a simple uniaxial tensile dilatation perpendicular to the habit plane. A strain of this type is termed: an invariant plane strain, because a shear parallel to the habit plane, or an extension or contraction perpendicular to it, cannot change the positions or magnitude of vectors lying in the plane. We shall now try to answer the question of whether the fcc lattice can be homogeneously deformed to generate the bct structure.

6.2.1 Bain Model of the fcc → bct Transformation

In 1924, Bain[5] demonstrated how the bct lattice could be obtained from the fcc structure with the minimum of atomic movement, and the minimum of strain to the parent lattice. To illustrate this we shall use the convention that x, y, z and x', y', z' represent the original and final axes of the fcc and bcc unit cells as illustrated in Figure 6.7. As shown by this figure, an elongated unit cell of the bcc structure can be drawn within two fcc cells. Transformation to a bcc unit cell is achieved by contracting the cell by 20% in the z-direction and expanding the cell by 12% along the x- and y-axes. In the case of steels, the carbon atoms fit into z'-axis of the bcc cell at $\frac{1}{2}\langle 100 \rangle$ positions causing the lattice to elongate in this direction. In 1 atomic % C steel, for example, carbon occupies one position along the z'-axis for every 50 iron unit cells. The positions occupied by the carbon atoms in the bct structure do not exactly correspond to the equivalent octahedral positions in the parent fcc structure, and it is assumed that small shuffles of the C atoms must take place during the transformation.

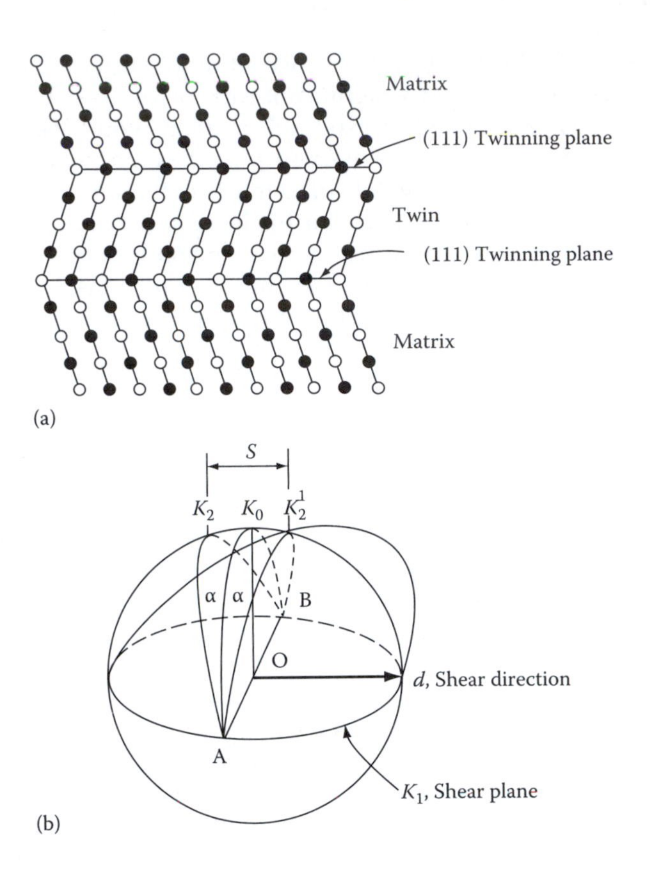

FIGURE 6.6
(a) Showing the twinning of an fcc structure. Black and while circles represent atoms on different levels. (From Reed-Hill, R.E., *Physical Metallurgy Principles*, 2nd edn., Van Nostrand, 1973. With permission.) (b) Graphical representation of a twinning shear occurring on a plane K_1 in a direction *d*. (From Wayman, C.M., *Introduction to the Crystallography of Martensite Transformations*, MacMillan, New York, 1964. With permission.)

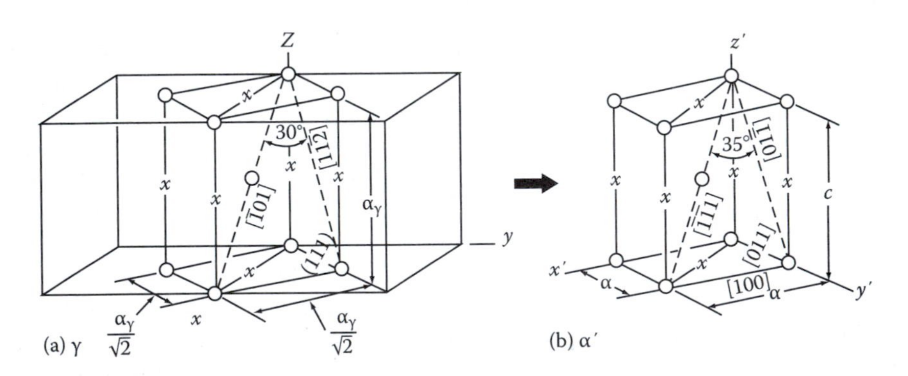

FIGURE 6.7
Bain correspondence for the $\alpha \rightarrow \alpha'$ transformation. Possible interstitial sites for carbon are shown by crosses. To obtain α' the γ unit cell is contracted about 20% on the *c*-axis and expanded about 12% on the α-axis.

It is an interesting fact that the Bain deformation involves the absolute minimum of atomic movements in generating the bcc from the fcc lattice. Examination of Figure 6.7 shows that the Bain deformation results in the following correspondence of crystal planes and directions:

$$(111)_\gamma \rightarrow (011)_{\alpha'}$$
$$[\bar{1}01]_\gamma \rightarrow [\bar{1}\,\bar{1}1]_{\alpha'}$$
$$[1\bar{1}0]_\gamma \rightarrow [100]_{\alpha'}$$
$$[11\bar{2}]_\gamma \rightarrow [01\bar{1}]_{\alpha'}$$

Experimental observations of orientation relationship between austenite and martensite show that $\{111\}_\gamma$ planes are approximately parallel to $\{011\}_{\alpha'}$ planes, and that the relative directions can vary between $\langle\bar{1}01\rangle_\gamma \parallel \langle 1\bar{1}1\rangle_{\alpha'}$ (the Kurd-jumov–Sachs relation) and $\langle 1\bar{1}0\rangle_\gamma \parallel \langle 101\rangle_{\alpha'}$ (the Nishiyama–Wasserman relation). These two orientations differ by $\sim 5°$ about $[111]_\gamma$.

By using the sphere → ellipsoid transformation applied earlier to demonstrate the twinning shear (Figure 6.6) we can now test whether the Bain deformation also represents a pure deformation in which there is an undeformed (invariant) plane. If a sphere of unit radius represents the fcc structure then after the Bain distortion it will be an ellipsoid of revolution with two axes (x' and y') expanded by 12% and the third axis (z') contracted by 20%. The $x'z'$ section through the sphere before and after distortion is shown in Figure 6.8. In this plane, the only vectors that are not shortened or elongated by the Bain distortion are OA or O′A′. However, in order to find a plane in the fcc structure that is not distorted by the transformation requires that the vector OY′ (perpendicular to the diagram) must also be undistorted. This is clearly not true and therefore the Bain transformation does not fulfill the requirements of bringing about a transformation with an undistorted plane.

Hence, the key to the crystallographic theory of martensitic transformations is to postulate an additional distortion which, in terms of Figure 6.8, reduces the extension of y' to zero (in fact a slight rotation, θ, of the AO plane should also be made as shown in the figure). This second deformation can be

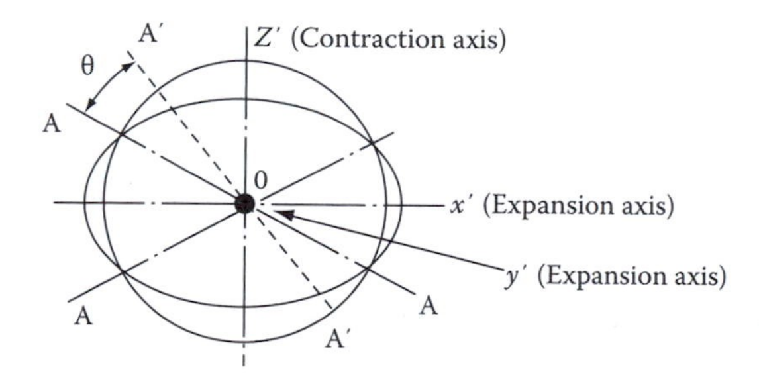

FIGURE 6.8
Bain deformation is here simulated by the pure deformation in compressing a sphere elastically to the shape of an oblate ellipsoid. As in the Bain deformation, this transformation involves two expansion axes and one contraction axis.

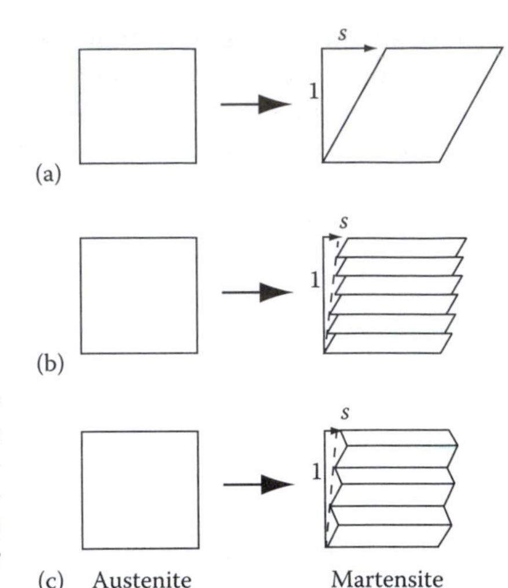

FIGURE 6.9
Schematic illustration of how dislocation glide or twinning of the martensite can compensate for a pure lattice deformation such as a Bain deformation and thereby reduce the strain of the surrounding austenite. The transformation shear (s) is defined. Note how s can be reduced by slip or twinning.

in the form of dislocation slip or twinning as illustrated in Figure 6.9. Applying the twinning analogy to the Bain model, we can see that an internally twinned martensite plate can form by having alternate regions in the austenite undergo the Bain strain along different contraction axes such that the net distortions are compensated. By also adjusting the width of the individual twins, the habit plane of the plate can even be made to adopt any desired orientation. These features of twinned martensite plates are illustrated in Figure 6.10. In this figure, ϕ defines the angle between some reference plane in the austenite and the martensite habit plane. It is seen that ϕ is a function of twin widths I, II (see, e.g., Figure 6.9c). On this basis, the habit plane of the martensite plate can be defined as a plane in the

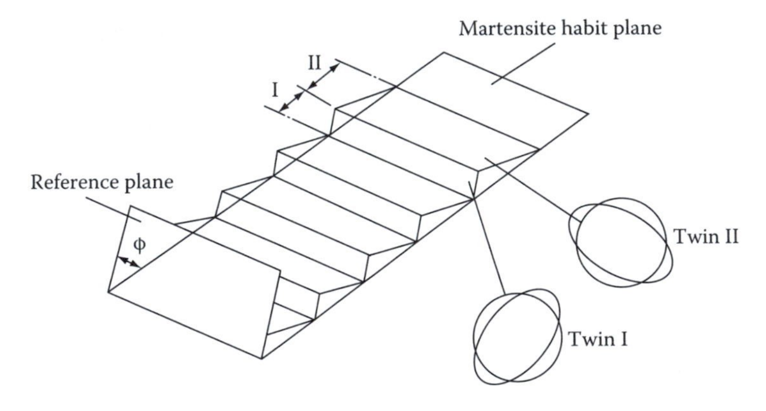

FIGURE 6.10
Twins in martensite may be self-accommodating and reduce energy by having alternate regions of the austenite undergo the Bain strain along different axes.

austenite which undergoes no net (macroscopic) distortion. By net distortion, it is meant that the distortion when averaged over many twins is zero. There will of course be local regions of strain energy associated with the α'/γ interface of the twins at the edge of the plate. However, if the plate is very thin (a few atomic spacings) this strain can be relatively small.

In the crystallographic theory, it is assumed that slip or twinning occurs on suitable $\langle 111\rangle\{11\bar{2}\}_\alpha$ systems, corresponding to equivalent $\langle 1\bar{1}0\rangle\{110\}_\gamma$ planes in the austenite. Since the $\{112\}_\alpha\langle 111\rangle_\alpha$ system is that commonly adopted for bcc slip or twinning, the physical requirements of the theory are satisfied.

6.2.2 Comparison of Crystallographic Theory with Experimental Results

Some plots of experimental measurements typical of habit planes in steel martensites are shown in Figure 6.11. These results indicate that there is a fairly wide scatter in experimental measurements for a given type of steel, and that alloying additions can have a marked effect on the habit plane. It appears that on reaching a critical carbon content, martensite in steel changes its habit plane, these transitions being approximately $\{111\} \rightarrow \{225\} \rightarrow \{259\}_\gamma$ with increasing C content (there is overlap of these transitions in practice). As a general rule, the {111} martensites are associated with a high dislocation density lath morphology, or consist of bundles of needles lying on $\{111\}_\gamma$ planes, while the $\{225\}_\gamma$ and $\{259\}_\gamma$ martensites have a mainly twinned plate or lens morphology. However, any exact morphological description of martensite is not possible since, after thickening and growth, the shapes of the martensites are often quite irregular. Twinning is more predominant at high carbon or nickel contents and is virtually complete for $\{259\}_\gamma$ martensites. In stainless steel, the habit plane is thought to be near $\{11\bar{2}\}$, which has been explained in terms of a lattice invariant shear on $\{101\}\langle 10\bar{1}\rangle_{\alpha'}$ corresponding to $\{111\}\langle 1\bar{2}1\rangle_\gamma$. Transmission electron micrographs of lath and twinned steel martensites are shown in Figure 6.12, which also illustrates the classical definition of lath and plate morphologies.

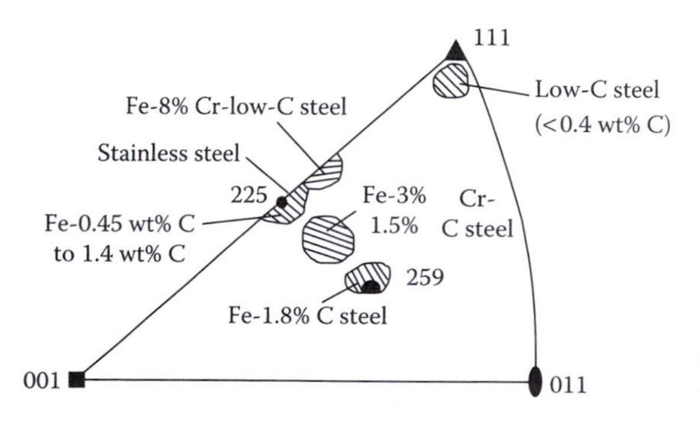

FIGURE 6.11
Martensite habit planes in various types of steel.

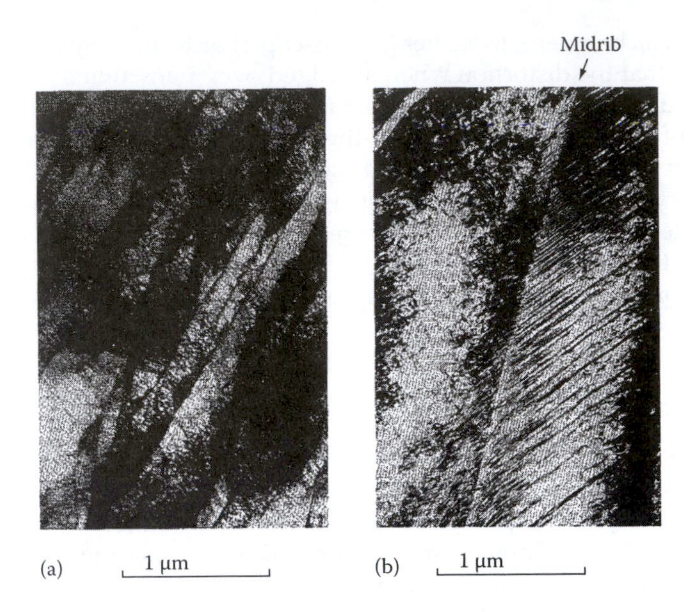

FIGURE 6.12
Transmission electron micrographs of (a) lath martensite and (b) twinned martensite. Note the midrib in the twinned martensite, which is thought to be the first part of the plate to grow.

The notable success of the crystallographic theory is that it can predict the fine substructure (twinning or slip) of martensite before it was actually observed in the electron microscope. For a typical (high-carbon) steel for example, to achieve a $\{259\}_\gamma$ habit plane, twins having a spacing of only 8–10 atomic planes, or ~3 nm, are predicted. Twin thicknesses of this order of magnitude are observed in electron micrographs of high-carbon martensites. On the other hand, it is usually difficult to predict exactly the habit plane of a given alloy on the basis of known lattice parameters, dilatations, etc., and apart from a few cases, the theory is mainly of qualitative interest. The theory is essentially phenomenological, and should not be used to interpret the kinetics of the transformation. Attempts at combining the crystallographical aspects of the transformation with the kinetics have, however, recently been made and will be discussed later.

6.3 Theories of Martensite Nucleation

A single plate of martensite in steel grows in 10^{-5} to 10^{-7} s to its full size, at velocities approaching the speed of sound. Using resistivity changes to monitor the growth of individual plates of martensite in, e.g., Fe–Ni alloys, speeds of 800–1100 m s^{-1} have been measured (Nishiyama[6]). The nucleation

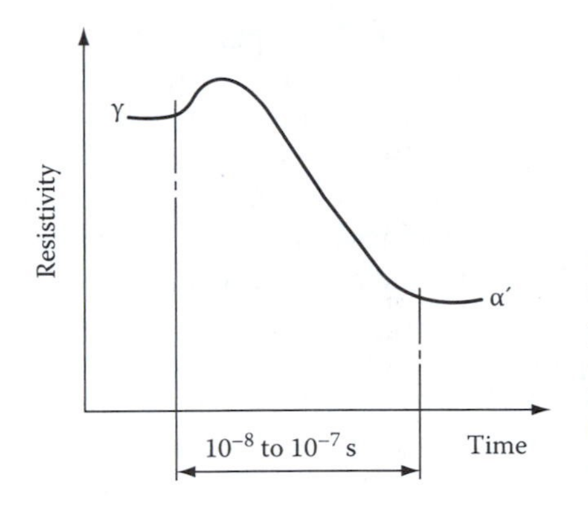

FIGURE 6.13
Resistivity changes during the growth of single plates of martensite across a grain in a Fe–Ni alloy. From this it can be calculated that the velocity of growth is about 1000 m s^{-1}. (After Bunshah, R. and Mehl, R.F., *Trans. AIME*, 197, 1251, 1953.)

event is thus very important in martensitic transformations because of its likely influence on the final form of the full-grown plate. This implies that the nucleation of martensite influences the strength and toughness of martensitic steels, since for a given austenite grain size, if the number of nuclei is large, then the final grain size of the martensite will be finer and hence the steel may be stronger.

Because of the great speed of growth of martensite, it is extremely difficult to study this transformation experimentally. An example of Bunshah and Mehl's resistivity measurements[7] is shown schematically in Figure 6.13, indicating that α' gives a lower resistivity than γ. The small initial increase in resistivity is explained in terms of the initial strain of the austenite lattice by the martensite nucleus. This suggests in turn that the initial nucleus should be coherent with the parent austenite. This factor could be an important starting point when considering how nucleation occurs.

6.3.1 Formation of Coherent Nuclei of Martensite

The total increase in Gibbs free energy associated with the formation of a fully coherent inclusion of martensite in a matrix of austenite can be expressed as

$$\Delta G = A\gamma + V\Delta G_s - V\Delta G_v \tag{6.5}$$

where

γ is the interfacial free energy
ΔG_s is the strain energy
ΔG_v is the volume free energy release
V is the volume of the nucleus
A is the surface area

This expression does not account for possible additional energies that may be available due, e.g., to thermal stresses during cooling, externally applied stresses, and stresses produced ahead of rapidly growing plates. As with other nucleation events there is a balance between surface and elastic energy on the one side, and chemical (volume) free energy on the other. However, in martensitic transformations the strain energy of the coherent nucleus is much more important than the surface energy, since the shear component of the pure Bain strain is as high as $s \simeq 0.32$ which produces large strains in the surrounding austenite. On the other hand, the interfacial (surface) energy of a fully coherent nucleus is relatively small.

Consider the nucleation of a thin ellipsoidal nucleus, with radius a, semi-thickness c, and volume V, as illustrated in Figure 6.14. In agreement with experimental observations, we assume that nucleation does not necessarily occur at grain boundaries. We also assume to begin with that nucleation occurs homogeneously without the aid of any other types of lattice defects. As seen from Figure 6.14, the nucleus forms by a simple shear, s, parallel to the plane of the disk, and complete coherency is maintained at the interface. On this basis, Equation 6.5 can be written as

$$\Delta G = 2\pi a^2\gamma + 2\mu V(s/2)^2 \frac{2(2-\nu)}{(1-\nu)} \pi c/a - \frac{4}{3}\pi a^2 c \cdot \Delta G_v \tag{6.6}$$

where

γ is the coherent interfacial energy of the coherent nucleus
ν is the Poissons ratio of the austenite
μ is the shear modulus of the austenite

If $\nu = (1/3)$, Equation 6.6 can be simplified to

$$\Delta G = \underbrace{2\pi a^2\gamma}_{\text{Surface}} + \frac{16\pi}{3}\underbrace{(s/2)^2\ \mu a c^2}_{\text{Elastic}} - \frac{4\pi}{3}\underbrace{a^2 c \cdot \Delta G_{\text{V}}}_{\text{Volume}} \tag{6.7}$$

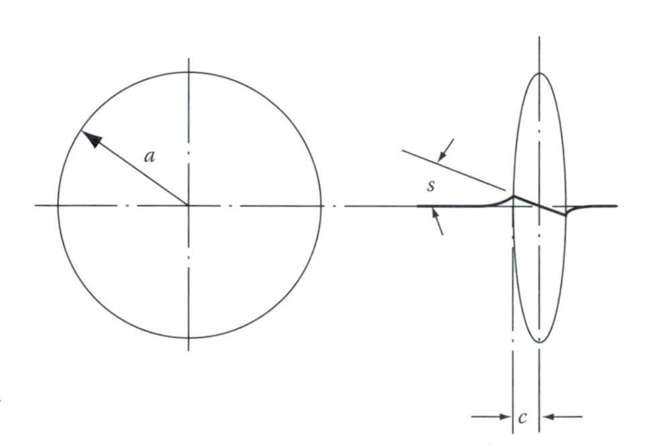

FIGURE 6.14
Schematic representation of a martensite nucleus.

In this expression, the negative term ΔG_v is the free energy difference at the M_s temperature between the austenite and martensite phases and is defined in Figure 6.3a. The middle term referring to the strain energy is due to the shear component of strain only and neglects the small additional strain due to the dilatation which is assumed to occur normal to the disk. As pointed out by Christian,[1] the most favorable nucleation path is given by the condition that the habit plane is exactly an invariant plane of the shape transformation (to reduce coherency energy) although this is not necessarily realized in practice. The minimum free energy barrier to nucleation is now found by differentiating Equation 6.7 with respect to a and c, and by subsequent substitution we obtain

$$\Delta G^* = \frac{512}{3} \cdot \frac{\gamma^3}{(\Delta G_v)^4} \cdot \left(\frac{s}{2}\right)^4 \mu^2 \pi \text{ J nucleus}^{-1} \tag{6.8}$$

This expression is thus the nucleation barrier to be overcome by thermal fluctuations of atoms if classical homogeneous nucleation is assumed. It is seen that the energy barrier is extremely sensitive to the values chosen for γ, ΔG_v, and s. The critical nucleus size (c^* and a^*) is also highly dependent upon these parameters. It can be shown that:

$$c^* = \frac{2\gamma}{\Delta G_v} \tag{6.9}$$

and

$$a^* = \frac{16\gamma\mu(s/2)^2}{(\Delta G_v)^2} \tag{6.10}$$

Typically $\Delta G_v = 174$ MJ m^{-3} for steel. s varies according to whether the net shear of a whole plate (e.g., as measured from surface markings) or the shear of a fully coherent plate (as measured from lattice fringe micrographs) is considered. Here, we shall assume a value of 0.2 which is the macroscopic shear strain in steel. We can only guess at the surface energy of a fully coherent nucleus, but a value of ~20 mJ m^{-2} seems reasonable. These values gives $c^*/a^* \simeq 1/40$, and $\Delta G^* \simeq 20$ eV, which in fact is too high for thermal fluctuations alone to overcome (at 700 K, $kT = 0.06$ eV). Indeed, there is plenty of experimental evidence to show that martensite nucleation is in fact a heterogeneous process. Perhaps the most convincing evidence of heterogeneous nucleation is given by small particle experiments.[8–10]

In these experiments, small single-crystal spheres of Fe–Ni of a size range from submicron to a fraction of a millimeter were cooled to various temperatures below the M_s, and then studied metallographically. These experiments showed that:

1. Not all particles transformed even if cooled down to +4 K, i.e., ~300°C below the M_s of the bulk material; this appears to completely rule out homogeneous nucleation, since this should always occur at a certain undercooling. Indeed, the maximum undercooling for certain alloys reached as much as 600°C–700°C.
2. The average number of nuclei (based on plate counts) was of the order of 10^4 per mm^3; this is less than expected for purely homogeneous nucleation.
3. The number of nuclei increases substantially with increasing supercooling prior to transformation; on the other hand, the average number of nuclei is largely independent of grain size, or even whether the particles (of a given size) are single crystals or polycrystalline.
4. The surface does not appear to be a preferred site for nucleation.

On the basis of (3) and (4), it is thought that since surfaces and grain boundaries are not significantly contributing to nucleation, then the transformation is being initiated at other defects within the crystal. The most likely types of defect which could produce the observed density of nuclei are individual dislocations, since an annealed crystal typically contains $\sim 10^5$ or more dislocations per mm^2.

6.3.2 Role of Dislocations in Martensite Nucleation

A number of researchers have considered the possible ways in which dislocations may contribute to martensite nucleation. It is instructive to consider some of these ideas and see how they can fit in with the various features of martensitic transformations already discussed.

Zener[11] demonstrated how the movement of $\langle 112 \rangle_\gamma$ partial dislocations during twinning could generate a thin bcc region of lattice from an fcc one and as illustrated in Figure 6.15. In this figure, the different layers of the close-packed planes of the fcc structure are denoted by different symbols and numbered 1, 2, 3, from bottom to top layer. As indicated, in the fcc lattice the normal twinning vector is $\bar{b}_1$, which can be formed by the dissociation of an $\frac{a}{2}\langle 110 \rangle$ dislocation into two partials:

$$\bar{b} = \bar{b}_1 + \bar{b}_2 \tag{6.11}$$

i.e.,

$$\frac{a}{2}[110] = \frac{a}{6}[\bar{2}11] + \frac{a}{6}[\bar{1}2\bar{1}]$$

In order to generate the bcc structure it requires that all the triangular (Level 3) atoms jump forward by $\frac{1}{2}\bar{b}_1 = \frac{a}{12}[\bar{2}11]$. In fact, the lattice produced

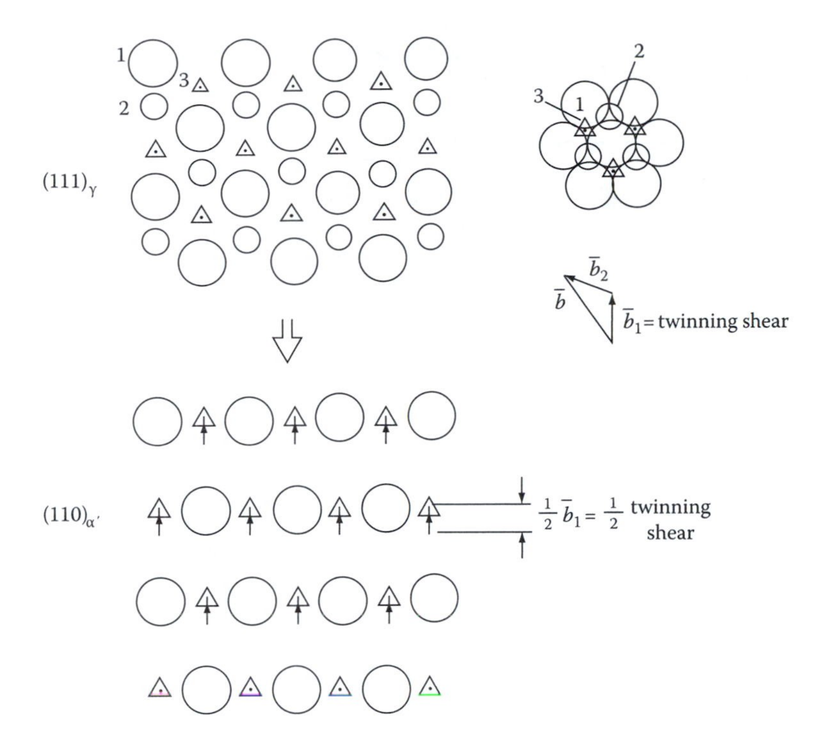

FIGURE 6.15
Zener's model of the generation of two-atom-thick martensite by a half-twinning shear (some additional minor adjustments are also needed).

is not quite the bcc one after this shear, but requires an additional dilatation to bring about the correct lattice spacings. As pointed out by Christian,[1] however, this reaction produces a bcc lattice only of two atom layers thick. Recent electron microscopy work by Brooks et al.[12] indicated that thicker nuclei could form by this mechanism at dislocation pileups, where the partial dislocations are forced closer together thereby reducing the slip vectors such that the core structures correspond to a bcc stacking. Pileups on nearby planes can hence interact such as to thicken the pseudo-bcc region.

An alternative suggestion was made earlier by Venables[13] also in connection with the formation of martensite in stainless steels, i.e., in the case of alloys of low stacking fault energy. Venables proposed that α' forms via an intermediate (hcp) phase which he termed epsilon martensite, thus

$$\gamma \rightarrow \varepsilon' \rightarrow \alpha' \tag{6.12}$$

Using the same atomic symbols as before, Venables' transformation mechanism is shown in Figure 6.16. The ε'-martensite structure thus thickens by inhomogeneous half-twinning shears on every other $\{111\}_\gamma$ plane. Such faulted regions have been observed to form in conjunction with martensite

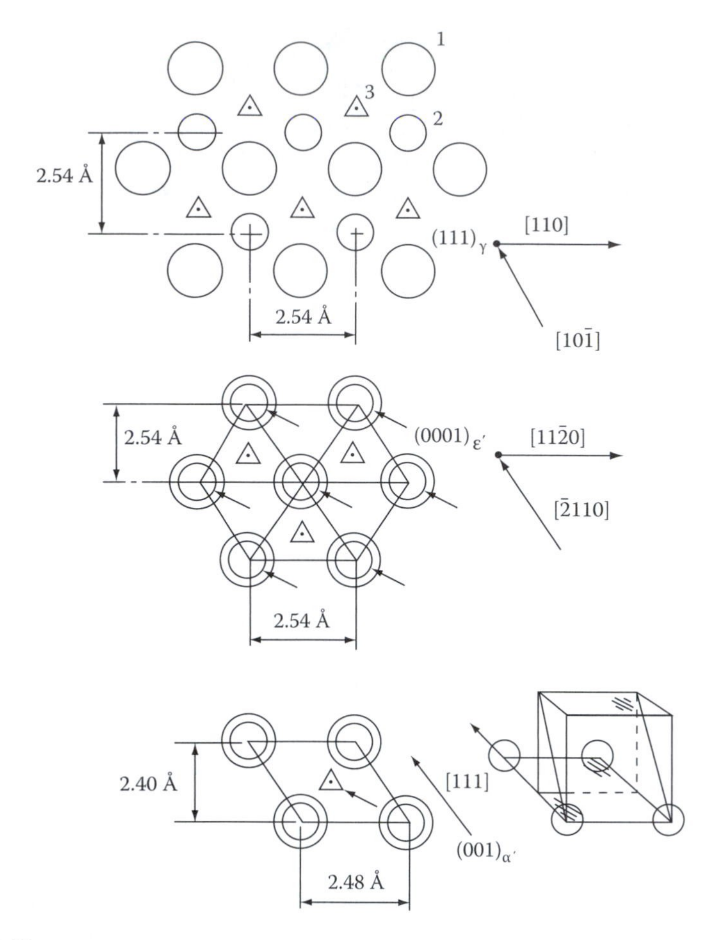

FIGURE 6.16
Venables' model for the $\gamma \rightarrow \varepsilon' \rightarrow \alpha'$ transformation in stainless steel.

and an example is given in Figure 6.17. On the other hand, there is no direct evidence of the $\varepsilon \rightarrow \alpha$ transition, and recent electron microscopy work indicates that the ε' and α' phases in martensitic stainless steels form independently of each other by different mechanisms, i.e., the transformation reactions in stainless steel are of the type $\gamma \rightarrow \varepsilon$ or $\gamma \rightarrow \alpha'$.[12,14] Other detailed models of how dislocations may bring about the martensitic transformation in iron alloys have been given, e.g., by Bogers and Burgers[15] and more recently by Olson and Cohen.[16]

Another example of the fact that the half-twinning shear in fcc material can induce a martensitic transformation is in cobalt.[17] In this case there is an fcc → cph transformation at around 390°C. The generation of large numbers of $\frac{a}{6}\langle 11\bar{2}\rangle_\gamma$ partial dislocations on $\{111\}_\gamma$ planes has been observed directly in the transmission electron microscope using a hot stage as shown in Figure 6.18. The stacking faults in this case appeared to initiate at grain boundaries.

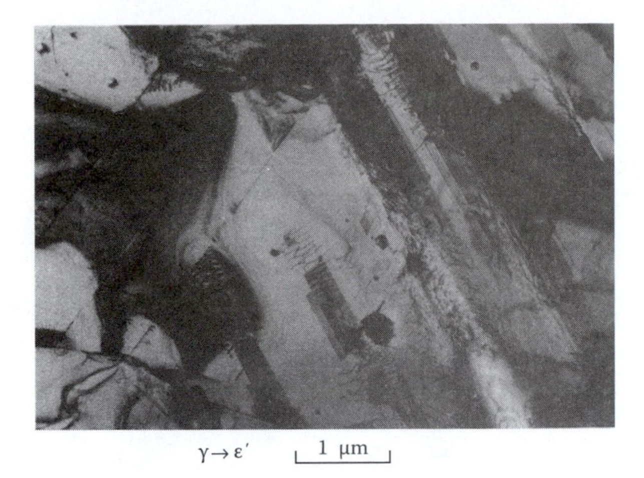

FIGURE 6.17
Dislocation-assisted martensite transformation in a plastically deformed 17% Cr–8% Ni stainless steel. (Courtesy of B. Lehtinen, Institute for Metals Research, Stockholm, Sweden.)

The habit plan is $\{111\}_\gamma$ and the orientation relationship is $(111)_\gamma \| (0001)_\alpha$. The transformation is reversible (at ~430°C) and the cph → fcc reaction occurs by the following dissociation on the hcp basal plane:

$$\frac{1}{3}[\bar{1}2\bar{1}0] \rightarrow \frac{1}{3}[01\bar{1}0] + \frac{1}{3}[\bar{1}100] \tag{6.13}$$

As before, the reaction has to occur on every other hcp plane in order to generate the fcc structure.

It is thus seen that some types of martensite can form directly by the systematic generation and movement of extended dislocations. It is as if the M_s temperature of these alloys marks a transition from positive to negative stacking fault energy. It appears, however, that this type of transformation can neither occur in high stacking fault energy nor in thermoelastic martensites, and it is thus necessary to consider alternative ways in which dislocations can nucleate martensite other than by changes at their cores. It is also difficult to understand twinned martensite, merely on the basis of dislocation core changes.

6.3.3 Dislocation Strain Energy Assisted Transformation

We now consider the possibility that the nucleation barrier to form coherent nuclei can be reduced by the help of the elastic strain field of a dislocation. This theory[19] thus differs fundamentally from the other dislocation-assisted transformation theories discussed earlier, all of which were based on atomic shuffles within the dislocation core. We also note that in this case it is unnecessary that the habit plane of the martensite corresponds to the glide planes of austenite. Furthermore, it is assumed that coherent nuclei are generated by a pure Bain strain, as in the classical theories of nucleation.

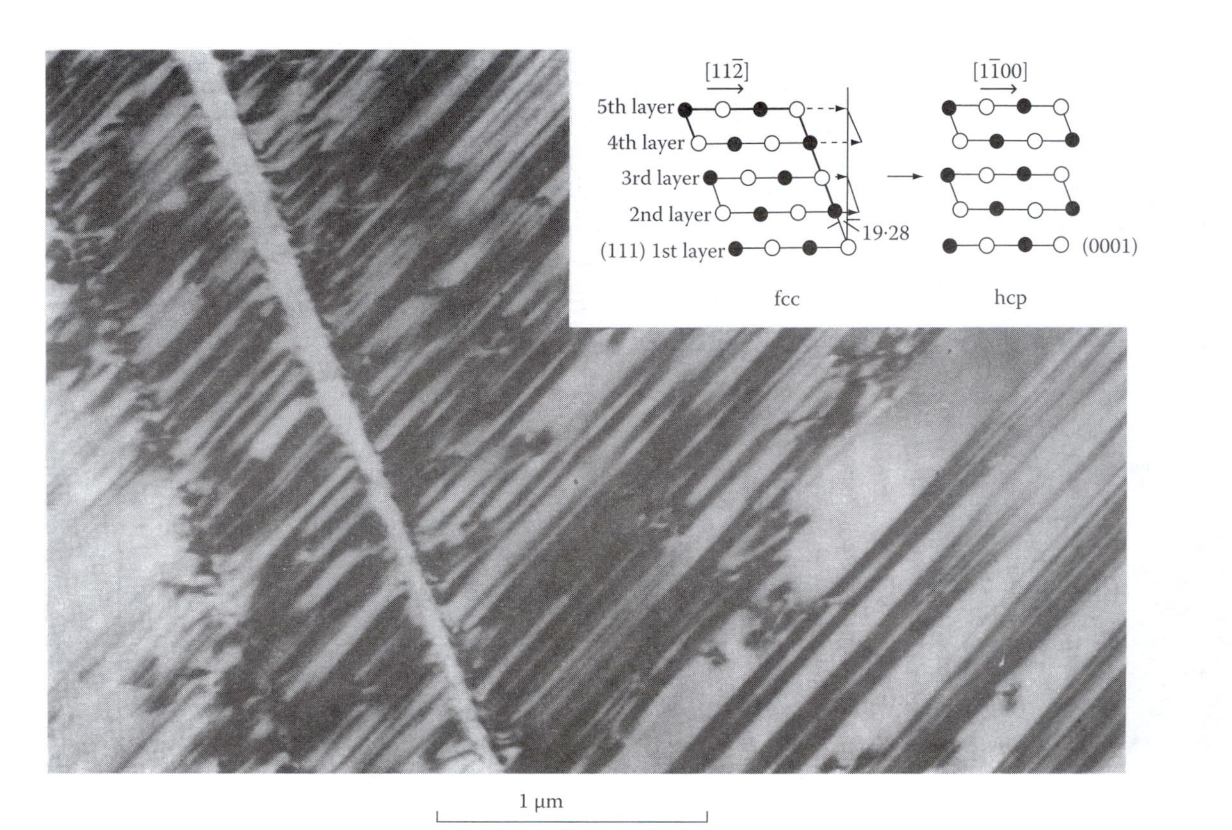

FIGURE 6.18
Dislocation-assisted martensite transformation in cobalt. Inset illustrates the way stacking fault formation induces the fcc → hcp transformation. (After Nishiyama, Z., *Martensitic Transformation*, Academic Press, New York, 1978.)

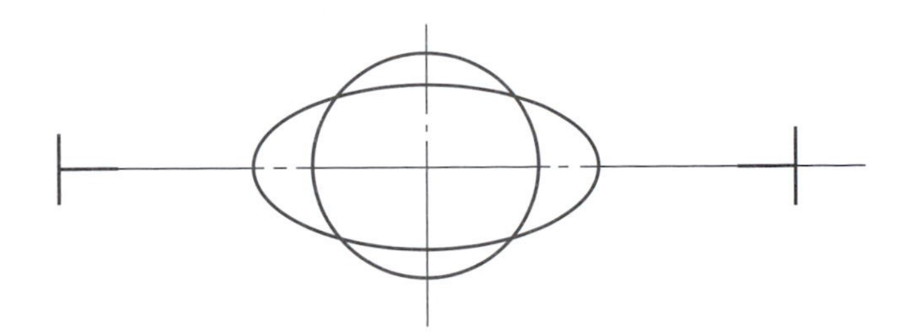

FIGURE 6.19
Illustrating how one of the strain components of the Bain deformation may be compensated for by the strain field of a dislocation which in this case is tending to push atom planes together.

It can be shown that the strain field associated with a dislocation can in certain cases provide a favorable interaction with the strain field of the martensite nucleus, such that one of the components of the Bain strain is neutralized thereby reducing the total energy of nucleation. This interaction is illustrated schematically in Figure 6.19, in which it is seen that the dilatation associated with the extra half plane of the dislocation contributes to the Bain strain. Alternatively the shear component of the dislocation could be utilized.

Such an interaction thus modifies the total energy of Equation 6.5 to

$$\Delta G_{\mathrm{d}} = 2\mu s\pi \cdot ac \cdot \bar{b} \tag{6.14}$$

where ΔG_{d} represents the dislocation interaction energy which reduces the nucleation energy barrier. It can be shown that this interaction energy is given by the expression

$$\Delta G_{\mathrm{d}} = 2\mu s\pi \cdot ac \cdot \bar{b} \tag{6.15}$$

where
$\bar{b}$ refers to the Burgers vector of the dislocation
s refers to the shear strain of the nucleus

The interaction energy used in Equation 6.15 assumes that a complete loop is interacting with the nucleus. In practice, it is likely that only a part of a dislocation will be able to react wit the nucleus in this way.

Equation 6.14 may now be written in full (see Equation 6.7) as

$$\Delta G = 2\pi a^2\gamma + \frac{16\pi}{3}(s/2)^2\mu ac^2 - \Delta G_{\mathrm{v}}\frac{4\pi}{3}a^2c - 2\mu s\pi ac \cdot \bar{b} \tag{6.16}$$

By summing the various components of this expression it is possible to compute the total energy of a martensite nucleus as a function of its diameter and thickness (a, c), whether it is twinned or not (this affects s, see, e.g., Figure 6.9) and the degree of assistance from the strain field of a dislocation (or group of dislocations). This result is shown schematically in Figure 6.20a. It has been calculated that a fully coherent nucleus can reach a

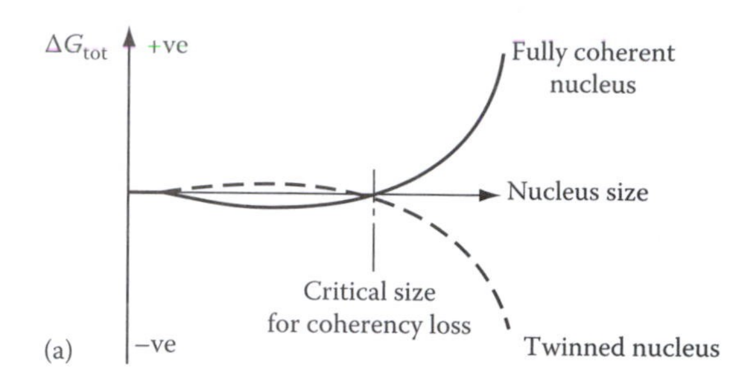

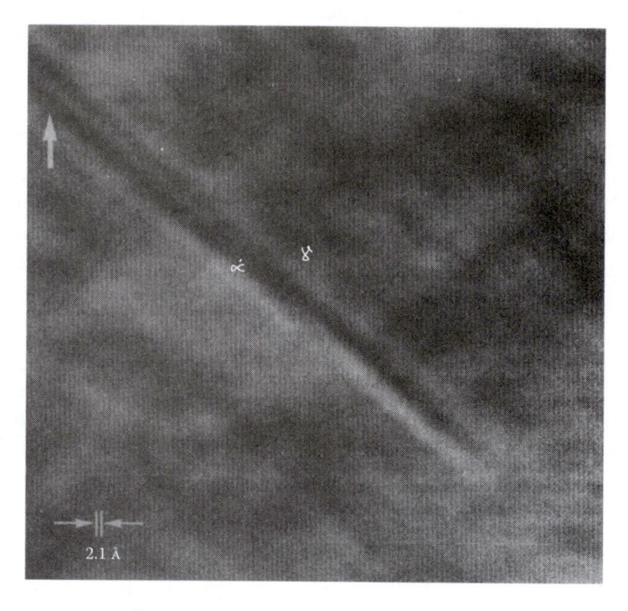

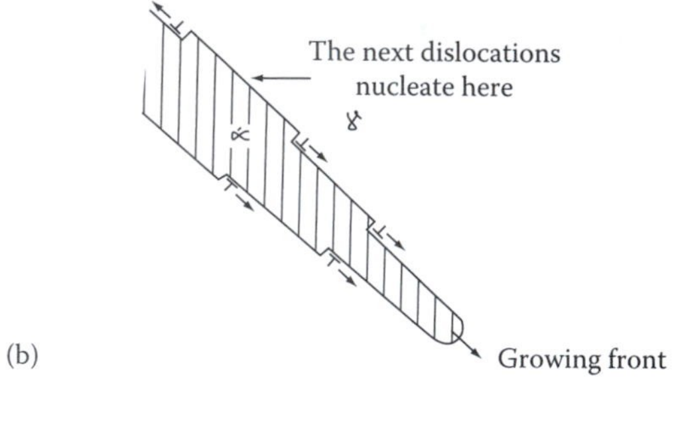

FIGURE 6.20
(a) Schematic diagram based on Equation 6.16, illustrating the need for the nucleus to twin if it is to grow beyond a certain critical size. (b) Lattice image of the tip of a martensite plate in a Ti–Ni alloy. The first interfacial dislocation behind the growing front is indicated. (After Sinclair, R. and Mohamed, H.A., *Acta Metall.*, 26, 623, 1978.)

size of about 20 nm diameter and two to three atoms in thickness by this partial interaction with the strain field of a dislocation. However, it will not be able to thicken or even grow larger unless twins form or slip occurs to further reduce strain energy. The attractive feature of this theory is that it essentially combines the crystallographic characteristics of the inhomogeneous shear and the Bain strain in terms of total strain energy at nucleation. It is thus in line with the majority of the known characteristics of martensite, including the initial straining of the lattice due to the coherent nucleus (see Figure 6.13) and the fact that an inhomogeneous shear is necessary for growth. It even shows that in principle nucleation can occur in the vicinity of any dislocation, thus underlining the statistical correlation between dislocation density in well-annealed austenite and martensite formation indicated by the small particle experiments.

The M_s temperature is thus associated with the most potent nuclei, perhaps depending on the orientation or configuration of the dislocation, or groups of dislocations with respect to the potential martensite nucleus. The large undercoolings below bulk M_s as observed from the small particle experiments thus reflect the statistical probability that ideally oriented dislocations are relatively few and far between, so that high chemical driving forces are needed in most cases. The burst phenomenon, in which an autocatalytic process of rapid, successive plate formation occurs over a small temperature range in, e.g., Fe–Ni alloys, is explicable on this basis by the large elastic stresses set up ahead of a growing plate. In this case, the elastic strain field of the plate acts as the necessary interaction term in Equation 6.10. The question of whether slip or twinning occurs at the critical nucleus size in order to assist growth of the nucleus appears to be a function of the alloy content and M_s temperature, and this factor will be taken up in more detail in Section 6.4 on martensite growth.

In summary, we have not dealt with all the theories of martensite nucleation in this section as recorded in the literature, or even with all alloys exhibiting martensitic transformations. Instead, we have attempted to illustrate some of the difficulties associated with explaining a complex event which occurs at such great speeds as to exclude experimental observation. A general, all-embracing theory of martensite nucleation has still evaded us, and may not even be feasible.

6.4 Martensite Growth

Once the nucleation barrier has been overcome, the chemical volume free energy term in Equation 6.10 becomes so large that the martensite plate grows rapidly until it hits a barrier such as another plate, or a high-angle grain boundary. It appears from observations, that very thin plates first form with a very large a/c ratio (see Figure 6.14) and then thicken afterward. In high-carbon martensites this often leaves a so-called "midrib" of fine twins, and an outer less-well-defined region consisting of fairly regular arrangements of dislocations. In low-carbon lath martensite, transmission microscopy reveals a high dislocation density, sometimes arranged in cellular networks in the case of very low C content, but no twins (see Figure 6.12). In very high-carbon martensite (259 type), only twins are observed.

In view of the very high speeds of growth, it has been conjectured that the interface between austenite and martensite must be a glissile semicoherent boundary consisting of a set of parallel dislocations or twins with Burgers vector common to both phases, i.e., transformation dislocations. The motion of the dislocations brings about the required lattice invariant shear transformation. As noted in Section 3.4.5, the motion of this interface may or may not generate an irrational habit plane.

The habit plane transition in steels and Fe–Ni alloys as a function of alloying content of $\{111\}_\gamma$ lath → $\{225\}_\gamma$ mixed lath/twins → $\{259\}_\gamma$ twinned martensite, is not properly understood. An important factor is thought to be that increased alloying lowers the M_s temperature and that it is the temperature of transformation that dictates the mode of lattice invariant shear. Qualitatively, the slip-twinning transition in a crystal at low temperatures is associated with the increased difficulty of nucleating whole dislocations needed for slip. It is thought that the critical stress needed for the nucleation of a partial twinning dislocation is not so temperature dependent as the Peierls stress for a perfect dislocation.[18] On the other hand, the chemical energy available for the transformation is largely independent of M_s temperature. This implies that as the M_s temperature is lowered the mechanism of transformation chosen is governed by the growth process having least energy. The other factor affecting mode of growth, as discussed in Section 6.3, is how the nucleus forms. If the nucleus forms by the generation of a homogeneous Bain deformation, the orientation of the nucleus in the austenite is again dependent upon finding the lowest energy. This may not coincide with a normal glide plane in the austenite—and in highly alloyed systems it evidently does not. On the other hand, the inhomogeneous shear during growth has to be dictated by the normal modes of slip or twinning available. This suggests that if the habit plane of the martensite is irrational, it may have to grow in discrete steps which are themselves developed by conventional modes of deformation. The resulting plate would then be, for example, likened to a sheared-over pack of cards (see, e.g., Figure 6.9b). We now consider the two main cases of rational (lath) and irrational (plate) martensite growth in steel in more detail.

6.4.1 Growth of Lath Martensite

The morphology of a lath with dimensions $a > b \gg c$ growing on a $\{111\}_\gamma$ plane (see Figure 6.20b) suggests a thickening mechanism involving the nucleation and glide of transformation dislocations moving on discrete ledges behind the growing front. This picture of growth is suggested, e.g., in the work of Sinclair and Mohamed[19] studying NiTi martensite and Thomas and Rao[20] in the case of steel martensite.

It seems possible that due to the large misfit between the bct and fcc lattices dislocations could be self-nucleated at the lath interface. The criterion to be satisfied for dislocation nucleation in this case is that the stress at the interface exceeds the theoretical strength of the material.

It can be shown using Eshelby's approach[21] that for a thin ellipsoidal plate in which $a \gg c$ the maximum shear stress at the interface between the martensite and austenite due to a shear transformation is given by the expression

$$\sigma \simeq 2\mu sc/a \tag{6.17}$$

where μ is the shear modulus of the austenite. It is seen in this simple model that the shear stresses are sensitive to particle shape as well as angle of shear. Of course, in practice, it is very difficult to define the morphology of martensite in such simple c/a terms, but this gives us at least a qualitative idea of what may be involved in the growth kinetics of martensite.

Kelly[22] has calculated a theoretical shear strength for fcc materials of 0.025 μ at ambient temperature, and this can be used as a minimum, or threshold stress for nucleating dislocations. Equation 6.17 is plotted in Figure 6.21 in terms of different $a{:}c$ ratios, assuming $s = 0.2$ which is typical of bulk lath and plate martensite. An approximate range of morphologies representative of lath or plate martensite is given in the figure. It is seen that Kelly's threshold stress for dislocation nucleation may be exceeded in the case of lath martensite, but seems unlikely in the case of the thinner plate martensite. It is interesting to note from Figure 6.21, however, that shear loop nucleation in plate martensite is feasible if $s \gtrsim 0.32$, which is the shear associated with a pure Bain strain (Figure 6.9a). In other words, coherency loss of the initial coherent nucleus is energetically possible.

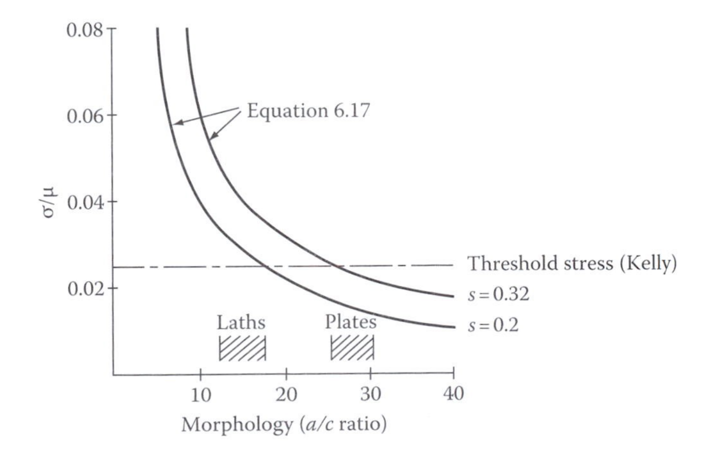

FIGURE 6.21
Equation 6.17 plotted for two values of shear corresponding to a pure Bain deformation (0.32) and a twinned plate (0.2).

The assumption of shear loop nucleation in fact seems reasonable and likely in conjunction with lath growth. The same mechanism of dislocation generation during growth could even be applied to bainite where the morphology appears to be fairly similar to lath martensite, although in this case some diffusion of carbon also occurs. It is thus seen that by nucleating dislocations at the highly strained interface of the laths, the misfit energy can be reduced and the lath can continue to grow into the austenite.

Internal friction measurements have shown that in lath martensite the density of carbon is slightly higher at cell walls than within cells, suggesting that limited diffusion of carbon takes place following or during the transformation. The transformation could also produce adiabatic heating which may affect diffusion of carbon and dislocation recovery, at least at higher M_s temperatures. In this respect, there appears to be a certain relationship between lower bainite and martensite. The higher M_s temperatures associated with lath martensite may be sufficient to allow dislocation climb and cell formation after the transformation, although the high growth speeds suggest an interface of predominantly screw dislocations. The volume of retained austenite between laths is relatively small in lath martensite (these small amounts of retained austenite are now thought to be important to the mechanical properties of low-carbon steels[23]), suggesting that sideways growth, and transformation between laths occurs without too much difficulty.

6.4.2 Plate Martensite

In medium- and high-carbon steels, or high-nickel steels, the morphology of the martensite appears to change from a lath to a roughly platelike product. This is associated with lower M_s temperatures and more retained austenite, as illustrated by Figure 6.22. However, as mentioned earlier, there is also a

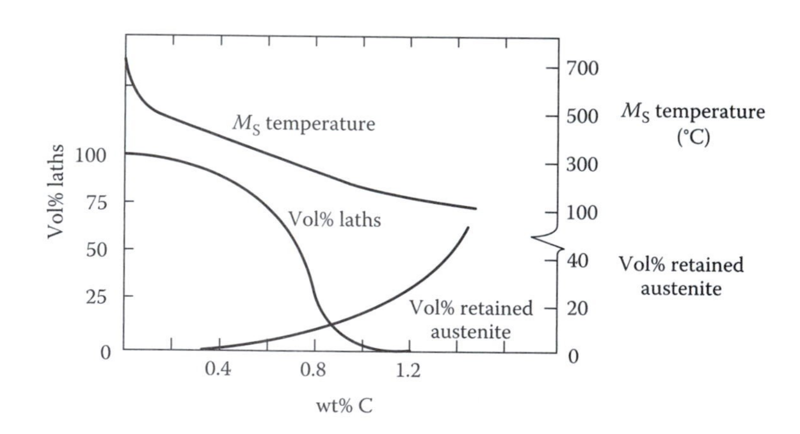

FIGURE 6.22
Approximate relative percentages of lath martensite and retained austenite as a function of carbon content in steels. (Data from Speich, G.R., *Metall. Trans.*, 3, 1045, 1972.)

transition from plates growing on $\{225\}_\gamma$ planes to $\{259\}_\gamma$ planes with increasing alloy content. The lower carbon or nickel $\{225\}_\gamma$ martensite often consists of plates with a central twinned midrib, the outer regions of the plate being free of twins. It appears that the twinned midrib forms first and the outer (dislocation) region which is less well defined than the midrib, grows afterward. The high-carbon or nickel {259} martensite on the other hand is completely twinned and the habit plane measurements have less scatter than the mixed structures.

Typical morphologies for plate martensite are usually thought to be much thinner than lath martensite or bainite. On the basis of Figure 6.21, it appears that there is likely to be a problem in nucleating whole dislocations in the case of growing plate martensite when $s \simeq 0.2$, but that partial twinning dislocations evidently are able to nucleate. Once nucleated, twinned martensite grows extremely rapidly, but the mechanism by which this occurs has not been clarified as yet. It is clear from the work on low-temperature deformation of fcc metals, that twinning can be an important deformation mechanism. However, the problem in martensite transformations is to explain the extremely rapid rates of plate growth based on twinning mechanisms. The pole mechanism seems inadequate in this respect, although mechanisms based on dislocation reflection processes may be more realistic.[26] Alternatively, it may be necessary to invoke theories in which standing elastic waves may nucleate twinning dislocations[24] as an aid to very rapid plate growth.

The transition from twinning → dislocations in midrib martensite is intriguing and could be the result of a change in growth rate after the midrib forms (see, e.g., Shewmon[25]). In other words, martensite formed at higher temperatures or slower rates grows by a slip mechanism, while martensite formed at lower temperatures and higher growth rates grows by a twinning mode. Indeed, in the case of ferritic steels, the normal mode of plastic deformation is very much a function of strain rate and temperature.

An elegant model for a dislocation-generated $\{225\}_\gamma$ martensite has been postulated by Frank.[26] Frank has basically considered the way to interface the fcc austenite lattice with that of the bcc martensite such as to reduce lattice misfit to a minimum. He finds that this can be achieved quite well with the help of a set of dislocations in the interface. In this model, the close-packed planes of the fcc and bcc structures are envisaged to meet approximately along the martensite habit plane as shown in Figure 6.23a. Since the (111) and $(101)_{\alpha'}$ planes meet edge-on at the interface, the close-packed direction are parallel and lie in the interface plane. The reason for the rotation, ψ, shown in Figure 6.23a, is to equalize the atomic spacings of the (111) and $(101)_{\alpha'}$ planes at the interface. However, in spite of this, there is still a slight misfit along the $[01\bar{1}]_{\gamma'}[11\bar{1}]_{\alpha'}$ direction where the martensite lattice parameter is ~2% less than that of austenite. Frank therefore proposed that complete matching can be achieved by the insertion of an array of screw dislocations with a spacing of six atom planes in the interface which has the effect of matching the two lattices and thus removing the misfit in this

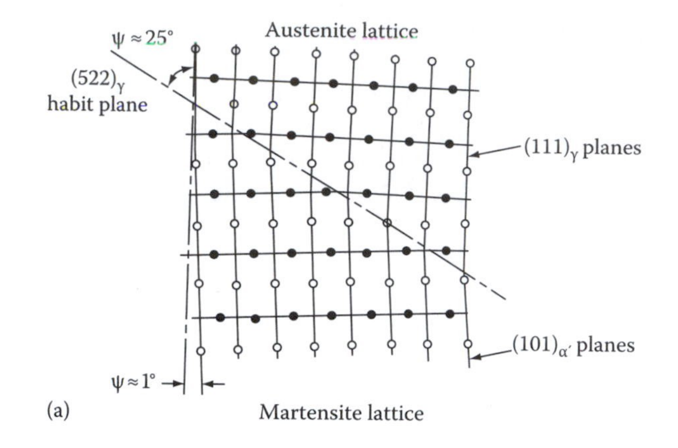

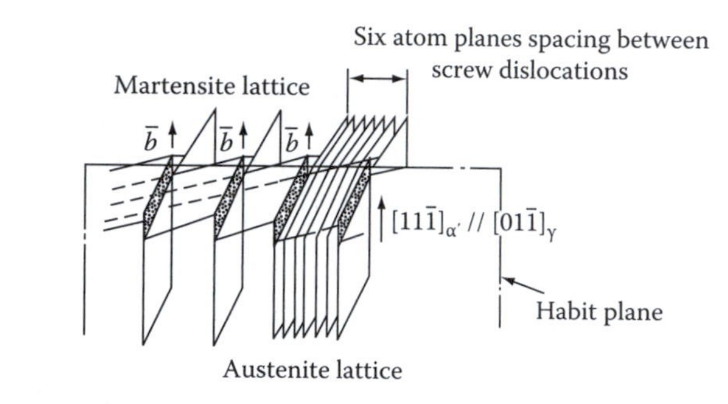

FIGURE 6.23
Model for the {225} habit austenite/martensite interface in steel. (Based on data by Frank, F.C., *Acta Metall.*, 1, 15, 1953.)

direction. This also brings about the required lattice-invariant shear on the $(112)_{\alpha'}$ plane as the interface advances. The resulting interface is illustrated in Figure 6.23b.

In terms of the minimum shear stress criterion (Figure 6.21), the further expansion and thickening of a $\{225\}_\gamma$ twinned midrib by a Frank dislocation interface could occur when the midrib reaches some critical a/c ratio. However, there have been no detailed models developed as to how the Frank interface can be generated from the nucleation event. Assuming a coherent nucleus with $s \simeq 0.32$, it is seen from Figure 6.21 that it is theoretically possible for dislocation nucleation to occur at this stage to relieve coherency. Qualitatively, the larger amount of chemical free energy available after the critical size for growth has been exceeded, may be sufficient to homogeneously nucleate dislocations[27] particularly in the presence of the large strain energy of the rapidly growing plate.[28]

Other factors known to affect the growth of martensite are grains size, external stresses, and the phenomenon of stabilization. We now briefly consider these effects.

6.4.3 Stabilization

This is a phenomenon associated with samples cooled to some temperature intermediate between M_s and M_f, held there for a period of time and then cooled again. In such a case, transformation does not immediately continue, and the total amount of transformed martensite is less than that obtained by continuous cooling throughout the transformation range. It has even been observed that existing plates do not continue to grow after stabilization, but new plates are nucleated instead. The degree of stabilization is a function of the time held at temperature. This phenomenon is not properly understood, although it seems conceivable that carbon has time to diffuse to the interface under the influence of the high stresses associated with plate growth. There could also be local atomic relaxation at the interface, thereby increasing the nucleation barrier for dislocation generation.

6.4.4 Effect of External Stresses

In view of the dependence of martensite growth on dislocation nucleation, it is expected that an externally applied stress will aid the generation of dislocations and hence the growth of martensite. It is well established, for example, that external stress lowers the nucleation barrier for coherency loss of second phase precipitates. External stresses can also aid martensite nucleation if the external elastic strain components contribute to the Bain strain. This will provide yet another interaction term in Equation 6.14. It has been shown in such cases that the M_s temperature can be raised.[29] However, if plastic deformation occurs, there is an upper limiting value of M_s defined as the M_d temperature, by, e.g., holding the sample being transformed under hydrostatic compression. This is because increasing pressure stabilizes the phase with the smaller atomic volume, i.e., the close-packed austenite, thereby lowering the driving force ΔG_v for the transformation to martensite. On the other hand, the presence of a large magnetic field can raise the M_s temperature on the grounds that it favors the formation of the ferromagnetic phase.

Plastic deformation of samples can aid both nucleation and growth of martensite, but too much plastic deformation may in some cases suppress the transformation. Qualitatively it could be expected that increase in dislocation density by deformation should raise the number of potential nucleation sites, but too much deformation may introduce restraints to nuclei growth.

The effect of plastically deforming the austenite prior to transformation on increasing the number of nucleation sites and hence refining plate size is of course the basis of the ausforming process. The high strength of ausformed steels is thus due to the combined effect of fine plate size, solution hardening (due to carbon), and dislocation hardening.

6.4.5 Role of Grain Size

Since martensite growth relies on maintaining a certain coherency with the surrounding austenite, a high-angle grain boundary is an effective barrier to plate growth. Thus, while grain size does not affect the number of martensite nuclei in a given volume, the final martensite plate size is a function of grain size. Another important feature of grain size is its effect on residual stress after transformation is completed. In large grain-sized material, the dilatational strain associated with the transformation causes large residual stresses to be built up between adjacent grains and this can even lead to grain-boundary rupture (quench cracking) and substantially increase the dislocation density in the martensite. Fine grain-sized metals tend to be more self-accommodating and this, together with the smaller martensite plate size, provides for stronger, tougher material.

In summary, theories of martensite nucleation and growth are far from developed to a state where they can be used in any practical way—such as helping to control the fine structure of the finished product. It does appear that nucleation is closely associated with the presence of dislocations and the process of ausforming (deforming the austenite prior to transformation) could possibly be influenced by this feature if we knew more of the mechanism of nucleation. Growth mechanisms, particularly by twinning, are still far from clarified, however.

6.5 Premartensite Phenomena

This is a subject that has provoked considerable attention in recent years from researchers, and is mainly concerned with ordered compounds exhibiting order $\rightarrow$ order martensitic transformations. A useful summary of this phenomenon has been given recently by Wayman.[30] The effect has been observed in the form of anomalous diffraction effects or even diffuse streaking as well as a resistivity anomaly, e.g., in TiNi alloys. In β-brass thin foils, a mottled contrast has been observed giving rise to side band reflections in diffraction patterns. In CuAu alloys the phenomenon occurs in the form of a streaming or shimmering effect in bright field images of thin foils. The latter observation, first noted by Hunt and Pashley in 1962,[31] has even been interpreted as possible evidence for the appearance of fluctuating strain fields due, e.g., to a Bain deformation. Recent work in Wayman's laboratory suggests, however, that while the appearance of local reordering reactions above the M_s temperature is possible, there is still little direct evidence that the observed phenomena can be related to the initial stage of the martensitic transformation. Nevertheless, the effect is an intriguing one, particularly bearing in mind the relatively low undercoolings associated with ordered alloys (see, e.g., Table 6.1), and in this respect it could be conjectured that some process is occurring which very effectively aids the transformation of these alloys.

6.6 Tempering of Ferrous Martensites

Although the diffusionless martensite transformation is fundamental to the hardening of steel, most (if not all) technological steels have to be heat treated after the transformation in order to improve toughness and in some cases even strength. Recent years have seen notable developments in these steels, achieving in some cases very high degrees of sophistication in the form of carbide dispersions and various types of substructure strengthening. For useful reviews see, e.g., Speich[32] and Honeycombe.[33]

The martensitic transformation usually results in a ferritic phase which is highly supersaturated with carbon and any other alloying elements that remain locked into the positions they occupied in the parent austenite. On aging, or tempering, therefore, there is a strong driving force for precipitation. As is usual with low-temperature aging, the most stable precipitate, as indicated by the equilibrium phase diagram, is not the first to appear. The ageing sequence is generally $\alpha' \rightarrow \alpha + \varepsilon$-carbide or $\alpha + Fe_3C$, depending upon the tempering temperature. It is not thought that ε-carbide ($Fe_{2.4}C$) decomposes directly to Fe_3C, but that the transition only occurs by the ε-carbide first dissolving. When strong carbide-forming alloying elements such as Ti, Nb, V, Cr, W, or Mo are present, the most stable precipitate can be an alloy carbide instead of cementite. See, for example, the Fe–Mo–C phase diagram in Figure 6.24. However, these ternary additions are dissolved substitutionally in the ferrite lattice and are relatively immobile in comparison to interstitial carbon. The precipitation of these more stable carbides is therefore preceded by the formation of ε-carbide and Fe_3C which can occur solely by the diffusion of carbon. Alloying elements are only incorporated into the precipitate in proportion to their overall concentration in the ferrite.

The various changes that can take place during the tempering of ferrous martensites are summarized in Table 6.3. In practice, heat-treatment times are limited to a few hours and the phases that appear within these time periods depend on the temperature at which tempering occurs. Therefore, Table 6.3 gives a summary of the new phases that appear within the various temperature ranges, and provides details of other microstructural changes that take place. It should be noted, however, that the temperature ranges given are only approximate and that there is a great deal of overlap between the various ranges.

Table 6.4 summarizes the observed precipitation sequences in a few selected steel compositions. These compositions are experimental alloys that have been studied to avoid the complications that arise with commercial alloys where many interacting alloying elements are present. The crystal structures, shapes, and orientation relationships for some of these precipitates are listed in Table 6.5.

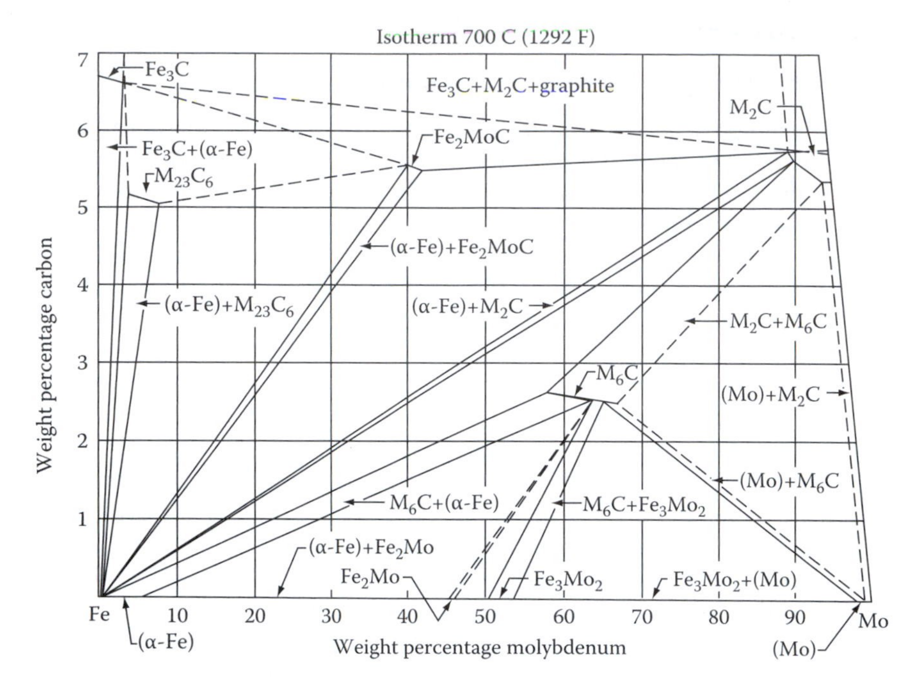

FIGURE 6.24
An isothermal section through the Fe–Mo–C phase diagram at 700°C. (After Wada, T., *Metals Handbook*, 8th edn., Vol. 8, American Society for Metals, 1973, p. 409.)

TABLE 6.3
Transformations Occurring during Tempering of Ferrous Martensites

Temperature (°C)	Transformation	Remarks
25–100	Carbon segregation to dislocations and boundaries; preprecipitation clustering and ordering	Clustering predominant in high-carbon steels
100–200	Transition-carbide precipitation, diameter 2 nm (first stage of tempering)	Carbides may be n(Fe_2C) or ε($Fe_{2.4}C$)
200–350	Retained austenite transforms to ferrite and cementite (second stage)	Associated with tempered martensite embrittlement
250–350	Lath-like Fe_3C precipitation (third stage)	
350–550	Segregation of impurity and alloying elements	Responsible for temper embrittlement
400–600	Recovery of dislocation substructure; lath-like Fe_3C agglomerates to form spheroidal Fe_3C	Lath structure maintained
500–700	Formation of alloy carbides (secondary hardening or fourth stage)	Occurs only in steels containing Ti, Cr, Mo, V, Nb, or W; Fe_3C may dissolve
600–700	Recrystallization and grain growth; coarsening of spheroidal Fe_3C	Recrystallization inhibited in medium-carbon and high-carbon steels; equiaxed ferrite formed

TABLE 6.4

Carbide Precipitation Sequences

Alloy (wt%)	Precipitation Sequence
Fe–C	ε-Carbide[a] → Fe_3C (→graphite)
Fe–2 V–0.2 C	Fe_3C → VC or V_4C_3
Fe–4 Mo–0.2 C	FeC → Mo_2C → M_6C[b]
Fe–6 W–0.2 C	Fe_3C → W_2C → $M_{23}C_6$ → M_6C
Fe–12 Cr–0.2 C	Fe_3C → Cr_7C_3 → $Cr_{23}C_6$

[a] Does not form when C $\leq$ 0.2%.

[b] M stands for a mixture of substitutional alloying elements, in this case Fe and Mo.

On the basis of the data given in Tables 6.3 through 6.5, we note the following features:

Carbon segregation

As a result of the large distortion caused by the carbon atoms in the martensitic lattice there is an interaction energy between carbon and the strain fields around dislocations. In lath martensite, for example, carbon tends to diffuse to sites close to dislocations in order to lower its chemical potential. In plate martensite, however, the martensite is internally twinned and there are relatively few dislocations. In this case carbon-rich clusters or zones tend to form instead. In low-carbon low-alloy steels, martensite starts to form at relatively high temperatures and there can be sufficient time during the quench for carbon to segregate or even precipitate as ε-carbide or cementite.

TABLE 6.5

Data Concerning Carbides Precipitated during Tempering of Martensite

Carbide	Crystal Structure	Shape	Orientation Relationship	Temperature of Formation (°C)
ε-Carbide ($Fe_{2-3}C$)	hcp	Laths	$(10\bar{1}1)_\varepsilon \parallel (101)_\alpha$ $[0001]_\varepsilon \parallel [011]_\alpha$	100–250
Cementite (Fe_3C)	Orthorhombic	Laths	$(001)_c \parallel (211)_\alpha$ $[100]_c \parallel [0\bar{1}1]_\alpha$	250–700
$VC–V_4C_3$	Cubic (NaCl structure)	Plates	$(100)_c \parallel (100)_\alpha$ $[011]_c \parallel [010]_\alpha$	~550
Mo_2C	hcp	—	$(0001)_c \parallel (011)_\alpha$ $[11\bar{2}0]_c \parallel [100]_\alpha$	~550
W_2C	hcp	Needles	As Mo_2C	~600
Cr_7C_3	Hexagonal	Spheres	—	~550
$Cr_{23}C_6$ ($M_{23}C_6$)	Cubic	Plates	$(100)_c \parallel (100)_\alpha$ $[010]_c \parallel [010]_\alpha$	
M_6C (Fe_3Mo_3C, Fe_3W_3C)	Cubic	—	—	~700

FIGURE 6.25
ε-Carbide (dark) precipitated from martensite in Fe–24 Ni–0.5 C after 30 min at 250°C. Thin foil electron micrograph (×90,000). (After Speich, G.R., *Metals Handbook*, 8th edn., Vol. 8, American Society for Metals, Metals Park, OH, 1973, p. 202.)

ε-Carbide

The reason for the 0.2% C limit (Table 6.4) is thought to be due to the fact that the M_s temperatures of very low-carbon martensites are high enough to allow considerable carbon diffusion to lath boundaries during cooling (see, e.g., Figure 6.3c). There is thus no carbon left in solution to precipitate out on reheating. ε-Carbide has a hexagonal crystal structure and precipitates in the form of laths with an orientation relationship as shown in Table 6.5 (see Figure 6.25). This orientation relationship provides good matching between the $(101)_{\alpha'}$ and $(10\bar{1}1)_{\varepsilon}$ planes.

Cementite

Cementite forms in most carbon steels on tempering between 250°C and 700°C. The precipitate is initially lath-like with a $\{011\}_{\alpha'}$ habit plane, Figure 6.26. It has an orthorhombic crystal structure and forms with the orientation relationship given in Table 6.5. At high temperatures cementite rapidly coarsens into a spheroidal form as shown in Figure 6.32. In alloy steels the cementite composition can often be represented as $(FeM)_3C$ where M is a carbide-forming alloying element. The composition may however be metastable if sufficient alloying elements are present.

Alloy carbides

In steels containing sufficient carbide-forming elements alloy carbides are formed above ~500°C where substitutional diffusion becomes significant. These carbides replace the less stable cementite which dissolves as a finer alloy carbide dispersion forms. Some typical precipitation sequences are listed in Table 6.4. There are two ways in which the $Fe_3C \rightarrow$ alloy carbide transformation can take place:

1. By in situ transformation—the alloy carbides nucleate at several points at the cementite/ferrite interfaces, and grow until the cementite disappears and is replaced by a finer alloy carbide dispersion, see, e.g., Figure 6.27.

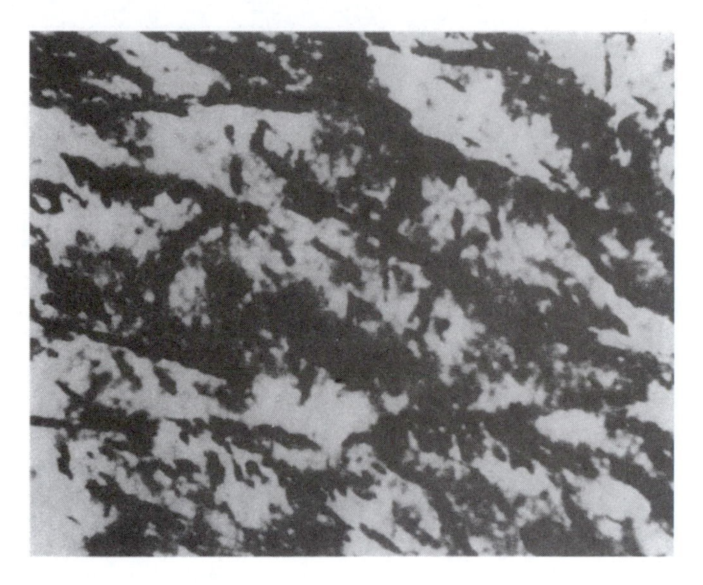

FIGURE 6.26
Cementite (dark laths) formed during tempering a 0.42 C steel for 1 h at 300°C. Thin foil electron micrograph (×39,000). (After Speich, G.R., *Metals Handbook*, 8th edn., Vol. 8, American Society for Metals, Metals Park, OH, 1973, p. 202.)

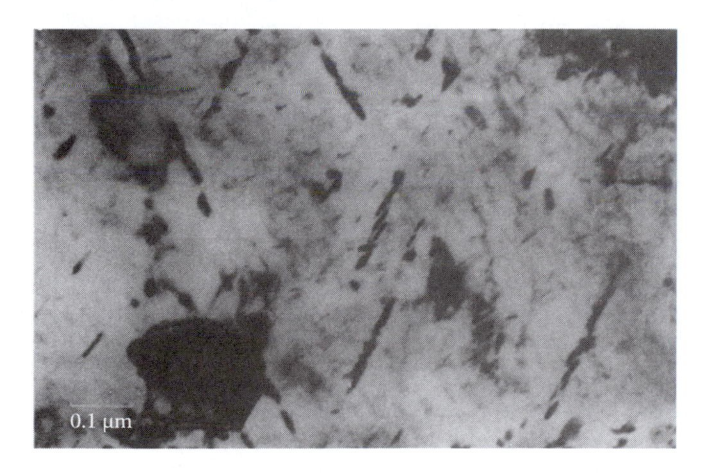

FIGURE 6.27
W_2C needles lying along the sites of former Fe_3C precipitates in Fe–6.3 W–0.23 C quenched and tempered for 20 h at 600°C. (After Honeycombe, R.W.K., *Structure and Strength of Alloy Steels*, Climax Molybdenum, London, 1973; after A.T. Davenport.)

2. By separate nucleation and growth—the alloy carbides nucleate heterogeneously within the ferrite on dislocations, lath boundaries, and prior austenite grain boundaries. The carbides then grow at the expense of cementite.

Either or both mechanisms can operate depending on the alloy composition.

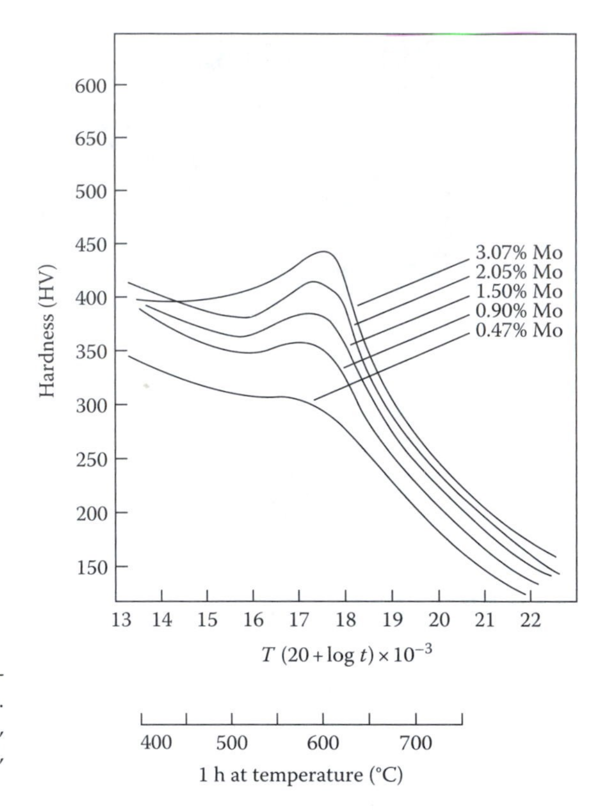

FIGURE 6.28
Effect of molybdenum on the tempering of quenched 0.1% C steels. (After Irvine, K.J. and Pickering, F.B., *J. Iron Steel Inst.*, 194, 137, 1960.)

The formation of alloy carbides is an important strengthening mechanism in high-speed tool steels that must operate at dull red heat without losing their cutting ability. The phenomenon is usually referred to as secondary hardening. Figure 6.28 shows the effect of tempering molybdenum steels for various times and temperatures. The hardness of plain carbon martensites usually decreases with increasing temperature due to recovery and overageing effects. The replacement of a coarse cementite dispersion by a finer alloy carbide that is more resistant to coarsening, however, can produce an increase in hardness at around 550°C–600°C.

The effectiveness of these carbides as strengtheners depends on the fineness of the dispersion and the volume fraction precipitated. The fineness of the dispersion depends on ΔG^* for nucleation which in turn is influenced by the free energy of formation of the carbide, the interfacial energy, and the misfit. A guide to the relative free energies of formation is given by Figure 6.29 which shows the heats of formation (ΔH_f) of various nitrides, carbides, and borides relative to that of cementite which is taken as $\Delta H_f = 0$. The finest precipitate dispersions are generally obtained from VC, NbC, TiC, TaC, and HfC. These are all close-packed intermetallic compounds. On the other hand, the carbides with complex crystal structures and low heats of formation, e.g., M_7C_3, M_6C, and Mn_3C_6, generally form relatively coarse dispersions.

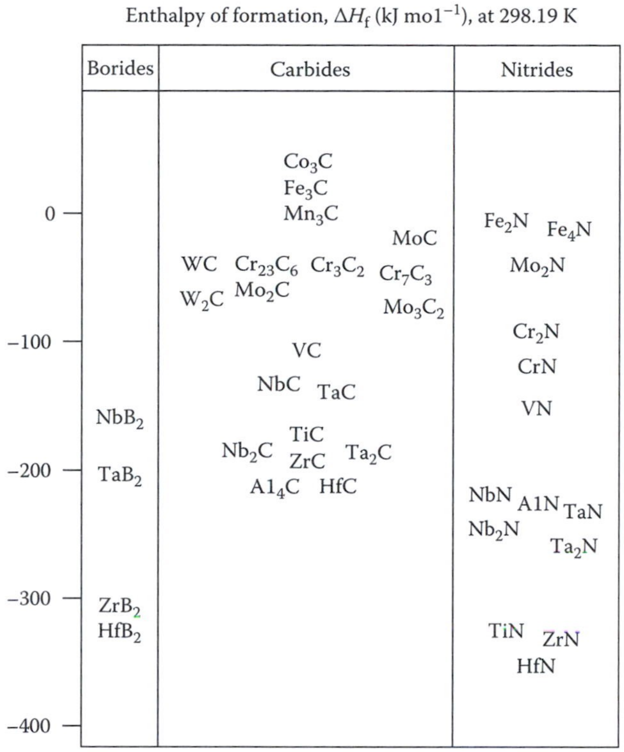

Enthalpy of formation, ΔH_f (kJ mol^{-1}), at 298.19 K

Borides | Carbides | Nitrides

0, −100, −200, −300, −400

Co_3C Fe_3C Mn_3C MoC WC $Cr_{23}C_6$ Cr_3C_2 Cr_7C_3 W_2C Mo_2C Mo_3C_2 VC NbC TaC TiC Nb_2C ZrC Ta_2C Al_4C HfC

Fe_2N Fe_4N Mo_2N Cr_2N CrN VN NbN AlN TaN Nb_2N Ta_2N TiN ZrN HfN

NbB_2 TaB_2 ZrB_2 HfB_2

FIGURE 6.29
Enthalpies of formation of carbides, nitrides, and borides. (Data from Schick, H.L., *Thermodynamics of Certain Refractory Compounds*, Academic Press, New York, 1966.)

The volume fraction of carbide precipitated depends on the solubility of the alloy carbide in the austenite prior to quenching, relative to the solubility in ferrite at the tempering temperature. Note that the solubility of a β-phase in a terminal solid solution of α was considered in Chapter 1 for binary alloys. It can similarly be shown that in ternary Fe–C–M alloys the concentrations of M and C in Fe in equilibrium with a carbide M_mC_n are approximately given by the relation:[36]

$$[M]^m[C]^n = K \tag{6.18}$$

where [M] and [C] are the atomic percentages or mole fractions of M and C in solution and K is the solubility product which can be expressed as

$$K = K_0 \exp \frac{-\Delta H}{RT} \tag{6.19}$$

where
K_0 is a constant
ΔH is the enthalpy of formation of M_mC_n from M and C in solution

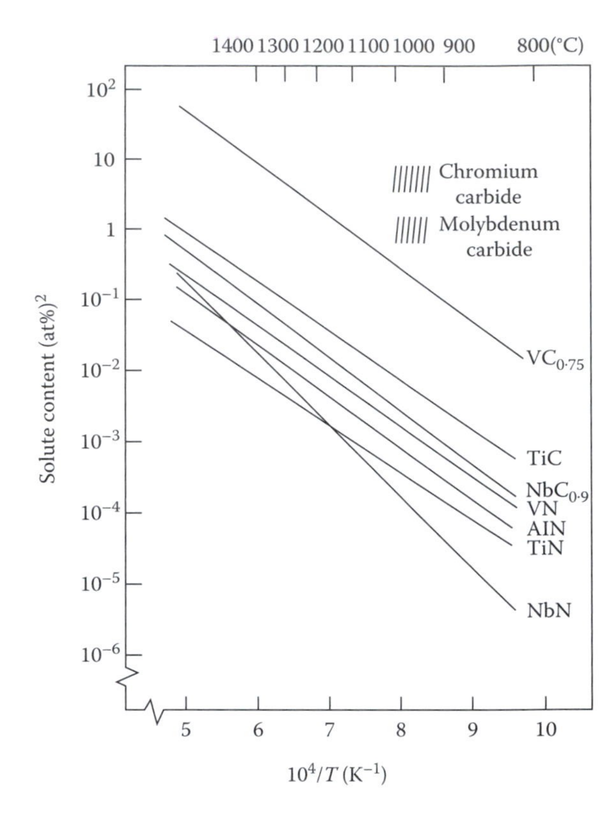

FIGURE 6.30
Solubility products (in atomic percent) of carbides and nitrides in austenite as a function of temperature. (After Honeycombe, R.W.K., *Structure and Strength of Alloy Steels*, Climax Molybdenum, London, 1973.)

Figure 6.30 shows the solubility products of various carbides and nitrides in austenite as a function of temperature. The solubilities of these compounds in ferrite are very much lower and to a first approximation can be considered to be approximately equal. It is clear therefore that chromium, molybdenum, and vanadium with the highest solubilities in austenite should precipitate in the highest volume fractions in the ferrite.

Effect of retained austenite
In most steels, especially those containing more than 0.4% C, austenite is retained after quenching. On aging in the range 200°C–300°C this austenite decomposes to bainite. In some high-alloy steels austenite can be stabilized to such low temperatures that the martensite partially reverts into austenite on heating. Very thin regions of retained austenite may even be present between laths in low-carbon steel, and this is thought to improve the toughness of these steels independently of tempering treatments.

Recovery, recrystallization, and grain growth
As-quenched lath martensite contains high-angle lath boundaries, low-angle cell boundaries within the laths, and dislocation tangles within the cells. Recovery usually occurs above 400°C and leads to the elimination of both the dislocation tangles and the cell walls. The lath-like structure, however,

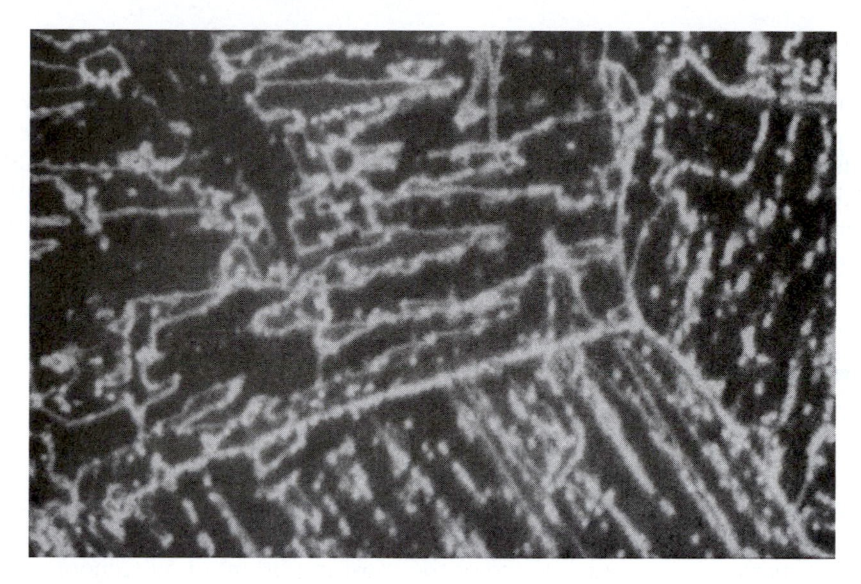

FIGURE 6.31
A recovered lath martensite showing that the lath boundaries are retained. Fe–0.18 C tempered for 10 min at 600°C (×2000). (After Speich, G.R., *Metals Handbook*, 8th edn., Vol. 8, American Society for Metals, Metals Park, OH, 1973, p. 202.)

remains as shown in Figure 6.31. The ferrite can recrystallize at high temperatures in low-carbon steels (see, e.g., Figure 6.32), but the process is inhibited in medium- to high-carbon steels by the grain boundary pinning caused by carbide precipitates. In the latter steels recovery is followed by grain growth.

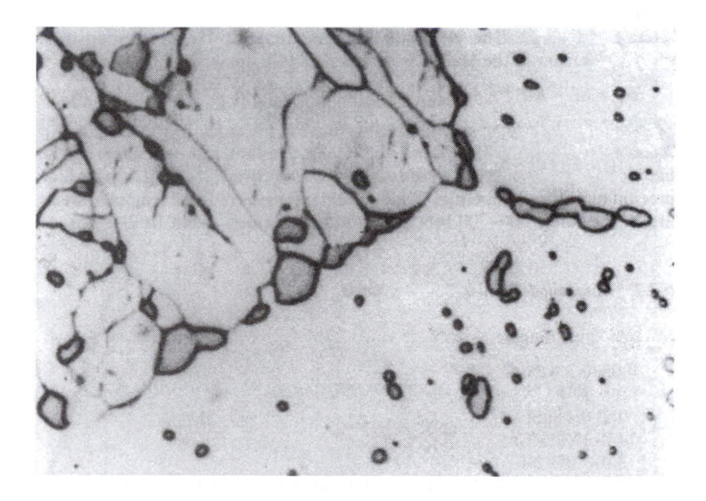

FIGURE 6.32
A partially recrystallized structure. Top left, recovered but not recrystallized; bottom left, a new recrystallized grain. Coarse spheroidal cementite is also apparent. Fe–0.18 C quenched and tempered for 96 h at 600°C (×2000). (After Speich, G.R., *Metals Handbook*, 8th edn., Vol. 8, American Society for Metals, Metals Park, OH, 1973, p. 202.)

Temper embrittlement
As pointed out in the introduction to this section the aim of tempering martensite is to improve ductility. However, in some steels tempering, or slow cooling through, in the range 350°C–575°C can lead to embrittlement. This has been attributed to the segregation of impurity atoms such as P, Sb, or Sn to prior austenite grain boundaries. Some steels also show an embrittlement on tempering between 230°C and 370°C. This may be caused by the formation of carbides with a critical platelike shape.

6.7 Case Studies

It is clear from the foregoing theory of martensite that much work remains to be done before we can fully understand this complex transformation, particularly in steels. In spite of this, the hardening of steel by quenching to obtain martensite is arguably one of the most important of all technological processes. In this section, we illustrate four examples of technological alloys based on the martensite transformation. These are a quenched and tempered structural steel, some controlled transformation steels including transformation-induced plasticity (TRIP) steels, dual-phase steel, and a TiNi memory metal possessing a unique shape-memory property based on a diffusionless transformation.

6.7.1 Carbon and Low-Alloy Quenched and Tempered Steels

Composition range:

0.1–0.5 wt% C; (C $<$ 0.3%: weldable without preheat)
0.6%–1.3% Mn with or without small alloying additions,
e.g., Si, Ti, Mo, V, Nb, Cr, Ni, W, etc.

Special properties: High strength, weldable constructional steels.

Relevant phase diagrams: Fe–C, in conjunction with appropriate CCT, TTT diagrams (see, e.g., Figure 6.3).

Microstructures: See Figures 6.1 and 6.12.

Comments: The compositions of these steels are chosen with respect to (a) hardenability; (b) weldability; (c) tempering properties, e.g., resistance to tempering, or increased tempering strength due to secondary hardening. Typically, lath or mixed (lath plus twinned) structures contain high densities of dislocations ($0.3–0.9 \times 10^{10}$ mm^{-2}), equivalent to a heavily worked steel. There is normally very little retained austenite associated with these steels see, e.g., Figure 6.22. The lattice structure is bct, at least for carbon contents greater than ~0.2%. Below this composition, it is suspected that due to the higher M_s temperature, some carbon segregates to dislocations or lath boundaries during the quench, as measured by resistivity and internal frictional measurements.

These results indicate that only in steels containing more than 0.2% C is carbon retained in solution. Curiously, this effect is not reflected by hardness changes and therefore the main strengthening mechanism in these steels is thought to be the fine lath or cellular structures and not so much due to carbon in solution. The yield strength can therefore be represented by a modified version of the Hall–Petch equation (Gladman et al.[34]):

$$\sigma_{\gamma} = \sigma_i + Kd^{-1/2} + \sigma_{disl}(+\sigma_{ppt}) \tag{6.20}$$

where

σ_i is the friction stress (due to alloying elements in solution)
d refers to the mean cell or lath size
K is a constant
σ_{disl} refers to the hardening contribution due to dislocations and/or twins
σ_{ppt} refers to carbide precipitation after tempering

Typical yield strengths of these tempered steels are in the range 500–700 MN m^{-2}, for a mean lath width of 2–3 μm.

6.7.2 Controlled Transformation Steels

Compositions ranges:

0.05–0.3 wt% C
0.5%–2.0% Mn
0.2%–0.4% Si
14.0%–17.0% Cr
3.0%–7.0% Ni
~2% Mo

Other possible additions: V, Cu, Co, Al, Ti, etc.

Special properties: Very high strength, weldable, good corrosion resistance; used, e.g., as skin for high-speed aircraft and missiles.

Relevant phase diagrams: See Figure 6.33.

Microstructures: Fine lath martensite with possible fine network of δ-ferrite.

Comments: Since it is required to form, or work this material at ambient temperatures prior to hardening and tempering, elements that stabilize the austenite are used in significant amounts, e.g., Ni, Cu, etc. On the other hand, the M_s–M_f range should not be depressed too far, and the relative effects of

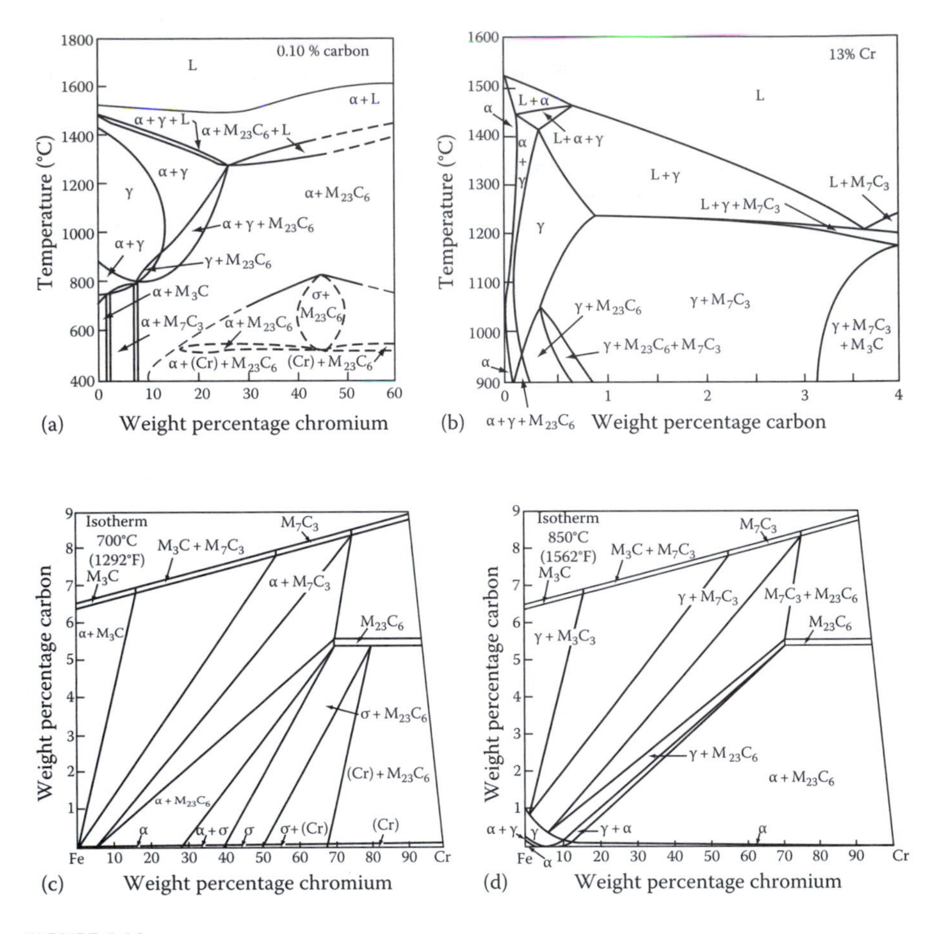

FIGURE 6.33
Phase diagram for Fe–Cr–C steels. ([a, c, and d] After Forgeng, W.D., *Metals Handbook*, 8th edn., Vol. 8, American Society for Metals, Metals Park, OH, 1973, p. 402; [b] from *A Guide to the Solidification of Steels*, Jernkontoret, Stockholm, 1977.)

alloying elements on M_s temperature are shown in Table 6.6. It is seen that in practice very strict control over composition of these steels must be made, balancing the amount of ferrite formers (e.g., Mn) with C content. Such amounts of δ-ferrite are sometimes retained in order to improve weldability

TABLE 6.6
Effect of Alloying Elements on M_s in Steels

Element	N	C	Ni	Co	Cu	Mn	W	Si	Mo	Cr	V	Al
Change in M_s (°C per wt%)	−450	−450	−20	+10	−35	−30	−36	−50	−45	−20	−46	−53

Source: From Pickering, F.B., *Int. Met. Rev.*, 21, 227, 1976.

(see Section 4.6.3). This also requires a careful balance in analyses, using, e.g., modified Schaeffer diagrams as a guide. Cold working is either carried out in the purely austenitic range (i.e., above M_d) the steel then being quenched to obtain martensite, or worked below M_d in which case deformation induces the transformation to occur without the need for refrigeration. The latter steels are known as TRIP steels. Since the M_s–M_f range is about 100°C–140°C, the M_s temperature should not lie too far below the working temperature, or refrigeration will have to be carried out at such low temperatures that it may become too expensive. Retained austenite is undesirable in these steels because of its adverse effect on strength. The fine martensitic structure, in combination with work hardening and tempering give these steels strengths up to ~1500 MN m^{-2}. In Table 6.7, typical properties of various controlled transformation steels are shown as a function of the type of heat treatment and transformation.

The mechanical properties given in Table 6.7 show that samples transformed by refrigeration generally give the higher strengths. It is also seen that the austenitizing temperatures may change from alloy to alloy. Choice of austenitizing temperature is critical with regard to solution treatment, resolution of carbides, and M_s temperature. For example, the lower the solution temperature, the more $M_{23}C_6$ will remain during austenitizing; this in turn reduces the Cr and C content of the austenite which raises the M_s temperature. The example given of a TRIP steel in Table 6.7 shows that this material has exceptional high strength and toughness (50% elongation).

6.7.3 TRIP-Assisted Steels

TRIP steels exhibit an excellent strength–ductility balance due to the continuous and progressive retained austenite transformation into martensite, a transformation induced by deformation in this case. From a technological point of view, this class of steels is very attractive for the automotive industry where the use of high-strength steels enables the use of thinner steel sections, saving weight, yet without compromising safety.[37] But usually high-strength steels perform poorly during metal forming which is not the case with TRIP steels.

A typical composition of such steels is Fe-0.15C-1.5Si-1.5Mn, in wt%. Note the high silicon content which is intended to suppress cementite precipitation during bainite transformation. A typical microstructure would be composed of, in volume fractions, 0.2 bainitic ferrite, 0.1 retained austenite with the balance being allotriomorphic ferrite.[38] Several publications reported studies on the microstructure–property relationships of these steels,[39–41] mainly associating the good uniform elongation obtained during deformation to the martensitic transformation of the retained austenite. The strength achieved in these steels was explained in terms of dislocation strengthening in the nearby ferrite caused by the transformation of retained austenite particles embedded in the microstructure.[42]

TABLE 6.7

Typical Properties of Controlled Transformation Steels

Composition	Heat Treatment	Tensile Strength (MN m^{-2})	0.2% Proof Stress (MN m^{-2})	Elongation (%)
0.1 C–17 Cr–4 Ni–3 Mn (showing effect of cold working)	925°C, tempered at 450°C for 3 h	950	620	10
	925°C, cold worked, 20% reduction, tempered at 450°C for 3 h	1340	1300	5
	925°C, cold worked, 40% reduction, tempered at 450°C for 3 h	1700	1670	3.5
0.1 C–17 Cr–4 Ni–3 Mn	950°C, 700°C for 2 h, tempered at 450°C for 2 h	1280	1110	15
	950°C, –78°C, tempered at 400°C for 1 h	1440	1200	19
0.1 C–17.5 Cr–3 Ni–3 Mn–2 Cu	1050°C, 700°C for 2 h, tempered at 400°C for 24 h	1120	940	5
	975°C, –78°C, tempered at 450°C for 24 h	1500	1260	11
0.1 C–17.5 Cr–4 Ni–3 Mn–1 A1	950°C, –78°C, tempered at 450°C for 4 h	1450	1260	13
0.1 C–17 Cr–4 Ni–2 Mo–2 Co	1000°C, –78°C, tempered at 450°C for 4 h	1340	1150	23
0.06 C–16.5 Cr–2 Mn–5 Ni–1.5 Mo–2 Co–1 A1	1050°C, 700°C for 2 h, tempered at 450°C for 4 h	1430	1270	3
	950°C, –78°C, tempered at 450°C for 4 h	1520	1240	21
0.07 C–17.5 Cr–3 Ni–2 Mn–2 Mo–2 Co–1 Cu	1050°C, 700°C for 2 h, tempered at 450°C for 4 h	1250	1110	10
	950°C, –78°C, tempered at 450°C for 4 h	1360	1240	20
TRIP steel: 0.3 C–2 Mn–2 Si–9 Cr–8.5 Ni–4.0 Mo	Austenitized 425°C, 80% reduction	1500	1430	50

Source: After Pickering, F.B., *Int. Met. Rev.*, 21, 227, 1976.

Strain-induced austenite transformation

The volume fraction of retained austenite V_γ as a function of plastic strain ε may be expressed as

$$\ln V_\gamma^0 - \ln V_\gamma = k\varepsilon \tag{6.21}$$

where

V_γ^0 is the initial volume fraction at zero plastic strain
k is a fitting parameter

This relation is an oversimplification in the sense that neither material-dependent factors nor deformation temperature effect is accounted for. In addition, the value of the fitting parameter changes as the steel composition/deformation temperature changes, i.e., it is not universal.

If one assumes that the fitting parameter k is proportional to the driving force for transformation ($\Delta G^{\gamma\alpha'} = G^{\alpha'} - G^{\gamma}$) then

$$k = k_1 \Delta G^{\alpha'\gamma} \tag{6.22}$$

where k_1 is a new fitting parameter. Therefore

$$\ln V_\gamma^0 - \ln V_\gamma = k_1 \Delta G^{\alpha'\gamma}\varepsilon \tag{6.23}$$

To verify the above hypothesis, the equation was tested against experimental data collected from the literature which represented a wide range of compositions and deformation temperatures.[38] The driving force for the transformation was calculated thermodynamically on the basis of the carbon-enriched austenite composition in conjunction with the temperature at which austenite was deformed. The model predictions are presented against collected experimental data in Figure 6.34.

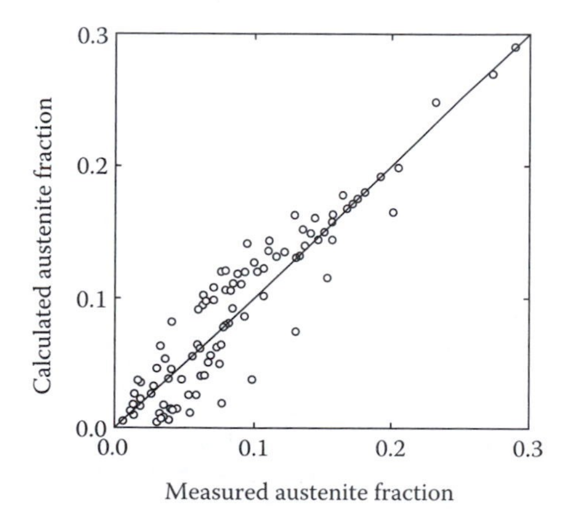

FIGURE 6.34
Predicted versus measured values of retained austenite content. (After Sherif, M.Y., Garcia Mateo, C., Sourmail, T., and Bhadeshia, H.K. D.H., *Mater. Sci. Tech.-Lond.*, 20, 319, 2004.)

As can be seen in Figure 6.34, the fitting was satisfactory with a correlation coefficient of 0.93 and a standard error of 0.023. k_1 was found to be 0.00446 mol J^{-1}. The advantage of the current model is that it enables a quantitative estimation of the strain-induced transformation of retained austenite in TRIP steels with a parameter k_1 which is composition and deformation temperature independent.

Model applicability to retained austenite transformation in fully bainitic steels

It is interesting now to see if the above model would hold if applied to a different microstructure. Figure 6.35 presents the calculated versus measured volume fractions of retained austenite as a function of engineering strain in a fully bainitic steel.[43] The steel transformed into bainite at different temperatures (300°C, 250°C, and 200°C) which resulted in different strength levels. Phase fraction measurements were carried out using x-ray diffraction.

Figure 6.35a, where the steel was transformed at 300°C, shows adequate prediction except at fracture strain where the model underestimated the austenite content. As can be seen in Figures 6.35b and c—transformation temperatures were 250°C and 200°C respectively—the model was less successful in its predictions.

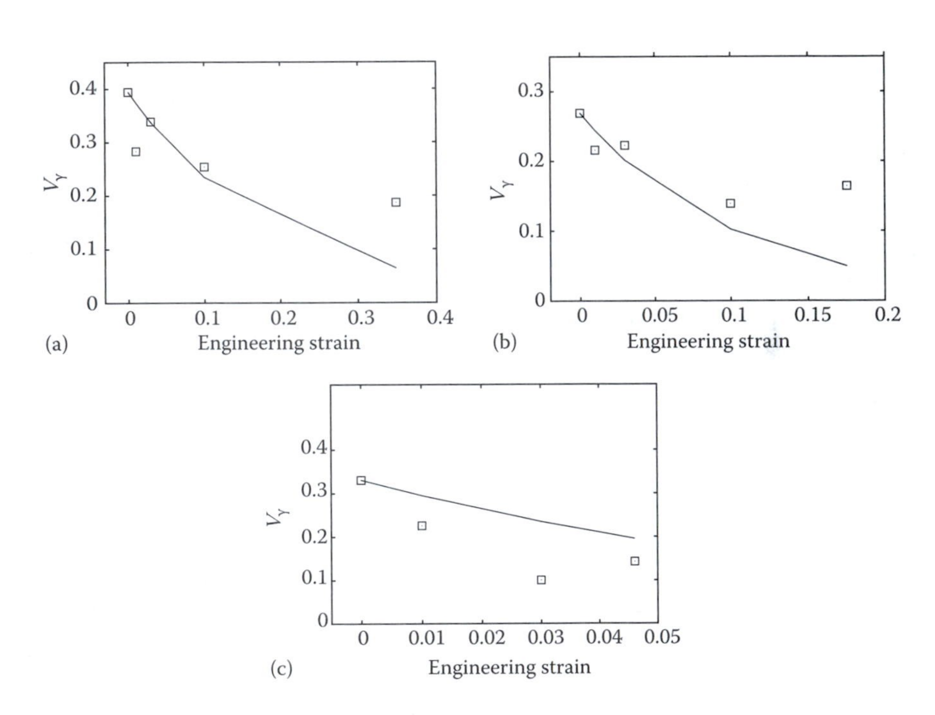

FIGURE 6.35
Measured (points) versus calculated (line) austenite phase fractions as a function of engineering strain. Steels were fully bainitic, carbide-free, transformed into bainite prior to straining at (a) 300°C, (b) 250°C, and (c) 200°C.

The inaccuracy in model predictions when bainitic steels were transformed at relatively low temperatures stems from the fact that the model does not account for the heterogeneity in the mechanical behavior of the phases in the microstructure. Accordingly, the model will tend to overestimate austenite content when austenite grains are embedded within a stronger matrix as the austenite in such a case sustains more plastic strain.[44]

When the model is applied to TRIP steels, where the allotriomorphic ferrite matrix is softer than austenite, austenite content could be underestimated. On the other hand, at a particular plastic strain level, the austenite should be subjected to greater stress which is not accounted for in the model in the free energy term. This missing aspect in the model would contribute to an overestimation of the austenite content predicted. As such, it seems both these two inaccuracies in the model more or less cancel each other out.

6.7.4 "Shape-Memory" Metal: Nitinol

Composition range: 55–55.5 wt% Ni–44.5%–45% Ti.
Possible additions: Small amounts of Co (to vary M_s).

Phase diagram: See Figure 6.36.
Phase transitions:

Ordered TiNi (I) bcc A2 structure
↓ {650°C–700°C diffusion controlled
TiNi (II): complex CsC1-type structure
↓ {170°C martensitic
TiNi (III): complex structure

Special properties: The TiNi (II ⇄ III) transformation is reversible and effectively enables the alloy to be deformed by a shear transformation, i.e., without irreversible plastic deformation occurring, by up to 16% elongation/contraction. Thus forming operations can be made below M_s which may be unformed simply by reheating to above the M_s. These unique properties are used in such applications as, e.g., toys, self-erecting space antennae, special tools, self-locking rivets, etc.

Microstructure: Very fine twinned martensite.

Comments: This transformation is interesting for two reasons: First, it involves a diffusionless transformation from one ordered structure to another. It is course fundamental to this type of transformation that if the austenitic phase is ordered, the martensitic product must also be ordered. Second, the mode of the transformation is such that very extensive deformation (up to 16%) can occur as a thermoelastic (nonplastic) martensitic shear mechanism, i.e., the transformation is reversible. Although this essentially involves an alloy of nominally stoichiometric composition, small additional increases in Ni (max ~55.6 wt%) can be tolerated. This increase in Ni content has the important effect of decreasing the M_s temperature. The M_s temperature as a function of Ni content is shown in Figure 6.36b. However, it is advisable not to exceed 55.6

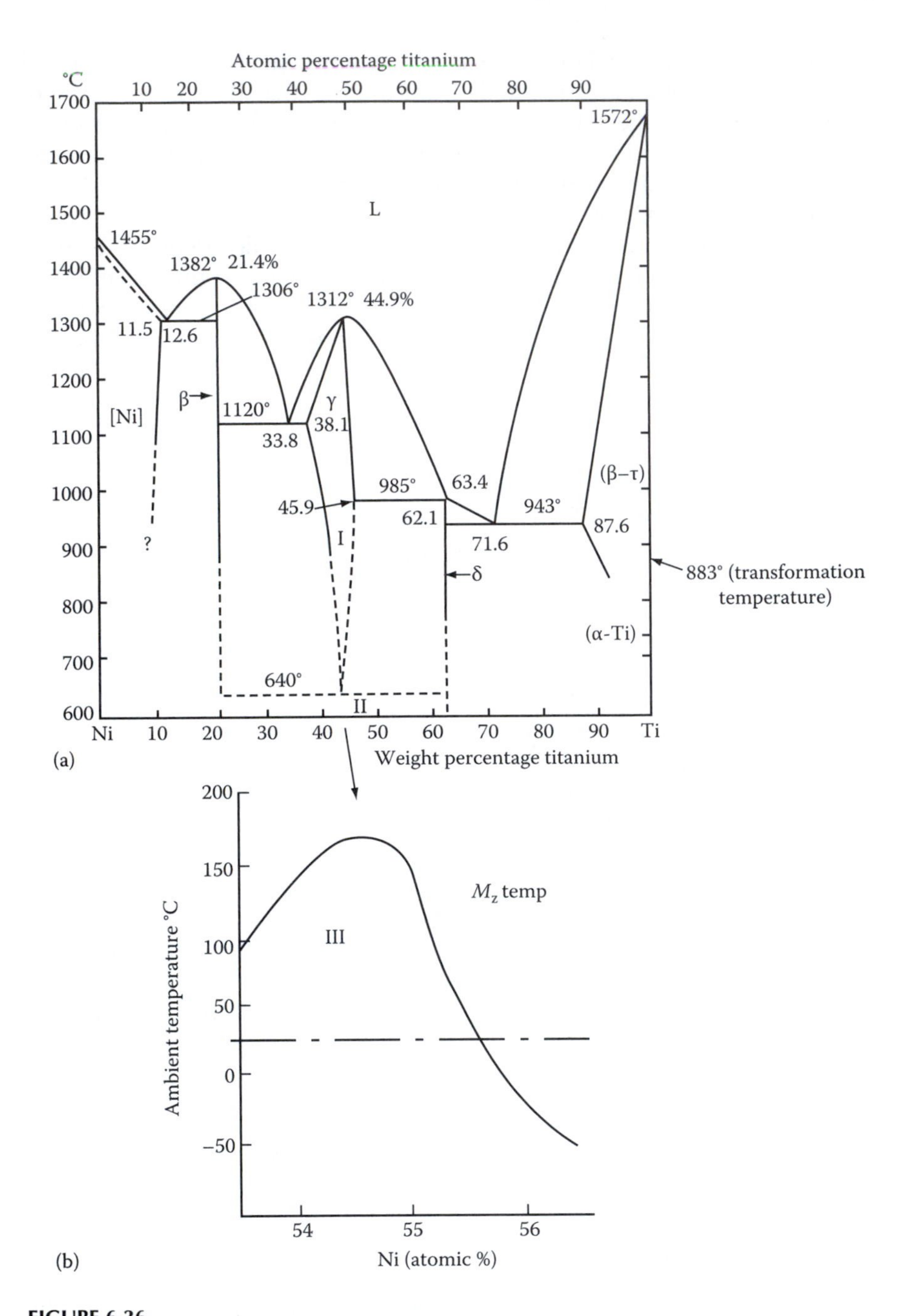

FIGURE 6.36
(a) Ti–Ni phase diagram. (After Hawkins, D.R., *Metals Handbook*, 8th edn., Vol. 8, American Society for Metals, Metals Park, Ohio, 1973, p. 326.) (b) M_s temperature as function of Ni content.

wt% Ni, to avoid the precipitation of the $TiNi_3$ phase. To avoid this problem, small amounts of Co can be added virtually on a 1:1 basis as a substitute for Ni.

While the martensitic phase is described as a complex CsCl ordered lattice, as a first approximation the transformation is related to the type bcc ⇄ hcp. The habit plane of the twinned martensite plate is irrational and close to $\{551\}_{bcc}$ (Sinclair and Mohamed[19]).

Another interesting feature of this transformation is that it appears to bring about an abrupt change in Young's modulus and yield strength. This change in Young's modulus also results in distinct changes in the modulus of resilience, the damping properties of the material being much greater in the martensitic (TiNi III) form.

The large amount of deformation that this alloy can undergo due its special transformation characteristics is utilized commercially. For example, it may be formed in a fully reversible way simply by deforming below the M_s temperature. Subsequent heating above the transition temperature then changes the deformed (sheared) structure back to its original form. This unique feature of alloys such as this has made them known as "memory metals." There are a number of so-called memory metals known today,[35] although none so commercially useful as TiNi alloys.

Exercises

6.1 Use free energy–composition diagrams to illustrate the driving force for the Fe–Ni martensitic transformation at $T > T_0$; $T = T_0$; $T = M_s$. Show how this chemical free energy can be estimated if the undercooling is known. Explain why the driving force for the nucleation of martensite at the M_s temperature is independent of carbon concentration in Fe–C steels.

6.2 What are the possible nonchemical energy terms in the martensitic transformation? Derive equations for the critical size and volumes of a martensite nucleus using classical nucleation theory. What evidence is there that martensite nucleates heterogeneously?

6.3 Evaluate Equations 6.8 through 6.10 for Fe–C martensite assuming $\Delta G_v = 174$ MJ m^{-3}, $\gamma = 20$ mJ m^{-2}, $s = 0.2$, $\mu = 80$ GN m^{-2}

6.4 Give an exact definition of the habit plane of martensite. Describe how this habit plane might be measured experimentally. Give possible reasons why there is so much scatter of habit plane measurements in a given sample.

6.5 In the phenomenological approach to martensitic transformations there are two different but equivalent ways of producing the lattice invariant shear. Show exactly what is meant by this. What is the experimental proof of both types of shear?

6.6 Draw a diagram to illustrate Bain's homogeneous deformation model for the fcc → bcc diffusionless transformation. Assuming $\alpha_\gamma = 3.56$ Å and $\alpha_\alpha = 2.86$ Å, and that c/a for martensite is 1.15 calculate the maximum movement experienced by atoms during the transformation. Assume that $c/a = 1.1$.

6.7 What are the essential differences in martensite nucleation models based (a) on changes at the core of a dislocation; (b) on dislocation strain field interaction? Discuss the advantages and disadvantages of both models in terms of the known characteristics of martensitic transformations.

6.8 Give possible reasons why the habit plane of martensite changes as a function of alloying content in steels and Fe–Ni alloys. What factors influence the retention of austenite in these alloys?

6.9 What is the role of austenitic grain size in martensitic transformations? Is austenitic grain size important to the strength of martensite? What others factors are important to the strength and toughness in technological hardened steels?

6.10 Suggest possible alloying and heat treatment procedures needed to design the following steels: (a) a quenched and tempered steel; (b) a dual phase steel; (c) a maraging steel; (d) a TRIP steel.

6.11 How would you characterize the unique properties of alloys which can be utilized as "memory metals." How would you design a TiNi alloy for use as, e.g., a self-locking rivet? Give instructions on how it is to be used.

References

1. J.W. Christian, *Theory of Transformations in Metals and Alloys*, Pergamon, Oxford, 1965.
2. Z. Nishiyama, *Martensitic Transformation*, Academic Press, New York, 1978.
3. A.H. Cottrell, *Theoretical Structural Metallurgy*, Edward Arnold, London, 1962.
4. For a comprehensive review see C.M. Wayman, *Introduction to the Crystallography of Martensitic Transformations*, Macmillan, New York, 1964.
5. E.C. Bain, *Trans. AIME*, 70, 25, 1924.
6. Z. Nishiyama, *Martensitic Transformation*, Academic Press, 1978, p. 232.
7. R. Bunshah and R.F. Mehl, *Trans. AIME*, 197: 1251, 1953.
8. R.E. Cech and D. Turnbull, *Trans. AIME*, 206: 124, 1956.
9. R. Huizing and J.A. Klostermann, *Acta Metall.*, 14: 1963, 1966.
10. K.E. Easterling and P.R. Swann, *Acta Metall.*, 19: 117, 1971.
11. C. Zener, *Elasticity and Anelasticity of Metals*, University of Chicago Press, Chicago, 1948.
12. J.W. Brooks, M.H. Loretto, and R.E. Smallman, *Acta Metall.*, 27: 1839, 1979.
13. J.A. Venables, *Phil. Mag.*, 7: 35, 1962.
14. B. Lehtinen and K.E. Easterling, in: P.R. Swann et al. (Eds.), *High Voltage Electron Microscopy*, Academic Press, London, 1974, p. 211.

15. A.J. Bogers and W.G. Burgers, *Acta Metall.*, 12: 255, 1964.
16. G.B. Olson and M. Cohen, *Met. Trans.*, 7A: 1897, 1905; 1915, 1976.
17. C.R. Houska, B.L. Averback, and M. Cohen, *Acta Metall.*, 8: 81, 1960.
18. J. Hirth and J. Lothe, *Theory of Dislocations*, McGraw-Hill, New York, 1968.
19. R. Sinclair and H.A. Mohamed, *Acta Metall.*, 26: 623, 1978.
20. G. Thomas and B.V.N. Rao, *Proc. Conf. on Martensitic Transformations*, Academy of Sciences, Kiev, 1978, pp. 57–64.
21. J.W. Eshelby, *Proc. Roy. Soc. A*, 241: 376, 1957.
22. A. Kelly, *Strong Solids*, Clarendon, Oxford, 1966.
23. J.Y. Koo and G. Thomas, *Mater. Sci. Eng.*, 24: 187, 1976.
24. A.R. Thölén, *Phys. Stat. Sol.*, 60: 153, 1980.
25. P.G. Shewmon, *Transformations in Metals*, McGraw-Hill, New York, 1969, p. 337.
26. F.C. Frank, *Acta Metall.*, 1: 15, 1953.
27. J. Friedel, *Dislocations*, Pergamon, Oxford, 1964.
28. K.E. Easterling and A.R. Thölén, *Acta Metall.*, 28: 1229, 1980.
29. J.R. Patel and M. Cohen, *Acta Metall.*, 1: 531, 1953.
30. C.M. Wayman, *Phase Transformations*, Series 3, No. 11, vol. 1, Institute of Metallurgists, London, 1979, pp. IV-1–IV-15.
31. A.M. Hunt and D.W. Pashley, *J. Phys. Rad.*, 23: 846, 1962.
32. G.R. Speich, *Metals Handbook*, vol. 8, ASM, Cleveland, Ohio, 1973, p. 202.
33. R.W.K. Honeycombe, *Structure and Strength of Alloy Steels*, Climax Molybdenum, London, 1973.
34. T. Gladman, D. Dulieu, and I.D. McIvor, *Microalloying* 75, Session 1, Metals Soc., London, 1975, p. 25.
35. J. Perkins (Ed.), *Shape Memory Effects in Alloys*, Plenum Press, New York, 1978.
36. M. Hillert, Calculation of phase equilibria, in: *Phase Transformations,* Chapter 5, American Society for Metals, Ohio, 1970.
37. O. Matsumura, Y. Sakuma, and H. Takechi, *Trans. ISIJ*, 27: 570, 1987.
38. M.Y. Sherif, C. Garcia Mateo, T. Sourmail, and H.K.D.H. Bhadeshia, Stability of retained austenite in TRIP-assisted steels, *Mater. Sci. Tech.-Lond.*, 20: 319, 2004.
39. Y. Tomota, K. Kuroki, T. Mori, and I. Tamura, *Mater. Sci. Eng.*, 24: 85, 1976.
40. H.K.D.H. Bhadeshia and D.V. Edmonds, *Met. Sci.*, 14: 41, 1980.
41. P. Jacques, E. Girault, Ph. Harlet, and F. Delannay, *ISIJ Int.*, 41: 1061, 2001.
42. P. Jacques, Q. Furnemont, A. Mertens, and F. Delannay, On the sources of work hardening in multiphase steels assisted by transformation-induced plasticity, *Philos. Mag. A*, 81: 1789, 2001.
43. M.Y. Sherif, Characterisation and development of nanostructured, ultrahigh strength, and ductile bainitic steels, PhD thesis, Cambridge University, 2006.
44. Q. Furnemont, P. Jacques, T. Pardoen, F. Lani, S. Godet, P. Harlet, K. Conlon, and F. Delannay, in: B.C. DeCooman (Ed.), *Int. Conf. on TRIP-Aided High Strength Ferrous Alloys*, Aachen, Germany, 2002, pp. 303–309. Wissenschaftsverlag Mainz GmbH.

Further Reading

J.W. Christian, *The Theory of Transformations in Metals and Alloys*, Pergamon, Oxford, 1965.

R.W.K. Honeycombe, *Steels—Microstructure and Properties*, Edward Arnold, London, 1981.

L. Kaufman and M. Cohen, Thermodynamics and kinetics of martensitic transformations, in: B. Chalmers and R. King (Eds.), *Progress in Metal Physics*, Vol. 7, Pergamon, New York, 1958, pp. 165–246.

Z. Nishiyama, *Martensitic Transformation*, Academic Press, New York, 1978.

F.B. Pickering, *Physical Metallurgy and the Design of Steels*, Applied Science Publishers, London, 1978.

A.K. Sinha, *Ferrous Physical Metallurgy*, Butterworths, London, 1989.

C.M. Wayman, *Introduction to the Crystallography of Martensitic Transformations*, Macmillan, New York, 1964.

C.M. Wayman, Phase transformations nondiffusive, in: R.W. Cahn and P. Haasen (Eds.), *Physical Metallurgy*, 2nd edn., Chapter 15, North-Holland, 1983.

*Solutions to Exercises**

Chapter 1

1.1 $C_p = 22.64 + 6.28 \times 10^{-3}T \text{ J mol}^{-1}\text{K}^{-1}$

$$\text{Entropy increase, } \Delta S = \int_{T_1}^{T_1} \frac{C_p}{T}\,dT$$

$$\begin{aligned}\Delta S_{300-1358} &= \int_{300}^{1358} \frac{22.64 + 6.28 \times 10^{-3}T}{T}\,dT \\ &= {}^{1358}_{300}\left[22.64 \ln T + 6.28 \times 10^{-3}T\right] \\ &= \underline{40.83 \text{ J mol}^{-1}\text{K}^{-1}}\end{aligned}$$

1.2

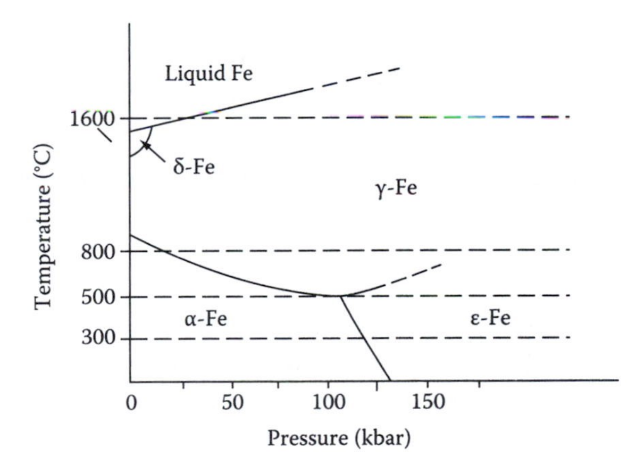

* Compiled by John C. Ion

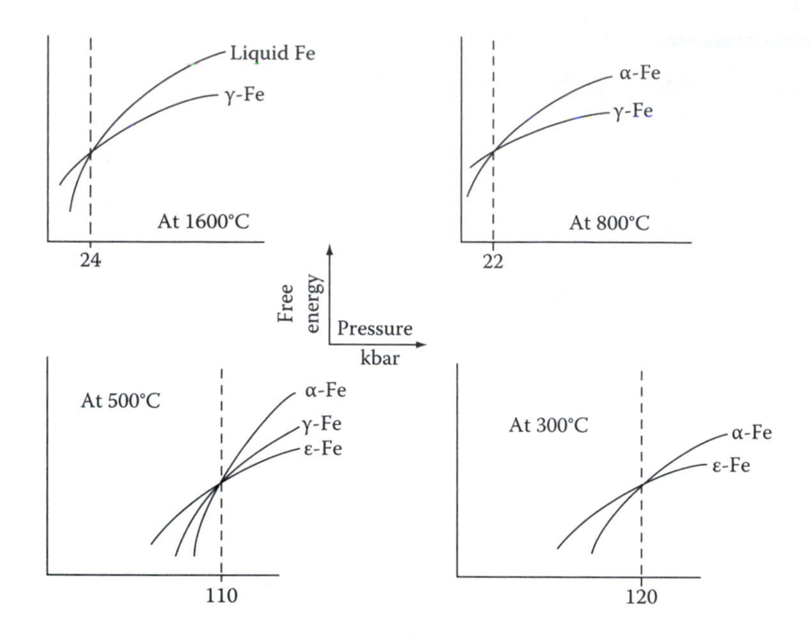

Schematic free energy–pressure curves for pure Fe.

1.3 From Equation 1.14

$$\left(\frac{\mathrm{d}P}{\mathrm{d}T}\right)_{\mathrm{eq}} = \frac{\Delta H}{T\Delta V}$$

Assuming ΔH and ΔV are independent of T and P for the range of interest, the equation may be rewritten as
where

$$\Delta H = H^{\mathrm{L}} - H^{\mathrm{S}} = 13{,}050 \text{ J mol}^{-1}$$

$$\Delta V = V^{\mathrm{L}} - V^{\mathrm{S}} = (8.0 - 7.6) \times 10^{-6} \text{ m}^3$$

$$T = (1085 + 273) \text{ K}$$

Thus if ΔP is 10 kbar, i.e., 10^9 N m^{-2}, the change in the equilibrium melting temperature is given by the above equation as

$$\underline{\Delta T = 42 \text{ K}}$$

1.4 Phases stable at low temperatures must have low enthalpies because the ($-TS$) term in the expression for G becomes negligible. Phases stable at high temperatures, on the other hand, have higher entropies to compensate for higher enthalpies.

1.5

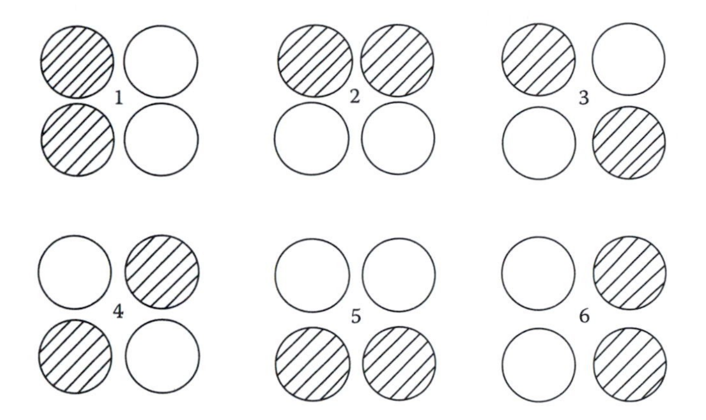

Six distinguishable configurations.

Theoretical number of distinguishable ways of arranging two black balls and two white balls in a square is

$$\frac{(N_B + N_W)!}{N_B!\,N_W!} = \frac{(2+2)!}{2!\,2!} = \underline{6}$$

1.6 Dividing both sides of Equation 1.30 by the number of moles of solution $(n_A + n_B)$ gives

$$\frac{dG'}{(n_A + n_B)} = \mu_A \frac{dn_A}{(n_A + n_B)} + \mu_B \frac{dn_B}{(n_A + n_B)}$$

The left-hand side of this equation is the free energy change per mole of solution and can therefore be written dG.

$$\frac{dn_A}{(n_A + n_B)} \quad \text{and} \quad \frac{dn_B}{(n_A + n_B)}$$

are the change in the mole fractions of A and B, dX_A and dX_B.

The above equation can therefore be written as

$$dG = \mu_A dX_A + \mu_B dX_B$$

Thus

$$\begin{aligned} \frac{dG}{dX_B} &= \mu_A \frac{dX_A}{dX_B} + \mu_B \\ &= -\mu_A + \mu_B \end{aligned} \tag{1.6.1}$$

But using Equation 1.31

$$G = \mu_A X_A + \mu_B X_B$$

gives

$$\mu_B = \frac{G - \mu_A X_A}{X_B}$$

giving

$$\frac{dG}{dX_B} = -\mu_A + \frac{G - \mu_A X_A}{X_B}$$

or

$$\mu_A = G - X_B \frac{dG}{dX_B}$$

From the figure

$$\mu_A = \overline{PR} - X_B \left(\frac{\overline{QR}}{X_B} \right) = \overline{PQ} = \overline{OS}$$

i.e., point S, the extrapolation of the tangent to point R on the G-curve represents the quantity μ_A

Equation 1.6.1 gives

$$\mu_B = \mu_A + \frac{dG}{dX_B}$$

i.e.,

$$\mu_B = \overline{OS} + \frac{\overline{UV}}{\overline{US}}$$

But

$$\overline{US} = \overline{OT} = 1$$

Thus

$$\mu_B = \overline{OS} + \overline{UV} = \overline{TV}$$

i.e., point V represents the quantity μ_B.

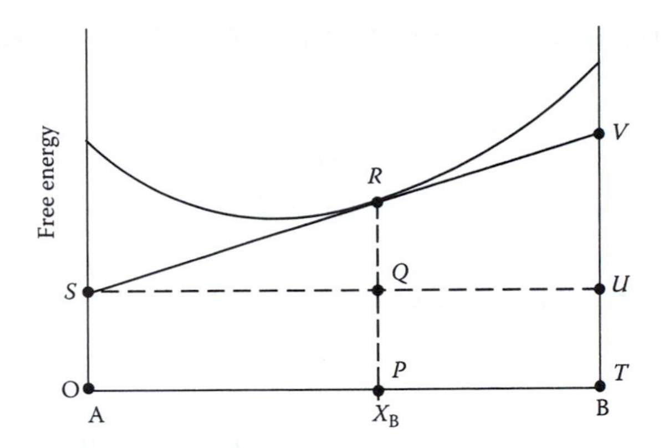

1.7 Equation 1.31: $G = \mu_A X_A + \mu_B X_B$

$$\begin{aligned}\text{Equation 1.39: } G &= X_A G_A + X_B G_B + \Omega X_A X_B + RT(X_A \ln X_A + X_B \ln X_B)\\ &= X_A G_A + X_B G_B + \Omega\left(X_A^2 X_B + X_B^2 X_A\right) + RT(X_A \ln X_A + X_B \ln X_B)\\ &= X_A\left[G_A + \Omega X_B^2 + RT \ln X_A\right] + X_B\left[G_B + \Omega X_A^2 + RT \ln X_B\right]\end{aligned}$$

Comparison with Equation 1.31 and using $X_A + X_B = 1$ gives

$$\mu_A = G_A + \Omega(1 - X_A)^2 + RT \ln X_A$$
$$\mu_B = G_B + \Omega(1 - X_B)^2 + RT \ln X_B$$

1.8 (a) Atomic weight of Au = 197
Atomic weight of Ag = 108

$$\text{No. of moles of Au} = \frac{15}{197} = 0.076$$
$$\text{No. of moles of Ag} = \frac{25}{108} = 0.231$$

$\therefore$ No. of moles of solution = $\underline{0.307}$

(b) $\text{Mole fraction of Au} = \dfrac{0.076}{0.307} = \underline{0.248}$

$$\text{Mole fraction of Ag} = \frac{0.231}{0.307} = \underline{0.752}$$

(c) Molar entropy of mixing, $\Delta S_{mix} = -R(X_A \ln X_A + X_B \ln X_B)$

$$\begin{aligned}\therefore \Delta S_{mix} &= -8.314(0.248 \cdot \ln 0.248 + 0.752 \cdot \ln 0.752)\\ &= \underline{4.66\ \text{J K}^{-1}\ \text{mol}^{-1}}\end{aligned}$$

(d) Total entropy of mixing = Molar entropy of mixing × no. of moles of solution

$$= 4.66 \times 0.307$$
$$= \underline{1.43\ \mathrm{J\ K^{-1}}}$$

(e) Molar free energy change at 500°C = ΔG_{mix}

$$= RT(X_A \ln X_A + X_B \ln X_B)$$

$$\therefore \Delta G_{mix} = -T\Delta S_{mix} = -773 \times 4.66 = \underline{-3.60\ \mathrm{kJ\ mol^{-1}}}$$

(f) $\mu_{Au} = G_{Au} + RT \ln X_{Au}$

$$= 0 + (8.314 \cdot 773 \cdot \ln 0.248)$$
$$= \underline{-8.96\ \mathrm{kJ\ mol^{-1}}}$$

$$\mu_{Ag} = G_{Ag} + RT \ln X_{Ag}$$
$$= 0 + (8.314 \cdot 773 \cdot \ln 0.752)$$
$$= \underline{-1.83\ \mathrm{kJ\ mol^{-1}}}$$

(g) For a very small addition of Au

$$dG' = \mu_{Au} \cdot dn_{Au}\ (T, P, n_B \text{ constant})$$

At 500°C, $\mu_{Au} = -8.96\ \mathrm{kJ\ mol^{-1}}$
Avogadro's number = 6.023×10^{23}

$$1\ \mathrm{eV} = 1.6 \times 10^{-19}\ \mathrm{J}$$

$$\therefore -8.96\ \mathrm{kJ\ mol^{-1}} = \frac{-8.96 \times 10^3}{1.6 \times 10^{-19} \times 6.023 \times 10^{23}}\ \mathrm{eV\ atom^{-1}}$$
$$= -0.1\ \mathrm{eV\ atom^{-1}}$$

∴ Adding one atom of Au changes the free energy of solution by $-\underline{0.1\ \mathrm{eV}}$.

1.9

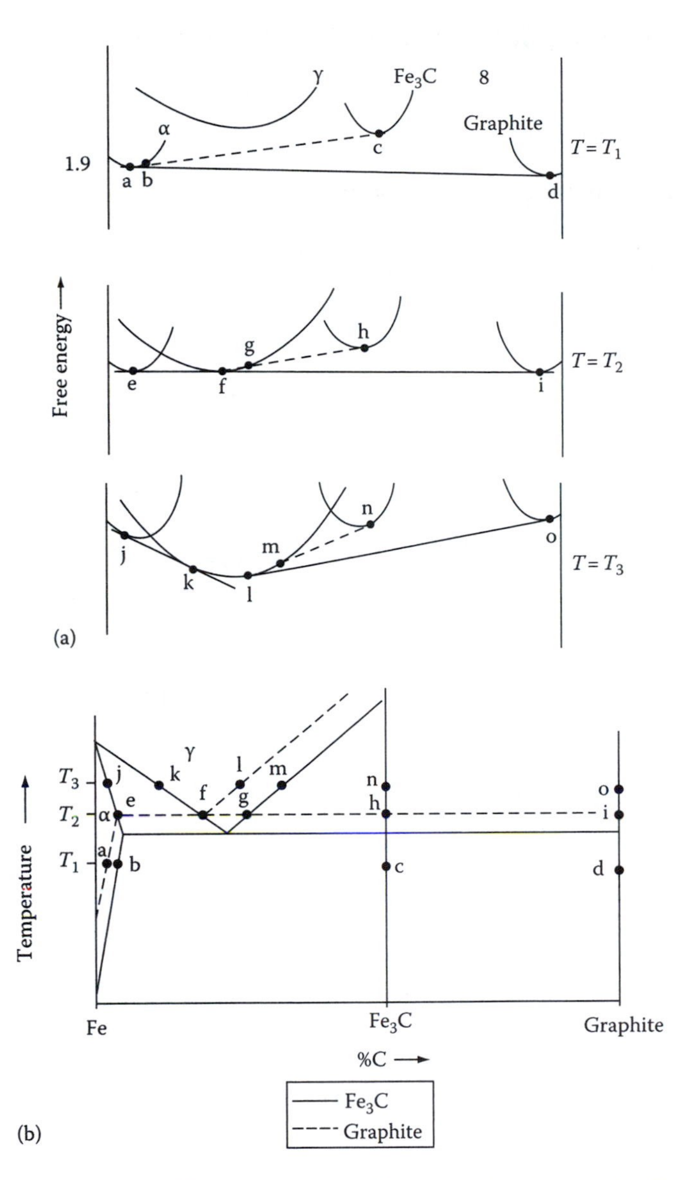

G-composition and T-composition diagrams for the Fe–Fe_3C and Fe–C systems (not to scale).

1.10

$$dG^{\alpha} = -\mu_A^{\alpha} dn_A$$

$$dG^{\beta} = +\mu_A^{\beta} dn_A$$

At equilibrium $dG^{\alpha} + dG^{\beta} = 0$
i.e., $-\mu_A^{\alpha} dn_A + \mu_A^{\beta} dn_A = 0$
i.e., $\mu_A^{\alpha} = \mu_A^{\beta}$
Similarly for B, C, etc.

1.11 Equilibrium vacancy concentration

$$X_v^e = \exp - \frac{\Delta G_v}{RT}$$
$$= \exp \frac{\Delta S_v}{R} \cdot \exp - \frac{\Delta H_v}{RT}$$

$1 \text{ eV} = 1.6 \times 10^{-19}$ J

$$\therefore R = 8.63 \times 10^{-5} \text{ eV atom}^{-1} \text{ K}^{-1}$$

$$\therefore X_v^c(933 \text{ K}) = \exp(2) \cdot \exp\left(\frac{-0.8}{8.63 \times 10^{-5} \times 933}\right)$$
$$= \underline{3.58 \times 10^{-4}}$$

$$X_v^c(298 \text{ K}) = \exp(2) \cdot \exp\left(\frac{-0.8}{8.63 \times 10^{-5} \times 298}\right)$$
$$= \underline{2.28 \times 10^{-13}}$$

1.12 Assume $X_{Si} = A \exp - (Q/RT)$

$$\therefore \ln X_{Si} = \ln A - \frac{Q}{RT}$$

At 550°C (823 K): $\ln 1.25 = \ln A - Q/(8.314 \times 823)$
which can be solved to give

$$Q = 49.45 \text{ kJ mol}^{-1}$$
$$A = 1721$$

Thus at 200°C (473 K)

$$X_{Si} = 1721 \cdot \exp - \left(\frac{49.450}{8.314 \times 473}\right)$$
$$\underline{= 0.006 \text{ atomic\%}}$$

According to the phase diagram, the solubility should be slightly under 0.001 atomic%. Reliable data is not available is such low temperatures due to the long times requires to reach equilibrium.

1.13 A sketch of the relevant phase diagram and free energy curves is helpful in solving this problem as shown in the figure. See the following two figures.

ΔG_A and ΔG_B are as defined in (b) and (c).

Since A and B are mutually immiscible, the tangent to the liquid curve G^L at $X_B = X_B^E$ will intercept the curves for the A and B phase as shown, i.e., $\mu_A^L = G_A^S$, $\mu_B^L = G_B^S$.

The liquid is assumed ideal, therefore from Figure 1.12

$$\Delta G_A \equiv -RT_E \ln X_A^E$$

But ΔG_A and ΔG_B can also be found from the relationships shown in figures (b) and (c).

If $C_P^L = C_P^S$, Equation 1.17 gives

$$\Delta G = \frac{L}{T_m} \cdot \Delta T$$

or

$$\Delta G = \Delta S_m \cdot \Delta T$$

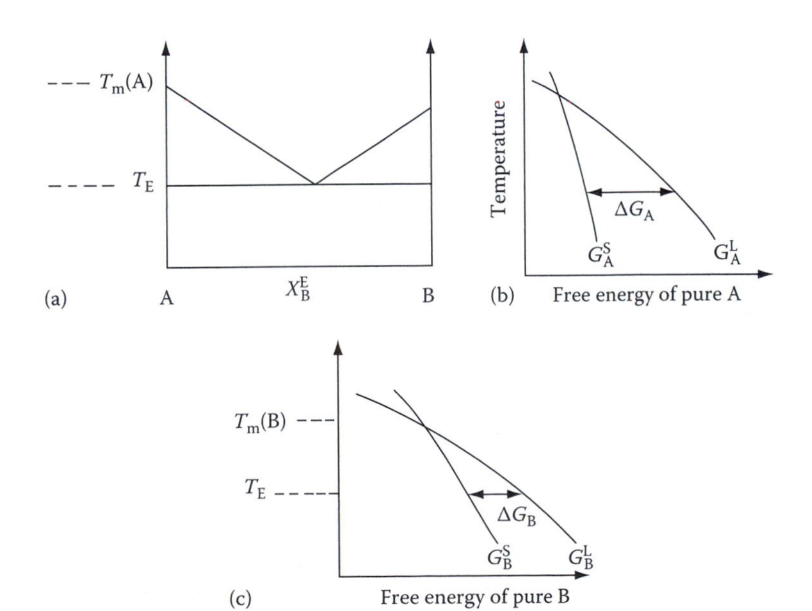

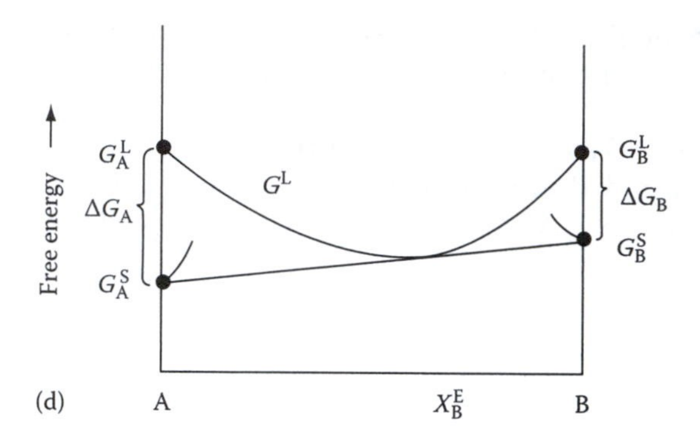

(a) Schematic phase diagram
(b) G–T curves for pure A
(c) G–T curves for pure B
(d) Free energy curves for the A–B system at T_E

$$\text{Thus } \Delta G_A = \Delta S_m(A) \cdot (T_m(A) - T_E)$$
$$\Delta G_B = \Delta S_m(B) \cdot (T_m(B) - T_E)$$

Finally therefore:

$$-RT_E \ln\ X_A^E = \Delta S_m(A) \cdot (T_m(A) - T_E)$$
$$-RT_E \ln\ X_B^E = \Delta S_m(B) \cdot (T_m(B) - T_E)$$

or

$$-\,8.314T_E \ln\ X_A^E = 8.4(1500 - T_E)$$
$$-\,8.314T_E \ln\left(1 - X_A^E\right) = 8.4(1300 - T_E)$$

Solving these equations numerically gives

$$X_A^E = 0.44$$
$$X_B^E = 0.56$$
$$T_E = 826\ \text{K}$$

1.14 If solid exists as a sphere of radius r within a liquid, then its free energy is increased by an amount

$$G_r^s - G_\infty^s = \frac{2\gamma V_m}{r} \quad \text{(from Equation 1.58)}$$

where
G_r^s is the molar free energy of the sphere
G_∞^s is the molar free energy in the absence of interfaces

Growth of the sphere must lead to a reduction of the total free energy of the system, i.e., growth can occur when

$$G_r^s < G^L$$

$$\text{i.e., } G^L - G_\infty^S > \frac{2\gamma V_m}{r}$$

See figure below

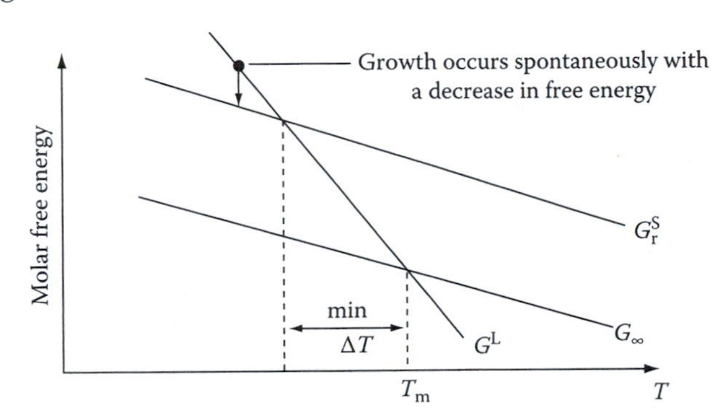

Substituting Equation 1.17 for $G^L - G_\infty^S$ gives

$$\frac{L\Delta T}{T_m} > \frac{2\gamma V_m}{r}$$

$$\text{i.e., } \Delta T > \frac{2\gamma V_m T_m}{rL}$$

Substituting the numerical values given

$$\Delta T(r = 1\ \mu\text{m}) > 0.2\ \text{K}$$
$$\Delta T(r = 1\ \text{nm}) > 200\ \text{K}$$

1.15 Composition = 40% A, 20%, B, 40% C

α = 80% A, 5% B, 15% C

β = 10% A, 70% B, 20% C

γ = 10% A, 20% B, 70% C

Let the mole fractions of α, β, and γ in the final microstructure be X_α, X_β, and X_γ, respectively.

$$\text{Balance on A: } 0.4 = 0.8X_\alpha + 0.1X_\beta + 0.1X_\gamma$$
$$\text{Balance on B: } 0.2 = 0.05X_\alpha + 0.7X_\beta + 0.2X_\gamma$$
$$\text{Balance on C: } 0.4 = 0.15X_\alpha + 0.2X_\beta + 0.7X_\gamma$$

Solving these equations gives: $\underline{X_\alpha = 0.43;\ X_\beta = 0.13;\ X_\gamma = 0.44}$

1.16 From Equations 1.41 and 1.43 we have

$$\mu_A = G_A + RT \ln \gamma_A X_A$$

where G_A is the free energy of pure A at temperature T and pressure P.
Suppose G_A is known for a given temperature and pressure T_0 and P_0

$$\text{i.e., } G_A(T_0, T_0) = G_A^0$$
$$\text{i.e., } G_A = -S_A \mathrm{d}T + V_m \mathrm{d}P$$

From Equation 1.9 for 1 mol of A

$$\mathrm{d}G_A = -S_A \mathrm{d}T + V_m \mathrm{d}P$$

Thus if S_A and V_m are independent of T and P, changing temperature from T_0 to T and pressure from P_0 to P will cause a total change in G_A of

$$\begin{aligned} \Delta G_A &= -S_A(T - T_0) + V_m(P - P_0) \\ \therefore G_A &= G_A^0 + \Delta G \\ &= G_A^0 + S_A(T_0 - T) + V_m(P - P_0) \end{aligned}$$

and

$$\mu_A = G_A^0 + S_A(T_0 - T) + V_m(P - P_0) + RT \ln \gamma_A X_A$$

The accuracy of this equation decreases as $(T - T_0)$ and $(P - P_0)$ increase.

Chapter 2

2.1 (a)

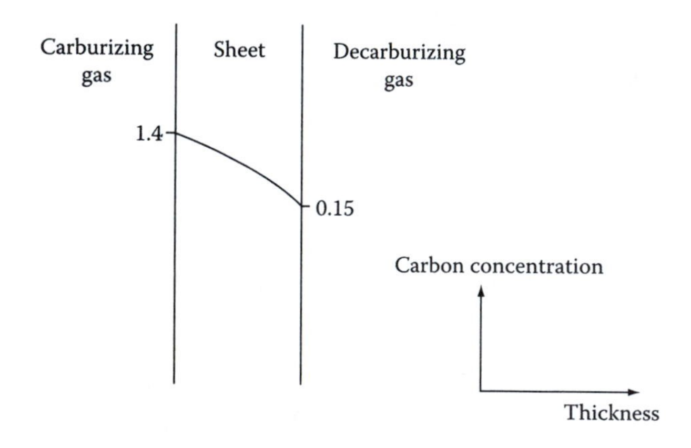

(b) Under steady-state conditions, flux of carbon atoms into one side = flux out of the other side = J.

$$J = -\frac{D_c dC}{dx}$$

$$\therefore D_{1.4}\left\{\frac{dC}{dx}\right\}_{1.4} = D_{0.15}\left\{\frac{dC}{dx}\right\}_{0.15}$$

$$\left(\frac{dC}{dx}\right)_{1.4} \Big/ \left\{\frac{dC}{dx}\right\}_{0.15} = \frac{D_{0.15}}{D_{1.4}} = \frac{2.5 \times 10^{-11}}{7.7 \times 10^{-11}} = \underline{0.32}$$

(c) Assume that the diffusion coefficient varies linearly with carbon concentration

$$D = a + bC$$

where a and b are constants that can be determined from the data given. Fick's first law then gives

$$J = -(a + bC)\frac{dC}{dx}$$

$$\text{or} \int J dx = -\int (a + bC) dC$$

$$\text{i.e.,} - Jx = aC + \frac{bC^2}{2} + d$$

where d is an integration constant. If we define $C = C_1$ at $x = 0$

$$d = -aC_1 - \frac{b}{2}C_1^2$$

Similarly, if $C = C_2$ at $x = l$, the thickness of the sheet gives

$$-Jl = aC_2 + \frac{b}{2}C_2^2 - aC_1 - \frac{b}{2}C_1^2$$

$$\text{i.e.,}\ J = \left\{a(C_1 - C_2) + \frac{b}{2}(C_1^2 - C_2^2)\right\} \Big/ l$$

The constant a and b can be determined from

$$D_1 = a + bC_1$$
$$D_2 = a + bC_2$$

$$\text{from which}\ a = D_2 - \left(\frac{D_1 - D_2}{C_1 - C_2}\right)C_2$$

$$\text{and}\ b = \frac{D_1 - D_2}{C_1 - C_2}$$

Substitution of these expressions into the equation for J gives after simplification

$$J = \left(\frac{D_2 + D_1}{2}\right)\frac{C_1 - C_2}{l}$$

$$\begin{aligned} \text{Substituting: } D_1 &= 7.7 \times 10^{-11}\ \text{m}^2\ \text{s}^{-1} \\ D_2 &= 2.5 \times 10^{-11}\ \text{m}^2\ \text{s}^{-1} \\ C_1 &= \frac{1.4}{0.8} \times 60\ \text{kg m}^{-3} \\ C_2 &= \frac{0.15}{0.8} \times 60\ \text{kg m}^{-3} \\ l &= 2 \times 10^{-3}\ \text{m} \end{aligned}$$

$$\text{gives} \quad \underline{J = 2.4 \times 10^{-6}\ \text{kg m}^{-2}\ \text{s}^{-1}}$$

2.2

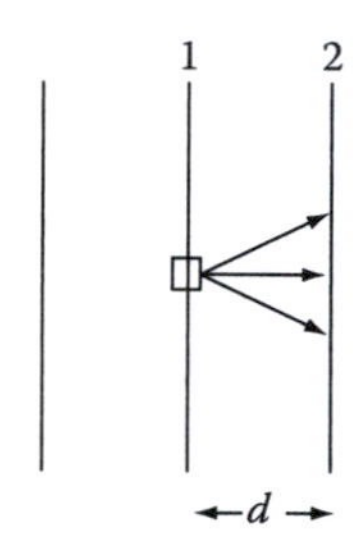

Consider two adjacent (111) planes in an fcc crystal. A vacancy in plane 1 can jump to one of three sites in plane 2. For the sake of generality, let this number of jumps be designed P (=3). In all there are 12 possible sites (nearest neighbors).

If n_1 and n_2 are the numbers of vacancies m^{-2} in planes 1 and 2, respectively, the number of jumps from 1 to 2 will be given by

$$\overrightarrow{J}_v = \frac{P}{12}\Gamma_v n_1\ \text{m}^{-2}\ \text{s}^{-1}$$

where Γ_v is the jump frequency of the vacancies.

Likewise

$$\overleftarrow{J}_v = \frac{P}{12}\Gamma_v n_2$$

Therefore following the same arguments as in Section 2.2.1 gives

$$J_v = -\left(\frac{P}{12}d^2\Gamma_v\right)\frac{\delta C_v}{\delta x}$$

where d is the perpendicular separation of the adjacent planes, i.e., we can write

$$D_v = \frac{Pd^2}{12} \cdot \Gamma_v$$

In fcc metals the jumps distance α is given by

$$\alpha = \frac{a}{\sqrt{2}}$$

where a is the lattice parameter.

For (111) planes

$$d = \frac{a}{\sqrt{3}} = \alpha\sqrt{\frac{2}{3}}$$

Putting $P = 3$ gives

$$\underline{D_v = \frac{1}{6}\alpha^2\Gamma_v}$$

For (100) planes, adjacent planes are in fact (200)

$$\therefore d = \frac{a}{2} = \frac{\alpha}{\sqrt{2}}$$

$$\therefore \underline{D_v = \frac{1}{6}\alpha^2\Gamma_v}$$

2.3 The activity along the bar is described by the following equation

$$\text{Activity} = \frac{A_0}{2} \cdot \exp\left(-\frac{x^2}{!4Dt}\right)$$

where

A_0 is the initial activity
D is the diffusion constant
t is the time
x is the distance along the bar

Thus by plotting in (activity) vs. x^2, a straight line of slope $-(4Dt)^{-1}$ is produced, enabling D to be found since t is known.

x (μm)	10	20	30	40	50
x (μm)	100	400	900	1600	2500
Activity	83.8	66.4	42.0	23.6	8.74
ln (activity)	4.43	4.20	3.74	3.16	2.17

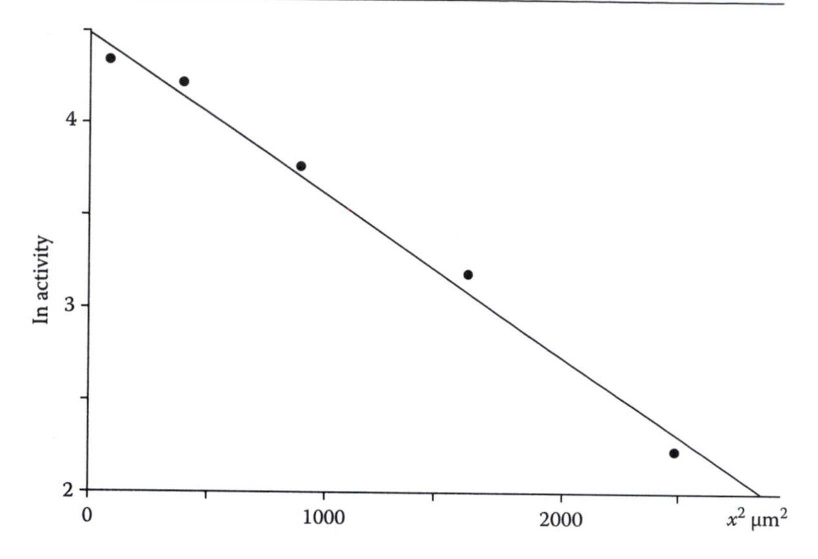

From the graph: slope $= -8.66 \times 10^{-4}$ μm^{-2}.

$$\text{Hence: } -\frac{1}{(4Dt)} = -8.66 \times 10^{-4}$$

Since $t = 24$ h

$$\underline{D = 3.34 \times 10^{-15}\ \text{m}^2\ \text{s}^{-1}}$$

2.4

$$C = \overline{C} + \beta_0 \sin\left(\frac{\pi x}{l}\right) \exp -\frac{t}{\tau}$$

For this equation to be a solution of Fick's second law, the following condition must be met

$$\frac{\delta C}{\delta t} = D_B \frac{\delta^2 C}{\delta x^2}$$

$$\frac{\delta C}{\delta t} = -\beta_0 \sin\left(\frac{\pi x}{l}\right) \cdot \exp -\frac{1}{\tau}$$

$$\frac{\delta C}{\delta x} = \beta_0 \exp -\frac{1}{\tau} \cdot \frac{\pi}{l} \cos\left(\frac{\pi x}{l}\right)$$

$$\frac{\delta^2 C}{\delta x^2} = -\beta_0 \exp - \frac{t}{\tau} \cdot \frac{\pi^2}{l^2} \sin\left(\frac{\pi x}{l}\right)$$

$$\therefore \tau \cdot \frac{\delta C}{\delta t} = \frac{\delta^2 C}{\delta x^2} \cdot \frac{l^2}{\pi^2}$$

$$\text{But } \tau = \frac{l^2}{\pi^2 D_B} \text{ (Equation 2.21)}$$

$$\therefore \underline{\frac{\delta C}{\delta t} = D_B \frac{\delta^2 C}{\delta x^2}}$$

2.5 (a)

$$C(x) = \frac{4C_0}{\pi} \sum_{i=0}^{\infty} \frac{1}{2i+1} \cdot \sin \frac{(2i+1)\pi x}{l}$$

where
l is the thickness of sheet
C_0 is the initial concentration

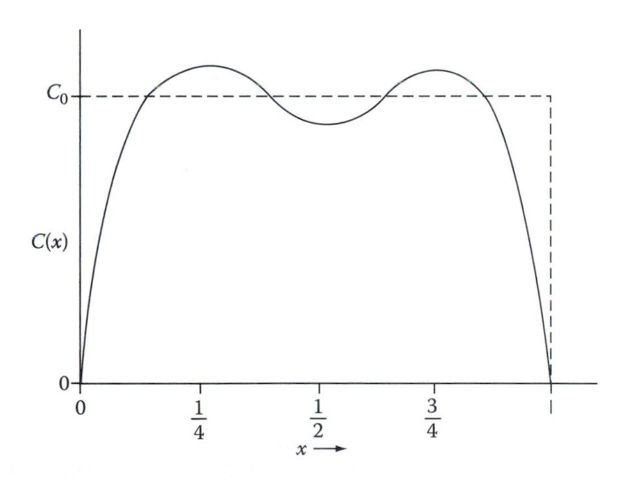

The first two terms of the series are

$$C(x) = \frac{4C_0}{\pi} \left[\sin \frac{\pi x}{l} + \frac{1}{3} \sin \frac{3\pi x}{l}\right]$$

Plotting the sum for the range $0 < x < l$ gives the curve plotted opposite.

(b) If the surface concentration is effectively zero, the solution to the diffusion equation becomes

$$C(x,t) = \frac{4C_0}{\pi} \sum_{i=0}^{\infty} \frac{1}{2i+1} \cdot \sin\left(\frac{(2i+1)\pi x}{l}\right) \cdot \exp\left[\frac{-(2i+1)^2 \pi^2 D t}{l^2}\right]$$

The amplitude of the first term (A_1) is obtained by putting $x = 1/2$ and $i = 0$, i.e.,

$$A_1 = \frac{4C_0}{\pi} \cdot \exp\left[\frac{-\pi^2 Dt}{l^2}\right]$$

The amplitude of the second term (A_2) is obtained by putting $x = l/6$ and $i = 1$, i.e.,

$$A_2 = \frac{4C_0}{3\pi} \cdot \exp\left[\frac{-9\pi^2 Dt}{l^2}\right]$$

If $A_2 < 0.05\ A_1$

$$\frac{4C_0}{3\pi} \cdot \exp\left[\frac{-9\pi^2 Dt}{l^2}\right] < 0.05 \cdot \frac{4C_0}{\pi} \cdot \exp\left[\frac{-\pi^2 Dt}{l^2}\right]$$

$$\text{which gives } \underline{t > 0.0240\frac{t^2}{D}}$$

(c) Assume that the time taken to remove 95% of all the hydrogen is so long that only the first term of the Fourier series is significant. The hydrogen concentration at this stage will then be given by

$$C(x,t) = \frac{4C_0}{\pi} \cdot \sin\frac{\pi x}{l} \cdot \exp\left(\frac{-\pi^2 Dt}{l^2}\right)$$

i.e., as shown in the following figure.

At the required time (t_1) the shaded area in the figure will be 5% of the area under the concentration line at $t = 0$, i.e.,

$$\int_0^l C(x_1 t_1)\mathrm{d}x = 0.05 C_0 l$$

$$\therefore \frac{4C_0}{\pi} \cdot \exp\left(\frac{-\pi^2 Dt_1}{l^2}\right) \int_0^l \sin\frac{\pi x}{l}\mathrm{d}x = 0.05 C_0 l$$

$$\text{which gives } t_1 = 0.282\frac{l^2}{D}$$

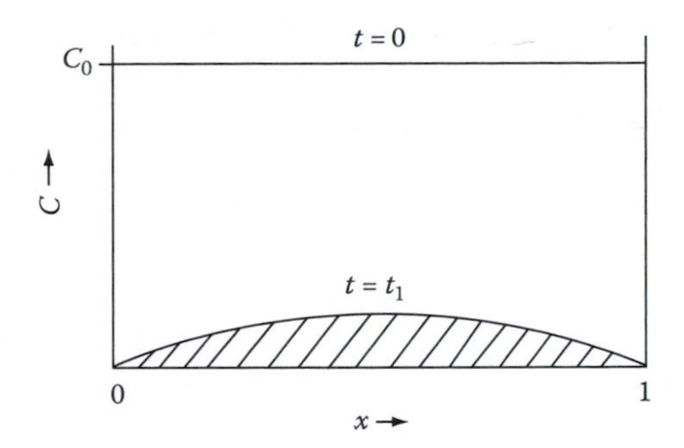

Note that this time is an order of magnitude larger than the time derived in part (b). Consequently it is clearly justified to ignore all terms of the Fourier series but the first.

From Table 2.1,

$$D = 0.1 \ \exp\left(\frac{-13.400}{RT}\right)$$

$$\text{i.e., } D(20°\text{C}) = 4.08 \times 10^{-4} \ \text{mm}^2 \ \text{s}^{-1}$$

Thus for $l = 10$ mm; $t_1 = 19.2$ h, for $l = 100$ mm; $t_1 = 1920$ h (80 days).

2.6

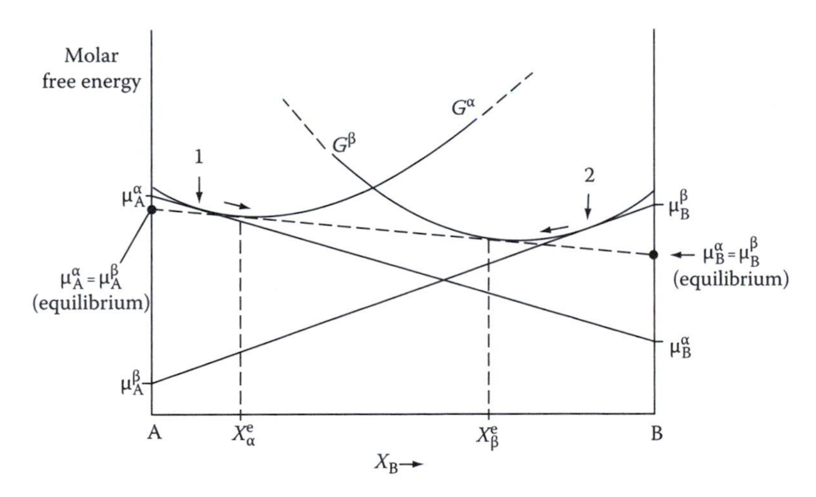

At the initial compositions 1 and 2 of α and β, respectively the chemical potentials of A and B atoms in each phase can be found by extrapolation of the tangents to the free energy curves at 1 and 2 to the corresponding sides of the free energy diagram, as just illustrated.

All atoms diffuse so as to reduce their chemical potential. Therefore, A atom will have a tendency to diffuse from α to $\beta(\mu_A^\alpha > \mu_A^\beta)$ and B atoms will have a tendency to diffuse from β to $\alpha(\mu_B^\beta > \mu_B^\alpha)$.

The resultant composition changes are indicated in the diagram. Diffusion stops, and equilibrium is reached, when $\mu_A^\alpha = \mu_A^\beta$ and $\mu_B^\alpha = \mu_B^\beta$. That this process results in a reduction in the total free energy of the diffusion couple can be seen from the diagram below. The initial free energy G_1 can be reduced to G_2 by a change in the compositions of the α- and β-phases to X_α^c and X_β^c, the equilibrium compositions (provided $X_\alpha^c < X_{Bulk} < X_\beta^c$).

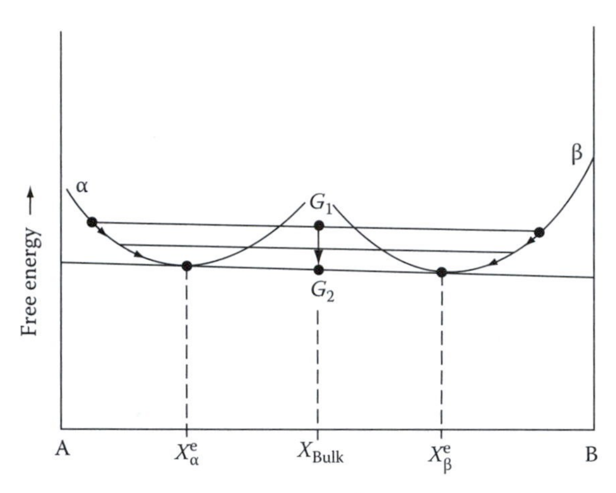

2.7 Substituting into Darken's equations (Equations 2.47 and 2.51)

$$v = \left(D_{Zn}^\alpha - D_{Cu}^\alpha\right)\frac{\delta X_{Zn}}{\delta x}$$
$$\overline{D}^\alpha = X_{Cu}D_{Zn}^\alpha + X_{Zn}D_{Cu}^\alpha$$

we obtain

$$0.026 \times 10^{-6} = \left(D_{Zn}^\alpha - D_{Cu}^\alpha\right) \times 0.089\,\text{mm s}^{-1}$$
$$4.5 \times 10^{-7} = 0.78D_{Zn}^\alpha + 0.22D_{Cu}^\alpha\,\text{mm}^2\,\text{s}^{-1}$$

From which

$$D_{Zn}^\alpha = 5.1 \times 10^{-7}\,\text{mm}^2\,\text{s}^{-1}$$
$$D_{Cu}^\alpha = 2.2 \times 10^{-7}\,\text{mm}^2\,\text{s}^{-1}$$

The expected variation of D^{α}_{Zn}, D^{α}_{Cu}, and $\tilde{D}^{\alpha}$ are shown schematically below

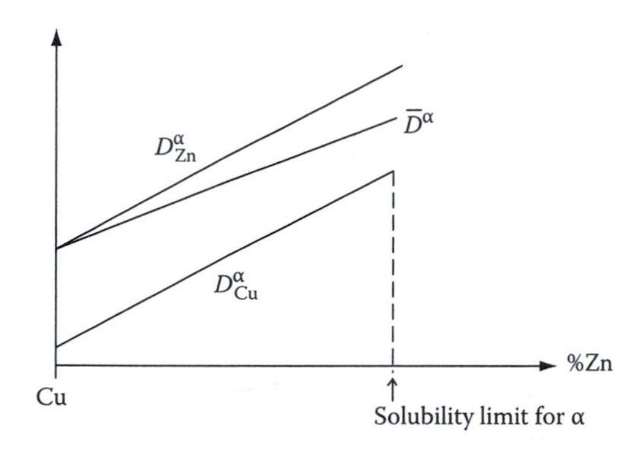

Since Zn has a lower melting point than Cu, it diffuses faster of the two, and since increasing the zinc content reduces the liquidus temperature, all diffusivities can be expected to increase with increasing Zn concentration.

2.8

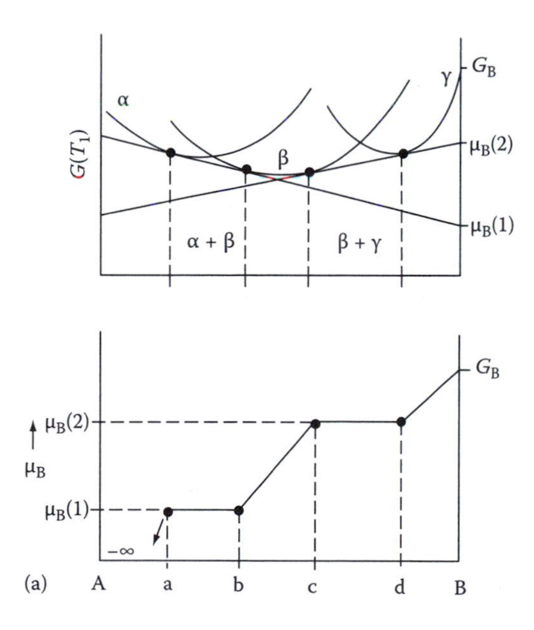

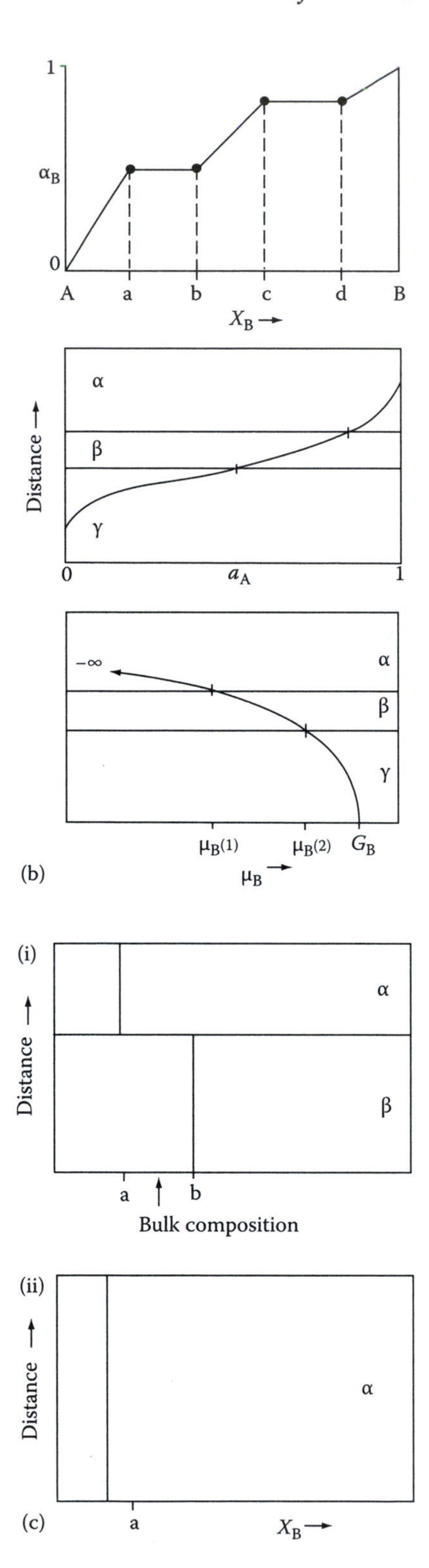
1
α_B
0
A
a
b
c
d
B
X_B
α
Distance
β
γ
0
a_A
1
−∞
α
β
γ
$\mu_B(1)$
$\mu_B(2)$
G_B
(b)
μ_B
(i)
α
Distance
β
a
b
Bulk composition
(ii)
Distance
α
(c)
a
X_B

2.9

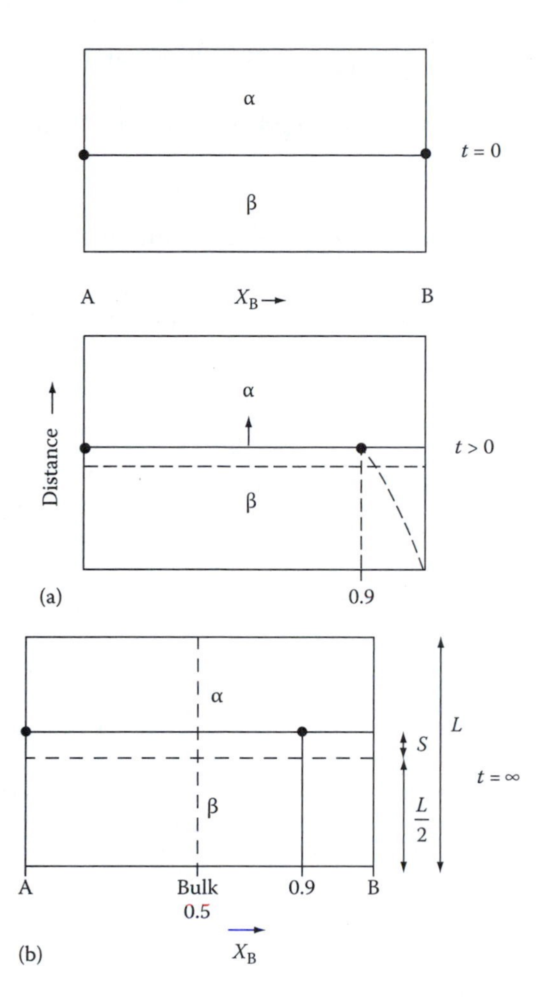

The total distance the interface moves, *s*, can be calculated in terms of the total couple thickness, *L*, by writing down an equation describing the conservation of B, i.e.,

$$0.9\left(\frac{L}{2}+s\right)+0\left(\frac{L}{2}-s\right)=0.5\,L$$

$$\therefore\ \underline{\frac{s}{L}=5.6\times10^{-2}}$$

Chapter 3

3.1 Considering only nearest neighbors, if a surface atom has *B* "broken" bonds, it will have an excess energy of B, $\varepsilon/2$, where ε is the bond energy.

For fcc crystals, each atom has 12 nearest neighbors in the bulk, so that $\varepsilon = L_s/6N_a$, where L_s is the molar latent heat of sublimation and N_a Avogadro's number (no. of atoms per mole).

$$\gamma'_{\text{sv}} = \frac{B}{12} \cdot \frac{L_s}{N_a} \text{ per surface atom}$$

If each surface atom is associated with a surface area A, the surface energy is

$$\gamma_{\text{sv}} = \frac{\text{B}}{12\text{A}} \cdot \frac{L_s}{N_a} \text{ per unit area}$$

A can be calculated in terms of the lattice parameter a:

$\{hkl\}$	**(111)**	**{200}**	**{220}**
	$\frac{a}{\sqrt{2}}$	a	a
A	$\frac{a}{\sqrt{2}} \cdot \frac{a}{\sqrt{2}} \cdot \frac{\sqrt{3}}{2}$	$\frac{a^2}{2}$	$a \cdot \frac{a\sqrt{2}}{2}$
B	3	4	5*
γ'_{SV}	$0.25\left[\frac{L_s}{N_a}\right]$	$0.33\left[\frac{L_s}{N_a}\right]$	$0.42\left[\frac{L_s}{N_a}\right]$
γ_{SV}	$0.58\left[\frac{L_s}{a^2 N_a}\right]$	$0.67\left[\frac{L_s}{a^2 N_a}\right]$	$0.59\left[\frac{L_s}{a^2 N_a}\right]$

* Each surface atom is connected to two nearest neighbors in the {220} plane. Therefore it must be connected to 10 others out of the plane. Since the atoms are symmetrically disposed about the {220} plane, there must be five bonds above the plane of the paper and five below (giving a total of 12).

It can be shown that in general for fcc metals.

$$A = \frac{a^2\sqrt{(h^2 + k^2 + l^2)}}{4}$$

For the simple cases above, however, A can be calculated directly from a sketch, as shown.

3.2

$$E_{\text{sv}} = (\cos\theta + \sin|\theta|) \cdot \frac{\varepsilon}{2a^2}$$

$$\text{i.e., } E_{\text{sv}} = (\cos\theta + \sin\theta) \cdot \frac{\varepsilon}{2a^2},\ \theta > 0$$

$$\frac{dE_{\text{sv}}}{d\theta} = (-\sin\theta + \cos\theta) \cdot \frac{\varepsilon}{2a^2}$$

$$\left(\frac{dE}{d\theta}\right)_{\theta=0} = \frac{\varepsilon}{2a^2}$$

$$\text{and } E_{sv} = (\cos\theta - \sin\theta) \cdot \frac{\varepsilon}{2a^2}, \ \theta < 0$$

$$\text{which gives } \left(\frac{dE_{sv}}{d\theta}\right)_{\theta=0} = -\frac{\varepsilon}{2a^2}$$

At $\theta = 0$ there is a cusp in the $E_{sv} - \theta$ curve with slopes $\pm(\varepsilon/2a^2)$.

3.3 For a two-dimensional rectangular crystal with sides of lengths l_1 and l_2 and surface energies γ_1 and γ_2, respectively, the total surface energy is given by

$$G = 2(l_1\gamma_1 + l_2\gamma_2)$$

The equilibrium shape is given when the differential of G equals zero, i.e.,

$$dG = 2(l_1 d\gamma_1 + \gamma dl_1 + l_2 d\gamma_2 + \gamma_2 dl_2) = 0$$

Assuming that γ_1 and γ_2 are independent of length gives

$$\gamma_1 dl_1 + \gamma_2 dl_2 = 0$$

But since the area of the crystal $A = l_1 l_2$

$$dA = l_1 dl_2 + l_2 dl_1 = 0$$

Giving

$$\underline{\frac{l_1}{l_2} = \frac{\gamma_2}{\gamma_1}}$$

3.4 (a) By measuring, the misorientation $\theta \approx 11$.

$$D = \frac{b}{\sin\theta}$$

$$D = \frac{1.53}{\sin 11^\circ} \approx 8.0 \text{ mm}$$

which is very close to the mean dislocation spacing in the boundary.

3.5 Like all other natural processes, grin-boundary migration always results in a reduction in total free energy.

Grain growth
During the process of grain growth all grains have approximately the same, low dislocation density, which remains unchanged during.

Grain boundaries move toward their centers of curvature in this case, because atoms tend to migrate across the boundaries in the opposite direction (from the high pressure side to the low pressure side), in order to reduce their free energy, or chemical potential.

The process also results in a reduction of the total number of grains by the growth of large grains at the expense of smaller ones. The net result is a reduction in the total grain-boundary area and total grain-boundary energy.

Recrystallization

In this case, grain-boundary energy is insignificant in comparison with the difference in dislocation energy density between recrystallized grain and surrounding deformed matrix. The small increase in total grain-boundary energy that accompanies growth of a recrystallization nucleus is more than compensated for by the reduction in total dislocation energy.

The boundaries of recrystallization nuclei can therefore migrate away from their centers of curvature.

3.6 (a) The pulling force acting on the boundary is equivalent to the free energy difference per unit volume of material.
If the dislocations have an energy of $(\mu b^2/4)\,\mathrm{J\ m^{-1}}$, and the dislocation density is $10^{16}\ \mathrm{m^{-2}}$, then the free energy per unit volume, G, is given by

$$G = 10^{16} \times \frac{10^{10} \times (0.28 \times 10^{-9})^2}{4} = 1.96\,\mathrm{MJ\ m^{-3}}$$

Thus the pulling force per unit area of boundary is $1.96\ \mathrm{MN\ m^{-2}}$.

(b) For nucleus growth, reduction in free energy due to annihilation of dislocations must be greater than or equal to the retarding force due to grain-boundary curvature.

Equating this with the driving force across a curved boundary

$$\text{i.e., } 1.96 \times 10^6 \geq \frac{2\gamma}{r}$$

$$\therefore\ r \geq \frac{2\gamma}{1.96 \times 10^6}$$

Thus the smallest diameter = 1.0 μm

3.7 From the phase diagrams, the limit of solid solubility of Fe in Al is 0.04 wt% Fe, whereas that of Mg in Al is 17.4 wt% Mg. If one element is able to dissolve another only to a small degree, the extent of grain-boundary enrichment will be large. (See for example Figure 3.28). Thus, grain-boundary enrichment of Fe in dilute Al–Fe alloys would be expected to be grater than that of Mg in Al–Mg alloys.

3.8 See Figure 3.35.

If $d_\alpha < \Delta_\beta$, then in general the dislocation spacing (D) will span n atom planes in the β-phase and $(n + 1)$ planes in the α-phase, i.e.,

$$D = nd_\beta = (n+1)d_\alpha$$

From the definition of δ we have

$$\delta = \frac{d_\alpha - d_\beta}{d_\alpha}(-\text{ve})$$
$$\therefore\ d_\beta = (1+\delta)d_\alpha$$

Substitution into the first equation gives

$$n(1+\delta)d_\alpha = (n+1)d_\alpha$$
$$\text{i.e., } n = \frac{1}{\delta}$$
$$\text{and } \underline{D = \frac{d_\beta}{\delta}}$$

3.9

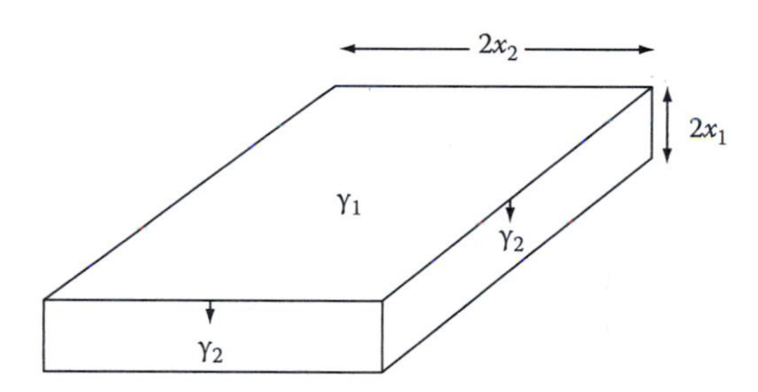

The edges of the plate exert a force on the periphery of the broad face equal to $\gamma_2 \cdot 4 \cdot 2x_2$

$$\therefore\ \Delta P = \frac{\gamma_2 \cdot 4 \cdot 2x_2}{(2x_2)^2} = \frac{2\gamma_2}{x_2}$$

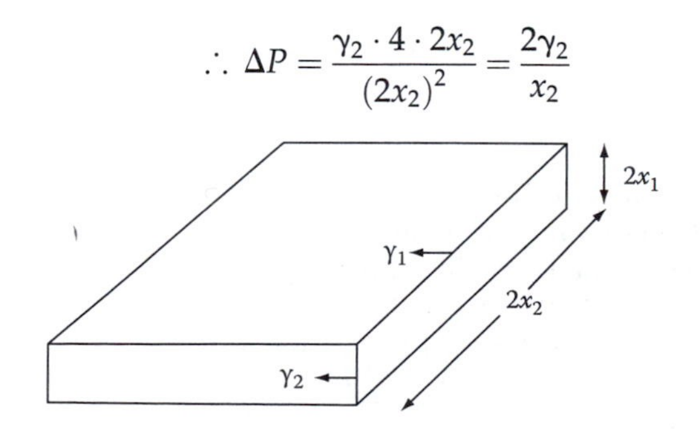

The periphery of each edge is acted on by a force of magnitude

$$2 \cdot \gamma_2 \cdot 2x_1 + 2 \cdot \gamma_1 \cdot 2x_2$$

The area of each edge is $2x_1 \cdot 2x_2$

$$\therefore \ \Delta P = \frac{2\gamma_2 \cdot 2x_1 + 2\gamma_1 \cdot 2x_2}{2x_1 \cdot 2x_2} = \frac{\gamma_2}{x_2} + \frac{\gamma_1}{x_1}$$

From the Wulff theorem (p. 116) for an equilibrium plate shape:

$$\frac{x_1}{x_2} = \frac{y_1}{y_2} \text{ (see also Exercise 3.3)}$$

$$\therefore \ \Delta P = \frac{2\gamma_1}{x_1} = \frac{2\gamma_2}{x_2}$$

3.10 See Section 3.4.1.

3.11 Atomic radius of Al = 1.43 Å
Atomic radius of Fe = 1.26 Å

$$\text{Hence zone misfit} = \frac{1.26 - 1.43}{1.43} \times 100\% = -11.89\%.$$

When the misfit is less than 5%, strain energy effects are less important than interfacial energy effects and spherical zones minimize the total free energy. However, when the misfit is greater than 5%, the small increase in interfacial energy caused by choosing a disk shape is more than compensated by the reduction in coherency strain energy.

Thus the zones in Al–Fe alloys would be expected to the disk-shaped.

3.12 Assuming that the matrix is elastically isotropic, that both Al and Mg atoms have equal elastic moduli, and taking a value of 1/3 for Poisson's ratio, the total elastic strain energy ΔG_s is given by

$$\Delta G_s = 4\mu\delta^2 V$$

where
μ is the shear modulus or matrix
δ is the unconstrained misfit
V is the volume of an Al atom

$$\delta = \frac{1.60 - 1.43}{1.43} = 0.119$$

$$V = 4/3 \cdot \pi \cdot \left(1.43 \times 10^{-10}\right)^3 = 1.225 \times 10^{-29}\ \text{m}^3$$
$$\mu_{\text{Al}} = 25\ \text{GPa} = 25 \times 10^9\ \text{N m}^{-2}$$
$$\therefore\ \Delta G_s = 4 \times 25 \times 10^9 \times (0.119)^2 \times 1.225 \times 10^{-29}\ \text{J atom}^{-1}$$
$$= 1.735 \times 10^{-20}\ \text{J atom}^{-1}$$

In 1 mol there are 6.023×10^{-23} atoms

$$\therefore \Delta G_s = \frac{1.735 \times 10^{-20} \times 6.023 \times 10^{23}}{1000}\ \text{kJ mol}^{-1}$$
$$= \underline{10.5\ \text{kJ mol}^{-1}}$$
$$1\ \text{eV} = 1.6 \times 10^{-19}\ \text{J, thus}$$
$$\Delta G_s = \frac{1.735 \times 10^{-20}}{1.6 \times 10^{-19}}\ \text{eV atom}^{-1}$$
$$= \underline{0.1\ \text{eV atom}^{-1}}$$

It is also implicitly assumed that individual Mg atoms are separated by large distances, so that each atom can be considered in isolation, i.e., dilute solutions.

The use of Equation 3.39 is also based on the assumption that the matrix surrounding a single atom is a continuum.

3.13 See Section 3.4.4.

3.14 When a Shockley partial dislocation passes through an fcc crystal the atoms above the glide plane in positions A are shifted to B position, B into C positions, etc.

i.e., A → B → C → A

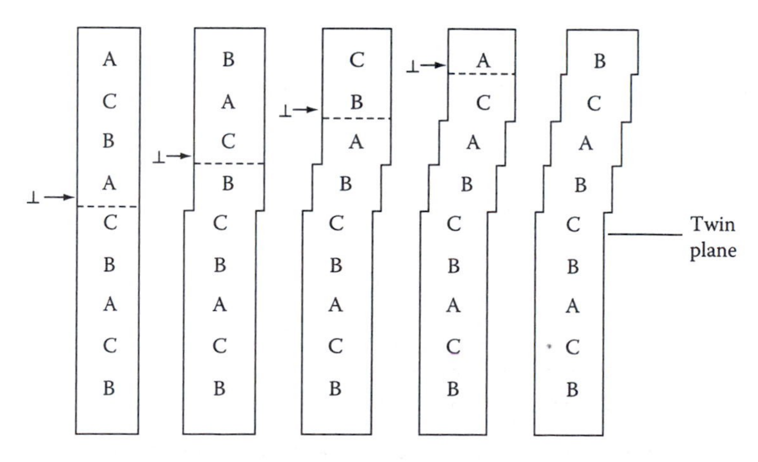

The above series of diagrams shows the twinning process.

3.15

Let the interface CD move with a velocity v perpendicular to the interface.

Consider unit area of interface perpendicular to both BC and CD.

$$\text{Mass flow perpendicular to BC} = u \times h.$$
$$\text{Mass flow perpendicular to CD} = v \times l.$$

From the conservation of mass: $u \times h = v \times l$.

$$v = \frac{u \times h}{l}$$

3.16

```
      B     B     B            B     B     B
      C     C     C            C     C     C
      A     A     A            A     A     A      fcc I
                                   ////////
      B     B     B            B  ┌--A--┐   B
--- C --- C --- C --------- C--┘   C   └--C---
      B     B     B            B     B     B
      A     A     A            A     A     A
      C     C     C            C     C     C      fcc II
      B     B     B            B     B     B
      A     A     A            A     A     A
```

If a single atom in crystal I attempts to jump into a crystal II position a ring of dislocation and an unstable A upon A situation results.

A Shockley partial dislocation in every {111} slip plane creates a glissile interface between two twinned crystals:

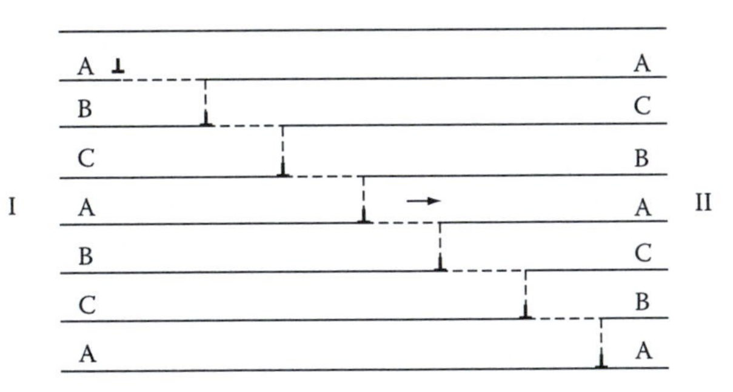

Note, however, that as a result of the shape change produced by the transformation large coherency stresses will be associated with the interface (see Figure 3.62a).

Similar coherency stresses will arise as a result of the fcc/hcp interface in Figure 3.61. Strictly speaking, Figure 3.60 is an incorrect representation of the stacking sequence that results from the passage of the partial dislocations. In layer 10 for example, there will not be a sudden change across the "extra half-plane" of A to B or B to C, but rather a gradual change associated with long-range strain fields in both crystals.

3.17 Solid/vapor interfaces and solid/liquid interfaces in nonmetals are faceted and therefore migrate by ledge mechanisms.

Solid/liquid interfaces in metals are diffuse and migration occurs by random atom jumps.

3.18 See Section 3.3.4.

3.19

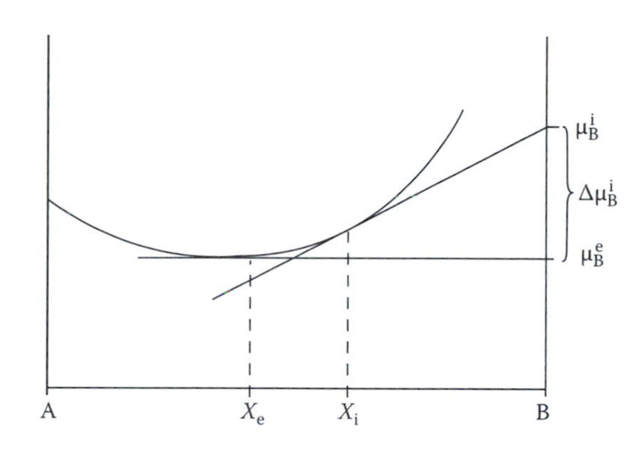

From Equations 1.41 and 1.43 we can write

$$\mu_B^i = G_B + RT \ln \gamma_i X_i$$
$$\mu_B^c = G_B + RT \ln \gamma_c X_c$$
$$\therefore \Delta\mu_B^i \mu_B^i - \mu_B^c = RT \ln \frac{\gamma_i X_i}{\gamma_c X_c}$$

For ideal solutions: $\gamma_i = \gamma_c = 1$

For dilute solutions: ($X_i \ll 1$): $\gamma_i = \gamma_c =$ constant (Henry's law) such that in both cases

$$\Delta\mu_B^i = RT \ln \frac{X_i}{X_e}$$

This can also be written as

$$\Delta\mu_B^i = RT \ln\left(1 + \frac{X_i - X_e}{X_e}\right)$$

If the supersaturation is small, i.e.,

$$(X_i - X_e) \ll X_e,$$

then

$$\Delta\mu_B^i = RT\left(\frac{X_i - X_e}{X_e}\right)$$

3.20

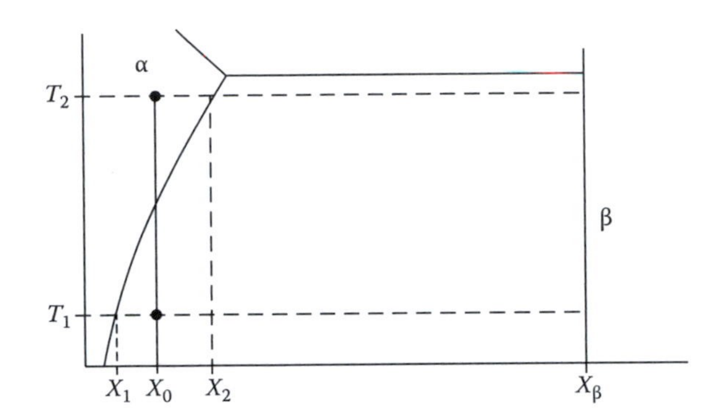

Suppose the alloy had reached equilibrium at a temperature T_1 and consists of long platelike precipitates. The bulk alloy composition is X_0, the equilibrium concentrations at T_1 and T_2 are X_1 and X_2, respectively, where T_2 is the temperature to which the alloy is heated.

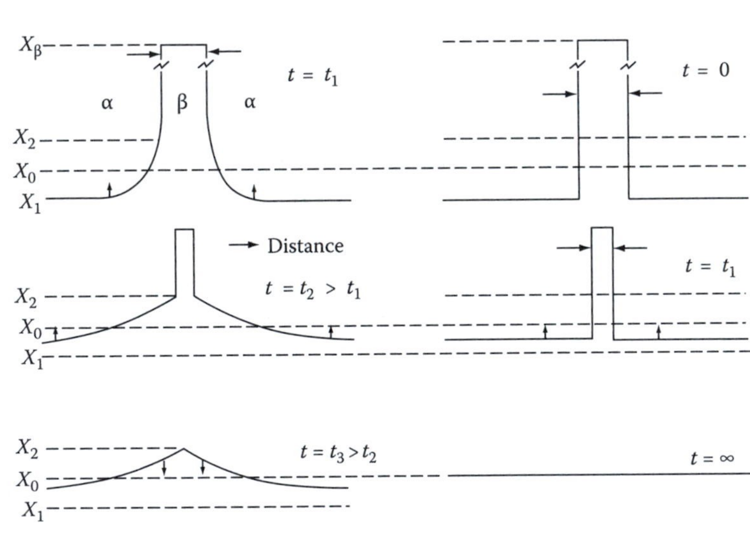

(i) Diffusion control, (ii) interface control, and (iii) mixed control: similar to diffusion control except the interface concentration in the a-matrix will be less than X_2, the equilibrium concentration at T_2.

Chapter 4

4.1

$$\Delta G_{\rm r} = \frac{-4\pi r^3}{3}\Delta G_{\rm v} + 4\pi r^2 \gamma_{\rm SL}$$

Differentiating this equation with respect to r:

$$\Delta G_r = \frac{-4\pi r^3}{3}\Delta G_{\rm v} + 4\pi r^2 \gamma_{\rm SL}$$

At the critical radius, r^*, this expression is equal to zero

$$0 = -4\pi r^{*2} \cdot \Delta G_{\rm v} + 8\pi r^* \gamma_{\rm SL}$$
$$\therefore\ r^* = \underline{\frac{2\gamma_{\rm SL}}{\Delta G_{\rm v}}}$$

In order to calculate the critical value of ΔG, ΔG^* at this radius, the value of r^* is substituted into the original equation

$$\Delta G^* = \frac{-4\pi}{3}\left(\frac{2\gamma_{\rm SL}}{\Delta G_{\rm v}}\right)^3 \cdot \Delta G_{\rm v} + 4\pi\left(\frac{2\gamma_{\rm SL}}{\Delta G_{\rm v}}\right)\gamma_{\rm SL}$$
$$= \underline{\frac{16\pi\gamma_{\rm SL}^3}{3(\Delta G_{\rm v})^2}}$$

4.2 From Equation 4.10, at the equilibrium melting temperature $T_{\rm m}$

$$n_{\rm r} = n_0 \exp\left(\frac{-\Delta G_{\rm r}}{kT_{\rm m}}\right)$$

At the equilibrium melting temperature $DG_{\rm v} = 0$, so that Equation 4.4 becomes

$$\Delta G_{\rm r}(T = T_{\rm m}) = 4\pi r^2 \gamma_{\rm SL}$$

For a cluster containing $n_{\rm c}$ atoms with an atomic volume Ω, we have

$$\frac{4\pi r^3}{3} = n_{\rm c}\,\Omega$$

Therefore the expression for ΔG_r becomes

$$\Delta G_r = 4\pi\left(\frac{3\Omega n_c}{4\pi}\right)^{2/3\gamma}$$

Substituting $\Omega = 1.6 \times 10^{-29}$ m^3 and $\gamma = 0.177$ J m^{-2} gives

$$\Delta G_r = (5.435 \times 10^{-20})n_c^{2/3}$$

For 1 mm^3, $n_0 = 6.25 \times 10^{19}$ atoms

Therefore when $n_c = 10$ atoms, $n_r = 9 \times 10^{13}$clusters mm^{-3}; and when $n_c = 60$ atoms, $n_r = 3$ clusters mm^{-3}; when $n_c = 100$ atoms, $n_r = 4 \times 10^{-8}$ clusters mm^{-3}; or, alternatively, 1 cluster in 2.5×10^7 mm^3

4.3 As the undercooling (ΔT) is increased, there is an increasing contribution from ΔG_v in the equation

$$\Delta G_r = -\frac{4}{3}\pi r^3 \Delta G_v + 4\pi r^2 \gamma_{SL}$$

whereas the interfacial energy is independent of ΔT. Consequently, for a given r, ΔG_r decreases with increasing ΔT, and the "maximum" cluster size increases somewhat.

4.4 From Equation 4.13

$$N_{hom} = f_0 C_0 \exp\left\{\frac{-16\pi\gamma_{SL}^3 T_m^2}{3L_v^2 kT} \cdot \frac{1}{\Delta T^2}\right\}$$

where $T = T_m - \Delta T$

From which the following values are obtained:

ΔT (K)	N_{hom} (m^{-3} s^{-1})	N_{hom} (cm^{-3} s^{-1})
180	0.7	7×10^{-7}
200	8×10^6	8
220	1×10^{12}	1×10^6

Note the large change in N over the small temperature range (see Figure 4.6).

4.5 $\Delta G^* = \frac{1}{2} \cdot V^* \cdot \Delta G_v$

For homogeneous nucleation, it has been shown (see Exercise 4.1) that

$$r^* = \frac{2\gamma_{SL}}{\gamma G_v}$$

Thus for a spherical nucleus

$$V^* = \frac{4\pi r^{*3}}{3} = \frac{32\pi\gamma_{SL}^3}{3(\Delta G_v)^3}$$

$$\Delta G^* = \frac{1}{2}\cdot V^*\cdot \Delta G_v = \frac{16\pi\gamma_{SL}^3}{3\Delta G_v^2}$$

This is identical to that derived in Exercise 4.1, and so the equation holds for homogeneous nucleation.

For heterogeneous nucleation, it can be shown that

$$r^* = \frac{2\gamma_{SL}}{\Delta G_v}$$

The volume of a spherical cap on a flat mould surface is given by

$$V = \pi r^3 \frac{(2+\cos\theta)(1-\cos\theta)^2}{3}$$

Thus

$$V^* = \pi\left(\frac{2\gamma_{SL}}{\Delta G_v}\right)^3 \cdot \frac{(2+\cos\theta)(1-\cos\theta)^2}{3}$$

where θ is the wetting angle.

Substituting into the given equation

$$\Delta G^* = \frac{1}{2}\cdot V^*\cdot \Delta G_v = \frac{4}{3}\pi\frac{\gamma_{SL}^3}{\Delta G_v^2}\cdot(2+\cos\theta)(1-\cos\theta)^2$$

Writing the normal free energy equation for heterogeneous nucleation in terms of the wetting angle θ and the cap radius r

$$\Delta G_{het} = \left\{-\frac{4}{3}\pi r^3 \Delta G_v + 4\pi r^2 \gamma_{SL}\right\}\frac{(2+\cos\theta)(1-\cos\theta)^2}{4}$$

But from Equations 4.17 and 4.19 we have

$$\Delta G^* = \frac{16\pi\gamma_{SL}^3}{3\Delta G_v^2} \cdot \frac{(2+\cos\theta)(1-\cos\theta)^2}{4}$$

which is identical to that obtained using

$$\Delta G^* = \frac{1}{2} V^* \Delta G_v$$

4.6 See Section 4.1.3.

4.7 Consider a cone-shaped crevice with semiangle α as shown below:

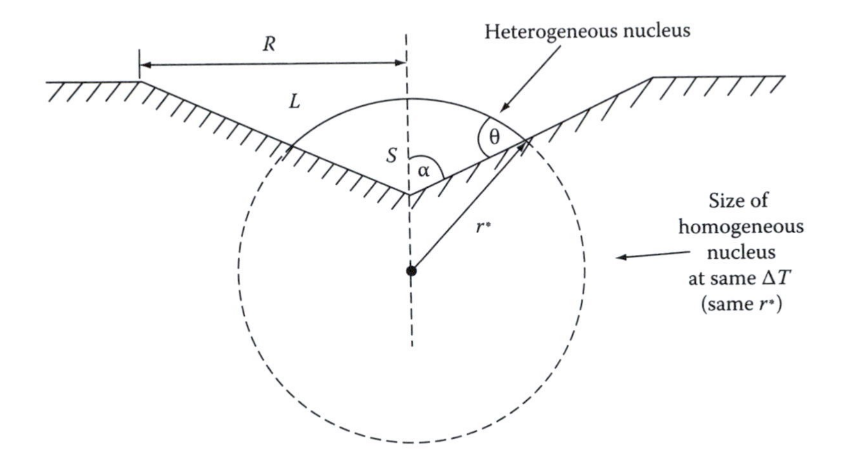

The wetting angle between nucleus and mold wall (θ) is fixed by the balance of surface tension forces (Equation 4.14). The activation energy barrier (ΔG^*_{het}) depends on the shape of the nucleus as determined by angles α and θ.

From Equation 4.23, for a given undercooling (ΔT), ΔG_v and r^* are constant, such that the following equalities apply

$$S = \frac{\Delta G^*_{het}}{\Delta G^*_{hom}} = \frac{V^*_{het}}{V^*_{hom}}$$

i.e., $S = \dfrac{\Delta G^*_{het}}{\Delta G^*_{hom}}$

$$= \frac{\text{Volume of the heterogeneous nucleous}}{\text{Volume of a sphere with the same nucleus/liquid interfacial radius}}$$

It can be seen that the shape factor (S) will decrease as α decreases, and on cooling below T_m the critical value of ΔG^* will be reached at progressively lower values of ΔT, i.e., nucleation becomes easier.

When $\alpha \leq 90 - \theta$, $S = 0$ and there is no nucleation energy barrier. (It can be seen that $\alpha = 90 - \theta$ gives a planar solid/liquid interface, i.e., $r = \infty$ even for a negligibly small nucleus volume.)

Once nucleation has occurred, the nucleus can grow until it reaches the edge of the conical crevice. However, further growth into the liquid requires the solid/liquid interface radius to pass through a minimum or R (the maximum radius of the cone). This requires an undercooling given by

$$\frac{2\gamma_{SL}}{R} = \frac{L\Delta T}{T_m}$$

$$\text{i.e., } \Delta T = \frac{2\gamma_{SL}T_m}{RL}$$

4.8 For conical crevices with $\alpha < 90 - \theta$ the solid/liquid interface can maintain a negative radius of curvature which stabilizes the solid above the equilibrium melting temperature (T_m):

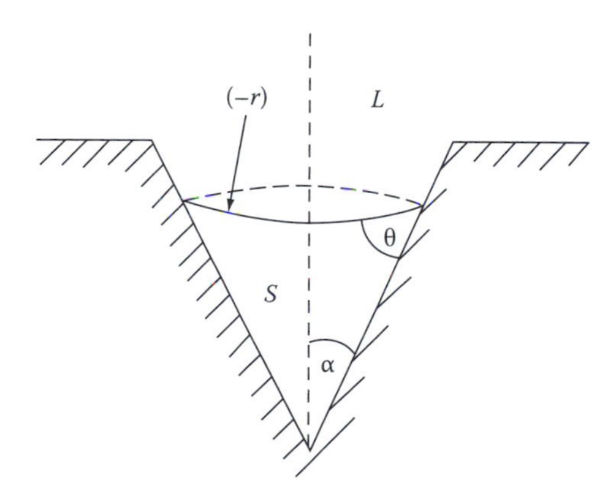

As the temperature is raised above T_m the solid will melt back into the crevice to maintain equilibrium with a radius given by

$$r = \frac{2\gamma_{SL}T_m}{L\Delta T}$$

where $(-\Delta T)$ is now the superheat above T_m.

4.9 If the situation described above is realized in practice it would explain the observed phenomena.

4.10 (a) The values of the three interfacial energies are as follows:

$$\text{Solid–liquid} = 0.132 \text{ J m}^{-2}$$

$$\text{Liquid–vapor} = 1.128 \text{ J m}^{-2}$$

$$\text{Solid–vapor} = 1.400 \text{ J m}^{-2}$$

Thus the sum of the solid–liquid and liquid–vapor interfacial free energies is less than the solid–vapor free energy, and there is no increase of free energy in the early stages of melting. Therefore, it would be expected that a thin layer of liquid should form on the surface below the melting point, because the difference in free energies could be used to convert solid into liquid.

(b) Imagine the system I below. The free energy of this system is given by

$$G(\mathrm{I}) = G^{\mathrm{S}} + \gamma_{\mathrm{SV}}$$

System II contains a liquid layer of thickness δ and solid reduced to a height (1 − δ). (The difference in molar volume between liquid and solid has been ignored.)

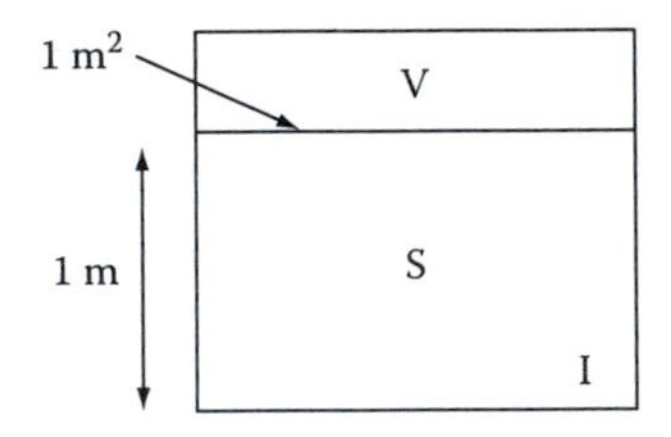

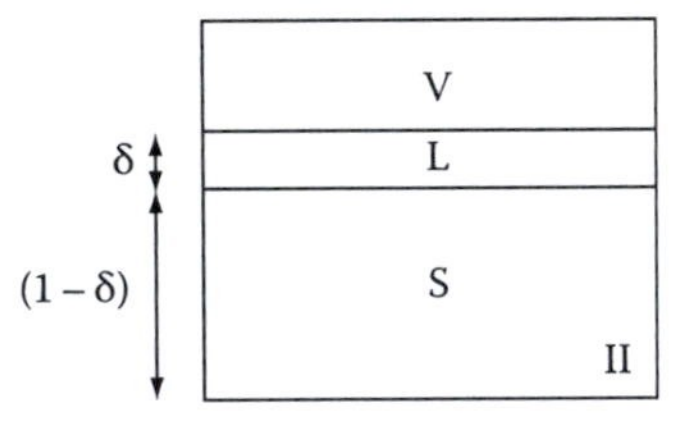

Thus

$$G(\mathrm{II}) = G^{\mathrm{S}}(1 - \delta) + G^{\mathrm{L}}\delta + \gamma_{\mathrm{SL}} + \gamma_{\mathrm{LV}} \quad (\delta > 0)$$

$$G(\mathrm{II}) = G^{\mathrm{S}} + \delta\left(G^{\mathrm{L}} - G^{\mathrm{S}}\right) + \gamma_{\mathrm{SL}} + \gamma_{\mathrm{LV}} \quad (\delta > 0)$$

At an undercooling ΔT below T_{m},

$$G^{\mathrm{L}} - G^{\mathrm{S}} = \frac{L\Delta T}{T_{\mathrm{m}}}$$

i.e.,

$$G(\mathrm{II}) = G^{\mathrm{S}} + \frac{L\Delta T}{T_{\mathrm{m}}}\delta + \gamma_{\mathrm{SL}} + \gamma_{\mathrm{LV}}$$

or

$$G(\text{II}) = G(\text{I}) - \Delta\gamma + \frac{L\Delta T}{T_m} \cdot \delta$$

where

$$\Delta\gamma = \gamma_{SV} - \gamma_{SL} - \gamma_{LV}$$

This is shown in the figure below:

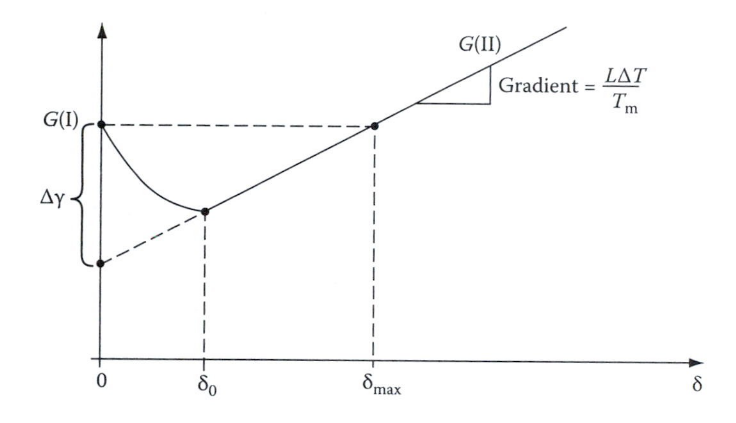

Note that as $\delta \rightarrow 0$, $G(\text{II}) \rightarrow G(\text{I})$, which means that in practice $\gamma_{SL} + \gamma_{LV} \rightarrow \gamma_{sv}$ as a result of an interaction between the S/L and L/V interfaces as they approach to within atomic dimension of each other.

The optimum liquid layer thickness (δ_0) will be that giving a minimum free energy as shown. We cannot calculate this value without a knowledge of the above interaction. However, it is reasonable to assume that the minimum will occur at a separation of a few atom diameters, provided δ_{max} in the above diagram is at least a few atom diameters. δ_{max} is defined by $G(\text{I}) = G(\text{II})$, i.e.,

$$G(\text{I}) = G(\text{I}) - \Delta\gamma + \frac{L\Delta T}{T_m} \cdot \delta_{max}$$

where

$$\delta_{max} = \frac{\Delta\gamma T_m}{L\Delta T}$$

Alternatively,

$$\Delta T = \frac{\Delta \gamma T_m}{L \delta_{max}}$$

If $\delta_{max} = 10$ nm (say), then $\Delta T = 16$ K.

It seems therefore that surface melting is theoretically possible a few degrees below T_m.

4.11 (a) Repeated surface nucleation (see Section 4.2.2).

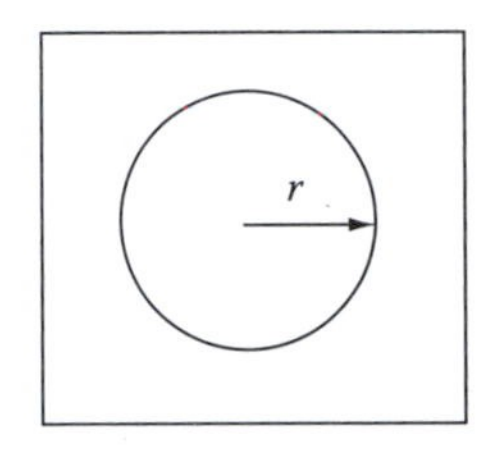

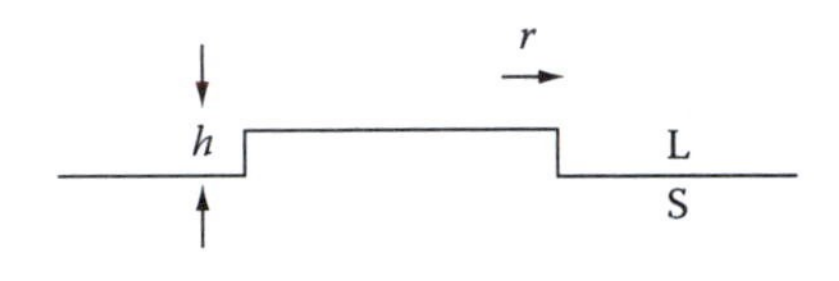

Suppose the edge of the cap nucleus is associated with an energy e (J m^{-1}). Formation of such a cap will cause a free energy change given by

$$\Delta G = -\pi r^2 h \Delta G_v + 2\pi r e$$

The critical cap radius r^* is given by

$$\frac{d(\Delta G)}{dr} = 0$$

i.e.,

$$r^* = \frac{e}{h\Delta G_V}$$

and

$$\Delta G^* = \frac{\pi e^2}{h \Delta G_v}$$

The rate at which caps nucleate on the surface should be proportional to $\exp(-\Delta G^*/kT)$, i.e.,

$$N \propto \exp\left(\frac{-\pi e^2}{h \Delta G_v k T}\right)$$

But, $\Delta G \propto \Delta T_i$, the undercooling at the interface, so for small undercooling we have

$$N \propto \exp\left(\frac{-k}{\Delta T_i}\right) \text{ nuclei m}^{-2} \text{ s}^{-1}$$

where k is approximately constant.

Each time a cap is nucleated, it should grow rapidly across the interface to advance a distance h. It seems reasonable to suppose therefore that the growth rate will be proportional to N, i.e.,

$$\underline{v \propto \exp\frac{-k}{\Delta T_i}}$$

(b) Very roughly, Equation 4.28 can be seen to be reasonable as follows: First, it is reasonable to suppose that the distance between successive turns of the spiral (L) will be linearly related to the minimum radius at the centre (r^*). Thus we have

$$L \propto r^* \propto \Delta T_i^{-1}$$

Second, for small undercoolings, the lateral velocity of the steps (u) should be proportional to the driving force, which in turn is proportional to ΔT_i

$$u \propto \Delta T_i$$

Thus the velocity normal to the interface v is given by

$$v = \frac{uh}{L} \propto \Delta T_i^2$$

where h is the step height.

4.12 Equilibrium solidification (see Figures 4.19 and 4.20)

From Figure 4.19 the lever rule gives the mole fraction solid (f_s) at T_2 as

$$f_S = \frac{X_L - X_0}{X_L - X_S} = \frac{(X_S/k) - X_0}{(X_S/k) - X_S}$$

$$\therefore\ \underline{X_S = \frac{kX_0}{1 - (1-k)f_S}}$$

This expression relates the composition of the solid forming at the interface at T_2 to the fraction already solidified. For the case shown in Figure 4.19, it will be roughly as shown below:

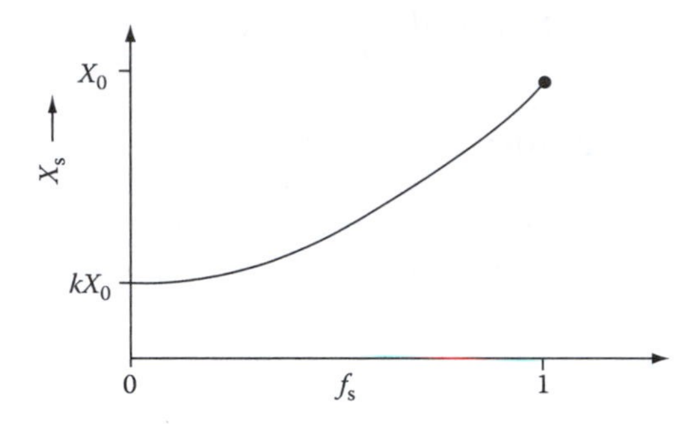

The temperature of the interface (T_2) as a function of the fraction solidified can be obtained using the following relationship which is apparent from Figure 4.19

$$\frac{T_2 - T_3}{T_1 - T_3} = \frac{X_0 - X_S}{X_0 - kX_0}$$

Substituting for X_S gives

$$\frac{T_2 - T_3}{T_1 - T_3} = \left\{1 + k\left(\frac{f_S}{1 - f_S}\right)\right\}^{-1}$$

This will be a curved line roughly as shown below for the case described for Figure 4.19 ($k \sim 0.47$).

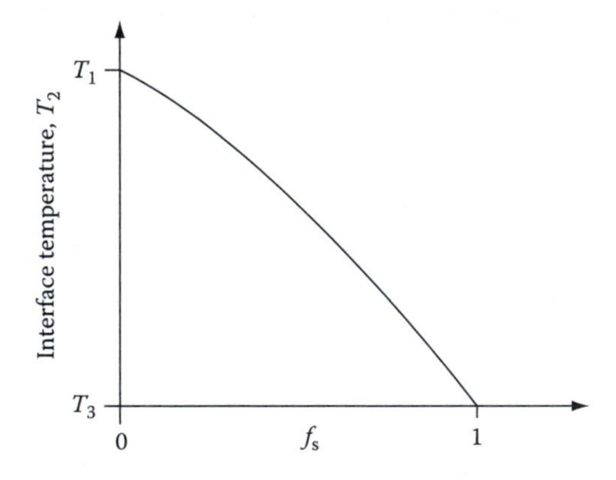

No diffusion in solid, perfect mixing in liquid (see Figure 4.21).

Again, we have

$$\frac{T_2 - T_3}{T_1 - T_3} = \frac{X_0 - X_S}{X_0 - kX_0}$$

X_S is now given by Equation 4.33 such that

$$\frac{T_2 - T_3}{T_1 - T_3} = \frac{X_0 - kX_0(1 - f_S)^{(k-1)}}{X_0 - kX_0}$$

$$\therefore \frac{T_2 - T_3}{T_1 - T_3} = \frac{1 - k(1 - f_S)^{(k-1)}}{(1 - k)}$$

where $T_1 > T_2 > T_E$. For the phase diagram in Figure 4.21a, the following variation is therefore obtained ($k - 0.47$, the exact form of the curve depends on k, of course)

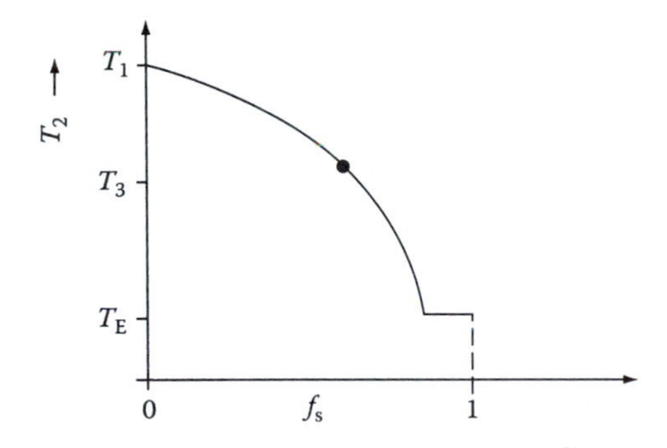

No diffusion in solid, no stirring in liquid (see Figure 4.22)

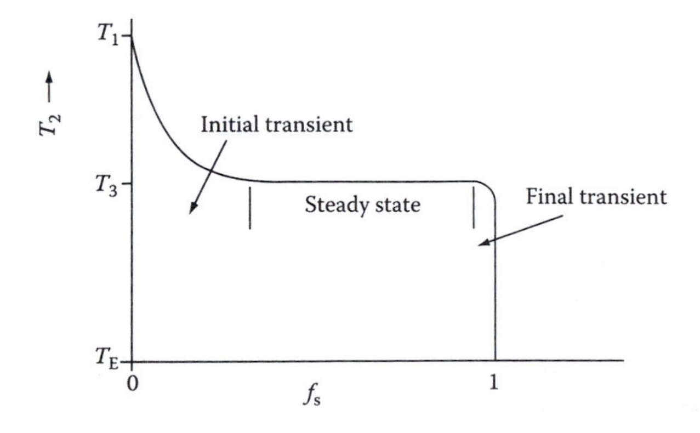

4.13 No diffusion in solid, complete mixing in liquid

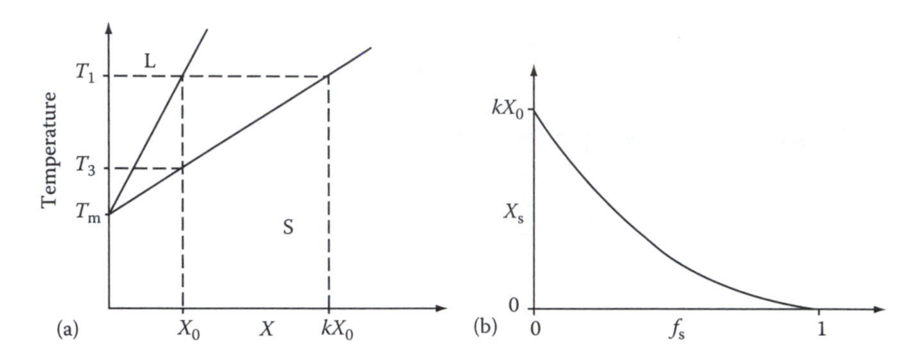

Diagram (a), above, is a typical phase diagram for $k>1$. (In this case, $k=3$.) The variation of composition along the bar can be calculated using Equation 4.33, i.e.,

$$X_S = kX_0(1 - f_S)^{(k-1)}$$

The result for $k=3$ is shown in diagram (b). f_S is proportional to distance along the bar. Note that the final composition to solidify is pure solvent ($X_S=0$).

No diffusion in solid, no stirring in liquid.

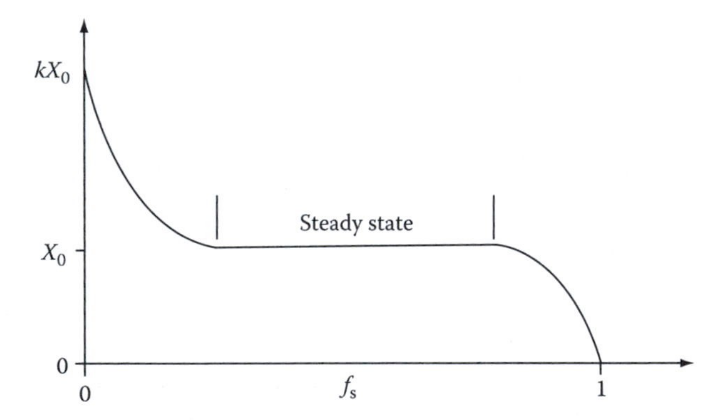

4.14 During steady-state growth the concentration profile in the liquid must be such that the rate at which solute diffuses down the concentration gradient away from the interface is balanced by the rate at which solute is rejected from the solidifying liquid, i.e.,

$$-DC_L' = v(C_L - C_S)$$

Assuming the molar volume is independent of composition, this becomes

$$\frac{-D\,\mathrm{d}X_L}{\mathrm{d}x} = v\left(\frac{X_0}{k} - X_0\right) \text{ at the interface}$$

The concentration profile in the liquid is given by

$$X_L = X_0\left\{1 - \frac{1-k}{k}\cdot\exp - \frac{x}{(D/v)}\right\}$$

$$\therefore\ \frac{\mathrm{d}X_L}{\mathrm{d}x} = X_0\left(\frac{1-k}{k}\right)\cdot\frac{v}{D}\cdot\exp - \frac{x}{(D/v)}$$

$$= \frac{v}{D}(X_0 - X_L)$$

Substituting this expression into the solute equation

$$-D\cdot\frac{v}{D} = (X_0 - X_L) = v\left(\frac{X_0}{k} - X_0\right)$$

Since $X_L = X_0/k$ at the interface, the expressions are equivalent, and the profile satisfies the solute balance.

4.15

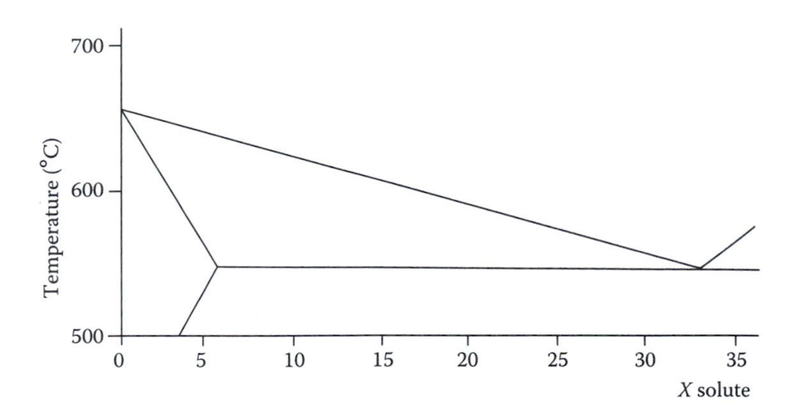

For an Al-0.5 wt% Cu alloy:

(a) Interface temperature in the steady state is given by the solidus temperature for the composition concerned.

$$\therefore \text{ Interface temperature} = \underline{650.1°\text{C}}$$

(b) Diffusion layer thickness is equivalent to the characteristic width of the concentration profile,

$$\therefore \text{ Thickness} = \frac{D_L}{v} = \frac{3 \times 10^{-9}}{5 \times 10^{-6}} = \underline{6 \times 10^{-4}\ \text{m}}$$

(c) A planar interface is only stable if there is no zone of constitutional undercooling ahead of it. Under steady-state growth, consideration of the temperature and concentration profiles in the liquid ahead of the interface gives that the critical gradient, T'_L, can be expressed as follows

$$T'_L = \frac{T_1 - T_3}{D/v}$$

where
T_1 is the liquidus temperature at X_0
T_3 is the solidus temperature at X_0

Thus

$$T'_L = \frac{(658.3 - 650.1)}{6 \times 10^{-4}}\ ^\circ\text{C m}^{-1}$$
$$= 13.7\ \text{K mm}^{-1}$$

(d) For an Al-2 wt% Cu alloy:

$$\text{Interface temperature} = \underline{620.4^\circ\text{C}}$$

$$\text{Diffusion layer thickness} = \frac{3 \times 10^{-9}}{5 \times 10^{-6}} = \underline{6 \times 10^{-4}\ \text{m}}$$

$$\text{Temperature gradient} = \frac{(653.2 - 620.4)}{6 \times 10^{-4}} = \underline{54.7\ \text{K mm}^{-1}}$$

4.16 (a) Scheil equation: $X_L = X_0 f_L^{(k-1)}$

Since it is assumed that the solidus and liquidus lines are straight, k is constant over the solidification range, and may be calculated using X_{max} and X_E as follows

$$k = \frac{X_S}{X_L} \text{ at a given temperature}$$

At the eutectic temperature, $X_S = X_{max}$ and $X_L = X_E$. Thus,

$$k = \frac{5.65}{33.0} = 0.17$$

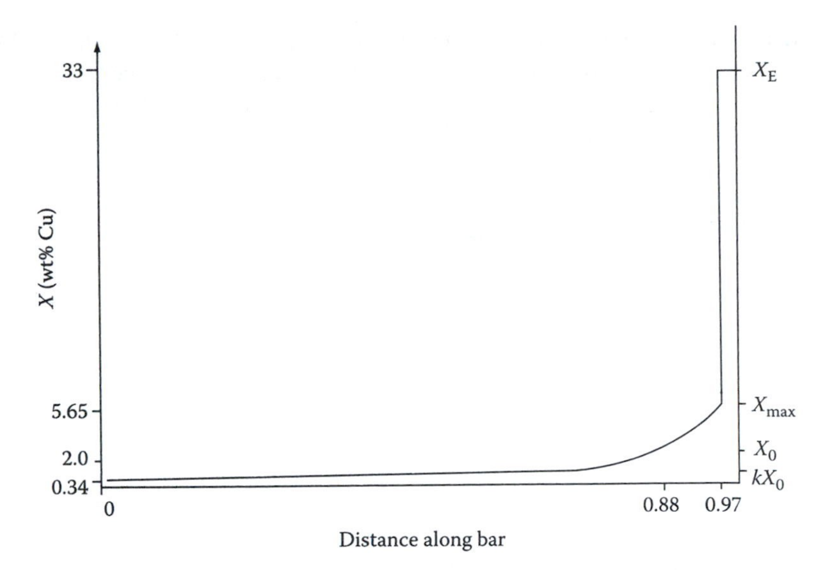

The above plot may be constructed by considering the composition of the initial solid formed (kX_0), the position at which the solid has the compositions X_0 and X_{max}, and the eutectic composition, X_E.

$$\begin{aligned}\text{Initial solid formed} &= kX_0\\ &= 0.17 \times 2 \text{ wt\% Cu}\\ &= 0.34 \text{ wt\% Cu}\end{aligned}$$

The volume fraction of liquid remaining, f_L when the solid deposited has a composition X_0 is found from the Scheil equation.

Thus when $X_S = X_0$; $X_L = X_0/k$, and the Scheil equation becomes

$$\frac{X_0}{k} = X_0 f_L^{(k-1)}$$

$$\therefore \frac{1}{0.17} = f_L^{(-0.83)}$$

$$\therefore f_L = 0.12, \text{ hence } f_s(\text{position along bar}) = \underline{0.88}$$

Similarly, when $X_S = X_{max} = 5.65$ wt% Cu, Scheil equation becomes

$$\frac{5.65}{0.17} = 2 \times f_L^{(-0.83)}$$

$$\therefore f_L = \left(\frac{0.17 \times 2}{5.65}\right)^{1/0.83}$$

$$= 0.03$$

Hence f_S (position along bar) = $\underline{0.97}$

From the information given, $X_E = 33$ wt% Cu for positions along the bar between 0.97 and 1.

(b) From the diagram, the fraction solidifying as a eutectic, $\underline{f_E} = \underline{0.03}$.

(c) For an Al-0.5 wt% Cu alloy solidified under the same conditions, the fraction forming eutectic may be found from the Scheil equation as before by putting X_S equal to X_{mas}:

$$X_L = X_0 f_L^{(k-1)}$$

$$\therefore \frac{X_s}{k} = X_0 f_L^{(k-1)}$$

$$\therefore \frac{X_{max}}{k} = X_0 f_L^{(k-1)}$$

$$\therefore \frac{5.65}{0.17} = 0.5 \times f_E^{(-0.83)}$$

$$\therefore f_E = \left(\frac{0.17 \times 0.5}{5.65}\right)^{1/0.83}$$

$$= \underline{0.006}$$

4.17 Cells grow in the direction of maximum temperature gradient, which is upstream in a convection current.

4.18

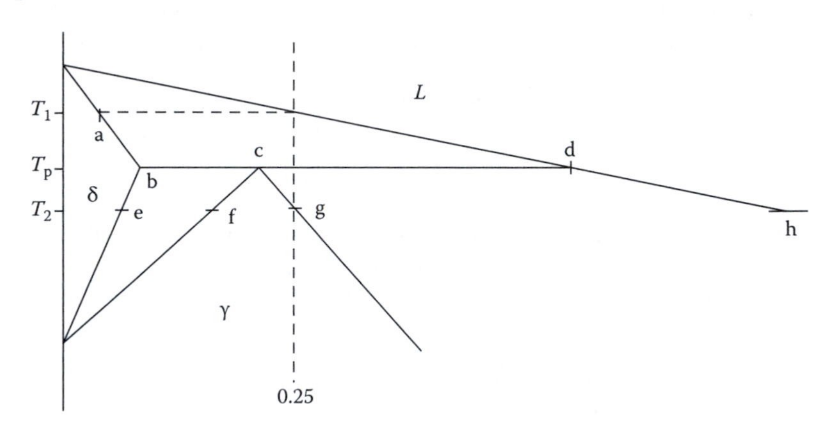

Assume equilibrium conditions between δ and L

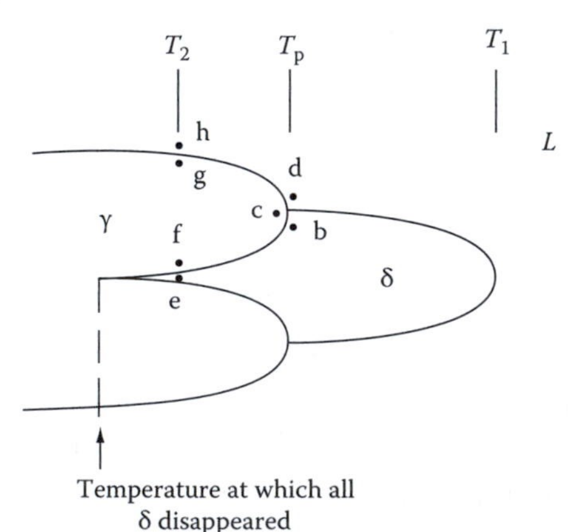

4.19 It can be sown that the growth rate of a lamellar eutectic v is given by the following equation

$$v = kD\Delta T_0 \cdot \frac{1}{\lambda}\left(1 - \frac{\lambda^*}{\lambda}\right)$$

where

k is the proportionality constant
D is the liquid diffusivity
ΔT_0 is the interface undercooling
λ is the lamellar spacing
λ^* is the minimum possible value of λ

(i) When the undercooling is fixed, k, D, and ΔT_0 may be combined to form a constant c, thus

$$v = c \cdot \frac{1}{\lambda} \cdot \left(1 - \frac{\lambda^*}{\lambda}\right)$$

Differentiating this equation with respect to λ

$$\frac{dv}{d\lambda} = \frac{-c}{\lambda^2} + \frac{2c\lambda^*}{\lambda^3}$$

Differentiating a second time:

$$\frac{d^2v}{d\lambda^2} = \frac{2c}{\lambda^3} - \frac{6c\lambda^*}{\lambda^4}$$

Hence

$$\frac{c}{\lambda^2} = \frac{2c\lambda^*}{\lambda^3}$$

at the maximum or minimum growth rate.

$$\therefore \underline{\lambda = 2\lambda^*}$$

Substituting this value into the equation of the second differential

$$\frac{d^2v}{d\lambda^{*3}} = \frac{2c}{\lambda^3} - \frac{6c\lambda^*}{\lambda^4}$$

$$= \frac{2c}{8\lambda^{*3}} - \frac{6c\lambda^*}{16\lambda^{*4}}$$

$$= \frac{-c}{8\lambda^{*3}}$$

Thus when $\lambda = 2\lambda^*$, the growth rate is a maximum.

(ii) When the growth rate is fixed, the original equation may be rewritten as follows:

$$a = \Delta T_0 \cdot \frac{1}{\lambda}\left(1 - \frac{\lambda^*}{\lambda}\right)$$

where

$$a = \frac{v}{kD}$$

Thus

$$\frac{a}{\Delta T_0} = \frac{1}{\lambda} - \frac{\lambda^*}{\lambda^2}$$

Differentiating with respect to λ

$$\frac{-a}{\Delta T_0^2} \cdot \frac{\mathrm{d}\Delta T_0}{\mathrm{d}\lambda} = -\frac{1}{\lambda^2} + \frac{2\lambda^*}{\lambda^3}$$

$$\therefore \quad \frac{\mathrm{d}\Delta T_0}{\mathrm{d}\lambda} = \frac{\Delta T_0^2}{a}\left(\frac{1}{\lambda^2} - \frac{2\lambda^*}{\lambda^3}\right)$$

Differentiating second time

$$\frac{-a}{\Delta T_0^2} \cdot \frac{\mathrm{d}^2\Delta T_0}{\mathrm{d}\lambda^2} + \frac{\mathrm{d}\Delta T_0}{\mathrm{d}\lambda} \cdot \frac{2a}{\Delta T_0^3} = \frac{2}{\lambda^3} - \frac{6\lambda^*}{\lambda^4}$$

$$\therefore \quad \frac{-a}{\Delta T_0^2} \cdot \frac{\mathrm{d}^2\Delta T_0}{\mathrm{d}\lambda^2} + \frac{2}{\Delta T_0}\left(\frac{1}{\lambda^2} - \frac{2\lambda^*}{\lambda^3}\right) = \left(\frac{2}{\lambda^3} - \frac{6\lambda^*}{\lambda^4}\right)$$

$$\therefore \quad \frac{\mathrm{d}^2\Delta T_0}{\mathrm{d}\lambda^2} = \frac{2\Delta T_0}{a}\left(\frac{1}{\lambda^2} - \frac{2\lambda^*}{\lambda^3}\right) = \frac{\Delta T_0^2}{a}\left(\frac{2}{\lambda^3} - \frac{6\lambda^*}{\lambda^4}\right)$$

Substituting $\lambda = 2\lambda^*$

$$\frac{\mathrm{d}^2\Delta T_0}{\mathrm{d}\lambda^2} = \frac{2\Delta T_0}{a}\left(\frac{1}{4\lambda^{*2}} - \frac{1}{4\lambda^{*2}}\right) - \frac{\Delta T_0^2}{a}\left(\frac{1}{4\lambda^{*3}} - \frac{1}{4\lambda^{*3}}\right)$$

Thus $\mathrm{d}^2\Delta T_0/\mathrm{d}\lambda^2$ is positive

4.20 The total change in molar free energy when liquid transforms into lamella $\alpha + \beta$ with a spacing λ is given by Equation 4.37, i.e.,

$$\Delta G(\lambda) = -\Delta G(\infty) + \frac{2\gamma_{\alpha\beta} V_m}{\lambda}$$

The equilibrium eutectic temperature T_E is defined by $\lambda = \infty$ and $G(\infty) = 0$.

We can define a metastable equilibrium eutectic temperature at $(T_E - \Delta T_E)$ such that at this temperature there is no change in free energy $L \rightarrow \alpha + \beta$ with a spacing λ, i.e., at $T_E - \Delta T_E$, $\Delta G(\lambda) = 0$.

Also from Equation 4.38 at an undercooling of ΔT_E

$$\Delta G(\infty) = \frac{\Delta H \Delta T_E}{T_E}$$

Finally, then combing these equations gives

$$\Delta T_E = \frac{2\gamma_{\alpha\beta} V_m T_E}{\Delta H \lambda}$$

Substituting: $\gamma_{\alpha\beta} = 0.4$ J m^{-2}, $T_E = 1000$ K, and $(\Delta H/V_m) = 8 \times 10^8$ J m^{-3} gives

$$\Delta T_E = \frac{10^{-6}}{\lambda}$$

i.e., for $\lambda = 0.2$ mm, $\Delta T_E = 5$ K; $\lambda = 1.0$ μm, $\Delta T_E = 1$ K.

Note that if these eutectics grow at the optimum spacing of $2\lambda^*$ the total undercooling at the interface during growth (ΔT_0) will be given by Equation 4.39 such that

$$\text{for } \lambda = 0.2\ \mu\text{m},\ \lambda^* = 0.1\ \mu\text{m},\ \Delta T_0 = 10\ \text{K}$$
$$\text{for } \lambda = 1.0\ \mu\text{m},\ \lambda^* = 0.5\ \mu\text{m},\ \Delta T_0 = 2\ \text{K}$$

4.21

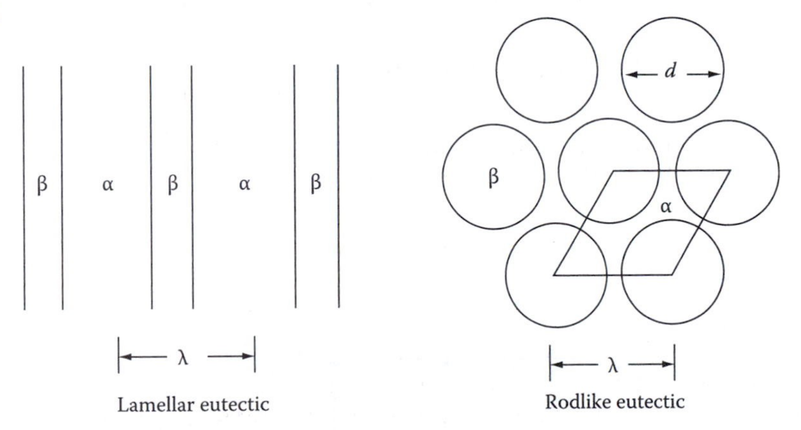

For a lamellar eutectic the total interfacial area per unit volume of eutectic is given by $2/\lambda$, irrespective of volume faction of β.

For the rod eutectic, considering rods of unit length, and diameter d, the area of α/β interface per unit volume of eutectic is given by

$$\frac{\pi d}{\lambda \cdot \lambda\sqrt{3}/2} = \frac{2\pi d}{\lambda^2\sqrt{3}}$$

For the rod eutectic to have the minimum interfacial energy, then

$$\frac{2\pi d}{\lambda^2\sqrt{3}} < \frac{2}{\lambda}, \quad \text{i.e.,} \quad \frac{d}{\lambda} < \frac{\sqrt{3}}{\pi}$$

d depends on the volume fraction of B, (f)

$$f = \frac{\pi d^2}{4} \bigg/ \frac{\lambda^2\sqrt{3}}{2}$$

From which $f < f_c - = \frac{\sqrt{3}}{2\pi} = \underline{0.28}$.

4.22 See Sections 4.4 and 4.5.

4.23 See Section 4.5.

Chapter 5

5.1

$$\Delta G_0 = RT\left[X_0 \ln\frac{X_0}{X_e} + (1 - X_0)\ln\frac{(1 - X_0)}{(1 - X_c)}\right] - \Omega(X_0 - X_e)^2$$

(a) By direct substitution into the above equation

$$\Delta G_0 = \underline{420.3 \text{ J mol}^{-1}}$$

(b) Applying the lever rule to the system at equilibrium

$$\text{Mole fraction of precipitate} = \frac{(X_0 - X_e)}{(X_\beta - X_e)} = 0.08$$

Assuming the molar volume is independent of composition, this will also be the volume fraction.

(c)

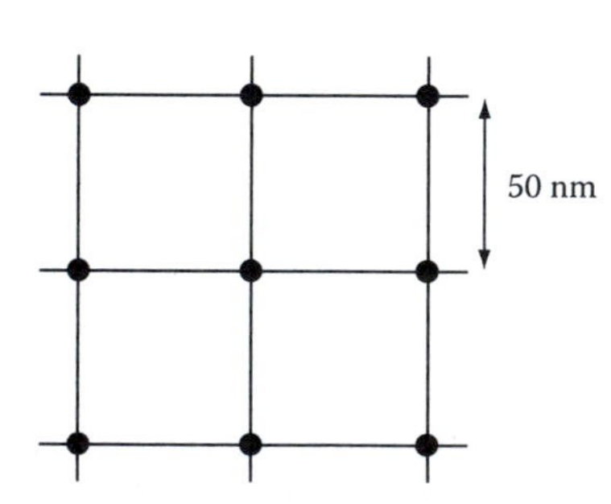

Assuming a regular cubic array with a particle spacing of 50 nm, the number of particles per cubic meter of alloy $= \frac{1}{(50\times10^{-9})^3} = 8 \times 10^{21}$

Let all the particles be of equal volume and spherical in shape with a radius r. Then the total volume of particles in 1 m^3 of alloy $= 8 \times 10^{21} \times (4/3)\pi r^3$

Equating this with the volume fraction of precipitate

$$8 \times 10^{21} \times \frac{4}{3}\pi r^3 = 0.08 \text{ m}$$

$$\therefore r = 13.4 \text{ nm}$$

Thus in 1 m^3 of alloy the total interfacial area $= 8 \times 10^{21} \times 4\pi r^2 = \underline{1.8 \times 10^7 \text{ m}^2}$

(d) If $\gamma_{\alpha\beta} = 200 \text{ mJ m}^{-2}$

$$\begin{aligned} \text{Total interfacial energy} &= 200 \times 1.8 \times 10^7 \text{ mJ m}^{-3} \text{ alloy} \\ &= 3.6 \times 10^6 \text{ J m}^{-3} \text{ alloy} \\ &= \underline{36 \text{ J mol}^{-1}} \end{aligned}$$

(e) The fraction remaining as interfacial energy $= \frac{36}{420.3} = \underline{9\%}$

(f) When the precipitate spacing is 1 μm,

$$\begin{aligned} \text{No. of particles per m}^3 &= \frac{1}{(1 \times 10^{-6})^3} \\ &= 1 \times 10^{18} \text{ m}^{-3} \end{aligned}$$

Using the same method as in (c), the particle radius is found to be 267 nm.

Thus in 1 m^3 of alloy the total interfacial area $= 1 \times 10^{18} \times 4\pi \times (2.67 \times 10^{-7})^2 = 8.96 \times 10^5 \text{ m}^2$

$$\begin{aligned} \text{Total interfacial energy} &= \underline{1.8 \times 10^5 \text{ J m}^{-3} \text{ alloy}} \\ &= \underline{1.8 \text{ J mol}^{-1}} \end{aligned}$$

Fraction remaining as interfacial energy $= \underline{0.4\%}$

5.2

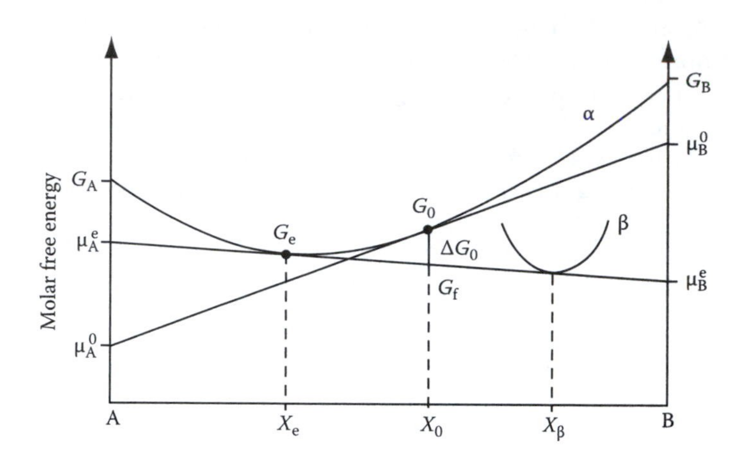

$$\Delta G_0 = G_0 - G_f$$
$$G_0 = X_0\mu_B^0 + (1 - X_0)\mu_A^e$$
$$G_f = X_0\mu_B^e + (1 - X_0)\mu_A^e$$

From Equation 1.40

$$\mu_B^0 = G_B + RT\ln X_0 + \Omega(1 - X_0)^2$$
$$\mu_B^e = G_B + RT\ln X_e + \Omega(1 - X_e)^2$$
$$\mu_A^0 = G_A + RT\ln(1 - X_0) + \Omega X_0^2$$
$$\mu_A^e = G_A + RT\ln(1 - X_e) + \Omega X_e^2$$

Combining the above equations gives

$$\Delta G_0 = RT\left[X_0\ln\frac{X_0}{X_e} + (1 - X_0)\ln\frac{(1 - X_0)}{(1 - X_e)}\right] - \Omega(X_0 - X_e)^2$$

5.3 (a) $\Delta G_n = RT\ln\frac{X_0}{X_e}$ per mole of precipitate
Thus for a precipitate with $X_0 = 0.1$ and $X_e = 0.02$ at 600 K:

$$\underline{\Delta G_n = 8.0 \text{ kJ mol}^{-1}}$$

(b) Assuming that the nucleus is spherical with a radius r, and ignoring strain energy effects and the variation of γ with interface orientation, the total free energy change associated with nucleation may be defined as

$$\Delta G = -\frac{4}{3}\pi r^3 \cdot \Delta G_v + 4\pi r^2\gamma$$

where ΔG_v is the free energy released per unit volume. Differentiation of this equation yields the critical radius r^*

$$r^* = \frac{2\gamma}{\Delta G_v} = \frac{2\gamma V_m}{\Delta G_n} = \underline{0.50\ \text{mm}}$$

(c) The mean precipitate radius for a particle spacing of 50 nm was calculated as 13.4 nm = $27r^*$. For a 1 μm dispersion the precipitate radius, 267 nm = $534r^*$.

5.4

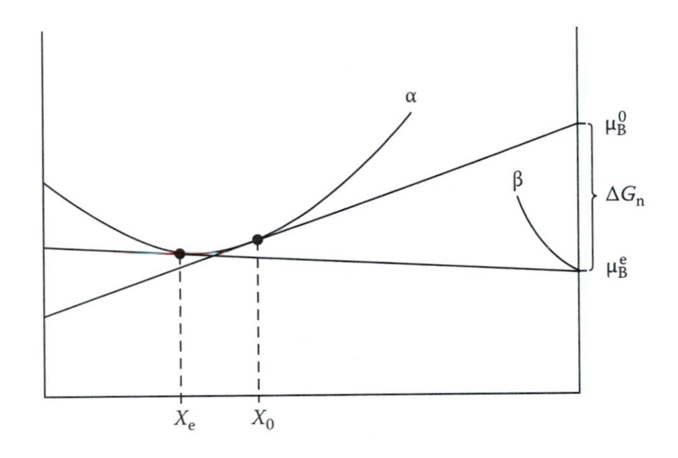

From Equation 1.68

$$\mu_B^0 = G_B + RT \ln \gamma_0 X_0$$
$$\mu_B^c = G_B + RT \ln \gamma_e X_e$$

where γ_0 and γ_e are the activity coefficients for alloy compositions X_0 and X_e, respectively,

$$\therefore \Delta G_n = \mu_B^0 - \mu_B^e = RT \ln \frac{\gamma_0 X_0}{\gamma_e X_e}$$

For ideal solutions $\gamma_0 = \gamma_e = 1$
For dilute solutions $\gamma_0 = \gamma_e =$ constant (Henry's law)
In both cases

$$\Delta G_n = RT \ln \frac{X_0}{X_e}$$

5.5 (a) Consider equilibrium of forces at the edge of the precipitate:

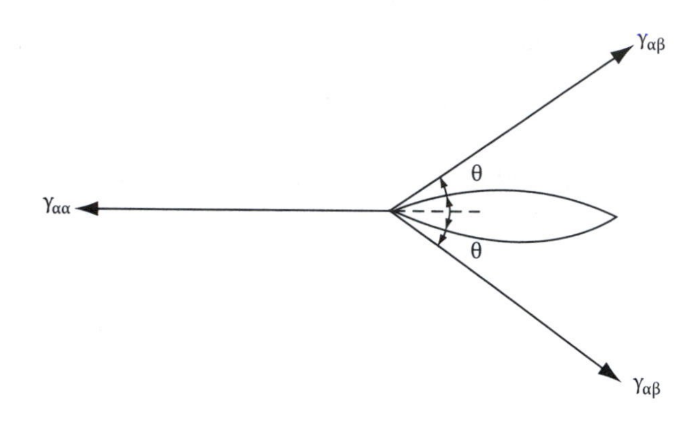

For unit area of interface

$$\gamma_{\alpha\alpha} = 2\gamma_{\alpha\beta}\cos\theta$$
$$\theta = \cos^{-1}\frac{\gamma_{\alpha\alpha}}{2\gamma_{\alpha\beta}} = \underline{53.1^\circ}$$

(b) The shape factor $S(\theta)$ is defined as

$$S(\theta) = \frac{1}{2}(2+\cos\theta)(1-\cos\theta)^2 = \underline{0.208}$$

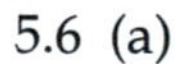
5.6 (a)

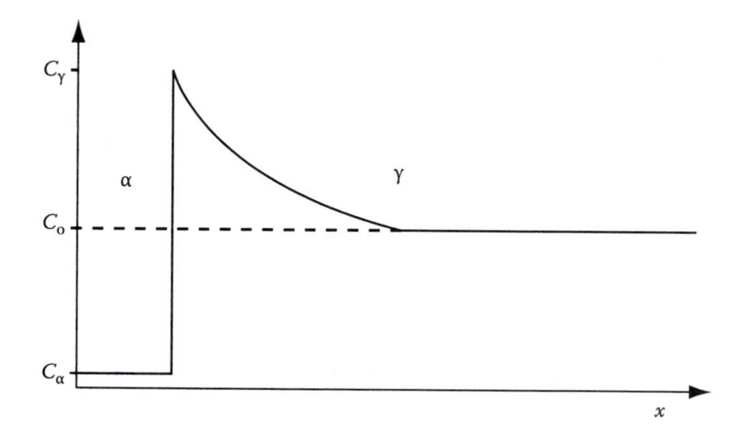

(b)

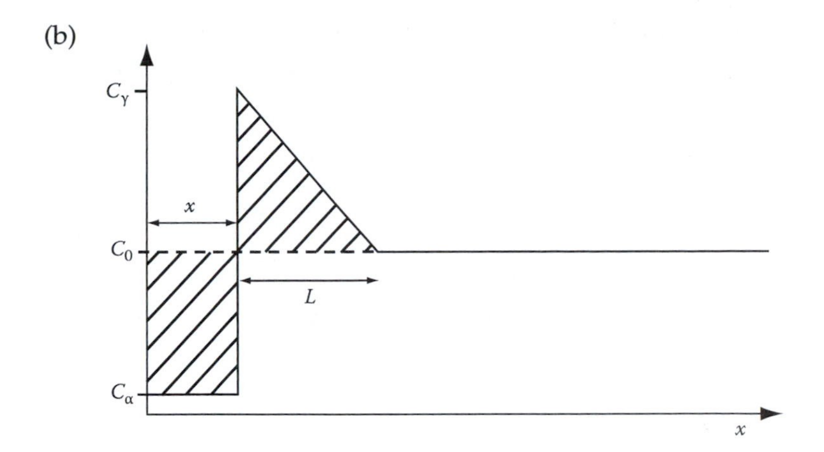

Using the simplified approach, above, the carbon concentration gradient in the austenite, $\frac{dC}{dx}$ may be expressed as

$$\frac{C_\gamma - C_0}{L}$$

For unit area of interface to advance a distance dx, a volume of material 1. dx must be converted from γ containing C_γ to a containing C_α moles of carbon per unit volume, i.e., $(C_\gamma - C_\alpha)dx$ moles of carbon must be rejected by diffusion through the γ. The flux of carbon through unit area in time dt is given by $D(dC/dx)dt$, where D is the diffusion coefficient. Equating the two expressions gives

$$(C_\gamma - C_\alpha)dx = D\left(\frac{dC}{dx}\right)dt$$

$$\therefore \; \frac{dx}{dt} = D\left(\frac{dC}{dx}\right)\cdot\frac{1}{(C_\gamma - C_\alpha)}$$

Thus using the simple concentration profile obtained earlier

$$\frac{dx}{dt} = D\left(\frac{C_\gamma - C_0}{L}\right)\cdot\frac{1}{(C_\gamma - C_\alpha)}$$

The width of the diffusion zone L may be found by noting that conservation of solute requires the two shaded areas in the diagram to be equal

$$(C_0 - C_\alpha)x = \frac{L(C_\gamma - C_0)}{2}$$
$$\therefore\ L = \frac{2(C_0 - C_\alpha)}{(C_\gamma - C_0)}$$

Substituting for L in the rate equation

$$\frac{\mathrm{d}x}{\mathrm{d}t} = \frac{D(C_\gamma - C_0)^2}{2(C_0 - C_\alpha)(C_\gamma - C_\alpha)x}$$

Assuming that the molar volume is constant, the concentrations may be replaced by mole fractions ($X = CV_m$). Integration of the rate equation gives the half-thickness of the boundary slabs as

$$X = \frac{(X_\gamma - X_0)\sqrt{(Dt)}}{(X_0 - X_\alpha)^{1/2}(X_\gamma - X_\alpha)^{1/2}}$$

(c) The mole fractions in the above equation can be replaced approximately by weight percentages. For ferrite precipitation from austenite in an Fe-0.15 wt% C alloy at 800°C, we have

$$X_\gamma = 0.32;$$
$$X_0 = 0.15;$$
$$X_\alpha = 0.02;$$
$$D_C^\alpha = 3 \times 10^{-12}\ \mathrm{m^2\ s^{-1}}$$

giving

$$x = 1.49 \times 10^{-6}\ t^{1/2}$$

(d) The previous derivation of $x(t)$ only applies for short times. At longer times the diffusion fields of adjacent slabs begin to overlap reducing the growth rate. The lever rule can be used to calculate the maximum half-thickness that is approached for long times.

Assume the grains are spherical with diameter D. When the transformation is complete the half-thickness of the ferritic slabs (x_{max}) is given by

$$\frac{(D - 2x_{max})^3}{D^3} = f_\gamma$$

where f_γ is the volume fraction of austenite.

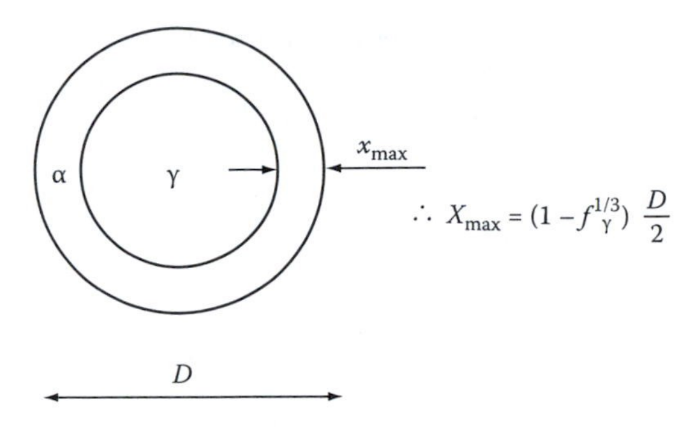

$$\therefore X_{max} = \left(1 - f_\gamma^{1/3}\right)\frac{D}{2}$$

(The same answer is obtained for any polyhedron.) Approximately, f_γ is given by

$$f_\gamma = \frac{X_0 - X_\alpha}{X_\gamma - X_\alpha}$$

In the present case $f_\gamma = 0.43$, such that for $D = 300\ \mu m$;

$$\underline{x_{max} = 36.5\ \mu m}$$

This value will be approached more slowly than predicted by the parabolic equation, as shown schematically in the diagram below.

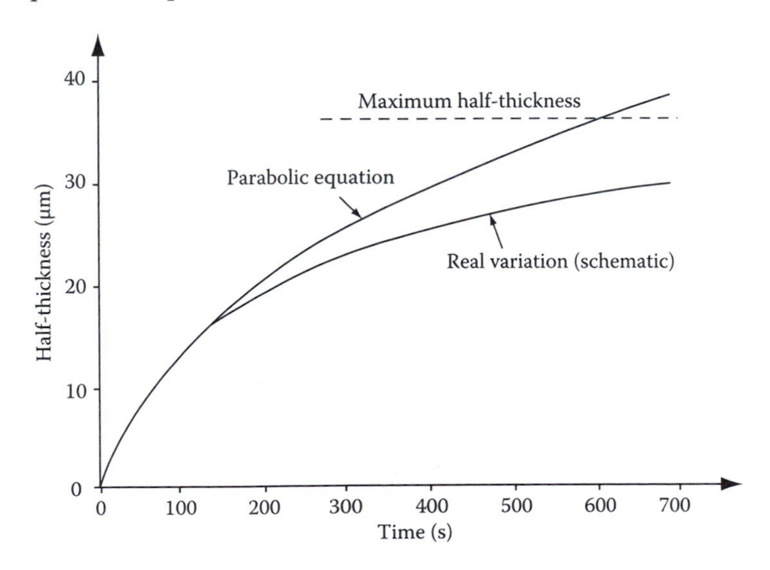

The next variation would require a more exact solution to the diffusion problem. However, the approximate treatment leading to the parabolic equation should be applicable for short times.

5.7

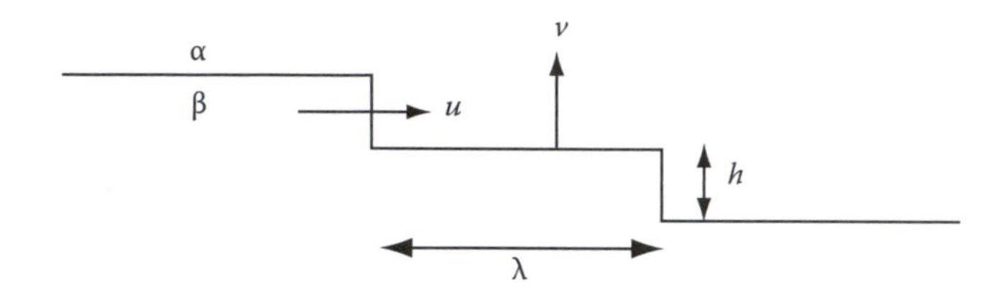

Consider unit area of interface perpendicular to the diagram:

Mass flow in the direction of $u = u \times h$;
Mass flow in the direction of $v = v \times \lambda$.

From the conservation of mass: $u \times h = v \times \lambda$

$$v = \frac{u \times h}{\lambda}$$

5.8 $f = 1 - \exp(-Kt^n)$
At short times this equation becomes

$$f = Kt^n$$

(a) Pearlitic nodules grow with a constant velocity, v. The volume fraction transformed after a short time t is given by

$$f = \frac{4\pi(vt)^3}{3d^3} = \left(\frac{4\pi v^3}{3d^3}\right)t^3$$

i.e.,

$$K = \frac{4\pi v^3}{3d^3}, \quad n = 3$$

(b) For short times, slabs growing in from the cube walls will give

$$f = \frac{6d^2 \cdot vt}{d^3} = \left(\frac{6v}{d}\right)t$$

i.e.,

$$K = \frac{6v}{d}, \quad n = 1$$

5.9

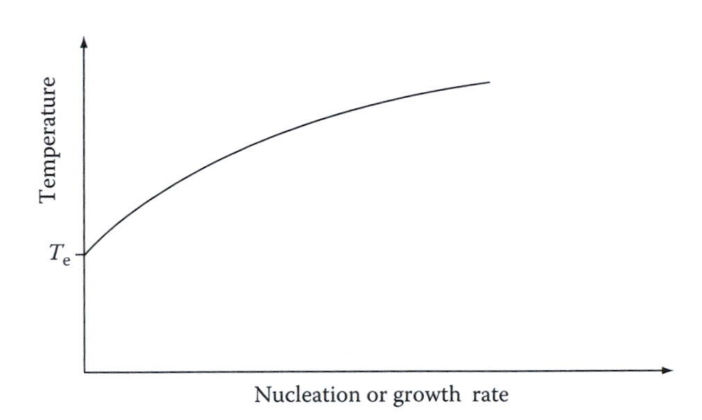

Civilian transformations that are induced by an increase in temperature show increasing nucleation and growth rates with increasing superheat above the equilibrium temperature (T_e). This is because both driving force and atom mobility (diffusivity) increase with increasing ΔT.

5.10 (a)

$$G = X_A G_A + X_B G_B + \Omega X_A X_B + RT(X_A \ln X_A + X_B \ln X_B)$$
$$G_A = G_B = 0 \quad \text{gives}$$
$$G = \Omega X_A X_B + RT(X_A \ln X_A + X_B \ln X_B)$$
$$dG = \Omega X_A dX_B + \Omega X_B dX_A + RT[dX_A + dX_B + \ln X_A dX_A + \ln X_B dX_B]$$

but

$$X_A + X_B = 1$$
$$dX_A + dX_B = 0$$
$$\therefore \frac{dG}{dX_B^2} = \Omega(X_A - X_B) + RT(\ln X_B - \ln X_A)$$
$$\frac{d^2G}{dX_B^2} = -2\Omega + RT\left(\frac{1}{X_B} + \frac{1}{X_A}\right)$$
$$\therefore \underline{\frac{d^2G}{dX_B^2} = \frac{RT}{X_A X_B} - 2\Omega}$$

(b) This system has a symmetrical miscibility gap with a maximum at $X_A = X_B = 0.5$ for which

$$\frac{d^2G}{dX_B^2} = 4RT - 2\Omega$$

It can be seen that as T increases d^2G/dX_B^2 changes from negative to positive values. The maximum of the solubility gap ($T = T_c$) corresponds to $\left(d^2G/dX_B^2\right) = 0$, i.e.,

$$\underline{T_c = \frac{\Omega}{2R}}$$

(c) Equating dG/dX_B to zero in the equations gives

$$\Omega(X_A - X_B) + RT(\ln X_B - \ln X_A) = 0$$

Putting $\Omega = 2RT_c$ gives

$$\underline{\frac{T}{T_c} = \frac{2(1 - 2X_B)}{\ln\left(\frac{1-X_B}{X_B}\right)}}$$

This equation can be used to plot the coordinates of the miscibility gap as shown blow:

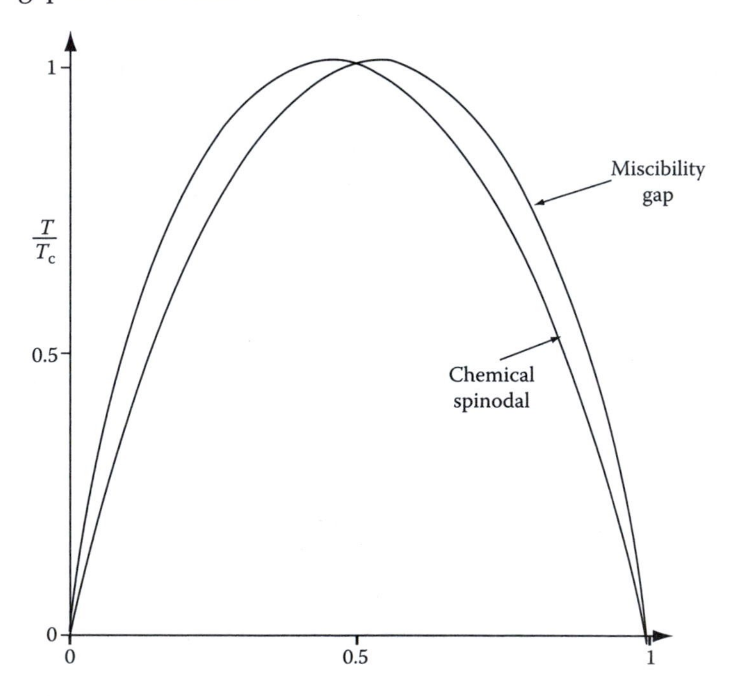

(d) The locus of the chemical spinodal is given by

$$\frac{d^2G}{dX_B^2} = 0$$

i.e.,

$$\frac{RT}{X_A X_B} - 2\Omega = 0$$

$$\underline{\frac{T}{T_c} = 4X_B(1 - X_B)}$$

This is also shown in the figure.

5.11

$$G(X_0 + \Delta X) = G(X_0) + \frac{dG}{dX}(\Delta X) + \frac{d^2G}{dX^2}\frac{(\Delta X)^2}{2} + \cdots$$

$$G(X_0 - \Delta X) = G(X_0) + \frac{dG}{dX}(-\Delta X) + \frac{d^2G}{dX^2}\frac{(\Delta X)^2}{2} + \cdots$$

$\therefore$ Total free energy of an alloy with parts of composition $(X_0 + \Delta X)$ and $(X_0 - \Delta X)$ is given by

$$\frac{G(X_0 + \Delta X)}{2} + \frac{G(X_0 + \Delta X)}{2} = \frac{1}{2}2G(X_0) + \frac{d^2G}{dX^2}(\Delta X)^2$$

$$= G(X_0) + \frac{1}{2}\frac{d^2G}{dX^2}(\Delta X)^2$$

$$\text{Original free energy} = G(X_0)$$

$$\therefore \text{Change in free energy} = \frac{1}{2}\frac{d^2G}{dX^2}(\Delta X)^2$$

5.12 Equation 5.50 gives the minimum thermodynamically possible wavelength λ_{min} as

$$\lambda_{min}^2 = \frac{-2K}{\left(\frac{d^2G}{dX^2} + 2\eta^2 E' V_m\right)}$$

$2\eta^2 E'V_m$ is a positive constant, while d^2G/dX^2 varies with composition X_B as shown below:

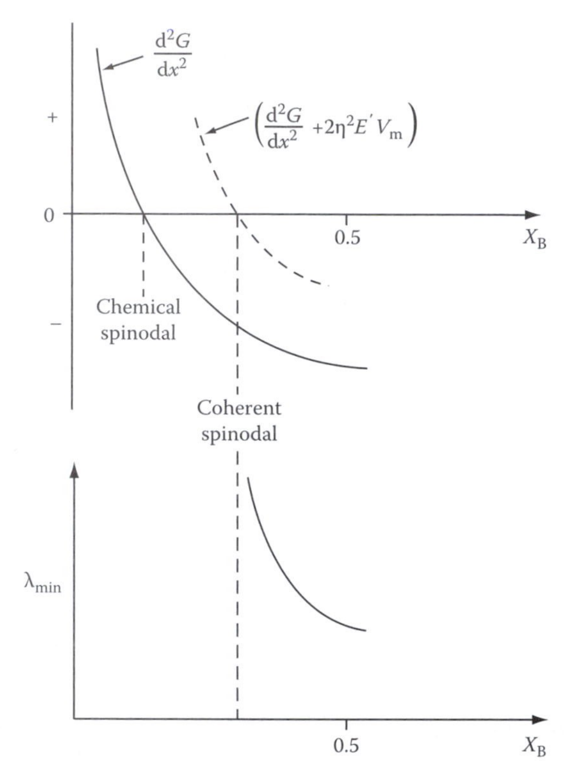

Thus $\lambda_{min} = \infty$ at the coherent spinodal, but decreases as X_B increases toward 0.5, as shown schematically above. The wavelength that forms in practice will be determined by a combination of thermodynamic and kinetic effects, but qualitatively it will vary in the same way as λ_{min}.

5.13 (a) Massive transformations are classified as civilian nucleation and growth transformations which are interface controlled. This is because massive transformations do not involve long-range diffusion, but are controlled by the rate at which atoms can cross the parent/product interface (see also Section 5.9).

(b) Precipitation reactions can occur at any temperature below that marking the solubility limit, whereas massive transformations cannot occur until lower temperatures at least lower than T_0 (Figure 5.4). Massive transformations therefore occur at lower temperatures than precipitation reactions. However, at low temperatures diffusion is slow, especially the long-range diffusion required for precipitation. Massive transformations have the advantage that only short-range atom jumps across the parent/product interface are needed. Thus it is possible for massive transformations to achieve higher growth rates than precipitation reactions despite the lower driving force.

Chapter 6

6.1

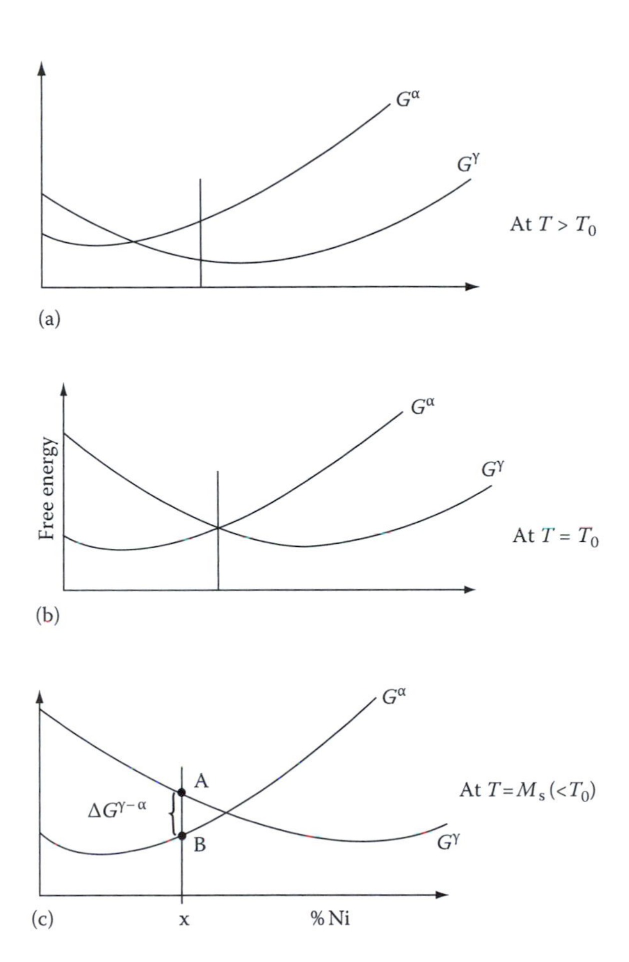

For an alloy of composition X, at $T > T_0$, the free energy curve for α lie above that for γ, thus austenite is stable at this composition and temperature, and the martensitic transformation is unable to occur.

At a temperature $T = T_0$ the α and α' free energy curve coincides with that for γ, and so at this temperature and composition both the martensite and austenite have equal free energy, and there is no driving force for the martensitic transformation.

At a temperature $T = M_s$ the γ free energy curve lies above that for α, therefore γ is thermodynamically unstable, and there is a driving force for the martensitic transformation to the length AB. The significance of the M_s temperature is that it is the maximum temperature for which the driving force is sufficient to cause the martensitic transformation. No such driving force is present at temperatures above M_s.

At the equilibrium temperature, ΔG for the transformation is zero, thus

$$\Delta G^{\gamma-\alpha'} = \Delta H^{\gamma-\alpha'} - T_0 \Delta S = 0$$

$$\text{at } T_0, \quad \Delta S = \frac{\Delta H^{\gamma-\alpha'}}{T_0}$$

For small undercoolings ΔH and ΔS may be considered to be independent of temperature, thus the free energy change may be expressed in terms of the undercooling as follows:

$$\Delta G^{\gamma-\alpha'} = \Delta H^{\gamma-\alpha'} \frac{(T_0 - M_s)}{T_0}$$

At the M_s temperature.

The driving force for the martensitic transformation has been shown to be proportional to the undercooling $(T_0 - M_s)$, where T_0 is the temperature at which austenite and martensite have the same free energy, and M_s is the temperature at which martensite starts to form. In the Fe–C system both T_0 and M_s fall with increasing carbon content, with an equal and linear rate. Thus the difference $(T_0 - M_s)$ remains constant for different carbon contents, which means that the driving force must remain constant.

6.2 See Section 6.3.1.

6.3

$$\Delta G^* = \frac{512}{3} \cdot \frac{\gamma}{(\Delta G_v)^4} \cdot \left(\frac{S}{2}\right)^4 \cdot \mu^2 \pi \text{ J nucleus}^{-1}$$

$$c^* = \frac{2\gamma}{\Delta G_v}$$

$$a^* = \frac{16\gamma\mu(S/2)^2}{(\Delta G_v)^2}$$

Substitution of the values given gives

$$\underline{\Delta G^* = 3.0 \times 10^{-18} \text{ J nucleus}^{-1}}$$

$$\underline{c^* = 0.23 \text{ nm}}$$

$$\underline{a^* = 8.5 \text{ nm}}$$

6.4 The habit plane of martensite is a common plane between martensite and the phase from which it forms which is undistorted and unrotated during transformation. Thus all directions and angular separations in the plane are unchanged during the transformation.

The martensitic habit plane may be measured using x-ray diffraction and constructing pole figures. The figures are analyzed and the plane index may be determined by measuring the positions of diffraction spots from martensite crystals produced from austenite crystals.

The main reason for the scatter in the measurement of habit planes is that the martensite lattice is not perfectly coherent with the parent lattice, and so a strain is inevitably caused at the interface. This may act to distort the habit plane somewhat. Internal stress formed during the transformation depends on transformation conditions. Habit plane scatter has been observed to increase when the austenite has been strained plastically prior to transformation, indicating that prior deformation of the austenite is an important factor.

Another reason for the scatter is that during the formation of twinned martensite, the twin width may be varied to obtain adjacent twin widths with low coherency energies. Experimental studies have shown that the lowest energy troughs are very shallow and quite extensive, enabling the production of habit planes which may vary by several degrees in a given alloy.

6.5 The key to the phenomenological approach to martensitic transformations is to postulate an additional distortion which reduces the elongation of the expansion axis of the austenite crystal structure to zero. This second deformation can occur in the form of dislocation slip or twinning as shown below:

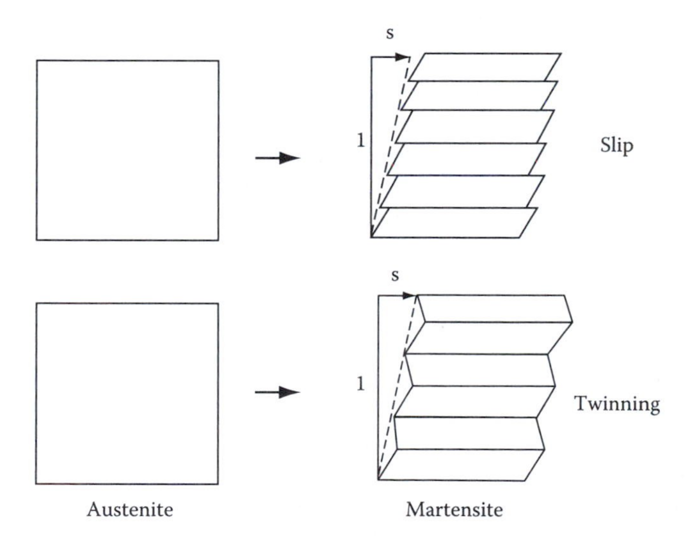

Dislocation glide or twinning of the martensite reduces the strain of the surrounding austenite. The transformation shear is shown as S.

Both types of shear have been observed under transmission electron microscopy.

6.6

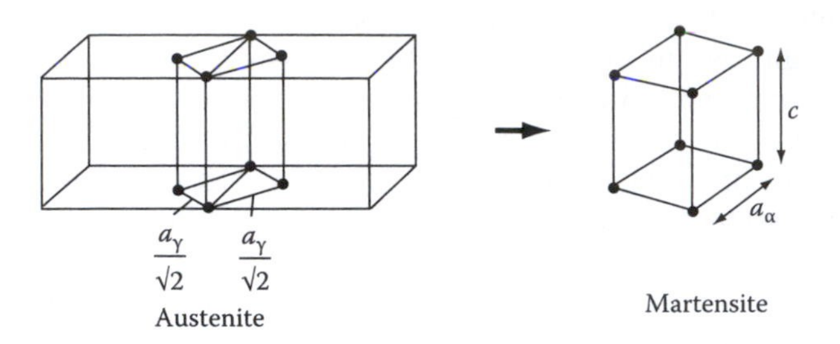

Assuming that $a_\gamma = 3.56$ Å and $a_\alpha = 2.86$ Å, and that c/a for martensite is equal to 1.1, the movements of atoms the c and a directions may be calculated

$$a_\alpha = 2.86A \quad \therefore \; c_a = 3.15A$$

$$a_\alpha = 3.56A \quad \therefore \; \frac{a}{\sqrt{21}}\gamma = 2.52A$$

Vertical movement of atoms = 3.56 − 3.15 Å = 0.41 Å
Horizontal movement of atoms = 2.86 − 2.52 Å = 0.34 Å
Thus by vector addition, the maximum movement is found to be 0.53 Å.

6.7 See Sections 6.3.2 and 6.3.3.

6.8 The habit plane of martensite is found to change with carbon and nickel contents in FeC and FeNi alloys, respectively. This may be explained by considering the nature and the method of formation of the martensite which is dependent on alloy content.

In low-carbon steels the M_S temperature is high and martensite forms with a lath morphology growing along a {111} plane. Growth occurs by the nucleation and glide of transformation dislocations. However, as the carbon content is increased the morphology changes to a plate structure which forms in isolation. The degree of twinning is higher in this type of martensite. An important difference in this process is that the M_S temperature is lowered with increasing alloy content which means that the austenite is not as uniformly for as efficiently eliminated as with lath martensites. Plate martensite is formed by a burst mechanism, this factor contributing to the fact that the habit plane changes to {225}, and to {259} with even higher carbon content.

Similar arguments may be used to explain the change in habit plane with increasing Ni content in FeNi alloys, since Ni acts in a similar way to C, lowering the M_S temperature and influencing martensite morphology and amount of retained austenite.

The amount of retained austenite is also influenced by the austenitizing temperature since this influences the amount of dissolved iron carbide. The quenching rate is also important, an oil quench will produce more retained austenite than a water quench.

6.9 See Section 6.4.5.

6.10 See Section 6.7.

6.11 See Section 6.7.4.

Index

C

E

K

L

M

N

R

S